# Probability and Statistical Inference

**CHAPMAN & HALL/CRC**

**Texts in Statistical Science Series**



**Recently Published Titles**

**Surrogates**
Gaussian Process Modeling, Design, and Optimization for the Applied Sciences
*Robert B. Gramacy*

**Statistical Analysis of Financial Data**
With Examples in R
*James Gentle*

**Statistical Rethinking**
A Bayesian Course with Examples in R and STAN, Second Edition
*Richard McElreath*

**Statistical Machine Learning**
A Model-Based Approach
*Richard Golden*

**Randomization, Bootstrap and Monte Carlo Methods in Biology**
Fourth Edition
*Bryan F. J. Manly, Jorje A. Navarro Alberto*

**Principles of Uncertainty, Second Edition**
*Joseph B. Kadane*

**Beyond Multiple Linear Regression**
*Applied Generalized Linear Models and Multilevel Models in R*
*Paul Roback, Julie Legler*

**Bayesian Thinking in Biostatistics**
*Gary L. Rosner, Purushottam W. Laud, and Wesley O. Johnson*

**Modern Data Science with R,** Second Edition
*Benjamin S. Baumer, Daniel T. Kaplan, and Nicholas J. Horton*

**Probability and Statistical Inference**
*From Basic Principles to Advanced Models*

*Miltiadis C. Mavrakakis and Jeremy Penzer*

**For more information about this series, please visit: https://www.crcpress.com/Chapman--HallCRC-Texts-in-Statistical-Science/book-series/CHTEXSTASCI**

# Probability and Statistical Inference

## From Basic Principles to Advanced Models

Miltiadis C. Mavrakakis
Jeremy Penzer

CRC Press is an imprint of the
Taylor & Francis Group, an **informa** business
A CHAPMAN & HALL BOOK

First edition published 2021
by CRC Press
6000 Broken Sound Parkway NW, Suite 300, Boca Raton, FL 33487-2742

and by CRC Press
2 Park Square, Milton Park, Abingdon, Oxon, OX14 4RN



CRC Press is an imprint of Taylor & Francis Group, LLC



Library of Congress CataloginginPublication Data

ISBN: 9781584889397 (hbk)
ISBN: 9780367749125 (pbk)
ISBN: 9781315366630 (ebk)

Typeset in Computer Modern font
by KnowledgeWorks Global Ltd.

*For JP. . . this book is dedicated to Debbie, Lily, and Bella.*

*For MCM. . . this book is dedicated to Sitali, Kostas, and Liseli.*

# Contents

**Preface** **xv**

**About the authors** **xix**

**1 Introduction** **1**

**2 Probability** **7**

2.1 Intuitive probability 7

2.2 Mathematical probability 9

2.2.1 Measure 9

2.2.2 Probability measure 11

2.3 Methods for counting outcomes 16

2.3.1 Permutations and combinations 17

2.3.2 Number of combinations and multinomial coefficients 20

2.4 Conditional probability and independence 23

2.4.1 Conditional probability 23

2.4.2 Law of total probability and Bayes' theorem 25

2.4.3 Independence 27

2.5 Further exercises 31

2.6 Chapter summary 32

**3 Random variables and univariate distributions** **33**

3.1 Mapping outcomes to real numbers 33

3.2 Cumulative distribution functions 37

3.3 Discrete and continuous random variables 41

3.3.1 Discrete random variables and mass functions 43
3.3.2 Continuous random variables and density functions 51
3.3.3 Parameters and families of distributions 60
3.4 Expectation, variance, and higher moments 61
3.4.1 Mean of a random variable 61
3.4.2 Expectation operator 63
3.4.3 Variance of a random variable 65
3.4.4 Inequalities involving expectation 67
3.4.5 Moments 70
3.5 Generating functions 74
3.5.1 Moment-generating functions 74
3.5.2 Cumulant-generating functions and cumulants 79
3.6 Functions of random variables 81
3.6.1 Distribution and mass/density for $g(X)$ 82
3.6.2 Monotone functions of random variables 84
3.7 Sequences of random variables and convergence 88
3.8 A more thorough treatment of random variables 93
3.9 Further exercises 96
3.10 Chapter summary 99

**4 Multivariate distributions** **101**
4.1 Joint and marginal distributions 101
4.2 Joint mass and joint density 104
4.2.1 Mass for discrete distributions 104
4.2.2 Density for continuous distributions 107
4.3 Expectation and joint moments 116
4.3.1 Expectation of a function of several variables 116
4.3.2 Covariance and correlation 117
4.3.3 Joint moments 120
4.3.4 Joint moment-generating functions 121
4.4 Independent random variables 123

4.4.1 Independence for pairs of random variables 124
4.4.2 Mutual independence 125
4.4.3 Identical distributions 126
4.5 Random vectors and random matrices 127
4.6 Transformations of continuous random variables 130
4.6.1 Bivariate transformations 130
4.6.2 Multivariate transformations 133
4.7 Sums of random variables 134
4.7.1 Sum of two random variables 135
4.7.2 Sum of $n$ independent random variables 137
4.8 Multivariate normal distribution 138
4.8.1 Bivariate case 139
4.8.2 $n$-dimensional multivariate case 140
4.9 Further exercises 143
4.10 Chapter summary 145

**5 Conditional distributions 147**
5.1 Discrete conditional distributions 147
5.2 Continuous conditional distributions 150
5.3 Relationship between joint, marginal, and conditional 153
5.4 Conditional expectation and conditional moments 154
5.4.1 Conditional expectation 155
5.4.2 Conditional moments 158
5.4.3 Conditional moment-generating functions 161
5.5 Hierarchies and mixtures 162
5.6 Random sums 164
5.7 Conditioning for random vectors 166
5.8 Further exercises 167
5.9 Chapter summary 169

**6 Statistical models** **171**

6.1 Modelling terminology, conventions, and assumptions 171

6.1.1 Sample, observed sample, and parameters 171

6.1.2 Structural and distributional assumptions 172

6.2 Independent and identically distributed sequences 173

6.2.1 Random sample 173

6.2.2 Error sequences 173

6.3 Linear models 174

6.3.1 Simple linear regression 174

6.3.2 Multiple linear regression 175

6.3.3 Applications 176

6.4 Generalised linear models 177

6.4.1 Motivation 177

6.4.2 Link function 178

6.5 Time-to-event models 181

6.5.1 Survival function and hazard function 182

6.5.2 Censoring of time-to-event data 186

6.5.3 Covariates in time-to-event models 187

6.6 Time series models 188

6.6.1 Autoregressive models 191

6.6.2 Moving-average models 191

6.6.3 Autocovariance, autocorrelation, and stationarity 192

6.7 Poisson processes 197

6.7.1 Stochastic processes and counting processes 197

6.7.2 Definitions of the Poisson process 199

6.7.3 Thinning and superposition 203

6.7.4 Arrival and interarrival times 203

6.7.5 Compound Poisson process 205

6.7.6 Non-homogeneous Poisson process 205

6.8 Markov chains 207

6.8.1 Classification of states and chains 210

6.8.2 Absorption 211
6.8.3 Periodicity 213
6.8.4 Limiting distribution 214
6.8.5 Recurrence and transience 215
6.8.6 Continuous-time Markov chains 216
6.9 Further exercises 217
6.10 Chapter summary 220

**7 Sample moments and quantiles 223**
7.1 Sample mean 223
7.1.1 Mean and variance of the sample mean 224
7.1.2 Central limit theorem 225
7.2 Higher-order sample moments 228
7.2.1 Sample variance 229
7.2.2 Joint sample moments 233
7.3 Sample mean and variance for a normal population 236
7.4 Sample quantiles and order statistics 240
7.4.1 Sample minimum and sample maximum 242
7.4.2 Distribution of $i^{\text{th}}$ order statistic 243
7.5 Further exercises 248
7.6 Chapter summary 249

**8 Estimation, testing, and prediction 251**
8.1 Functions of a sample 251
8.1.1 Statistics 251
8.1.2 Pivotal functions 252
8.2 Point estimation 255
8.2.1 Bias, variance, and mean squared error 257
8.2.2 Consistency 258
8.2.3 The method of moments 261
8.2.4 Ordinary least squares 263
8.3 Interval estimation 267

8.3.1 Coverage probability and length 269

8.3.2 Constructing interval estimators using pivotal functions 272

8.3.3 Constructing interval estimators using order statistics 275

8.3.4 Confidence sets 278

8.4 Hypothesis testing 278

8.4.1 Statistical hypotheses 279

8.4.2 Decision rules 280

8.4.3 Types of error and the power function 281

8.4.4 Basic ideas in constructing tests 282

8.4.5 Conclusions and $p$-values from tests 283

8.5 Prediction 285

8.6 Further exercises 289

8.7 Chapter summary 290

**9 Likelihood-based inference 293**

9.1 Likelihood function and log-likelihood function 293

9.2 Score and information 296

9.3 Maximum-likelihood estimation 302

9.3.1 Properties of maximum-likelihood estimates 304

9.3.2 Numerical maximisation of likelihood 308

9.3.3 EM algorithm 310

9.4 Likelihood-ratio test 316

9.4.1 Testing in the presence of nuisance parameters 317

9.4.2 Properties of the likelihood ratio 318

9.4.3 Approximate tests 320

9.5 Further exercises 323

9.6 Chapter summary 326

**10 Inferential theory** **327**

10.1 Sufficiency 327

10.1.1 Sufficient statistics and the sufficiency principle 327

10.1.2 Factorisation theorem 329

10.1.3 Minimal sufficiency 334

10.1.4 Application of sufficiency in point estimation 338

10.2 Variance of unbiased estimators 341

10.3 Most powerful tests 345

10.4 Further exercises 350

10.5 Chapter summary 352

**11 Bayesian inference** **355**

11.1 Prior and posterior distributions 355

11.2 Choosing a prior 357

11.2.1 Constructing reference priors 357

11.2.2 Conjugate priors 360

11.3 Bayesian estimation 363

11.3.1 Point estimators 363

11.3.2 Absolute loss 364

11.3.3 0-1 loss 364

11.3.4 Interval estimates 366

11.4 Hierarchical models and empirical Bayes 370

11.4.1 Hierarchical models 370

11.4.2 Empirical Bayes 371

11.4.3 Predictive inference 373

11.5 Further exercises 376

11.6 Chapter summary 378

**12 Simulation methods** **379**

12.1 Simulating independent values from a distribution 379

12.1.1 Table lookup 380

12.1.2 Probability integral 381

12.1.3 Box-Muller method 382

12.1.4 Accept/reject method 382

12.1.5 Composition 385

12.1.6 Simulating model structure and the bootstrap 387

12.2 Monte Carlo integration 390

12.2.1 Averaging over simulated instances 390

12.2.2 Univariate vs. multivariate integrals 391

12.2.3 Importance sampling 394

12.2.4 Antithetic variates 397

12.3 Markov chain Monte Carlo 400

12.3.1 Discrete Metropolis 400

12.3.2 Continuous Metropolis 402

12.3.3 Metropolis-Hastings algorithm 402

12.3.4 Gibbs sampler 404

12.4 Further exercises 408

12.5 Chapter summary 410

**A Proof of Proposition 5.7.2** **411**

**Index** **415**

# Preface

## About this book

**Probability and Statistical Inference: From Basic Principles to Advanced Models** covers aspects of probability, distribution theory, and inference that are fundamental to a proper understanding of data analysis and statistical modelling. It presents these topics in an accessible manner without sacrificing mathematical rigour, bridging the gap between the many excellent introductory books and the more advanced, graduate-level texts. The book introduces and explores techniques that are relevant to modern practitioners, while being respectful to the history of statistical inference. It seeks to provide a thorough grounding in both the theory and application of statistics, with even the more abstract parts placed in the context of a practical setting.

## Who is it for?

This book is for students who have already completed a first course in probability and statistics, and now wish to deepen and broaden their understanding of the subject. It can serve as a foundation for advanced undergraduate or postgraduate courses. Our aim is to challenge and excite the more mathematically able students, while providing explanations of statistical concepts that are more detailed and approachable than those in advanced texts. This book is also useful for data scientists, researchers, and other applied practitioners who want to understand the theory behind the statistical methods used in their fields. As such, it is intended as an instructional text as well as a reference book.

## Chapter summary

Chapter 2 is a complete introduction to the elements of probability theory. This is the foundation for random variables and distribution theory, which are explored in Chapter 3. These concepts are extended to multivariate distributions in Chapter 4, and conditional distributions in Chapter 5. Chapter 6 is a concise but broad account of statistical modelling, covering many topics that are essential to modern statisticians, such as generalised linear models, survival analysis, time series, and random processes. Sampling distributions are first encountered in Chapter 7, through the concept of sample moments and order statistics. Chapter 8 introduces the key topics in classical statistical inference: point estimation, interval estimation, and hypothesis testing. The main techniques in likelihood-based inference are developed further in Chapter 9, while Chapter 10 examines their theoretical basis. Chapter 11 is an overview of the fundamental concepts of Bayesian statistics. Finally, Chapter 12 is an introduction to some of the main computational methods used in modern statistical inference.

**Teaching a course**

The material in this book forms the basis of our 40-lecture course, spread across two academic terms. Our students are mostly second- and third-year undergraduates who have already completed an introductory statistics course, so they have some level of familiarity with probability and distribution theory (Chapters 2–5). We use the opportunity to focus on the more abstract concepts – measure-theoretic probability (section 2.2), convergence of sequences of random variables (3.8), etc. – and introduce some useful modelling frameworks, such as hierarchical models and mixtures (5.5), and random sums (5.7).

We approach the section on statistical inference (Chapters 7–10) in a similar way. As our students have already encountered the fundamental concepts of sampling distributions, parameter estimation, and hypothesis tests, we spend more time on advanced topics, such as order statistics (7.4), likelihood-ratio tests (9.4), and the theoretical basis for likelihood-based inference (9.3 and 10.1).

In addition to these, we cover the first few sections in Chapter 6 (6.1–6.3) and a selection of topics from the remaining parts. These are almost entirely self-contained, so it is possible to focus on, say, generalised linear models (6.4) and survival models (6.5), and omit stochastic processes (6.7 and 6.8) – or vice versa. Note that the MCMC techniques in section 12.3 require at least a basic familiarity with Markov chain theory (6.8).

For most of our students, this course is their first encounter with Bayesian inference, and we spend quite a lot of time on the basic ideas in Chapter 11 (prior and posterior, loss functions and estimators). Finally, we demonstrate the main techniques in Chapter 12 by implementing the algorithms in R, and encouraging the students to do the same. For a shorter course, both chapters can be omitted entirely.

There are many other ways to structure a course from this book. For example, Chapters 2–5 could be taught as a standalone course in mathematical probability and distribution theory. This would be useful as a prerequisite for advanced courses in statistics. Alternatively, Chapters 6, 11, and 12 could form the basis of a more practical course on statistical modelling.

Exercises appear within each section and also at the end of each chapter. The former are generally shorter and more straightforward, while the latter seek to test the key learning outcomes of each chapter, and are more difficult as a result. Solutions to the exercises will appear in a separate volume.

### Jeremy's acknowledgements

This book started life as a set of study guides. A number of people helped enormously in the preparation of these guides; in particular, Chris Jones and Wicher Bergsma (who reviewed early versions), James Abdey (who proof read patiently and thoroughly), and Martin Knott and Qiwei Yao (who were generous with both their time and their advice) – thank you all. I would like to acknowledge that without Milt this would have remained, at best, a half-finished piece of work. I am very grateful to my wife Debbie whose love and support make all things possible.

### Milt's acknowledgements

I would like to thank Jeremy for starting this book and trusting me to finish it; Mark Spivack, for fighting my corner when it really mattered; Lynn Frank, for being the very first person who encouraged me to teach; my LSE students, for helping me polish this material; and my colleagues at Smartodds, particularly Iain and Stu, for their valuable suggestions which helped keep the book on track. I am forever grateful to my parents, Maria and Kostas, for their unconditional love and support. Finally, I would like to thank my wife, Sitali, for being my most passionate fan and fiercest critic, as circumstances dictated; this book would never have been completed without her.

# About the authors

**Jeremy Penzer's** first post-doc job was as a research assistant at the London School of Economics. Jeremy went on to become a lecturer at LSE and to teach the second-year statistical inference course (ST202) that formed the starting point for this book. While working at LSE, his research interests were time series analysis and computational statistics. After 12 years as an academic, Jeremy shifted career to work in financial services. He is currently Chief Marketing and Analytics Officer for Capital One Europe (plc). Jeremy lives just outside Nottingham with his wife and two daughters.

**Miltiadis C. Mavrakakis** obtained his PhD in Statistics at LSE under the supervision of Jeremy Penzer. His first job was as a teaching fellow at LSE, taking over course ST202 and completing this book in the process. He splits his time between lecturing (at LSE, Imperial College London, and the University of London International Programme) and his applied statistical work. Milt is currently a Senior Analyst at Smartodds, a sports betting consultancy, where he focuses on the statistical modelling of sports and financial markets. He lives in London with his wife, son, and daughter.

Chapter 1

# Introduction

Statistics is concerned with methods for collecting, analysing, and drawing conclusions from data. A clear understanding of the theoretical properties of these methods is of paramount importance. However, this theory is often taught in a way that is completely detached from the real problems that motivate it.

To infer is to draw general conclusions on the basis of specific observations, which is a skill we begin to develop at an early age. It is such a fundamental part of our intelligence that we do it without even thinking about it. We learn to classify objects on the basis of a very limited training set. From a few simple pictures, a child learns to infer that anything displaying certain characteristics (a long neck, long thin legs, large brown spots) is a giraffe. In statistical inference, our specific observations take the form of a data set. For our purposes, a data set is a collection of numbers. Statistical inference uses these numbers to make general statements about the process that generated the data.

Uncertainty is part of life. If we were never in any doubt about what was going to happen next, life would be rather dull. We all possess an intuitive sense that some things are more certain than others. If I knock over the bottle of water that is sitting on my desk, I can be pretty sure some of the contents will spill out; as we write this book, we might hope that it is going to be an international bestseller, but there are many other (far more likely) outcomes. Statistical inference requires us to do more than make vague statements like "I can be pretty sure" and "there are more likely outcomes". We need to be able to quantify our uncertainty by attaching numbers to possible events. These numbers are referred to as probabilities.

The theory of probability did not develop in a vacuum. Our aim in studying probability is to build the framework for modelling real-world phenomena; early work in the field was motivated by an interest in gambling odds. At the heart of the models that we build is the notion of a random variable and an associated distribution. Probability and distribution theory provide the foundation. Their true value becomes apparent when they are applied to questions of inference.

Statistical inference is often discussed in the context of a sample from a population. Suppose that we are interested in some characteristic of individuals in a given population. We cannot take measurements for every member of the population so instead we take a sample. The data set consists of a collection of measurements of the characteristic of interest for the individuals in the sample. In this context, we will often refer to the data set as the observed sample. Our key inferential question is then, "what can we infer about the population from our observed sample?".

Consider the following illustration. Zoologists studying sea birds on an island in the South Atlantic are interested in the journeys made in search of food by members of a particular species. They capture a sample of birds of this species, fit them with a small tracking device, then release them. They measure the distance travelled over a week by each of the birds in the sample. In this example, the population might be taken to be all birds of this particular species based on the island. The characteristic of interest is distance travelled over the week for which the experiment took place. The observed sample will be a collection of measurements of distance travelled. Inferential questions that we might ask include:

i. Based on the observed sample, what is a reasonable estimate of the population mean distance travelled?
ii. How sure can we be about this estimate?
iii. The zoological community has an established view on the mean distance that these birds travel in search of food. Does the observed sample provide any evidence to suggest that this view is incorrect?

Consider the first of these questions. The population mean is the value that we would get if we measured distance travelled for every member of the population, added up these values, and divided by the population size. We can use our observed sample to construct an estimate of the population mean. The usual way to do this is to add up all the observed sample values and divide by the sample size; the resulting value is the observed sample mean. Computing observed sample values, such as the mean, median, maximum, and so on, forms the basis of descriptive statistics. Although widely used (most figures given in news reports are presented as descriptive statistics), descriptive statistics cannot answer questions of inference adequately.

The first of our questions asks, "what is a *reasonable* estimate of the population mean?". This brings us to a key idea in statistical inference: it is the properties of the sample that determine whether an estimate is reasonable or not. Consider a sample that consists of a small number of the weakest birds on the island; it is clear that this sample will not generate a reasonable estimate of the population mean. If, on the other hand, we sample at random and our sample is large, the observed sample mean provides an intuitively appealing estimate. Using ideas from probability and distribution theory, we can give a precise justification for this intuitive appeal.

It is important to recognise that all statistical inference takes place in the context of a model, that is, a mathematical idealisation of the real situation. In practice, data collection methods often fall well short of what is required to satisfy the model's properties. Consider sampling at random from a population. This requires us to ensure that every member of the population is equally likely to be included in the sample. It is hard to come up with a practical situation where this requirement can be met. In the sea birds illustration, the sample consists of those birds that the zoologists capture and release. While they might set their traps in order to try to ensure that every bird in the population is equally likely to be captured, in practice they may be more likely to capture young, inexperienced birds or, if their traps are baited, the hungriest members of the population. In the sage, if well-worn, words of George Box, *"all models are wrong, but some are useful"*. The best we can hope for is a model that provides us with a useful representation of the process that generates the data.

A common mistake in statistical inference is to extend conclusions beyond what can be supported by the data. In the sea bird illustration, there are many factors that potentially have an impact on the length of journeys: for example, prevailing weather conditions and location of fish stocks. As the zoologists measure the distance travelled for one week, it is reasonable to use our sample to draw inferences for this particular week. However, if we were to try to extend this to making general statements for any week of the year, the presence of factors that vary from week to week would invalidate our conclusions. To illustrate, suppose the week in which the experiment was conducted was one with particularly unfavourable weather conditions. In this situation, the observed sample mean will grossly underestimate the distance travelled by the birds in a normal week. One way in which the zoologists could collect data that would support wider inferences would be to take measurements on randomly selected days of the year. In general, close attention should be paid to the circumstances in which the data are collected; these circumstances play a crucial role in determining what constitutes sensible inference.

Statistical inference is often divided into three topics: point estimation, interval estimation, and hypothesis testing. Roughly, these correspond to the three questions asked in the sea birds illustration. Finding a reasonable estimate for the population mean is a question of point estimation. In general, point estimation covers methods for generating single-number estimates for population characteristics and the properties of these methods. Any estimate of a population characteristic has some uncertainty associated with it. One way to quantify this uncertainty is to provide an interval that covers, with a given probability, the value of the population characteristic of interest; this is referred to as interval estimation. An interval estimate provides one possible mechanism for answering the second question, "how sure can we be of our estimate?". The final question relates to the use of data as evidence. This is an important type of problem whose relevance is sometimes lost in the formality of the third topic, hypothesis testing. Although these labels are useful, we will try not to get too hung up on them; understanding the common elements in these topics is at least as important as understanding the differences.

The ideas of population and sample are useful. In fact, so useful that the terminology appears in situations where the population is not clearly identified. Consider the following illustration. A chemist measures the mass of a chemical reagent that is produced by an experiment. The experiment is repeated many times; on each occasion the mass is measured and all the measurements are collected as a data set. There is no obvious population in this illustration. However, we can still ask sensible inferential questions, such as, “what is a reasonable estimate of the mass of the reagent we might expect to be produced if we were to conduct the experiment again?”. The advantage of thinking in terms of statistical models is now more apparent. It may be reasonable to build a model for the measurements of mass of reagent that shares the properties of the model of distance travelled by the sea birds. Although there is no obvious population, the expected mass of reagent described by this model may still be referred to as the population mean. Although we are not doing any sampling, the data set may still be referred to as the observed sample. In this general context, the model may be thought of as our mathematical representation of the process that generated the data.

So far, the illustrations we have considered are univariate, that is, we consider a single characteristic of interest. Data sets that arise from real problems often contain measurements of several characteristics. For example, when the sea birds are captured, the zoologists might determine their gender, measure their weight and measure their wingspan before releasing them. In this context, we often refer to the characteristics that we record as variables. Data sets containing information on several variables are usually represented as a table with columns corresponding to variables and rows corresponding to individual sample members (a format familiar to anyone who has used a spreadsheet). We may treat all the variables as having identical status and, by building models that extend those used in the univariate case, try to take account of possible relationships between variables. This approach and its extensions are often referred to as multivariate analysis.

Data sets containing several variables often arise in a related, though subtly different, context. Consider the chemistry experiment described above. Suppose that the temperature of the reaction chamber is thought to have an impact on the mass of reagent produced. In order to test this, the experimenter chooses two temperatures and runs the experiment 100 times at the first temperature, then 100 times at the second temperature. The most convenient way to represent this situation is as two variables, one being the mass of reagent produced and the other temperature of reaction chamber. An inferential question arises: do the results of the experiment provide evidence to suggest that reaction chamber temperature has an impact on the mass of reagent produced? As we think that the variations in reaction chamber temperature may explain variations in the production of reagent, we refer to temperature as an explanatory variable, and mass of reagent produced as the response variable. This setup is a simple illustration of a situation where a linear model may be appropriate. Linear models are a broad class that include as special cases models for regression, analysis of variance, and factorial design.

In the example above, notice that we make a distinction between the variable that we control (the explanatory variable, temperature) and the variable that we measure (the response variable, mass of reagent). In many situations, particularly in the social sciences, our data do not come from a designed experiment; they are values of phenomena over which we exert no influence. These are sometimes referred to as observational studies. To illustrate, suppose that we collect monthly figures for the UK average house price and a broad measure of money supply. We may then ask “what is the impact of money supply on house prices?”. Although we do not control the money supply, we may treat it as an explanatory variable in a model with average house price as the response; this model may provide useful inferences about the association between the two variables.

The illustrations in this preface attempt to give a flavour of the different situations in which questions of statistical inference arise. What these situations have in common is that we are making inferences under uncertainty. In measuring the wingspan of sea birds, it is sampling from a population that introduces the uncertainty. In the house price illustration, in common with many economic problems, we are interested in a variable whose value is determined by the actions of tens of thousands of individuals. The decisions made by each individual will, in turn, be influenced by a large number of factors, many of them barely tangible. Thus, there are a huge number of factors that combine to determine the value of an economic variable. It is our lack of knowledge of these factors that is often thought of as the source of uncertainty. While sources of uncertainty may differ, the statistical models and methods that are useful remain broadly the same. It is these models and methods that are the subject of this book.

Chapter 2

# Probability

This chapter provides an introduction to the fundamental ideas in mathematical probability. The subject involves some formalism that, at first reading, might seem somewhat detached from everyday statements about probability. We attempt to explain how the formal definition of probability has evolved out of our intuition about how probability should behave. Methods for calculating probabilities when the outcomes are very numerous are given. In the final section we introduce conditional probability; this provides a natural mechanism for tackling many real-life problems.

## 2.1 Intuitive probability

Every year, at the start of the first lecture, we ask students to put their hand up if they do not know what probability is; no-one puts their hand up. We then ask for volunteers to explain probability to their colleagues; no-one volunteers. Probability is something about which we all have some intuitive notions, but these are rather hard to explain. The following simple example is used to illustrate.

**Example 2.1.1** (Roll of two fair dice)
We roll two fair dice. What is the probability that the sum of the values on the dice is greater than 10? You should be able to work this out easily. The rest of this section is an attempt to give a thorough account of the reasoning you might have used to arrive at your answer.

The first thing to note is that probabilities are always associated with events. The probability of an event is a number between 0 and 1 (inclusive) providing an indication of how likely the event is; an event with probability 0 will not happen while an event with probability 1 is certain to happen. We can stretch our intuition a bit further. Some informal definitions are helpful at this stage.

**Definition 2.1.2** (Experiment, sample space, and events)

i. An **experiment** is a repeatable procedure that has a well-defined set of possible outcomes.

ii. The **sample space**, $\Omega$, is the set of all possible outcomes of an experiment. Thus, any sample outcome $\omega$ is a member of the sample space $\Omega$, that is, $\omega \in \Omega$.

iii. An **event**, $A$, is a set of outcomes that is of interest to us. An event is a subset of the sample space, $A \subseteq \Omega$.

iv. The **complement** of $A$, denoted $A^c$, is the set of all outcomes not contained in $A$, that is, $A^c = \{\omega \in \Omega | \omega \notin A\}$.

If all the outcomes in the sample space are equally likely and the sample space is finite, we can construct an intuitively appealing definition of probability of the event $A$,

$$\mathrm{P}(A) = \frac{|A|}{|\Omega|} \tag{2.1}$$

where $|A|$ is the number of outcomes that are in $A$, and $|\Omega|$ is the total number of possible outcomes. The statement that the sample space is finite means that there is a finite number of possible outcomes of the experiment, that is, $|\Omega| < \infty$.

It is important to remember that probability is a mathematical construct. When we apply probability ideas to real situations we always make assumptions. Thus, probability statements are statements about a mathematical model, not statements about reality.

**Example 2.1.1 (Revisited)** In our example, the experiment is the act of throwing two fair dice. Our sample space is the set of all possible outcomes of the experiment,

$$\begin{aligned}
\Omega = \{&(1,1) & (1,2) && (1,3) && (1,4) && (1,5) && (1,6)\\
&(2,1) & (2,2) && (2,3) && (2,4) && (2,5) && (2,6)\\
&(3,1) & (3,2) && (3,3) && (3,4) && (3,5) && (3,6)\\
&(4,1) & (4,2) && (4,3) && (4,4) && (4,5) && (4,6)\\
&(5,1) & (5,2) && (5,3) && (5,4) && (5,5) && (5,6)\\
&(6,1) & (6,2) && (6,3) && (6,4) && (6,5) && (6,6)\}.
\end{aligned}$$

Here outcomes are of the form (value on first die, value on second die). The event we are interested in is that the sum of the two values of the dice is greater than 10. Put more precisely, the event is the set of outcomes for which the sum of the values of the two dice is greater than 10, $A = \{(5,6),(6,5),(6,6)\}$. Note that we could also write this as $A = \{(5,6)\} \cup \{(6,5)\} \cup \{(6,6)\}$. It is clear that $A \subset \Omega$. If we assume that all the outcomes in the sample space are equally likely, it makes sense to use the definition of probability given by counting outcomes. In this instance, $|A| = 3$ and $|\Omega| = 6 \times 6 = 36$, so $\mathrm{P}(A) = 1/12$.

The probability calculated above is a statement about a model for rolling two dice in which each of the possible outcomes is equally likely. In order for this to hold, we need each die to be fair, meaning that the six outcomes from rolling each die are all equally likely. This is a reasonable assumption but will not hold in practice. For example, imperfections in the dice may make some numbers slightly more likely than others.

**Exercise 2.1**

1. (**De Morgan's laws**) For any two events $A, B \subset \Omega$, show that

$$(A \cup B)^c = A^c \cap B^c \quad \text{and} \quad (A \cap B)^c = A^c \cup B^c .$$

2. An urn contains five balls numbered 1 to 5. We conduct a sample of two balls with replacement.
   (a) Draw a diagram to represent the sample space for this experiment. Mark the following events on your diagram:
   $A_1$ first ball drawn is a 5,
   $A_2$ second ball drawn has a value less than 4,
   $A_3$ sum of the values of the two draws is greater than or equal to 8.
   (b) Evaluate $\text{P}(A_1)$, $\text{P}(A_2)$, and $\text{P}(A_3)$.
   (c) Repeat parts (a) and (b) assuming no replacement.
3. Six people stand in a circle. We choose two at random, say $A$ and $B$. What is the probability that there are exactly $r$ people between $A$ and $B$ in the clockwise direction? What if the six people stand in a line instead of a circle? Try replacing 6 by $n$ for general problems of the same sort.

## 2.2 Mathematical probability

In order to develop a more rigorous framework for probability, we take a brief detour into an area of mathematics known as measure theory. The ideas here may seem a bit esoteric at first. Later we will see how they relate to our intuition about how probability should behave.

### 2.2.1 *Measure*

Consider a set, $\Psi$, and a subset, $A \subseteq \Psi$. We want to get some idea of the size of $A$. If $A$ is finite, one obvious way to do this is just to count the number of elements in $A$. Measures are functions acting on subsets that give us an idea of their size and generalise the notion of counting elements. Since a measure acts on subsets of the sample space, the domain for a measure will be a collection of subsets. In order to ensure that the measure can be defined sensibly, we need this collection to have certain properties.

**Definition 2.2.1** ($\sigma$-algebra)
Let $\Psi$ be a set and let $\mathcal{G}$ be a collection of subsets of $\Psi$. We say that $\mathcal{G}$ is a **$\sigma$-algebra** defined on $\Psi$ when the following conditions hold:

i. $\emptyset \in \mathcal{G}$,

ii. if $A \in \mathcal{G}$ then $A^c \in \mathcal{G}$,

iii. if $A_1, A_2, \ldots \in \mathcal{G}$ then $\bigcup_{i=1}^{\infty} A_i \in \mathcal{G}$.

We will discuss some of the intuitive reasoning behind these properties in the context of probability in section 2.2.2. The following example shows two $\sigma$-algebras that may be constructed for any set that has a non-trivial subset.

**Example 2.2.2** (Small and large $\sigma$-algebras)
Consider a set $\Psi$ together with a non-trivial subset $A \subset \Psi$. Two examples of $\sigma$-algebras defined on $\Psi$ are given below.

1. The smallest non-degenerate $\sigma$-algebra contains 4 elements, $\mathcal{G} = \{\emptyset, A, A^c, \Psi\}$, where $A \subset \Psi$.
2. The $\sigma$-algebra with the largest number of members is given by including every subset of $\Psi$. We can write this as $\mathcal{G} = \{A : A \subseteq \Psi\}$. This is referred to as the power set of $\Psi$, and is sometimes written $\mathcal{P}(\Psi)$ or $\{0, 1\}^{\Psi}$.

The pair consisting of a set and a $\sigma$-algebra defined on that set, $(\Psi, \mathcal{G})$, is referred to as a **measurable space**. As the name suggests, we define measure on $(\Psi, \mathcal{G})$.

**Definition 2.2.3** (Measure)
Given a measurable space $(\Psi, \mathcal{G})$, a **measure** on $(\Psi, \mathcal{G})$ is a function, $m : \mathcal{G} \to \mathbb{R}^+$, such that,

i. $m(A) \geq 0$ for all $A \in \mathcal{G}$,
ii. $m(\emptyset) = 0$,
iii. if $A_1, A_2, \ldots \in \mathcal{G}$ are disjoint then $m(\bigcup_{i=1}^{\infty} A_i) = \sum_{i=1}^{\infty} m(A_i)$.

The triple $(\Psi, \mathcal{G}, m)$, consisting of a set, a $\sigma$-algebra defined on that set, and a measure, is referred to as a **measure space**. The three properties given by Definition 2.2.3 correspond with our intuition about how size behaves. The first two are rather trivial; i. says "size is a non-negative quantity" and ii. says "if we do not have anything in our set then the size of our set is zero". The third requires a bit more thought but is fundamentally an extension of the idea that, if two sets have no elements in common, when we put them together the size of the resulting combined set is the sum of the sizes of the individual sets. As mentioned at the start, measure generalises the notion of counting the number of elements of a set. This is the starting point for Example 2.2.5. Before we consider the example we define an indicator function. This function will prove useful in a number of different contexts.

**Definition 2.2.4** (Indicator function)
The **indicator function** for a set $S \subseteq \mathbb{R}$ is $\mathbf{1}_S(x)$ where

$$\mathbf{1}_S(x) = \begin{cases} 1 & \text{for } x \in S, \\ 0 & \text{otherwise.} \end{cases}$$

The set $S$ often takes the form of an interval such as $[0, 1]$ or $[0, \infty)$. The indicator function $\mathbf{1}_S$ is simply a function taking two values: 1 on the set $S$, and 0 outside the set $S$. For example, if $S = [0, \infty)$ we have

$$\mathbf{1}_{[0,\infty)}(x) = \begin{cases} 1 & \text{for } x \geq 0, \\ 0 & \text{otherwise.} \end{cases}$$

**Example 2.2.5** (Counts and indicators as measures)
Suppose that $(\Psi, \mathcal{G})$ is a measurable space.

1. If $\Psi$ is finite and we define $m$ by $m(A) = |A|$ for all $A \in \mathcal{G}$, then $m$ is a measure.
2. If $\mathbf{1}_A$ is the indicator function for set $A$, then the function defined by $m(A) = \mathbf{1}_A(x)$ for $x \in \Psi$ is a measure.

Showing that the functions defined in this example are measures is part of Exercise 2.2.

### 2.2.2 *Probability measure*

In this section we will show how the framework of section 2.2.1 allows us to develop a rigorous definition of probability. Measure gives us a sense of the size of a set. Probability tells us how likely an event is. We will put these two ideas together to define probability as a measure.

To define a measure we need a measurable space, that is, a set and a $\sigma$-algebra defined on the set. Our intuitive description of probability in section 2.1 introduces the idea of a sample space, $\Omega$, the set of all possible outcomes of our experiment. We also define events as subsets of $\Omega$ containing outcomes that are of interest. From this setup we can generate a measurable space, $(\Omega, \mathcal{F})$, where $\mathcal{F}$ is a $\sigma$-algebra defined on $\Omega$. Here $\mathcal{F}$ is a collection of subsets of $\Omega$ (as usual), and we interpret the elements of $\mathcal{F}$ as being events. Thus, if $A \in \mathcal{F}$ then $A$ is an event. Remember that probability is always associated with events so $\mathcal{F}$ will be the domain for probability measure.

**Definition 2.2.6** (Probability measure)
Given a measurable space $(\Omega, \mathcal{F})$, a **probability measure** on $(\Omega, \mathcal{F})$ is a measure $\mathrm{P} : \mathcal{F} \to [0, 1]$ with the property that $\mathrm{P}(\Omega) = 1$.

Note that, as we might expect, the definition restricts the codomain of P to be the unit interval, $[0, 1]$. The triple consisting of a sample space, a collection of events (forming a $\sigma$-algebra on the sample space), and a probability measure, $(\Omega, \mathcal{F}, \mathrm{P})$, is referred to as a **probability space**.

We give two examples of functions which satisfy the conditions for probability measures. Showing that these functions are probability measures is part of Exercise 2.2.

**Example 2.2.7** (Intuitive and not so intuitive probability measures)
Suppose that we have a measurable space $(\Omega, \mathcal{F})$. Two functions that we have encountered before satisfy the properties of probability measures.

1. Equation (2.1) defines a probability measure P by $\mathrm{P}(A) = |A|/|\Omega|$ for $A \in \mathcal{F}$, where $|A|$ is the number of outcomes in $A$. It is easy to show that this satisfies the conditions for a probability measure.

2. Not all probability measures are so intuitively appealing. If we define $\mathrm{P}(A) = \mathbf{1}_A(\omega)$ for some $\omega \in \Omega$, this is a perfectly valid probability measure; it takes the value 1 for any event containing outcome $\omega$ and 0 for all other events.

We now have a definition of probability based on the set-theoretic notion of measure. It is worth taking a bit of time to check that we fully understand the implications of this definition. First, note that events are sets but we use them a bit sloppily – this is fine because no ambiguity arises. For example, when we say "[event] $A$ occurs" we mean "the outcome of the experiment is a member of [set] $A$". When we write $\mathrm{P}(A)$ we refer to this as "the probability of [event] $A$" which can be interpreted as "the probability that the outcome of the experiment is in [set] $A$". Table 2.1 elaborates.

| Statement | Interpretation:<br>"probability that outcome of experiment is . . . |
|---|---|
| $\mathrm{P}(A)$ | ". . . in $A$" |
| $\mathrm{P}(A^c)$ | ". . . not in $A$" |
| $\mathrm{P}(A \cap B)$ | ". . . in both $A$ and $B$" |
| $\mathrm{P}(A \cup B)$ | ". . . in $A$ or $B$ or in both" |
| $\mathrm{P}(A \backslash B)$ | ". . . in $A$ but not in $B$" |

Table 2.1 *The interpretation of some probability statements thinking about events A and B as sets of outcomes.*

It is important to understand the distinction between outcomes and events. This is analogous to the distinction between elements and subsets in set theory: an outcome is an element of the sample space and an event is a subset of the sample space. For example, suppose

$$\Omega = \{\omega_1, \omega_2, \ldots\},$$

then $\omega_1, \omega_2, \ldots$ are outcomes, and subsets like $\{\omega_1, \omega_3, \omega_5\}$ and $\{\omega_7, \omega_{207}\}$ are events. In every practical example that we consider, we take the collection of events to be the power set, $\mathcal{F} = \{0, 1\}^\Omega$, that is, the collection of all possible subsets of $\Omega$. In this case, sets containing single outcomes will be events, $\{\omega_1\}, \{\omega_2\}, \ldots$ (note the enclosing curly brackets). The events containing single outcomes are, by definition, disjoint. In a slight abuse of terminology we may refer to "the probability of [outcome] $\omega_r$" when we mean "the probability of [event] $\{\omega_r\}$".

Events form a $\sigma$-algebra on the sample space, and probability is a measure on the resulting measurable space. This means that the events form a collection that satisfies the properties given in Definition 2.2.1 of a $\sigma$-algebra, and probability is a function that satisfies the properties given in Definition 2.2.3 of a measure. We will go through some of these properties in turn to try to establish that they correspond with our intuition about how events and probability should behave.

Suppose that we have a probability space $(\Omega, \mathcal{F}, \mathrm{P})$.

i. **Empty set and sample space**: By property i. of a $\sigma$-algebra, $\emptyset \in \mathcal{F}$. By definition $\Omega = \emptyset^c$, so by property ii. of a $\sigma$-algebra, $\Omega \in \mathcal{F}$. This means that the empty set is an event and the sample space is an event. By property i. of a measure, $\mathrm{P}(\emptyset) = 0$. A probability measure has the additional property that $\mathrm{P}(\Omega) = 1$. You should not think of the event $\emptyset$ as nothing happening; nothing happening might be an outcome with positive probability in some experiments. Rather, you should remember that, since $\mathrm{P}(\Omega) = 1$, the sample space includes all possible outcomes of the experiment, so $\Omega$ is often referred to as a **certain event** and $\emptyset$ is an **impossible event**.

ii. **Complement**: Property ii. of a $\sigma$-algebra says that if $A \in \mathcal{F}$ then $A^c \in \mathcal{F}$. This means that if $A$ is an event then $A^c$ is also an event. In other words, if we are interested in outcomes in $A$ then we are, by default, also interested in outcomes that are not in $A$. This ensures that we can always sensibly talk about the probability of events not happening; $A^c \in \mathcal{F}$ means that $\mathrm{P}(A^c)$ is well defined.

iii. **Union**: Property iii. of a $\sigma$-algebra says that if $A_1, A_2, \ldots \in \mathcal{F}$ then $\bigcup_{i=1}^{\infty} A_i \in \mathcal{F}$. This means that the union of a collection of events is also an event. The property could be stated as "the collection of events is closed under union". It ensures that if $A$ and $B$ are events, the probability that $A$ or $B$ (or both) happens, that is, $\mathrm{P}(A \cup B)$, is well defined. In addition, property iii. of a measure tells us that if $A$ and $B$ do not have any outcomes in common then the probability that $A$ or $B$ happens is just the sum of the individual probabilities, $\mathrm{P}(A \cup B) = \mathrm{P}(A) + \mathrm{P}(B)$.

From the definition of probability as a measure we can derive a number of useful results. Some of these, such as the fact that the probability of an event not happening is one minus the probability that it happens, seem so obvious that you might have been taught them without proof. We now have the tools to establish some properties of probability from first principles.

**Lemma 2.2.8** (Basic properties of probability)
*Consider a probability space $(\Omega, \mathcal{F}, \mathrm{P})$ with $A, B \in \mathcal{F}$, that is, $A$ and $B$ are events.*

*i.* $\mathrm{P}(A^c) = 1 - \mathrm{P}(A)$.

*ii. If* $A \subseteq B$ *then* $\mathrm{P}(B \backslash A) = \mathrm{P}(B) - \mathrm{P}(A)$.

*iii.* $\mathrm{P}(A \cup B) = \mathrm{P}(A) + \mathrm{P}(B) - \mathrm{P}(A \cap B)$.

*Proof.*

i. By definition of $A^c$ we have $A \cap A^c = \emptyset$ and $A \cup A^c = \Omega$. In other words $A$ and $A^c$ are disjoint events whose union is the whole sample space. Thus, $\mathrm{P}(\Omega) = \mathrm{P}(A \cup A^c) = \mathrm{P}(A) + \mathrm{P}(A^c)$ by property iii. of measure. We know $\mathrm{P}(\Omega) = 1$ by the additional property of probability measure. Putting these together we get $\mathrm{P}(A) + \mathrm{P}(A^c) = 1$ and the result follows immediately.

ii. We use reasoning similar to i. above. Here, rather than partitioning the sample space into disjoint events, we partition the set $B$. Since $A \subseteq B$, we can write $B = A \cup (B \backslash A)$. By definition, $A \cap (B \backslash A) = \emptyset$. Thus, $\mathrm{P}(B) = \mathrm{P}(A) + \mathrm{P}(B \backslash A)$ by property iii. of measure and the result follows immediately.

iii. Again, our approach is to partition the set of interest, $A \cap B$, into disjoint events. We can write $A \cup B = A \cup (B \backslash A)$, thus, $\mathrm{P}(A \cup B) = \mathrm{P}(A) + \mathrm{P}(B \backslash A)$. The final step is to note that $B = (B \cap A) \cup (B \backslash A)$, which are also disjoint events, and so $\mathrm{P}(B) = \mathrm{P}(B \cap A) + \mathrm{P}(B \backslash A) \Leftrightarrow \mathrm{P}(B \backslash A) = \mathrm{P}(B) - \mathrm{P}(B \cap A)$. The result follows immediately.

□

**Corollary 2.2.9**
*Consider a probability space* $(\Omega, \mathcal{F}, \mathrm{P})$ *with* $A, B \in \mathcal{F}$. *If* $A \subseteq B$ *then* $\mathrm{P}(A) \leq \mathrm{P}(B)$.

The proofs of part iii. of Lemma 2.2.8 and Corollary 2.2.9 are part of Exercise 2.2. The proofs of the first two parts of Lemma 2.2.8 illustrate a useful general method. In each instance we partition an event of interest into disjoint events. Once we have disjoint events, we can add probabilities. We will return to this idea in section 2.4.2. Proposition 2.2.10 is sometimes referred to as the **inclusion–exclusion principle**.

**Proposition 2.2.10** (General Addition Law)
*Consider a probability space* $(\Omega, \mathcal{F}, \mathrm{P})$. *If* $A_1, \ldots, A_n \in \mathcal{F}$ *then*

$$\mathrm{P}\left(\bigcup_{i=1}^{n} A_i\right) = \sum_i \mathrm{P}(A_i) - \sum_{i<j} \mathrm{P}(A_i \cap A_j) + \sum_{i<j<k} \mathrm{P}(A_i \cap A_j \cap A_k) + \ldots$$
$$+ (-1)^{n+1} \mathrm{P}(A_1 \cap \ldots \cap A_n).$$

This result generalises part iii. of Lemma 2.2.8.

One of the properties inherited from measure is that the probability of a union of disjoint events is the sum of the individual probabilities. If the events are not disjoint, the relationship becomes an inequality.

**Proposition 2.2.11** (Boole's inequality)
*If* $(\Omega, \mathcal{F}, \mathrm{P})$ *is a probability space and* $A_1, A_2, \ldots \in \mathcal{F}$ *then* $\mathrm{P}(\bigcup_{i=1}^{\infty} A_i) \leq \sum_{i=1}^{\infty} \mathrm{P}(A_i)$.

*Proof.*

The result here is intuitively obvious: when we add up the probabilities of overlapping events, the overlaps get included repeatedly so the resulting probability is larger than the probability of the union. The basic method of proof is (as ever) to partition the event of interest, $\bigcup_{i=1}^{\infty} A_i$, into disjoint events. We do this recursively by removing from each event the part that overlaps with events we have already considered. This yields

$$B_1 = A_1,\ B_2 = A_2 \backslash B_1,\ B_3 = A_3 \backslash (B_1 \cup B_2), \ldots, B_i = A_i \backslash \bigcup_{j=1}^{i-1} B_j, \ldots$$

Then, $B_1, B_2, \ldots \in \mathcal{F}$ are disjoint with $\bigcup_{i=1}^{\infty} A_i = \bigcup_{i=1}^{\infty} B_i$, (see Exercise 2.2). The proof is completed by noting that $B_i \subseteq A_i$ so, by Corollary 2.2.9, $\mathrm{P}(B_i) \leq \mathrm{P}(A_i)$ for all $i$, and thus,

$$\mathrm{P}\left(\bigcup_{i=1}^{\infty} A_i\right) = \mathrm{P}\left(\bigcup_{i=1}^{\infty} B_i\right) = \sum_{i=1}^{\infty} \mathrm{P}(B_i) \leq \sum_{i=1}^{\infty} \mathrm{P}(A_i).$$

□

The notion of a sequence of sets is used in Chapter 3 to prove results associated with random variables. The following proposition establishes the connection between limits of probabilities and sequences of sets.

**Proposition 2.2.12** (Probability limit of a sequence of sets)

i. *If* $\{A_i : i = 1, 2, \ldots\}$ *is an increasing sequence of sets* $A_1 \subseteq A_2 \subseteq A_3 \subseteq \ldots$ *then* $\lim_{n\to\infty} \mathrm{P}(A_n) = \mathrm{P}\left(\bigcup_{i=1}^{\infty} A_i\right)$.

ii. *If* $\{A_i : i = 1, 2, \ldots\}$ *is a decreasing sequence of sets* $A_1 \supseteq A_2 \supseteq A_3 \supseteq \ldots$ *then* $\lim_{n\to\infty} \mathrm{P}(A_n) = \mathrm{P}\left(\bigcap_{i=1}^{\infty} A_i\right)$.

*Proof.*

We prove part i. The proof of part ii. is part of Exercise 2.2. Define $B_1 = A_1$ and $B_i = A_i \backslash A_{i-1}$ for $i \geq 2$. This definition ensures that the $B_i$s are disjoint and that $A_n = \bigcup_{i=1}^{n} B_i$ for any $n$. Thus, $\{B_1, \ldots, B_n\}$ forms a partition of $A_n$ and

$$\mathrm{P}(A_n) = \sum_{i=1}^{n} \mathrm{P}(B_i).$$

It is also clear from the definition of the $B_i$s that

$$\bigcup_{i=1}^{\infty} A_i = \bigcup_{i=1}^{\infty} B_i.$$

Putting these facts together yields

$$\lim_{n\to\infty} \mathrm{P}(A_n) = \lim_{n\to\infty} \sum_{i=1}^{n} \mathrm{P}(B_i) = \sum_{i=1}^{\infty} \mathrm{P}(B_i) = \mathrm{P}\left(\bigcup_{i=1}^{\infty} B_i\right) = \mathrm{P}\left(\bigcup_{i=1}^{\infty} A_i\right).$$

Note that $\mathrm{P}\left(\bigcup_{i=1}^{\infty} A_i\right)$ is well defined since $\bigcup_{i=1}^{\infty} A_i \in \mathcal{F}$, by definition of a $\sigma$-algebra. □

**Exercise 2.2**

1. (**Various facts about $\sigma$-algebras**)
   (a) What is the smallest possible $\sigma$-algebra on $\Psi$?
   (b) Show that $\{\emptyset, A, A^c, \Psi\}$, where $A \subset \Psi$, is a $\sigma$-algebra on $\Psi$.
   (c) If $|\Psi| < \infty$, how many elements are there in the power set $\{0, 1\}^{\Psi}$? Can you think of an explanation for the notation $\{0, 1\}^{\Psi}$?

(d) Use de Morgan's laws to show that a $\sigma$-algebra is closed under intersection, that is, if $A_1, A_2, \ldots \in \mathcal{G}$ then $\bigcap_{i=1}^{\infty} A_i \in \mathcal{G}$.

2. Consider a set $\Psi$ and two subsets $A, B \subset \Psi$, where $A \cap B \neq \emptyset$ but $A \cap B \neq A$ and $A \cap B \neq B$. What is the size of the smallest $\sigma$-algebra containing both $A$ and $B$? What about if $A \subset B$ or $A \cap B = \emptyset$. Generalise your results to 3 sets then to $N$ sets.

3. (**Counts and indicators as measures**) Consider a measurable space $(\Psi, \mathcal{G})$. Show that, if $A \in \mathcal{G}$, both of the following are measures:
   (a) $m(A) = |A|$,
   (b) $m(A) = \mathbf{1}_A(x)$ for given $x \in \Psi$.

4. (**Intuitive and not so intuitive probability measures**) Consider a sample space $\Omega$ and an event $A$. Show that both of the functions P defined below satisfy the conditions for a probability measure.
   (a) $\mathrm{P}(A) = |A|/|\Omega|$ where $|\Omega| < \infty$,
   (b) $\mathrm{P}(A) = \mathbf{1}_A(\omega)$ for some fixed $\omega \in \Omega$.

5. (**Properties of probability and events**)
   (a) Prove part iii. of Lemma 2.2.8 and Corollary 2.2.9 using only the definition of probability measure and parts i. and ii. of the lemma.
   (b) Let $A_1, A_2, \ldots \in \mathcal{F}$, and define $B_1 = A_1$ and $B_i = A_i \backslash \bigcup_{j=1}^{i-1} B_j$ for $i > 1$. Show that $B_i \in \mathcal{F}$ and $\bigcup_{j=1}^{i} B_j = \bigcup_{j=1}^{i} A_j$ for all $i$, then prove that $B_i \cap B_j = \emptyset$ for any $i \neq j$.
   (c) Prove part ii. of Proposition 2.2.12 using part i.

## 2.3 Methods for counting outcomes

Suppose that we have a sample space $\Omega$ which is finite, that is, $|\Omega| < \infty$, and a probability measure $\mathrm{P}(A) = |A|/|\Omega|$ for all $A \in \mathcal{F}$. Example 2.1.1 gives a simple illustration of throwing two fair dice, for which $|\Omega| = 6 \times 6 = 36$. In this case the entire sample space is easy to map out and the probabilities of events are readily calculated by counting. In practical experiments, the sample space is usually too large to be written down in its entirety.

**Example 2.3.1**

1. To play Lotto you select six numbers from $\{1, \ldots, 59\}$, pay £2 and receive a ticket with your numbers printed on it. The main Lotto draw on a Saturday night consists of six balls selected at random without replacement from an urn containing 59 balls labelled $1, \ldots, 59$. Initially, we would like to know the probability of winning with a single entry, that is, we would like to know the probability that the six numbers drawn match those on our ticket.

2. I have a lecture in a room with 100 people in it (99 students and me). What is the probability that there is at least one pair of people with the same birthday?

In order to tackle problems of the sort given in Example 2.3.1 we need to develop formal methods for counting outcomes. The basis of these methods is the following simple claim.

**Claim 2.3.2** (Multiplication rule)
*Suppose that $Q_1, \ldots, Q_k$ are experiments and that experiment $Q_i$ has $n_i$ possible outcomes for $i = 1, \ldots, k$. The experiment consisting of the ordered sequence $(Q_1, Q_2, \ldots, Q_k)$ has $n_1 \cdot n_2 \cdot \ldots \cdot n_k = \prod_{i=1}^{k} n_i$ possible outcomes.*

**Example 2.3.1 (Revisited I)**
We can use the multiplication rule to calculate the size of the sample space in each of our two cases.

1. In Lotto there are 59 ways to choose the first ball. Since this ball is not replaced, there are 58 ways to choose the second ball, and so on. Thus the number of outcomes of the form (ball 1, ..., ball 6) is $59 \times 58 \times 57 \times 56 \times 55 \times 54 = 3.244 \times 10^{10}$ (4 significant figures). We assume, on the basis that the balls are selected at random, that each of these outcomes is equally likely. Of course for the lottery the order in which the numbers appear is unimportant – we will consider this in more detail in section 2.3.1.
2. First of all note that the question is ill posed. Considering the birthdays of a given group of 100 people does not constitute an experiment – it is not repeatable. The real question is: if we were to take a random sample of 100 people from an entire population, what is the probability that at least one pair of people have the same birthday? In order to make the problem manageable we make assumptions. First we assume that there are 365 days in a year (apologies to those of you born on 29 February). This means that the number of outcomes of the form (my birthday, student 1 birthday, ..., student 99 birthday) is $365 \times 365 \times \ldots \times 365 = 365^{100} = 1.695 \times 10^{256}$. We assume that each of these outcomes is equally likely. Note that the assumptions we are making are clearly untrue; some years have 366 days and some birthdays are more common than others (in the US, September 9th is reported to be the most common birthday). However, the model that we propose is a reasonable approximation to reality so the probabilities will be meaningful.

### 2.3.1 *Permutations and combinations*

In this section we start by defining **permutations** and **combinations**. Our real interest lies in counting permutations and combinations. We will show how the same general counting rules can be applied to a number of apparently different problems. Our eventual aim is to use these rules to evaluate probabilities.

**Definition 2.3.3** (Permutations and combinations)
Suppose that we have a set $Q$ of $n$ distinct objects.

i. A permutation of length $k \leq n$ is an **ordered** subset of $Q$ containing $k$ elements.

ii. A combination of length $k \leq n$ is an **unordered** subset of $Q$ containing $k$ elements.

We distinguish between these two cases by using $(\ldots)$ to denote permutation and $\{\ldots\}$ to denote combination.

**Claim 2.3.4** (Number of permutations and number of combinations)

*i. If the number of permutations of length k that can be formed from n distinct elements is denoted* $^nP_k$*, then*

$$^nP_k = \frac{n!}{(n-k)!}.$$

*ii. If the number of combinations of length k that can be formed from n distinct elements is denoted* $^nC_k$*, then*

$$^nC_k = \frac{n!}{k!(n-k)!}.$$

The number of permutations, $^nP_k = n \times (n-1) \times \ldots \times (n-k+1)$, is a direct consequence of the multiplication rule. Note that one implication of this is that the number of ways of ordering all $n$ elements is $^nP_n = n!$. The general expression for the number of combinations requires a little more thought.

Suppose that we know the number of combinations of length $k$, that is, we know $^nC_k$. By the above argument, the number of ways of ordering each one of these combinations is $k!$. The multiplication rule then tells us that $^nP_k = k!^nC_k$. By rearranging we arrive at the general result, $^nC_k = {^nP_k}/k!$.

**Example 2.3.5** (Illustrating the difference between permutations and combinations)
Consider a set $Q$ consisting of the first 9 integers, $Q = \{1,2,\ldots,9\}$. An example of a permutation of length 4 from $Q$ is $q_1 = (6,3,4,9)$. The permutation $q_2 = (3,6,4,9)$ is distinct from $q_1$ because, although the elements are the same, the order is different. In contrast, the combinations $c_1 = \{6,3,4,9\}$ and $c_2 = \{3,6,4,9\}$ are the same since order is unimportant. The number of permutations of length 4 from a set of 9 objects is, by Claim 2.3.4,

$$^9P_4 = \frac{9!}{(9-4)!} = \frac{9!}{5!} = 9 \times 8 \times 7 \times 6 = 3024.$$

Justification is provided by the multiplication rule (Claim 2.3.2); there are 9 ways of choosing the first element, 8 ways of choosing the second, and so on. Now consider the number of combinations. From Claim 2.3.4, the number of combinations of length 4 from a set of 9 objects is

$$^9C_4 = \frac{9!}{4!5!} = {^9P_4}/4! = 126.$$

It is important not to become too fixed on the idea of permutations as being associated with ordered sequences and combinations with unordered sequences; in section 2.3.2 we will see that the number of combinations arises in several different contexts. In problems where order is unimportant, such as the Lotto problem, either method can be used to calculate probabilities. Permutations can be introduced rather artificially to solve problems such as the birthday problem. We are finally in a position to answer both of the questions posed in Example 2.3.1.

**Example 2.3.1 (Revisited II)**

1. As indicated above, although order is unimportant in a lottery, we can use either permutations or combinations to solve this problem – in this simple example, the equivalence of these approaches is obvious. Suppose $A$ is the event that the numbers in the draw match those on our ticket.
   (a) Permutations method: The sample space, $\Omega$, consists of all possible permutations of length 6 from $\{1,\ldots,59\}$. We have already calculated $|\Omega| = {}^{59}P_6 = 3.244 \times 10^{10}$. Since order is unimportant, any of the ${}^{6}P_6 = 6!$ outcomes with the same numbers as on our ticket will result in us winning the jackpot, so $|A| = 6!$. By intuitive probability measure (2.1), $\mathrm{P}(A) = 6!/{}^{59}P_6 = \frac{1}{45057474} = 2.219 \times 10^{-8}$ (4 significant figures).
   (b) Combinations method: Since order is unimportant, we may take $\Omega$ to be all possible combinations of length 6 from $\{1,\ldots,59\}$. Thus, $|\Omega| = {}^{59}C_6 = {}^{59}P_6/6!$. Only one of these outcomes corresponds to the numbers on our ticket so $\mathrm{P}(A) = 1/{}^{59}C_6 = 6!/{}^{59}P_6$, yielding exactly the same answer as the permutations approach.
2. We will solve the general problem of finding the probability that, for a random sample of $k(\leq 365)$ people, there is at least one pair with the same birthday. We have established that the sample space, $\Omega$, is every possible sequence of the form (person 1 birthday,..., person $k$ birthday). Assuming just 365 days in a year and using the multiplication rule we have $|\Omega| = 365^k$. Note that birthday sequences are not permutations since the birthdays do not have to be distinct objects. In fact, we are interested in the event, $A$, that at least two of the entries in our birthday sequence are the same. Calculating $\mathrm{P}(A)$ directly is rather a hard problem. However, calculating $\mathrm{P}(A^c)$, the complementary problem, is remarkably easy. Using (2.1), $\mathrm{P}(A^c) = |A^c|/|\Omega|$ and thus, by Lemma 2.2.8, $\mathrm{P}(A) = 1 - |A^c|/|\Omega|$. We have reduced our problem to calculating $|A^c|$, that is, calculating the number of birthday sequences in which all the birthdays are distinct. Since we are now considering distinct objects, this can now be thought of naturally in terms of permutations; there are 365 ways of choosing the first person's birthday, 364 ways of choosing the second person's birthday and so on. Thus, $|A^c| = 365\times 364\times\ldots\times(365-k+1) = {}^{365}P_k$ and $\mathrm{P}(A) = 1-{}^{365}P_k/365^k$. Note that when $k = 100$, as in the original question, $\mathrm{P}(A) = 0.9999999693$ (10 decimal places).

### 2.3.2 *Number of combinations and multinomial coefficients*

In Claim 2.3.4 we define ${}^nC_k = n!/(k!(n-k)!)$ as being the number of combinations of length $k$ from $n$ distinct objects. These numbers arise in a number of different – sometimes surprising – contexts and are worthy of consideration in their own right. A common notation for the number of combinations is

$$\binom{n}{k} = {}^nC_k = \frac{n!}{k!(n-k)!}. \tag{2.2}$$

This quantity is sometimes referred to as "$n$ **choose** $k$". We start by considering a property that is closely related to our original definition in terms of counting combinations.

**Proposition 2.3.6**
*Consider a collection of n objects, k of which are of type a and $(n-k)$ of which are of type b. The number of ways of arranging these objects into sequences of type a and type b is $\binom{n}{k}$.*

*Proof.*

Consider the problem as one of positioning $k$ things of type $a$ into $n$ slots (the remaining slots will be filled with things of type $b$). If we label the slots $1, \dots, n$, the problem is then equivalent to selecting a set of $k$ numbers from $\{1, \dots, n\}$; each number we choose will give us a position occupied by something of type $a$, so order is unimportant. By Claim 2.3.4, the number of ways of doing this is $\binom{n}{k}$. □

The number of combinations also appears in the expansion of expressions of the form $(a+b)^n$. In order to expand this type of expression we can write it out in full and multiply out the brackets; for example,

$$\begin{aligned}
(a+b) &= a+b, \\
(a+b)^2 &= (a+b)(a+b) = a^2 + ab + ba + b^2 = a^2 + 2ab + b^2, \\
(a+b)^3 &= (a+b)(a+b)(a+b) = \dots = a^3 + 3a^2b + 3ab^2 + b^3, \\
&\vdots
\end{aligned}$$

This becomes rather tedious. Fortunately, Proposition 2.3.6 allows us to write down a general closed form for $(a+b)^n$. Start by considering the coefficients in the expansion of $(a+b)^4 = (a+b)(a+b)(a+b)(a+b)$. We know that there is only one way to get $a^4$, since we have to take the $a$ term from each bracket. To get $a^3b$ we must take the $a$ term from 3 brackets and the $b$ term from 1. Since there are 4 brackets, there are 4 ways of selecting the $b$ term so the coefficient of $a^3b$ will be 4. For the $a^2b^2$ term we have to take the $a$ term from 2 brackets and the $b$ term from 2. This is analogous to arranging 2 things of type $a$ and 2 of type $b$, and we know that the number of ways in which this can be done is $\binom{4}{2} = 6$. In general, $\binom{n}{k}$ will be the coefficient of $a^k b^{n-k}$ in the expansion of $(a + b)^n$, and we can write

$$(a + b)^n = \sum_{k=0}^{n} \binom{n}{k} a^k b^{n-k}. \tag{2.3}$$

For this reason, quantities of the form $\binom{n}{k}$ are sometimes referred to as **binomial coefficients**.

**Proposition 2.3.7** (Basic properties of binomial coefficients)

*i.* $\binom{n}{k} = \binom{n}{n-k}$, *in particular* $\binom{n}{0} = \binom{n}{n} = 1$ *and* $\binom{n}{1} = \binom{n}{n-1} = n$.
*ii.* $\binom{n}{k} = \binom{n-1}{k} + \binom{n-1}{k-1}$.
*iii.* $\sum_{k=0}^{n} \binom{n}{k} = 2^n$.

*Proof.*

Part i. is a simple consequence of the definition. Part ii. can be established by substituting in the right-hand side from the definition given by (2.2) and adding the resulting fractions. Part iii. is a consequence of the multiplication rule; the sum represents every possible way of labelling $n$ slots as either type $a$ or type $b$. It is also given by taking $a = b = 1$ in (2.3). □

Part ii. of Proposition 2.3.7 says that each binomial coefficient is the sum of two coefficients of lower order. This forms the basis of a beautiful construction known as **Pascal's triangle**. This should really be called **Yang Hui's triangle**, as it was discovered five centuries before Pascal by the Chinese mathematician Yang Hui. Starting with 1s down the outside of the triangle, we sum adjacent entries to generate the next row. If we label the top row of the triangle as row 0 then, for $k = 0, \ldots, n$, the binomial coefficient $\binom{n}{k}$ is the $k^{\text{th}}$ entry in the $n^{\text{th}}$ row.

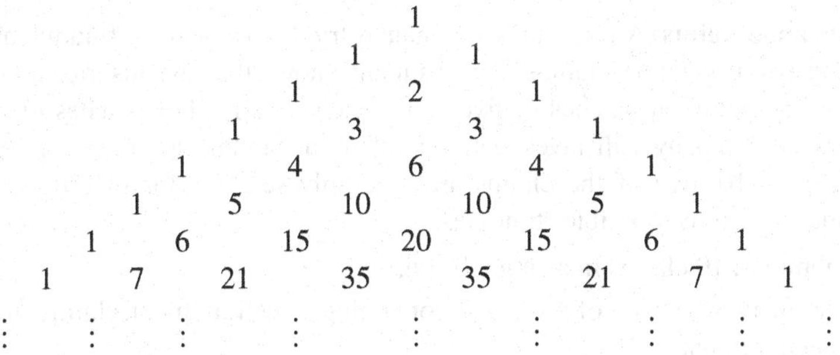


Figure 2.1 *Pascal's (or Yang Hui's) triangle.*

Part iii. of Proposition 2.3.7 can be thought of in the context of binary representation of integers. Suppose that we have $n$ binary digits ($n$ bits). Each one of these can take the value 0 or 1, so by the multiplication rule we can represent $2^n$ integers using $n$ bits. If we start at 0 then these numbers will be $0, 1, \ldots, 2^n - 1$.

We can generalize the notion of a binomial coefficient to the multinomial case.

**Proposition 2.3.8**
*Suppose that we have n objects where $n_1$ are of type 1, $n_2$ are of type 2, ..., $n_r$ are of type r. The number of ways of arranging these n objects is*

$$(n_1, n_2, \ldots, n_r)! = \frac{n!}{n_1! n_2! \ldots n_r!},$$

*where* $\sum_{i=1}^{r} n_i = n$.

*Proof.*

Suppose that we know the number of ways of arranging the objects, that is, we know $(n_1, n_2, \ldots, n_r)!$. If we treated all the objects of type 1 as being distinct this would increase the number of arrangements by a factor of $n_1!$. In general, if we treat the objects of type $j$ as being distinct, this increases the number of arrangements by a factor of $n_j!$. If we treat all the objects as being distinct, we know the number of arrangements is ${}^nP_n = n!$. Thus, by the multiplication rule, $n_1! n_2! \ldots n_r! (n_1, n_2, \ldots, n_r)! = n!$. The result follows by rearrangement. □

Quantities of the form $(n_1, n_2, \ldots, n_r)!$ are referred to as **multinomial coefficients.** The notation here is a little bit cumbersome; note that in the binomial case we have $(k, n-k)! = \binom{n}{k}$. The multinomial coefficients feature in the multinomial expansion

$$(a_1 + a_2 + \ldots + a_r)^n = \sum_{\substack{n_1, \ldots, n_r \geq 0 \\ n_1 + \ldots + n_r = n}} (n_1, n_2, \ldots, n_r)! a_1^{n_1} a_2^{n_2} \ldots a_r^{n_r}.$$

**Exercise 2.3**

1. (**Insurance claims**) An insurance claimant is trying to hide three fraudulent claims among seven genuine claims. The claimant knows that the insurance company processes claims in batches of five or in batches of 10. For batches of five, the insurance company will investigate one claim at random to check for fraud; for batches of 10, two of the claims are randomly selected for investigation. The claimant has three possible strategies:
   (a) submit all 10 claims in a single batch,
   (b) submit two batches of five, one containing two fraudulent claims, the other containing one,
   (c) submit two batches of five, one containing three fraudulent claims, the other containing 0.

   What is the probability that all three fraudulent claims will go undetected in each case? What is the optimal strategy for the fraudster?

2. **(Lotto)** The Lotto game described in example 2.3.1 is a slight simplification of the real game. In addition a seventh, bonus ball is drawn. There are five ways of winning a prize:

| | |
|---|---|
| Match six | £5 000 000 |
| Match five and bonus ball | £250 000 |
| Match five | £2 500 |
| Match four | £50 |
| Match three | £10 |

   In reality, prize amounts are variable but the above table is a reasonable approximation. Note that matching three means that three (and no more) of the numbers on your ticket match those of the main draw (not bonus ball). The same applies for matching four, matching five, and matching six. The bonus ball is only considered for the second prize. Calculate the expected return for a single ticket.

3. **(Poker)** Suppose that a poker hand is made up of five cards selected at random (without replacement obviously) from a deck of 52, made up of four suits each containing denominations $2, \dots, 10, J, Q, K, A$. Calculate the probability of each of the following hands:
   (a) A full house (three of one denomination, two of another).
   (b) One pair (two of one denomination, all others distinct).
   (c) A straight (five consecutive denominations, not all the same suit). Note that aces can be either low or high, so $A, 2, 3, 4, 5$ and $10, J, Q, K, A$ are both valid straights.

## 2.4 Conditional probability and independence

### 2.4.1 *Conditional probability*

We start with the by-now-familiar setup of an experiment, a probability space $(\Omega, \mathcal{F}, \mathrm{P})$, and an event of interest $A$. The probability $\mathrm{P}(A)$ gives us an indication of how likely it is that the outcome of the experiment, $\omega \in \Omega$, is in $A$, that is, how likely the event $A$ is to occur. Now suppose that we know that event $B$ has occurred. This will alter our perception of how probable $A$ is since we are now only interested in outcomes that are in $B$. In effect we have shrunk the sample space from $\Omega$ to $B$. In these circumstances the appropriate measure is given by conditional probability.

**Definition 2.4.1** (Conditional probability)
Consider the probability space $(\Omega, \mathcal{F}, \mathrm{P})$ and $A, B \in \mathcal{F}$ with $\mathrm{P}(B) > 0$. The **conditional probability** of $A$ given $B$ is the probability that $A$ will occur given that $B$ has occurred,

$$\mathrm{P}(A|B) = \frac{\mathrm{P}(A \cap B)}{\mathrm{P}(B)}.$$

Note the importance of the statement that $\mathrm{P}(B) > 0$. We cannot condition on events with zero probability. This makes sense intuitively; it is only reasonable to condition on events that have some chance of happening. The following example illustrates the use of Definition 2.4.1.

**Example 2.4.2** (Rolling two dice again)
Consider the setup in Example 2.1.1. We have shown that, if we roll two fair dice and $A$ is the event that the sum of the two values is greater than 10, then $\mathrm{P}(A) = \frac{1}{12}$. Now suppose that event $B$ is the event that the value on the second die is a 6. The situation is illustrated in Figure 2.2. By looking at the sample space, we can see that $|B| = 6$ and $|A \cap B| = 2$, so $\mathrm{P}(B) = \frac{1}{6}$ and $\mathrm{P}(A \cap B) = \frac{1}{18}$. By Definition 2.4.1, $\mathrm{P}(A|B) = \frac{1}{18}/\frac{1}{6} = \frac{1}{3}$.

| | | Second die roll | | | | | |
|---|---|---|---|---|---|---|---|
| | | 1 | 2 | 3 | 4 | 5 | 6 |
| First die roll | 1 | . | . | . | . | . | $B$ |
| | 2 | . | . | . | . | . | $B$ |
| | 3 | . | . | . | . | . | $B$ |
| | 4 | . | . | . | . | . | $B$ |
| | 5 | . | . | . | . | . | $A, B$ |
| | 6 | . | . | . | . | $A$ | $A, B$ |

Figure 2.2 *Sample space for experiment of throwing two fair dice, with events A and B marked.*

When we use the intuitive definition given by (2.1), the terms in $|\Omega|$ cancel out:

$$\mathrm{P}(A|B) = \frac{\mathrm{P}(A \cap B)}{\mathrm{P}(B)} = \frac{|A \cap B|}{|\Omega|}\frac{|\Omega|}{|B|} = \frac{|A \cap B|}{|B|}.$$

By conditioning on $B$, we are making $B$ take on the role of the whole sample space in our probability calculations. This makes sense since we know that the outcome of the experiment is in $B$.

**Proposition 2.4.3**
*The function* $\mathrm{P}(.|B)$ *in Definition 2.4.1 is a probability measure on* $(\Omega, \mathcal{F})$.

The proof of this proposition is part of Exercise 2.4.

**Claim 2.4.4** (Simple properties of conditional probability)

*i.* $\mathrm{P}(A \cap B) = \mathrm{P}(A|B)\mathrm{P}(B)$.
*ii. If A and B are disjoint then* $\mathrm{P}(A|B) = 0$.
*iii. If* $B \subseteq A$ *then* $\mathrm{P}(A|B) = 1$.

These properties are simple consequences of our definition of conditional probability. Part i. is the basis for a result that has a number of applications, in particular in dealing with sequences of correlated observations. We start by extending the result to the three-way intersection $A \cap B \cap C$. By transitivity of intersections we can write

$A \cap B \cap C = A \cap (B \cap C)$, so by Claim 2.4.4, $P(A \cap B \cap C) = P(A|B \cap C)P(B \cap C) = P(A|B \cap C)P(B|C)P(C)$. This generalises further as the following proposition.

**Proposition 2.4.5** (Multiplication law)
*Consider the probability space* $(\Omega, \mathcal{F}, P)$ *with* $A_1, \ldots, A_n \in \mathcal{F}$ *and* $P(\bigcap_{j=1}^n A_j) \neq 0$, *then*

$$P\left(\bigcap_{j=1}^{n} A_j\right) = \prod_{j=1}^{n} P\left(A_j \middle| \bigcap_{i=0}^{j-1} A_i\right),$$

*where we define* $A_0 = \Omega$.

Note that the multiplication *law* is something quite distinct from the multiplication *rule*.

**Notation 2.4.6** (Conditioning notation)
It is important to understand the distinction between $P(A|B)$ and $P(A \backslash B)$. Despite similar notation, the meanings are completely different.

- $P(A|B) = P(A \cap B)/P(B)$, the conditional probability of $A$ given $B$.
- $P(A \backslash B) = P(A \cap B^c)$, the probability of A occurring but not $B$.

### 2.4.2 *Law of total probability and Bayes' theorem*

In its simplest form, Bayes' theorem is a mechanism for reversing the order of conditioning. The following example illustrates the sort of situation in which this might be useful.

**Example 2.4.7** (Insurance fraud)
An insurance company processes claims for car theft on a monthly basis. An individual's probability of having their car stolen in any given month is 0.005, and 99% of those who have their car stolen make a claim. A small proportion of customers (1 in 10000) will put in a claim even though their car has not been stolen.

1. What is the probability that an individual customer makes a claim?
2. What is the probability that a claim made by a customer is fraudulent?

In section 2.2.2 we indicate that many problems in probability are made more tractable by dividing our sample space into disjoint events.

**Definition 2.4.8** (Partition)
Consider a probability space $(\Omega, \mathcal{F}, P)$. A set of events $\{B_1, \ldots, B_n\}$ is a **partition** if it has the following properties:

i. Exhaustive: $\bigcup_{i=1}^n B_i = \Omega$,
ii. Mutually exclusive: $B_i \cap B_j = \emptyset$ for $i \neq j$,
iii. Non-zero probability: $P(B_i) > 0$ for $i = 1, \ldots, n$.

Note that, since the elements of the partition are events, we have $B_i \in \mathcal{F}$ for all $i$. The exhaustive property means that the members of a partition cover the whole of the sample space, that is, for any $\omega \in \Omega$ we have $\omega \in B_i$ for some $i$. The term **mutually exclusive** is just another way of saying disjoint or non-overlapping. We know from Definition 2.2.3 of a measure that, for mutually exclusive events, the probability of the union is the sum of the probabilities. We can exploit this property to write the probability of an event in terms of the probabilities conditioning on elements of the partition.

**Proposition 2.4.9** (Law of total probability)
*Consider the probability space $(\Omega, \mathcal{F}, \mathrm{P})$ with a partition $\{B_1, \ldots, B_n\}$ of $\Omega$. For all $A \in \mathcal{F}$,*

$$\mathrm{P}(A) = \sum_{i=1}^{n} \mathrm{P}(A|B_i)\mathrm{P}(B_i).$$

*Proof.*

This is a simple consequence of the underlying set theory. Since intersection distributes over union, $\bigcup_{i=1}^{n}(A \cap B_i) = A \cap (\bigcup_{i=1}^{n} B_i)$. Now, since $\{B_1, \ldots, B_n\}$ is a partition, we know that $\bigcup_{i=1}^{n} B_i = \Omega$ by the exhaustive property. Thus, $\bigcup_{i=1}^{n}(A \cap B_i) = A$. The mutually exclusive property of the partition tells us that $(A \cap B_i) \cap (A \cap B_j) = A \cap (B_i \cap B_j) = A \cap \varnothing = \varnothing$, so $(A \cap B_i)$ and $(A \cap B_j)$ are also mutually exclusive. Putting these observations together and using Claim 2.4.4 we get

$$\mathrm{P}(A) = \mathrm{P}\left(\bigcup_{i=1}^{n}(A \cap B_i)\right) = \sum_{i=1}^{n} \mathrm{P}(A \cap B_i) = \sum_{i=1}^{n} \mathrm{P}(A|B_i)\mathrm{P}(B_i),$$

where the conditional probability is well defined since $\mathrm{P}(B_i) > 0$ for all $i$. □

Suppose that we know $\mathrm{P}(A|B)$ but what we would really like to know is $\mathrm{P}(B|A)$. Some simple manipulation yields

$$\mathrm{P}(B|A) = \frac{\mathrm{P}(B \cap A)}{\mathrm{P}(A)} = \frac{\mathrm{P}(A|B)\mathrm{P}(B)}{\mathrm{P}(A)}.$$

Although mathematically trivial, this is one possible statement of **Bayes' theorem**, a result with surprisingly wide applications. We will illustrate its use by finally answering the question posed in Example 2.4.7. First of all, we give three mutually consistent statements, any one of which might be labelled as Bayes' theorem.

**Claim 2.4.10** (Bayes' theorem)
*Consider the probability space $(\Omega, \mathcal{F}, \mathrm{P})$ with $A, B \in \mathcal{F}$ and a partition $\{B_1, \ldots, B_n\}$ of $\Omega$, then*

*i.* $\mathrm{P}(B|A) = \mathrm{P}(A|B)\mathrm{P}(B)/\mathrm{P}(A)$.

*ii.* $\mathrm{P}(B|A) = \mathrm{P}(A|B)\mathrm{P}(B)/(\mathrm{P}(A|B)\mathrm{P}(B) + \mathrm{P}(A|B^c)\mathrm{P}(B^c))$.

*iii.* $\mathrm{P}(B_j|A) = \mathrm{P}(A|B_j)\mathrm{P}(B_j)/(\sum_{i=1}^{n} \mathrm{P}(A|B_i)\mathrm{P}(B_i))$.

**Example 2.4.7** (Revisited)
We can use the law of total probability to answer the first part of the question. Let $A$ be the event that a customer makes a claim, $B_1$ be the event that their car is stolen, and $B_2$ be the event their car is not stolen. From the question,

$$\begin{aligned} \text{P}(B_1) &= 0.005, \quad \text{P}(A|B_1) = 0.99, \\ \text{P}(B_2) &= 0.995, \quad \text{P}(A|B_2) = 0.0001. \end{aligned}$$

Note that $B_2 = B_1^c$ so $\{B_1, B_2\}$ is a partition of the sample space. The probability that a customer makes a claim is then

$$\begin{aligned} \text{P}(A) &= \sum_{i=1}^{2} \text{P}(A|B_i)\text{P}(B_i) = (0.99 \times 0.005) + (0.0001 \times 0.995) \\ &= 0.0050495. \end{aligned}$$

Notice that, as the insurance company, we know whether or not a customer has made a claim, that is, we know whether or not event $A$ has occurred. The second question asks for the probability that a claim is fraudulent. Another way to say this is, given that a customer makes a claim, what is the probability that their car has not actually been stolen, that is, $\text{P}(B_2|A)$. Using Bayes' theorem and substituting values we already know yields

$$\text{P}(B_2|A) = \frac{\text{P}(A|B_2)\text{P}(B_2)}{\text{P}(A)} = \frac{0.0001 \times 0.995}{0.0050495} = 0.0197(4 \text{ d.p.}).$$

### *2.4.3 Independence*

Before we give a precise definition, it is worthwhile briefly exploring intuitive notions about independence. The Concise Oxford Dictionary's definition is *"free from outside control; not subject to another's authority"*. This is actually reasonably close to the mathematical use of the term. We will define independence in terms of events. An intuitively appealing definition is that events $A$ and $B$ are independent if the probability that $A$ occurs is unaffected by whether or not $B$ has occurred. Assuming $\text{P}(B) > 0$, we could write this as $\text{P}(A|B) = \text{P}(A)$. Substituting from Definition 2.4.1 of conditional probability, $\text{P}(A \cap B)/\text{P}(B) = \text{P}(A)$, and hence, $\text{P}(A \cap B) = \text{P}(A)\text{P}(B)$. This is the basis of our definition of independence.

**Definition 2.4.11** (Independence)
Consider a probability space $(\Omega, \mathcal{F}, \text{P})$ with $A, B \in \mathcal{F}$. $A$ and $B$ are said to be **independent**, denoted $A \perp\!\!\!\perp B$, if and only if $\text{P}(A \cap B) = \text{P}(A)\text{P}(B)$.

Note that this definition, unlike our conditional probability statements, does not exclude the possibility that $\text{P}(B) = 0$. It is important not to confuse the properties of mutual exclusivity with independence. If $A$ and $B$ are mutually exclusive then $\text{P}(A \cap B) = 0$. Thus, mutually exclusive events will not be independent unless $\text{P}(A) = 0$ or $\text{P}(B) = 0$ or both.

Consider the case where the sample space, $\Omega$, is made up of a finite number of equally likely events, so we use the probability measure (2.1). Assuming $P(B) > 0$, if $A$ is independent of $B$ then $P(A|B) = P(A)$ and so

$$\frac{|A \cap B|}{|B|} = \frac{|A|}{|\Omega|}.$$

This means that if $A$ and $B$ are independent then the proportion of the outcomes in $B$ that are also in $A$ is equal to the proportion of outcomes in $\Omega$ (that is, all outcomes) that are also in $A$.

**Proposition 2.4.12** (Basic properties of independent events)

*i. If $A \perp\!\!\!\perp B$ then $B \perp\!\!\!\perp A$.*

*ii. If $P(A) > 0$ then $P(B|A) = P(B) \iff A \perp\!\!\!\perp B$.*
*If $P(B) > 0$ then $P(A|B) = P(A) \iff A \perp\!\!\!\perp B$.*

*iii. If $A \perp\!\!\!\perp B$ then $A^c \perp\!\!\!\perp B^c$, $A^c \perp\!\!\!\perp B$, and $A \perp\!\!\!\perp B^c$.*

For more than two events we define independence recursively.

**Definition 2.4.13** (Mutual independence)
Consider a probability space $(\Omega, \mathcal{F}, P)$ and a set of events $\{A_1, \ldots, A_n\}$. We say that the events $\{A_1, \ldots, A_n\}$ are **mutually independent** if every subset of two or more elements are (mutually) independent and $P(A_1 \cap A_2 \cap \ldots \cap A_n) = P(A_1)P(A_2)\ldots P(A_n)$.

Independence is often inferred from the physics of an experiment. An assumption of independence can be used to construct an intuitively appealing probability measure in situations where we do not have a sample space of equally likely outcomes.

**Example 2.4.14** (Simple examples of independence)

1. **Throwing two dice yet again**: When throwing two fair dice, we assume the dice are rolled in such a way that the value shown on the first die does not have any impact on the value shown on the second. Events that are purely associated with the roll of die 1 will be independent of those purely associated with the roll of die 2. Consider, for example, events $A$ and $B$, where $A$ is the event that the value on the first die is odd and $B$ is the event that the value on the second die is bigger than 4. We can easily check that $P(A) = \frac{1}{2}$, $P(B) = \frac{1}{3}$, $P(A \cap B) = \frac{1}{6}$, and thus $A$ and $B$ are independent.

2. **Flipping a coin three times**: Suppose that we flip a coin three times, with probability of a head on any given flip $p$ and probability of a tail $(1 - p)$. If we assume independence of the flips then we use multiplication to work out probabilities of particular outcomes and, thus, work out probabilities associated with events. For example, suppose that we are interested in the number of heads in the sequence of

three tosses;

$$\begin{aligned}
P(\#\text{ heads} = 0) &= P(\{TTT\}) \\
&= (1-p) \times (1-p) \times (1-p) = (1-p)^3, \\
P(\#\text{ heads} = 1) &= P(\{TTH\} \cup \{THT\} \cup \{HTT\}) \\
&= P(\{TTH\}) + P(\{THT\}) + P(\{HTT\}) \\
&= (1-p)(1-p)p + (1-p)p(1-p) + p(1-p)(1-p) \\
&= 3p(1-p)^2, \\
P(\#\text{ heads} = 2) &= P(\{THH\} \cup \{HTH\} \cup \{HHT\}) \\
&= P(\{THH\}) + P(\{HTH\}) + P(\{HHT\}) \\
&= 3p^2(1-p), \\
P(\#\text{ heads} = 3) &= P(\{HHH\}) \\
&= p^3.
\end{aligned}$$

Each one of these probabilities is a term in the expansion of $(p + (1-p))^3$.

In the coin-flipping example above, we are not really interested in the outcomes themselves; we are interested in a property of the outcomes, namely the number of heads. Quantities that associate real numbers with outcomes are referred to as **random variables.** Random variables and the associated distributions of probability are the subject of the next chapter.

**Exercise 2.4**

1. (**Relative probabilities and conditioning**) Consider events $A$, $B$ and $C$ where $A \subseteq C$, $B \subseteq C$ and $P(B) > 0$. Show that the relative probabilities of $A$ and $B$ are unchanged by conditioning on $C$, that is, $P(A|C)/P(B|C) = P(A)/P(B)$.
2. (**Proof of multiplication law**) Prove the multiplication law (Proposition 2.4.5) using the definition of conditional probability. Take care to make sure any events that are being conditioned on have non-zero probability.
3. (**Conditions for mutual independence**) Consider a set of three events $\{A, B, C\}$. According to Definition 2.4.13, in order to establish independence we need to check that every subset of two or more elements is independent. This reduces to three conditions,
$$\begin{aligned}
P(A \cap B) &= P(A)P(B), \\
P(A \cap C) &= P(A)P(C), \\
P(B \cap C) &= P(B)P(C).
\end{aligned}$$
With the additional condition that $P(A \cap B \cap C) = P(A)P(B)P(C)$ we have a total of four conditions to check. Show that in order to establish that a set of four events is independent we need to check 11 conditions. Generalise to a set of $n$ events.
4. (**Conditional probability as measure**) Consider a probability space $(\Omega, \mathcal{F}, P)$ with $A, B \in \mathcal{F}$ and $P(B) > 0$. We define the conditional probability of $A$ given $B$ as $P(A|B) = P(A \cap B)/P(B)$.

(a) Show that P(.|$B$) defines a probability measure on $(\Omega, \mathcal{F})$.

(b) Define $\mathcal{G} = \{A \cap B : A \in \mathcal{F}\}$. Show that $\mathcal{G}$ is a $\sigma$-algebra in $B$. Hence show that, if $m(C) = \mathrm{P}(C)/\mathrm{P}(B)$ for all $C \in G$, then $m$ is a probability measure on $\{B, \mathcal{G}\}$.

5. (a) Consider the mutually exclusive and exhaustive events $A_0$, $A_1$, and $A_2$. Is it possible to have $\mathrm{P}(A_0 \cup A_1) = \frac{1}{2}$, $\mathrm{P}(A_1 \cup A_2) = \frac{1}{2}$, and $\mathrm{P}(A_2 \cup A_0) = \frac{2}{3}$?

(b) Consider three mutually exclusive and exhaustive events $A_0$, $A_1$, and $A_2$ where

$$\mathrm{P}(A_0 \cup A_1) = p_0,$$
$$\mathrm{P}(A_1 \cup A_2) = p_1,$$
$$\mathrm{P}(A_2 \cup A_0) = p_2.$$

What condition on $p_0$, $p_1$, and $p_2$ must hold? Now generalise to $n$ mutually exclusive and exhaustive events $A_0, \ldots, A_{n-1}$ where

$$\mathrm{P}\left(\bigcup_{i=r}^{r+k-1} A_{i(\mathrm{mod}\ n)}\right) = p_r,$$

for $r = 0, \ldots, n-1$ and $0 < k < n$. What condition on $p_0, \ldots, p_{n-1}$ must hold? [Note: $i(\mathrm{mod}\ n)$ is the remainder when $i$ is divided by $n$.]

6. Let $\Omega = \{abc, acb, cab, cba, bca, bac, aaa, bbb, ccc\}$ and let each element of $\Omega$ have equal probability. Let $A_k$ be the event that the $k^{\text{th}}$ letter in the outcome is $a$. Find $\mathrm{P}(A_1|A_2)$ and $\mathrm{P}(A_1|A_2^c)$. Show that the events $\{A_1, A_2, A_3\}$ are pairwise independent. Determine whether they are mutually independent.

7. Suppose that the genders of all children in a family are independent and that boys and girls are equally probable, that is, both have probability 0.5.

(a) For families of three children, calculate the probabilities of the events $A$, $B$, and $A \cap B$, where $A$ = "There are children of both genders" and $B$ = "Not more than one child is a girl".

(b) Do the same for families of four children.

(c) Are $A$ and $B$ independent events in parts (a) and (b)?

Let the probability $p_n$ that a family has exactly $n$ children be $(1-p)p^n$ for $n = 0, 1, \ldots$, with $p$ such that $0 < p < 1$.

(d) Show that the probability that a family contains exactly $k$ boys is given by

$$2(1-p)p^k(2-p)^{-(k+1)}.$$

[You might like to prove this for $k = 0$ and $k = 1$ before attempting general $k$. You will need to use a form of the negative binomial expansion for general $k$ – see equation (3.8)].

## 2.5 Further exercises

1. Consider a probability space $(\Omega, \mathcal{F}, \mathrm{P})$ and events $A, B, C \in \mathcal{F}$.
   (a) If $\mathrm{P}(A) = \frac{3}{4}$ and $\mathrm{P}(B) = \frac{1}{3}$, show that $\frac{1}{12} \leq \mathrm{P}(A \cap B) \leq \frac{1}{3}$. When do the two equalities hold?
   (b) Is it possible to find events for which the following four conditions hold: $A \cap B \subset C^c$, $\mathrm{P}(A) > 0.5$, $\mathrm{P}(B) > 0.5$, and $\mathrm{P}(C) > 0.5$?
   (c) If $A \cap B \subset C^c$, $\mathrm{P}(A) = 0.5$, and $\mathrm{P}(B) = 0.5$, what is the largest possible value for $\mathrm{P}(C)$?
2. Consider a probability space $(\Omega, \mathcal{F}, \mathrm{P})$ and events $A, B, C \in \mathcal{F}$. Starting from the definition of probability measure, show that
   (a) $\mathrm{P}((A \cup B) \cap (A^c \cup B^c)) = \mathrm{P}(A) + \mathrm{P}(B) - 2\mathrm{P}(A \cap B)$.
   (b) $\mathrm{P}(A \cup B \cup C) = \mathrm{P}(A) + \mathrm{P}(B) + \mathrm{P}(C) - \mathrm{P}(A \cap B) - \mathrm{P}(A \cap C) - \mathrm{P}(B \cap C) + \mathrm{P}(A \cap B \cap C)$.
3. (a) In 1995 an account on the LSE network came with a three letter (all uppercase Roman letter) password. Suppose a malicious hacker could check one password every millisecond. Assuming the hacker knows a username and the format of passwords, what is the maximum time that it would take to break into an account?
   (b) In a bid to improve security, IT services propose to either double the number of letters available (by including lowercase letters) or double the length (from three to six). Which of these options would you recommend? Is there a fundamental principle here that could be applied in other situations?
   (c) Suppose that, to be on the safe side, IT services double the number of letters, include numbers, and increase the password length to twelve. You have forgotten your password. You remember that it contains the characters $\{t, t, t, S, s, s, I, i, i, c, a, 3\}$. If you can check passwords at the same rate as a hacker, how long will it take you to get into your account?
4. $A$ and $B$ are events of positive probability. Supply a proof for each of the following.
   (a) If $A$ and $B$ are independent, $A$ and $B^c$ are independent.
   (b) If $A$ and $B$ are independent, $\mathrm{P}(A^c|B^c) + \mathrm{P}(A|B) = 1$.
   (c) If $\mathrm{P}(A \mid B) < \mathrm{P}(A)$, then $\mathrm{P}(B \mid A) < \mathrm{P}(B)$.
   (d) If $\mathrm{P}(B \mid A) = \mathrm{P}(B \mid A^c)$ then $A$ and $B$ are independent.
5. A fair coin is independently tossed twice. Consider the following events:
   $A$ = "The first toss is heads"
   $B$ = "The second toss is heads"
   $C$ = "First and second toss show the same side"
   Show that $A, B, C$ are pairwise independent events, but not independent events.
6. Show that, if $A$, $B$, and $C$ are independent events with $\mathrm{P}(A) = \mathrm{P}(B) = \mathrm{P}(C)$, then the probability that exactly one of $A$, $B$, and $C$ occurs is less than or equal to $4/9$.
7. Consider a probability space $(\Omega, \mathcal{F}, \mathrm{P})$ and events $A, B, C_1, C_2 \in \mathcal{F}$. Suppose, in addition, that $C_1 \cap C_2 = \emptyset$ and $C_1 \cup C_2 = B$. Show that

$$\mathrm{P}(A|B) = \mathrm{P}(A|C_1)\mathrm{P}(C_1|B) + \mathrm{P}(A|C_2)\mathrm{P}(C_2|B).$$

## 2.6 Chapter summary

On completion of this chapter you should be able to:

- calculate probabilities for simple situations by counting,
- write down the definition of a $\sigma$-algebra,
- derive properties of probability measures from axioms,
- calculate probabilities for problems with large sample spaces using combinations and permutations,
- explain the association between the number of combinations and multinomial coefficients,
- derive and use conditional probability,
- prove that conditional probability defines a valid probability measure,
- use Bayes' theorem to solve problems involving reversing conditioning, and
- define independent events and exploit independence to calculate probabilities.

Chapter 3

# Random variables and univariate distributions

This chapter introduces the core ideas of distribution theory: random variables, distributions, and expectation. The three are closely connected; random variables associate real numbers with outcomes, distributions associate probability with real numbers, and expectation gives us an idea of where the centre of a distribution is. Our starting point is the definition of a random variable. Understanding random variables and the associated (not always very transparent) notation is a fundamental part of understanding distribution theory.

## 3.1 Mapping outcomes to real numbers

At an intuitive level, the definition of a random variable is straightforward; a random variable is a quantity whose value is determined by the outcome of the experiment. The value taken by a random variable is always real. The randomness of a random variable is a consequence of our uncertainty about the outcome of the experiment. Example 3.1.1 illustrates this intuitive thinking, using the setup described in Example 2.4.14 as a starting point.

In practice, the quantities we model using random variables may be the output of systems that cannot be viewed as experiments in the strict sense. What these systems have in common, however, is that they are **stochastic**, rather than **deterministic**. This is an important distinction; for a deterministic system, if we know the input, we can determine exactly what the output will be. This is not true for a stochastic model, as its output is (at least in part) determined by a random element. We will encounter again the distinction between stochastic and deterministic systems in Chapter 12, in the context of random-number generation.

**Example 3.1.1** (Coin flipping again)
Define a random variable $X$ to be the number of heads when we flip a coin three times. We assume that flips are independent and that the probability of a head at each flip is $p$. We know that $X$ can take one of four values, 0, 1, 2, or 3. For convenience,

we say that $X$ can take any real value, but the probability of it taking a value outside $\{0, 1, 2, 3\}$ is zero. The probabilities evaluated in Example 2.4.14 can now be written as

$$P(X = x) = \begin{cases} (1-p)^3, & x = 0, \\ 3p(1-p)^2, & x = 1, \\ 3p^2(1-p), & x = 2, \\ p^3, & x = 3, \\ 0, & \text{otherwise.} \end{cases} \tag{3.1}$$

The final line in this statement is often omitted; it is assumed that we have zero probability for values not explicitly mentioned. We can write down an equivalent formulation in terms of probability below a point:

$$P(X \leq x) = \begin{cases} 0, & -\infty < x < 0 \\ (1-p)^3, & 0 \leq x < 1, \\ (1-p)^2(1+2p), & 1 \leq x < 2, \\ 1-p^3, & 2 \leq x < 3, \\ 1, & 3 \leq x < \infty. \end{cases}$$

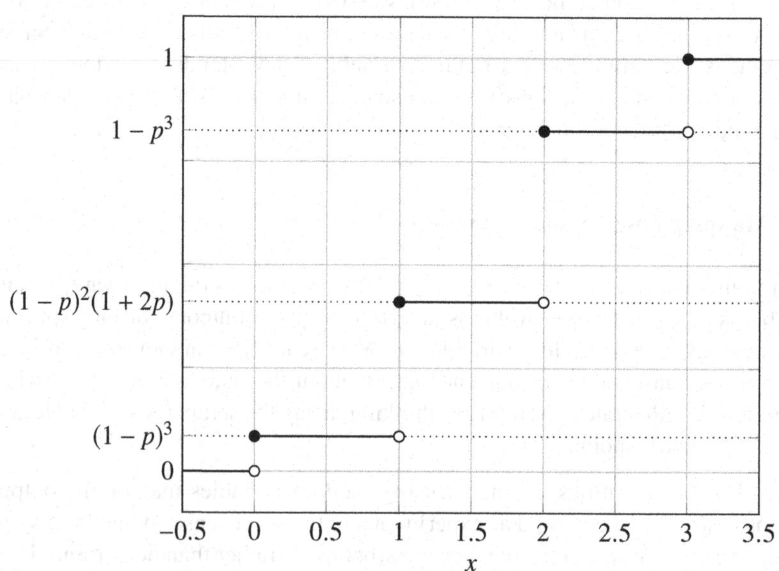


Figure 3.1 *The function* $P(X \leq x)$ *where* $X$ *is the number of heads when we flip a coin three times.*

In the statements $P(X = x)$ and $P(X \leq x)$, the random variable $X$ can be thought of as notational shorthand. Substituting the phrase "the number of heads when we flip a coin three times" for $X$ leads to sensible statements. For example, $P(X \leq 2)$ is interpreted as "the probability that the number of heads, when we flip a coin three times, is less than or equal to 2".

The definition of a random variable as a quantity whose value is determined by the outcome of the experiment is one that will serve us very well. However, there are situations in which it is important to understand how random variables fit into the rigorous framework set out in section 2.2. Recall that probability is defined as a measure that maps events to the unit interval. This means that whenever we refer to a probability $P(A)$, the argument, $A$, must be an event. Thus, if we are going to use $P(X \leq x)$, we must be able to interpret $\{X \leq x\}$ as an event. The key is to be more precise about exactly what sort of quantity a random variable is. We give the formal definition then explain how this makes sense both intuitively and in terms of the formal setup of section 2.2.

**Definition 3.1.2** (Random variable)
A random variable is a function $X : \Omega \longrightarrow \mathbb{R}$ with the property that, if

$$A_x = \{\omega \in \Omega : X(\omega) \leq x\}$$

then $A_x \in \mathcal{F}$ for all $x \in \mathbb{R}$. Thus, $A_x$ is an event, for every real-valued $x$.

The single most important thing to understand from Definition 3.1.2 is that random variables are functions that map outcomes to real numbers. Thus, a random variable $X$ is a function and, if $\omega$ is an outcome, $X(\omega)$ is a real number. This corresponds with our intuitive definition of a random variable as a quantity whose value is determined by the outcome of the experiment. The remaining part of Definition 3.1.2 ensures that we can talk about quantities such as $P(X \leq x)$. In fact, it should now be clear that $P(X \leq x)$ is an abuse of notation; $X \leq x$ compares a function with a real number. The probability we are referring to is more accurately described as $P(\{\omega \in \Omega : X(\omega) \leq x\})$. This is legitimate, since Definition 3.1.2 ensures that $\{\omega \in \Omega : X(\omega) \leq x\}$ is an event. Put another way, $P(X \leq x)$ refers to the probability of an event made up of all the outcomes whose images under $X$ are real numbers less than or equal to $x$. We will also use $\{X \leq x\}$ as shorthand for the event $\{\omega \in \Omega : X(\omega) \leq x\}$.

Suppose that, in Example 3.1.1, we are also interested in the number of tails $Y$. We can readily see that $Y(\omega) = 3 - X(\omega)$, for any outcome $\omega$. This is usually written as $Y = 3 - X$ and provides a simple illustration of a widely used concept, that is, a function of a random variable. We define notation as follows.

**Definition 3.1.3** (Function of a random variable)
If $g : \mathbb{R} \to \mathbb{R}$ is a well-behaved function, $X : \Omega \to \mathbb{R}$ is a random variable and $Y = g(X)$, then $Y$ is a random variable, $Y : \Omega \to \mathbb{R}$ with $Y(\omega) = g(X(\omega))$ for all $\omega \in \Omega$.

In Definition 3.1.3 we use the term "well-behaved function". The precise meaning of "well-behaved" in this context is given in section 3.8. For now we may assume that all the functions we consider are well behaved, in other words, for random variable $X$ and function $g$, we will assume $g(X)$ is also a random variable.

For many real phenomena we are interested in quantities that, by definition, cannot be negative. Examples include length of life, distance travelled or rain accumulated.

It is natural that random variables used as models for these phenomena should also be constrained to be positive.

**Definition 3.1.4** (Positive random variable)
A random variable $X$ is positive, often written $X \geq 0$, if it takes a positive value for every possible outcome, that is, if $X(\omega) \geq 0$ for all $\omega \in \Omega$.

Note that a more precise term would be *non-negative* random variables, as they are allowed to take the value 0. However, in keeping with standard convention, we will refer to such random variables as positive.

We return to Example 3.1.1 to clarify the definitions we have introduced.

**Example 3.1.1 (Revisited)** There are eight possible outcomes to the experiment of flipping a coin three times,

$$\Omega = \{TTT, TTH, THT, THH, HTT, HTH, HHT, HHH\}.$$

Suppose we take $\mathcal{F}$ to be the collection of all possible subsets of $\Omega$, that is $\mathcal{F} = \{0,1\}^{\Omega}$, so any set of outcomes is an event. The random variable $X$ is defined as the number of heads. Recall that $X$ is a function defined on the sample space $\Omega$. In this simple example, we can specify $X$ completely,

$$\begin{aligned}
&X(TTT) = 0,\\
&X(TTH) = X(THT) = X(HTT) = 1,\\
&X(THH) = X(HTH) = X(HHT) = 2,\\
&X(HHH) = 3.
\end{aligned}$$

We can now see how statements of the form $\mathrm{P}(X \leq x)$ refer to probabilities of events. For example,

$$\begin{aligned}
\mathrm{P}(X \leq 0) &= \mathrm{P}(\{\omega \in \Omega : X(\omega) \leq 0\}) = \mathrm{P}(\{TTT\}) = (1-p)^3,\\
\mathrm{P}(X \leq 1) &= \mathrm{P}(\{\omega \in \Omega : X(\omega) \leq 1\}) = \mathrm{P}(\{TTT, TTH, THT, HTT\})\\
&= (1-p)^2(1+2p).
\end{aligned}$$

This will work for any real numbers, for example,

$$\begin{aligned}
&\mathrm{P}(X \leq 0.141) = \mathrm{P}(\{\omega \in \Omega : X(\omega) \leq 0.141\}) = \mathrm{P}(\{TTT\}) = (1-p)^3,\\
&\mathrm{P}(X \leq -5.9265) = \mathrm{P}(\{\omega \in \Omega : X(\omega) \leq -5.9265\}) = \mathrm{P}(\emptyset) = 0,\\
&\mathrm{P}(X \leq \pi) = \mathrm{P}(\{\omega \in \Omega : X(\omega) \leq \pi\}) = \mathrm{P}(\Omega) = 1.
\end{aligned}$$

If $Y$ is the number of tails, we use $Y = 3 - X$ to mean $Y(\omega) = 3 - X(\omega)$ for any outcome $\omega$. We can see that this makes sense in our simple example,

$$\begin{aligned}
&Y(TTT) = 3 - X(TTT) = 3,\\
&Y(TTH) = 3 - X(TTH) = 2,
\end{aligned}$$

and so on. We can evaluate probabilities associated with $Y$ using what we know about probabilities associated with $X$, for example,

$$\begin{aligned}
&\mathrm{P}(Y = 2) = \mathrm{P}(3 - X = 2) = \mathrm{P}(X = 1) = 3p(1-p)^2,\\
&\mathrm{P}(Y \leq 2) = \mathrm{P}(X > 0) = 1 - \mathrm{P}(X \leq 0) = 1 - (1-p)^3.
\end{aligned}$$

As both $X$ and $Y$ are counts (number of heads and number of tails, respectively), both $X$ and $Y$ are positive random variables.

For most practical purposes, the true nature of random variables is suppressed. In most statistical applications the distribution is of paramount importance; thinking about an underlying sample space is often neither very natural nor particularly helpful. We will use notation such as $\mathrm{P}(X \leq x)$ and $g(X)$ in the usual way, and for the most part not be troubled by the fact that random variables are actually functions. However, you should keep this fact in mind; it is useful in ensuring that we always make the distinction between random variables (functions) and instances of random variables (real numbers).

So far, our discussion has focused on probabilities of the form $\mathrm{P}(X \leq x)$, that is, the probability that the variable of interest takes a value below a given point. By convention, this is the starting point for discussing the distribution of probability associated with a random variable. Of course, we will also be interested in other probabilities, such as $\mathrm{P}(X > x)$ and $\mathrm{P}(x < X \leq y)$. In the next section we show how all probabilities that may be of interest can be constructed from a function giving probability below a point: the cumulative distribution function.

**Exercise 3.1**

1. Suppose that we roll two fair dice. Let $X$ be the absolute value of the difference of the values taken by the two dice. Translate each of the following into an English statement and evaluate.
   i. $\{X \leq 2\}$,
   ii. $\{X = 0\}$,
   iii. $\mathrm{P}(X \leq 2)$,
   iv. $\mathrm{P}(X = 0)$.
2. We throw a die and toss a coin. Defining whatever random variables you feel necessary, give succinct mathematical expressions for each of the following events and probabilities:
   i. the value on the die is less than 3,
   ii. the probability that the value on the die is less than 3,
   iii. the coin shows a head,
   iv. the probability that the number of heads shown is less than 1,
   v. there are no heads and the die roll is a 6,
   vi. the probability that the number of heads is less than the value on the die.

## 3.2 Cumulative distribution functions

As mentioned above, we are usually more interested in probabilities associated with a random variable than in a mapping from outcomes to real numbers. The probability associated with a random variable is completely characterised by its **cumulative distribution function**.

**Definition 3.2.1** (Cumulative distribution function)
The cumulative distribution function (CDF) of a random variable $X$ is the function $F_X : \mathbb{R} \longrightarrow [0,1]$ given by $F_X(x) = \text{P}(X \leq x)$.

A couple of points to note about cumulative distribution functions.

1. We will use $F_X$ to denote the cumulative distribution function of the random variable $X$, $F_Y$ to denote the cumulative distribution function of the random variable $Y$, and so on.
2. Be warned; some texts use the argument to identify different distribution functions. For example, you may see $F(x)$ and $F(y)$ used, not to denote the same function applied to different arguments, but to indicate a value of the cumulative distribution function of $X$ and a value of the cumulative distribution function of $Y$. This can be deeply confusing and we will try to avoid doing it.

In our discussion of the properties of cumulative distribution functions, the following definition is useful.

**Definition 3.2.2** (Right continuity)
A function $g : \mathbb{R} \rightarrow \mathbb{R}$ is **right-continuous** if $g(x+) = g(x)$ for all $x \in \mathbb{R}$, where $g(x+) = \lim_{h \downarrow 0} g(x+h)$.

The notation $g(x+)$ is used for limit from the right. There is nothing complicated about this; it is just the limit of the values given by $g$ as we approach the point $x$ from the right-hand side. Right continuity says that we can approach any point from the right-hand side without encountering a jump in the value given by $g$. There is an analogous definition of left continuity in terms of the limit from the left; $g$ is left-continuous if $g(x-) = \lim_{h \downarrow 0} g(x-h) = g(x)$ for all $x$. Somewhat confusingly, the notation $\lim_{h \uparrow 0}$ is sometimes used. This is discussed as part of Exercise 3.2.

The elementary properties of cumulative distribution functions are inherited from their definition in terms of probability. It is true, but rather harder to show, that any function satisfying the three properties given in Proposition 3.2.3 is the distribution function of some random variable. We will only prove necessity of the three conditions.

**Proposition 3.2.3** (Properties of cumulative distribution functions)
*If $F_X$ is a cumulative distribution function then*

*i.* *$F_X$ is a non-decreasing function: if $x < y$ then $F_X(x) \leq F_X(y)$,*
*ii.* $\lim_{x \to -\infty} F_X(x) = 0$ *and* $\lim_{x \to \infty} F_X(x) = 1$,
*iii.* *$F_X$ is right-continuous: $F_X(x+) = F_X(x)$ for all $x \in \mathbb{R}$.*

*Proof.*

Throughout this proof we will use $A_x = \{\omega \in \Omega : X(\omega) \leq x\}$. This gives us three equivalent expressions for the cumulative distribution function,

$$F_X(x) = \text{P}(X \leq x) = \text{P}(\{\omega \in \Omega : X(\omega) \leq x\}) = \text{P}(A_x).$$

We will use each as appropriate.

i. This is a simple consequence of the fact that, if $x < y$, then $A_x \subseteq A_y$. Thus, by Corollary 2.2.9, $\mathrm{P}(A_x) \leq \mathrm{P}(A_y)$ and so $F_X(x) \leq F_X(y)$.

ii. Let $\{x_n : n = 1, 2, \ldots\}$ be a sequence of real numbers, such that $x_n \to -\infty$ as $n \to \infty$. We assume, without loss of generality, that $\{x_n\}$ is a decreasing sequence, that is, $x_n \geq x_{n+1}$ for $n = 1, 2, \ldots$. As a direct consequence of this definition, $A_{x_n} \supseteq A_{x_{n+1}}$ and $\cap_{n=1}^{\infty} A_{x_n} = \emptyset$. Thus,

$$\lim_{n\to\infty} F_X(x_n) = \lim_{n\to\infty} \mathrm{P}(A_{x_n}) = \mathrm{P}\left(\bigcap_{n=1}^{\infty} A_{x_n}\right) = \mathrm{P}(\emptyset) = 0, \tag{3.2}$$

where the second equality is a consequence of Proposition 2.2.12. Since equation (3.2) holds for any sequence $\{x_n\}$ such that $x_n \to -\infty$, we conclude that $\lim_{x\to-\infty} F_X(x) = 0$. Similar reasoning gives $\lim_{x\to\infty} F_X(x) = 1$; this proof is part of Exercise 3.2.

iii. The method is similar to that used in part ii. Now suppose we have a decreasing sequence such that $x_n \downarrow x$ as $n \to \infty$. By definition, $A_x \subseteq A_{x_n}$ for all $n$ and $A_x$ is the largest interval for which this is true. Thus, $\cap_{n=1}^{\infty} A_{x_n} = A_x$ and

$$\lim_{n\to\infty} F_X(x_n) = \lim_{n\to\infty} \mathrm{P}(A_{x_n}) = \mathrm{P}\left(\bigcap_{n=1}^{\infty} A_{x_n}\right) = \mathrm{P}(A_x) = F_X(x), \tag{3.3}$$

where, as before, the second equality is a consequence of Proposition 2.2.12. Since equation (3.3) holds for any sequence $\{x_n\}$ such that $x_n \downarrow x$, we conclude that $\lim_{h\downarrow 0} F_X(x+h) = x$ for all $x$, and so $F_X$ is right-continuous.

□

Cumulative distribution functions tell us about the probability below a point. Clearly there are other probabilities of interest. The following proposition indicates how these are calculated using the cumulative distribution function. As part of the proof we show how, for each of the probabilities, the argument can be interpreted as an event.

**Proposition 3.2.4** (Probabilities from cumulative distribution functions)
*Consider real numbers $x$ and $y$ with $x < y$. Then:*

*i.* $\mathrm{P}(X > x) = 1 - F_X(x)$.

*ii.* $\mathrm{P}(x < X \leq y) = F_X(y) - F_X(x)$.

*iii.* $\mathrm{P}(X < x) = \lim_{h\downarrow 0} F_X(x-h) = F_X(x-)$.

*iv.* $\mathrm{P}(X = x) = F_X(x) - F_X(x-)$.

*Proof.*

Consider a probability space $(\Omega, \mathcal{F}, \mathrm{P})$.

i. From the second property of a $\sigma$-algebra given by Definition 2.2.1, we know that if $A_x \in \mathcal{F}$, then $A_x^c \in \mathcal{F}$. By definition, $A_x^c = \{\omega \in \Omega : X(\omega) > x\}$, so it makes sense to talk about $P(X > x)$. From Lemma 2.2.8, we know that $P(A_x^c) = 1 - P(A_x)$. We conclude that

$$P(X > x) = P(A_x^c) = 1 - P(A_x) = 1 - P(X \leq x) = 1 - F_X(x).$$

ii. We can show that $\{\omega \in \Omega : x < X(\omega) \leq y\} \in \mathcal{F}$ by noting that $\{\omega \in \Omega : x < X(\omega) \leq y\} = A_y \cap A_x^c$, and $\mathcal{F}$ is closed under intersection. Consider partitioning $(-\infty, y]$ into disjoint subintervals $(-\infty, x]$ and $(x, y]$. Since the intervals are disjoint,

$$P(X \leq y) = P(X \leq x) + P(x < X \leq y).$$

The result follows immediately.

iii. Let $\{x_n : n = 1, 2, \ldots\}$ be an increasing sequence, such that $x_n \uparrow x$. We can write

$$\{\omega \in \Omega : X(\omega) < x\} = \bigcup_{n=1}^{\infty} A_{x_n},$$

and hence

$$\{\omega \in \Omega : X(\omega) < x\} \in \mathcal{F}.$$

Thus,

$$P(X < x) = P\left(\bigcup_{n=1}^{\infty} A_{x_n}\right) = \lim_{n\to\infty} F_X(x_n) = \lim_{h\downarrow 0} F_X(x - h) = F_X(x-).$$

iv. This is part of Exercise 3.2.

□

We will consider examples of cumulative distribution functions in section 3.3.

**Exercise 3.2**

1. (**Left continuity and right continuity**) Define left continuity and right continuity using the $\lim_{h\uparrow 0}$ notation.
2. (**Properties of cumulative distribution functions**) Let $F_X$ be a cumulative distribution function. Prove $\lim_{x\to\infty} F_X(x) = 1$.
3. (**Probability from cumulative distribution functions**) If $F_X$ is the cumulative distribution function of $X$, prove that $P(X = x) = F_X(x) - F_X(x-)$.
4. The claims received by an insurance company have CDF $F_X(x)$. Now suppose that the company decides not to pay out claims for amounts less than or equal to $c$. Work out the CDF of the claims paid in terms of $F_X(x)$. What is the CDF for the claims *not* paid? [*Hint*: Work out the probability that a claim is for an amount $\leq x$, conditional on it being $> c$.]

## 3.3 Discrete and continuous random variables

Any function satisfying the properties given in Proposition 3.2.3 is a cumulative distribution function. The general theory associated with functions of this type is rather technical. For the most part we restrict our attention to two classes of random variable and the associated distributions. These classes are referred to as **discrete** and **continuous**.

Discreteness and continuity are properties that we attribute to random variables and their distributions. In other words, these are properties of our models for real phenomena. The issue of whether the world is fundamentally discrete or continuous was a preoccupation of ancient philosophers, a preoccupation that led, among other things, to Zeno's paradoxes. If you ask a modern philosopher, they might say "the world is both continuous and discrete" which is to say "both are useful models"; this is a good answer. In the situations encountered by practising statisticians, it is usually fairly easy to determine whether a discrete or continuous variable should be used. In many instances, it is convenient to use a continuous random variable to model situations that are clearly discrete. The ideas developed for dealing with discreteness and continuity are also useful when the situation demands a variable that is neither continuous nor discrete. Before we give the formal definitions, it is useful to consider some examples illustrating the distinction between continuity and discreteness. As an initial attempt, we may define discreteness as a property that we attribute to things that we can count, and continuity as being associated with things that we measure.

**Example 3.3.1** (Appropriateness of discrete or continuous models)

1. **Discrete model**: Hurricanes develop in the Atlantic Basin and generally move westwards towards North America. We will assume that the rules for classification of hurricanes are clear cut, and that the equipment used to perform measurements is accurate (this is not too far from the truth). Suppose that we want a model for the number of hurricanes that develop each year. We know that the number of hurricanes will turn out to be a natural number, $0, 1, 2, \ldots$, and that we cannot have any values in between. This is a candidate for modelling using a discrete random variable.
2. **Continuous model**: Suppose that I snap a matchstick. By eye we could probably judge the distance between the bottom of the match and the break to within 5mm. To get a more accurate measurement, we could use a ruler. A ruler would be accurate to within, say, 0.5mm. If we were still not happy, we might use a more accurate device; for example, a micrometer would give us a reading accurate to around 0.01mm. The point here is that we treat length as a continuous scale; that is, we can measure length to any degree of accuracy, given a sufficiently sophisticated device. Thus, the length of a snapped match is a candidate for modelling using a continuous random variable.
3. **Continuous model for a discrete situation**: Consider claims received by an insurance company on a particular policy. These claims will take a range of values. For example the smallest claim might be £50 (that is, 5,000p) and the largest say £50,000 (that is, 5,000,000p). The values are discrete since currency

comes in discrete units (pence). However, as the number of possible distinct values is large, we may consider using a continuous variable as a model for the size of claims.

4. **Neither discrete nor continuous model**: Suppose that I have 15 identical matchsticks. I snap five of them, put all 20 pieces (10 whole and 10 pieces generated by snapping) into a bag and choose one at random. I want a model for the length of the match that I choose. An appropriate model has some discrete characteristics. At one level, I can count the number of possible outcomes; there are two, either I choose a whole match or a snapped match. At another level, the problem is continuous; the length of a snapped match is measured on a continuous scale. In fact, an appropriate model is one that is neither discrete nor continuous. A more realistic example follows.
5. **Neither discrete nor continuous model for a discrete situation**: Returning to the insurance claims example given above, suppose that we want to model the total value of claims made by an individual in a year. Many customers will not make any claims so there is a high probability that this value will be zero. If a customer makes one or more claims, we might choose to represent the total value on a continuous scale. The resulting model has some discrete characteristics (the customer either makes some claims or does not make any claims) and some continuous characteristics (we are measuring the total value on a continuous scale). Thus, we might choose to model this situation using a variable that is neither discrete nor continuous.

Before turning to a detailed discussion of discrete and continuous random variables, we introduce a piece of terminology. Many random variables only take values on a subset of $\mathbb{R}$. A useful concept for describing this situation is that of the **support**.

**Definition 3.3.2** (Support of a function)
The support of a positive real-valued function, $f$, is the subset of the real line where $f$ takes values strictly greater than zero, $\{x \in \mathbb{R} : f(x) > 0\}$.

We will regularly refer to the support of a random variable. This is just a slightly lazy way of referring to the support of the **mass function** for a discrete variable or the support of the **density function** for a continuous variable.

**Notation 3.3.3** (The $\sim$ notation)
We use $\sim$ in two distinct (though related) contexts.

i. $\sim$ means "is distributed as"; for example, $X \sim \text{Unif}[a, b]$ or $Y \sim \text{Bin}(n, p)$.
ii. $\sim$ means "has cumulative distribution function"; for example, $X \sim F_X$ is read as "$X$ is a random variable with cumulative distribution function $F_X$".

### 3.3.1 Discrete random variables and mass functions

**Definition 3.3.4** (Discrete random variable)
A random variable $X$ is **discrete** if it only takes values that are in some countable subset $\{x_1, x_2, \ldots\}$ of $\mathbb{R}$.

Another way of stating Definition 3.3.4 is that the probability of $X$ taking a value outside $\{x_1, x_2, \ldots\}$ is zero, that is, $\text{P}(X = x) = 0$ if $x \notin \{x_1, x_2, \ldots\}$. As with all random variables, we can characterise the distribution of probability associated with a discrete random variable using the cumulative distribution function. For a discrete random variable, it also makes sense to consider probability at a point. This leads to the following definition.

**Definition 3.3.5** (Probability mass function)
The **probability mass function** of a discrete random variable $X$ is the function $f_X : \mathbb{R} \longrightarrow [0, 1]$ given by $f_X(x) = \text{P}(X = x)$.

Probability mass functions are often referred to simply as mass functions. The properties of mass functions are easily stated. As mass is a probability, it must be between 0 and 1, and it must sum to 1 over all possible values. Claim 3.3.6 summarises these characteristics. Any function satisfying these properties is a valid mass function.

**Claim 3.3.6** (Properties of mass functions)
*If $f_X$ is a mass function then*

i. $0 \leq f_X(x) \leq 1$ *for all* $x$.
ii. $f_X(x) = 0$ *if* $x \notin \{x_1, x_2, \ldots\}$.
iii. $\sum_x f_X(x) = 1$.

It is clear that the support of the mass function of a discrete variable is the countable set $\{x_1, x_2, \ldots\}$ on which the variable can take values. In summations such as $\sum_x f_X(x)$, where no range of values is specified, the sum is assumed to be over the support of $f_X$. The relationship between mass and distribution functions is readily established. The following claim is a simple consequence of Proposition 3.2.4 and our definition of the mass function. You should satisfy yourself that you understand why each part is true.

**Claim 3.3.7** (Relationship between mass and distribution function)
*If $X$ is a random variable, $f_X$ is the mass function of $X$, and $F_X$ is the cumulative distribution function of $X$, then*

i. $f_X(x) = F_X(x) - F_X(x-)$.
ii. $F_X(x) = \sum_{u:u \leq x} f_X(u)$.

Note that the first part of this claim implies $F_X(x) = F_X(x-) + f_X(x)$. We know that $f_X(x) = 0$ if $x \notin \{x_1, x_2, \ldots\}$. Thus, $F_X(x) = F_X(x-)$ for $x \notin \{x_1, x_2, \ldots\}$. This indicates that the cumulative distribution function will be flat except for discontinuities at the points $\{x_1, x_2, \ldots\}$; this type of function is known as a **step function**. We conclude that non-zero probability at a given point is associated with a discontinuity in the cumulative distribution function. We will return to this point later in this section when we consider continuous random variables. The following examples introduce some well-known discrete distributions. These examples introduce the term **parameter**. Ideas associated with parameters are discussed in greater depth in subsection 3.3.3.

**Example 3.3.8** (Binomial and Bernoulli distributions)
Recall Example 3.1.1, in which we consider tossing a coin three times. Equation (3.1) gives us the mass function and can be summarised as

$$f_X(x) = \binom{3}{x} p^x (1-p)^{3-x} \text{ for } x = 0, 1, 2, 3.$$

We can generalise to the case where $X$ is the number of heads when we toss the coin $n$ times. The mass function is then

$$f_X(x) = \binom{n}{x} p^x (1-p)^{n-x} \text{ for } x = 0, \ldots, n. \tag{3.4}$$

The logic here is as follows. Any specific arrangement of $x$ heads has probability $p^x(1-p)^{n-x}$, since independence of the coin tosses means probabilities multiply. There are $\binom{n}{x}$ such arrangements. These arrangements are disjoint; for example, in three tosses you cannot have both HHT and THH. From the properties of probability measure, we know that the probability of a union of disjoint events is the sum of the individual probabilities; this leads to (3.4).

We can show that $f_X$ as defined by equation (3.4) is a valid mass function. It is clear from the form of the function that $0 \leq f_X(x) \leq 1$ for all $x$, and, by definition, $f_X(x) = 0$ if $x \notin \{0, \ldots, n\}$. The only real work that we have to do to establish that $f_X$ is a valid mass function is to show $\sum_x f_X(x) = 1$. We work directly using the specified mass function and support,

$$\begin{aligned}
\sum_x f_X(x) &= \sum_{x=0}^{n} \binom{n}{x} p^x (1-p)^{n-x} \\
&= [p + (1-p)]^n && \text{by equation (2.3)} \\
&= 1^n = 1.
\end{aligned}$$

Notice that when we replace $f_X(x)$ with the specific mass function, we also give the precise limits of the summation as dictated by the support.

Random variables with a mass function of the form given by equation (3.4) are referred to as having a **binomial distribution** with parameters $n$ and $p$. To indicate that a random variable $X$ has this distribution we use the notation $X \sim \text{Bin}(n, p)$. This is often described as the distribution of the number of successes in $n$ independent trials, where the probability of success at each trial is $p$.

The binomial distribution is used as a model for situations where we are repeating a procedure (the trials) with two outcomes that are of interest. The trials themselves can be modelled using a $\text{Bin}(1, p)$ distribution. This type of distribution is referred to as a **Bernoulli distribution**. If $Y$ has a Bernoulli distribution then the mass function is

$$f_Y(y) = \begin{cases} 1-p & \text{for } y = 0, \\ p & \text{for } y = 1, \\ 0 & \text{otherwise.} \end{cases}$$

The labels 0 and 1 for failure and success are convenient; in order to calculate the number of successes in $n$ trials, we just add up the outcomes of the underlying Bernoulli variables. In general, any random variable that can take just two values is referred to as a Bernoulli variable.

The term success is rather an emotive one – we do not always associate success with a desirable outcome. For example, suppose that we think that $n$ hurricanes will develop in the Atlantic Basin next year and that each one has probability $p$ of making landfall independent of all other hurricanes. The total number of hurricanes making landfall next year may then be modelled using a binomial distribution with parameters $n$ and $p$. The trials in this instance are the hurricanes that develop in the basin. Success is a hurricane hitting land (clearly not the outcome that we hope for).

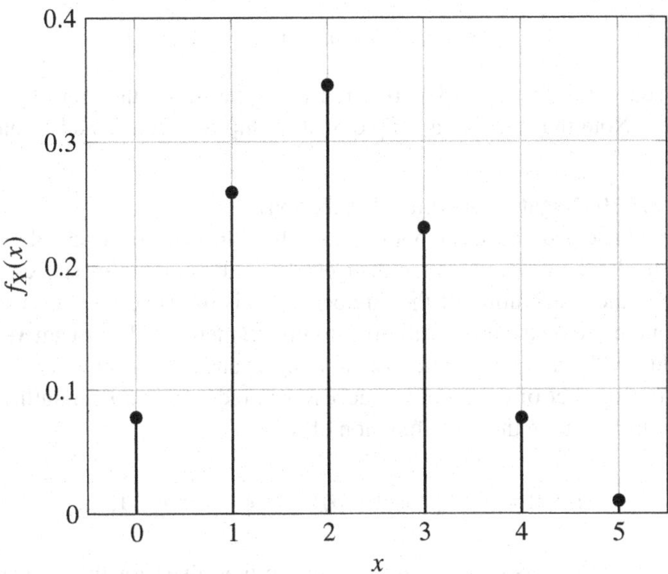


Figure 3.2 *The* Bin(5, 0.4) *mass function.*

**Example 3.3.9** (Geometric distribution)
Suppose that we flip a coin repeatedly until we get a head. Let $X$ be the number of times that we flip the coin. We want an expression for the mass function of $X$. We

know that we flip the coin at least once. The event $X = 1$ corresponds to getting a head on the first flip, the event $X = 2$ corresponds to getting a tail on the first flip but getting a head on the second, and so on. Thus, the probabilities are given by

$$\begin{aligned} f_X(1) &= \mathrm{P}(X = 1) = \mathrm{P}(\{H\}) = p, \\ f_X(2) &= \mathrm{P}(X = 2) = \mathrm{P}(\{TH\}) = (1 - p)p, \\ f_X(3) &= \mathrm{P}(X = 3) = \mathrm{P}(\{TTH\}) = (1 - p)^2 p, \\ &\vdots \qquad\qquad \vdots \end{aligned}$$

This is known as the **geometric distribution**. In general we can write

$$f_X(x) = (1 - p)^{x-1} p, \quad \text{for } x = 1, 2, \ldots \tag{3.5}$$

Showing that equation (3.5) is a valid mass function is part of Exercise 3.3. This is another situation that can be modelled as repeated independent Bernoulli trials. The expression given by (3.5) is the mass function of the number of trials up to and including the trial in which we get the first head. There is, rather confusingly, an alternative form. This alternative defines the geometric as the distribution of the number of trials *before* the first success. If $Y$ is a geometric distribution under this second definition, the mass function of $Y$ is given by

$$f_Y(y) = (1 - p)^y p, \quad \text{for } y = 0, 1, \ldots \tag{3.6}$$

You will see both (3.5) and (3.6) referred to as the mass function of a geometric distribution. Note that there is a simple relationship between $X$ and $Y$, namely $Y = X - 1$.

**Example 3.3.10** (Negative binomial distribution)
The binomial and geometric can both be thought of in terms of repeated independent Bernoulli trials; the binomial is the distribution of the number of successes and the geometric is the distribution of the number of trials until the first success. We will consider one more distribution that fits into this framework. The **negative binomial distribution** with parameters $r$ (a positive integer) and $p$ (a probability) is the distribution of the number of trials until $r$ successes in independent Bernoulli($p$) trials. If $X \sim \text{NegBin}(r, p)$, then the mass function of $X$ is

$$f_X(x) = \binom{x-1}{r-1} p^r (1 - p)^{x-r} \quad \text{for } x = r, r + 1, \ldots \tag{3.7}$$

The justification for this expression is similar to that given for the geometric mass. In order to have $r$ successes for the first time on the $x^{\text{th}}$ trial, we must have $r - 1$ successes in the first $x - 1$ trials. The number of ways of doing this is $\binom{x-1}{r-1}$ and the probability associated with each of these sequences is $p^{r-1}(1 - p)^{x-r}$. Noting that the final trial must be a success yields equation (3.7). It should be clear from this explanation that the negative binomial generalises the geometric distribution. In fact, if $X \sim \text{Geometric}(p)$, then $X \sim \text{NegBin}(1, p)$.

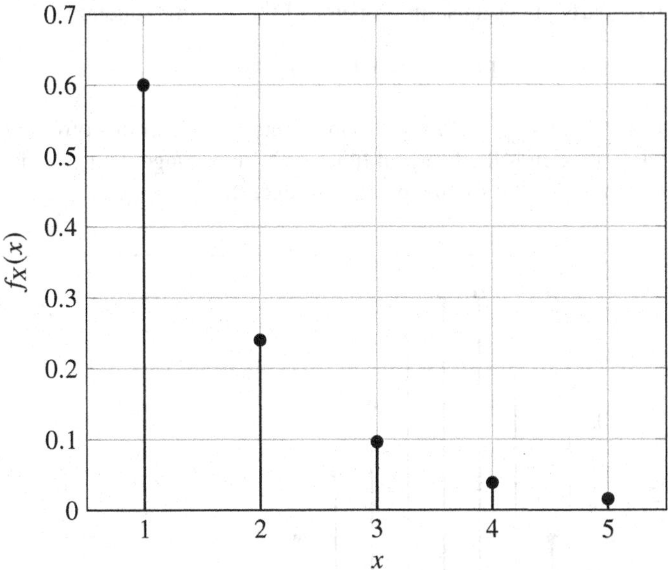


Figure 3.3 *The* Geometric(0.6) *mass function.*

Showing that equation (3.7) defines a valid mass function is part of Exercise 3.3. In order to do this, you will need the negative binomial expansion,

$$(1-a)^{-n} = \sum_{j=0}^{\infty} \binom{j+n-1}{n-1} a^j, \tag{3.8}$$

provided $|a| < 1$.

The negative binomial can also be formulated as the number of failures before we have $r$ successes. In this case the mass function is

$$f_Y(y) = \binom{y+r-1}{r-1} p^r (1-p)^y \quad \text{for} \quad y = 0, 1, \dots$$

It is easy to see that $Y = X - r$. We can generalise the negative binomial by allowing $r$ to take non-integer values. The resulting distribution is referred to as a **Polya distribution** (although it may also be called a negative binomial). If $Z \sim \text{Polya}(r, p)$, then the mass function of $Z$ is

$$f_Z(z) = \frac{\Gamma(r+z)}{z!\,\Gamma(r)} p^r (1-p)^z \quad \text{for} \quad z = 0, 1, \dots$$

Here, $\Gamma$ is the Gamma function,

$$\Gamma(t) = \int_0^{\infty} u^{t-1} e^{-u} du. \tag{3.9}$$

Integration by parts yields a useful property of the Gamma function:

$$\Gamma(t) = (t-1)\Gamma(t-1) \text{ for } t > 1. \tag{3.10}$$

Notice also that $\Gamma(1) = \int_0^\infty e^{-u} du = 1$. One interpretation of this property is that the Gamma function extends the factorial function to non-integer values. It is clear from (3.10) that $\Gamma(n) = (n-1)!$ for any positive integer $n$.

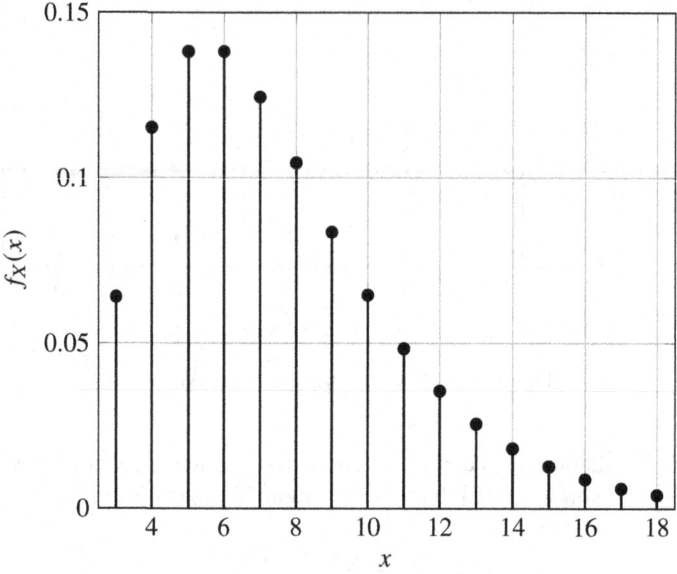


Figure 3.4 *The* NegBin(3, 0.4) *mass function.*

**Example 3.3.11** (Poisson distribution)
The **Poisson distribution** is often described as the distribution for the number of occurrences of rare events. It arises as the limit of the binomial distribution as the number of trials increases without bound, and the probability of success tends to zero; more on this shortly. The Poisson has a single parameter, $\lambda$, a positive real number often referred to as the **intensity** or **rate**. If $X \sim \text{Pois}(\lambda)$, the mass function of $X$ is given by

$$f_X(x) = \frac{\lambda^x e^{-\lambda}}{x!} \quad \text{for } x = 0, 1, \ldots$$

**Theorem 3.3.12** (Poisson limit theorem)
*Suppose that* $X \sim \text{Bin}(n, p)$. *If we let* $n \to \infty$ *and* $p \downarrow 0$ *while* $np = \lambda$ *(constant), then*

$$f_X(x) = \binom{n}{x} p^x (1-p)^{n-x} \to \frac{\lambda^x e^{-\lambda}}{x!},$$

*a* Pois($\lambda$) *mass function.*

*Proof.*

Begin by making the substitution $p = \lambda/n$. We have

$$\begin{aligned}\lim_{n\to\infty} f_X(x) &= \lim_{n\to\infty} \frac{n!}{x!(n-x)!}\left(\frac{\lambda}{n}\right)^x\left(1-\frac{\lambda}{n}\right)^{n-x} \\ &= \lim_{n\to\infty} \frac{n(n-1)\dots(n-x+1)}{n^x}\frac{\lambda^x}{x!}\left(1-\frac{\lambda}{n}\right)^n\left(1-\frac{\lambda}{n}\right)^{-x}.\end{aligned}$$

Now consider separately each term involving $n$,

$$\begin{aligned}&\lim_{n\to\infty} \frac{n(n-1)\dots(n-x+1)}{n^x} = \lim_{n\to\infty} \frac{n}{n}\frac{n-1}{n}\dots\frac{n-x+1}{n} = 1, \\ &\lim_{n\to\infty}\left(1-\frac{\lambda}{n}\right)^{-x} = 1^{-x} = 1, \\ &\lim_{n\to\infty}\left(1-\frac{\lambda}{n}\right)^{n} = e^{-\lambda}.\end{aligned}$$

Putting everything together, we obtain

$$\lim_{n\to\infty} f_X(x) = \frac{\lambda^x}{x!}e^{-\lambda},$$

the required mass function. □

We can give a more intuitive explanation of this result. Suppose that we are interested in the number of people, $Y$, visiting a website between, say, 8pm and 9pm on a weekday. We assume that potential visitors to the site behave independently of each other, and that the rate of arrivals is constant over this hour. Under these assumptions, if we subdivide the hour into one-minute intervals and define $X^{(60)}$ to be the number of these intervals with at least one visitor, then $X^{(60)} \sim \text{Bin}(60, p)$ for some probability $p$. If the website is popular, $X^{(60)}$ will be a poor approximation of $Y$, as we expect there to be many intervals with *more than one* visitor. Suppose that we halve the length of the intervals; the rate is constant, so the number of half-minute intervals with at least one visitor will be $X^{(120)} \sim \text{Bin}(120, p/2)$, which is a better approximation. Notice that the product $np$ remains constant. By subdividing the hour into smaller and smaller intervals, we obtain better and better approximations, that is, the binomial converges to a Poisson.

Early probability calculations were often associated with games of chance (in other words, gambling), and the binomial distribution arises frequently in these situations. Calculating probabilities associated with the binomial is tedious for anything more than a very small number of trials, while probabilities for the Poisson are readily computed, since

$$f_X(x) = \begin{cases} e^{-\lambda} & \text{for } x = 0, \\ \frac{\lambda}{x} f_X(x-1) & \text{for } x = 1, 2, \dots \end{cases}$$

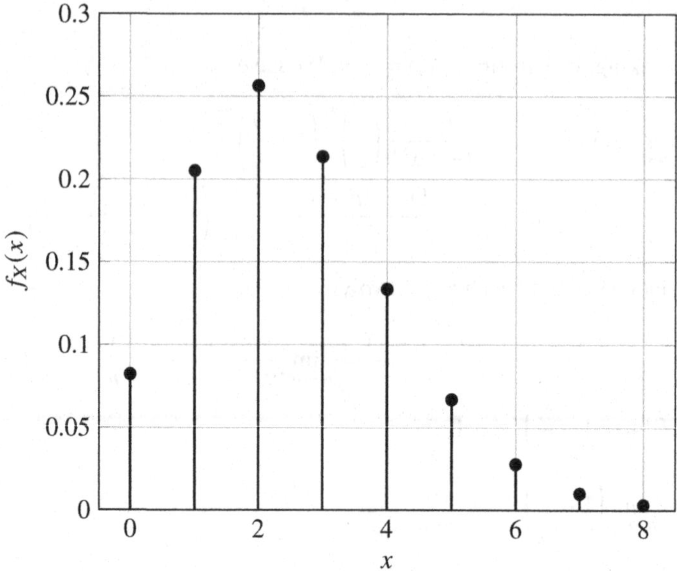


Figure 3.5 *The* Pois(2.5) *mass function.*

The Poisson thus provides a computationally convenient approximation to the binomial for the case where $n$ is large and $p$ is small. In an era when computation is fast and cheap, this fact is somewhat irrelevant. However, to early practitioners, the ease with which probabilities could be calculated made the Poisson an appealing distribution to work with.

**Example 3.3.13** (Discrete uniform distribution)
A **discrete uniform distribution** assigns equal probability to every member of its support. Thus, if $X$ is a discrete uniform with support $\{x_1, \dots, x_n\}$, the mass function of $X$ is

$$f_X(x) = \begin{cases} \frac{1}{n} & \text{for } x \in \{x_1, \dots, x_n\}, \\ 0 & \text{otherwise.} \end{cases}$$

The cumulative distribution function for the discrete uniform is a step function in which the steps are all of the same magnitude, $\frac{1}{n}$. Using the indicator function described by Definition 2.2.4, we can write the cumulative distribution function as

$$F_X(x) = \sum_{i=1}^{n} \frac{1}{n} \mathbf{1}_{(-\infty, x]}(x_i). \tag{3.11}$$

This function plays an important role in nonparametric statistics; the empirical cumulative distribution function for a sample $\{x_1, \dots, x_n\}$ has the same form as the cumulative distribution function given by equation (3.11).

### *3.3.2 Continuous random variables and density functions*

A continuous random variable can be defined as a variable whose cumulative distribution function is continuous. We use a stronger definition that ensures that a continuous random variable has a well-defined **(probability) density function**.

**Definition 3.3.14** (Continuous random variable)
A random variable $X$ is **continuous** if its distribution function can be expressed as

$$F_X(x) = \int_{-\infty}^{x} f_X(u)du \quad \text{for } x \in \mathbb{R},$$

for some integrable function $f_X : \mathbb{R} \longrightarrow [0, \infty)$. The function $f_X$ is called the **(probability) density function** of $X$.

Definition 3.3.14 shows how the distribution function for a random variable can be found by integrating the density. We can also work in the opposite direction: the density is found by differentiating the cumulative distribution function.

**Claim 3.3.15** (Density from cumulative distribution function)
*For a continuous random variable $X$ with cumulative distribution function $F_X$, the density function is given by*

$$f_X(x) = \frac{d}{du} F_X(u)\bigg|_{u=x} = F_X'(x) \textit{ for all } x \in \mathbb{R}.$$

The basic properties of density functions are given by the following claim.

**Claim 3.3.16** (Properties of continuous random variables)
*If $f_X$ is a density function then*

*i.* $f_X(x) \geq 0$ *for all* $x \in \mathbb{R}$.
*ii.* $\int_{-\infty}^{\infty} f_X(x)dx = 1$.

The first property is a direct consequence of Definition 3.3.14 and the second property can be seen directly from part ii. of Proposition 3.2.3.

We use the same notation (lowercase $f$) for density functions as we do for mass functions. This serves to emphasise that the density plays the same role for a continuous variable as mass does for a discrete variable. However, there is an important distinction; it is legitimate to have density function values that are greater than one since *values of a density function do not give probabilities*. Probability is not associated with the values of the density function but with the area beneath the curve that the density function defines. In order to work out the probability that a random variable $X$ takes a value between $a$ and $b$, we work out the area above the $x$-axis, beneath the density, and between the lines $x = a$ and $x = b$. As we might expect, given this interpretation, the total area beneath a density function is one, as stated in Claim 3.3.16. The general relationship between probability and density is given by the following proposition.

**Proposition 3.3.17** (Probability from a density function)
*If $X$ is a continuous random variable with density $f_X$, and $a, b \in \mathbb{R}$ with $a \leq b$, then*

$$P(a < X \leq b) = F_X(b) - F_X(a) = \int_a^b f_X(x)dx. \tag{3.12}$$

*Proof.*

This is an immediate consequence of part ii. of Proposition 3.2.4 and Definition 3.3.14 of a continuous random variable. □

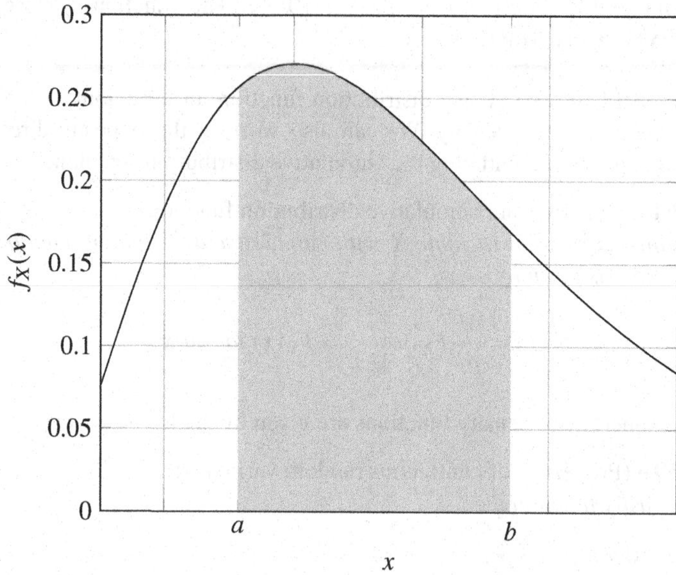


Figure 3.6 *Probability from density. The grey shaded region corresponds to* $P(a < X \leq b)$.

In fact, a more general result holds. If $B$ is a reasonably well-behaved subset of $\mathbb{R}$ then

$$P(X \in B) = \int_B f_X(x)dx.$$

Here, $\{X \in B\}$ is our shorthand for the event $\{\omega \in \Omega : X(\omega) \in B\}$. This relationship certainly holds if $B$ is an interval or a countable union of intervals.

Recall from part iv. of Proposition 3.2.4 that for any random variable $X$ we have $P(X = x) = F_X(x) - F_X(x-)$. In other words, we only have non-zero probability at points where there is a discontinuity in the cumulative distribution function. For a continuous random variable, there are no discontinuities in the distribution function. The rather disturbing consequence of this observation is summed up in the following corollary.

**Corollary 3.3.18** (Probability of a continuous random variable at a point)
*If $X$ is a continuous random variable then*

$$\mathrm{P}(X = x) = 0 \textit{ for all } x \in \mathbb{R}.$$

In other words, for any continuous random variable and any real value, the probability that the variable takes that value is zero. We can confirm the fact that $\mathrm{P}(X = x) = 0$ by taking $a = b = x$ in equation (3.12). Consider the following illustration. I throw a javelin; if $X$ is the length of the throw, it is reasonable to model $X$ using a continuous random variable. However, using this model, the probability that the length of the throw takes any real value is zero. But the javelin has to land somewhere, so how can this be? The apparent paradox is resolved by considering precisely what it means to measure something on a continuous scale. Suppose when we measure the throw it comes to 513cm (I'm not exactly an elite athlete). If our tape measure is accurate to within 0.5cm, our measurement actually indicates that the throw fell somewhere in the interval $[512.5, 513.5]$. This interval has some non-zero probability associated with it. No matter how much we improve our measuring device, there will always be limitations to its accuracy. Thus, any measurement on a continuous scale yields an interval rather than a single value.

The following examples introduce a number of well-known continuous distributions. We also discuss the idea of an exponential family, which provides a convenient representation for some commonly used distributions (both discrete and continuous).

**Example 3.3.19** (Continuous uniform distribution)
A **continuous uniform distribution** on an interval $[a, b]$ assigns equal probability to sub-intervals of equal length. Thus, if $X$ has a continuous uniform distribution on $[a, b]$, denoted $X \sim \mathrm{Unif}[a, b]$, the density function of $X$ is

$$f_X(x) = \begin{cases} \frac{1}{b-a} & \text{for } a \leq x \leq b, \\ 0 & \text{otherwise.} \end{cases}$$

Deriving the cumulative distribution function of a continuous uniform is part of Exercise 3.3. The uniform distribution arises in simulation problems. Instances of the continuous uniform on the unit interval, $\mathrm{Unif}[0, 1]$, are used as the basis for constructing samples from other distributions.

**Example 3.3.20** (Exponential distribution)
Consider the function

$$f(x) = \begin{cases} ke^{-\lambda x} & \text{for } 0 \leq x < \infty, \\ 0 & \text{otherwise.} \end{cases} \tag{3.13}$$

We would like to find the value of $k$ for which this is a valid density. We can integrate this function over the whole real line,

$$\int_{-\infty}^{\infty} f(x)dx = \int_{0}^{\infty} ke^{-\lambda x}dx = k\left[-\frac{1}{\lambda}e^{-\lambda x}\right]_0^{\infty} = \frac{k}{\lambda}. \tag{3.14}$$

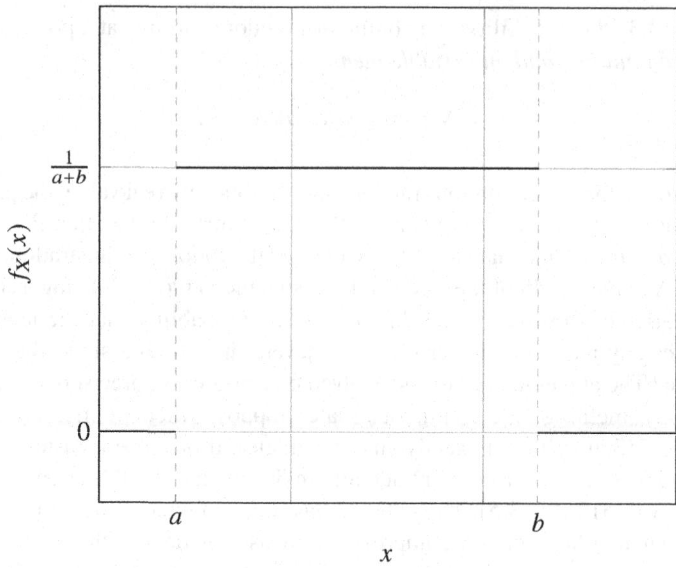


Figure 3.7 *The* Unif$[a, b]$ *density function.*

Since $\int_{-\infty}^{\infty} f(x)dx = 1$, we have $k = \lambda$. Notice that in equation (3.14) we write the general form $\int_{-\infty}^{\infty} f(x)dx$ first. When $f(x)$ is replaced by the specific function given in equation (3.13), the range of integration is replaced by the support of $f$, that is, $[0, \infty)$.

If $X$ has an **exponential distribution** with parameter $\lambda > 0$, this is denoted $X \sim \text{Exp}(\lambda)$ and the density function of $X$ is

$$f_X(x) = \begin{cases} \lambda e^{-\lambda x} & \text{for } 0 \leq x < \infty, \\ 0 & \text{otherwise.} \end{cases}$$

For $x \in [0, \infty)$ the cumulative distribution function of $X$ is given by

$$F_X(x) = \int_{-\infty}^{x} f_X(u)du = \int_{0}^{x} \lambda e^{-\lambda u} du = \left[-e^{-\lambda u}\right]_0^x = 1 - e^{-\lambda x}.$$

So, in full, the cumulative distribution function is

$$F_X(x) = \begin{cases} 1 - e^{-\lambda x} & \text{for } 0 \leq x < \infty, \\ 0 & \text{otherwise.} \end{cases}$$

The parameter $\lambda$ is known as the rate. The exponential distribution is also commonly expressed in terms of its scale parameter, $\theta$, where $\theta = 1/\lambda$. Confusingly, this is also denoted $X \sim \text{Exp}(\theta)$. In this book, we will always parameterise the exponential in terms of its rate.

One way in which the exponential arises is as a model for the length of time between events. It is used as a simple model for length of life (time between birth and death), although, as we will see in Example 6.5.5, it has a particular characteristic that makes it an unappealing model for human life length.

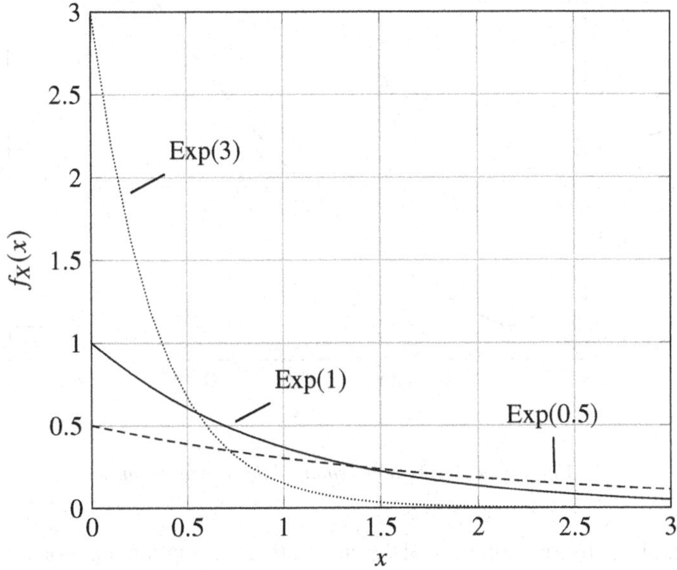


Figure 3.8 *The* Exp($\lambda$) *density function for various values of* $\lambda$.

**Example 3.3.21** (Simple polynomial density)
We can easily construct continuous density functions from polynomials. While these are not always particularly useful models for real-life phenomena, they provide a mechanism for illustrating methods associated with continuous variables while avoiding intractable integrals. Consider the function

$$f_X(x) = \begin{cases} \frac{3}{2}x^2 + x & \text{for } 0 \leq x \leq 1, \\ 0 & \text{otherwise.} \end{cases}$$

A plot of this function is shown in Figure 3.9. Proving that this is a valid density is part of Exercise 3.3. For $x \in [0, 1]$ the associated cumulative distribution function is given by

$$F_X(x) = \int_{-\infty}^{x} f_X(u)du = \int_{0}^{x} (\frac{3}{2}u^2 + u)du = \left[\frac{1}{2}u^3 + \frac{1}{2}u^2\right]_0^x = \frac{1}{2}(x^3 + x^2).$$

So, in full,

$$F_X(x) = \begin{cases} 0, & x < 0, \\ \frac{1}{2}(x^3 + x^2), & 0 \leq x \leq 1, \\ 1, & x > 1. \end{cases}$$

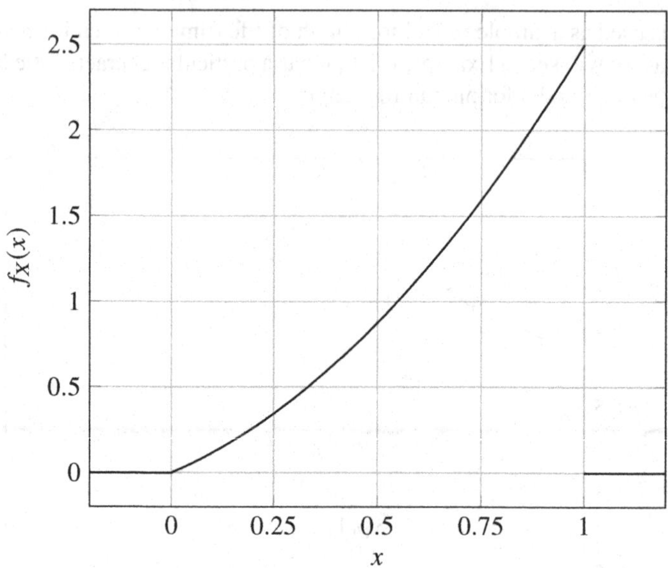


Figure 3.9 *A simple polynomial density function.*

We can readily go from a definition of the cumulative distribution function to density in the polynomial case. Suppose that $Y$ is a random variable with cumulative distribution function

$$F_Y(y) = \begin{cases} 0, & y < 0, \\ \frac{1}{2}y^2(1 - \frac{1}{8}y^2), & 0 \leq y \leq 2, \\ 1, & y > 2. \end{cases}$$

By differentiation we can see that, for $y \in [0, 2]$,

$$f_Y(y) = F'_Y(y) = y - \frac{1}{4}y^3.$$

So, in full,

$$f_Y(y) = \begin{cases} y - \frac{1}{4}y^3 & \text{for } 0 \leq y \leq 2, \\ 0 & \text{otherwise.} \end{cases}$$

You should check that this is a valid density.

**Example 3.3.22** (Normal distribution)
The **normal** or **Gaussian distribution** is fundamental in statistical inference. There are two main reasons. Firstly, the normal provides a reasonable model for many quantities that practitioners are interested in measuring. Examples abound; often quoted are quantities like height or, more controversially, intelligence for a human population. The second reason that the normal distribution is fundamental is that the distribution of the sample mean (and many other statistics) is approximately normal for large samples.

The normal distribution is characterised by two parameters: the mean, $\mu$, and the variance, $\sigma^2$. If $X \sim \mathrm{N}(\mu, \sigma^2)$, that is, if $X$ is normally distributed with mean $\mu$ and variance $\sigma^2$, the density of $X$ is given by

$$f_X(x) = \frac{1}{\sqrt{2\pi\sigma^2}} \exp\left\{-\frac{1}{2\sigma^2}(x-\mu)^2\right\} \quad \text{for} \quad -\infty < x < \infty.$$

An important special case is the normal distribution with mean 0 and variance 1. This distribution is referred to as a standard normal. If $Z \sim \mathrm{N}(0, 1)$, the density of $Z$ is

$$f_Z(z) = \frac{1}{\sqrt{2\pi}} \exp\left\{-\frac{1}{2}z^2\right\}.$$

A linear transformation relates general and standard normal distributions:

$$X \sim \mathrm{N}(\mu, \sigma^2) \Rightarrow \frac{X-\mu}{\sigma} \sim \mathrm{N}(0, 1).$$

Similarly,

$$Z \sim \mathrm{N}(0, 1) \Rightarrow \mu + \sigma Z \sim \mathrm{N}(\mu, \sigma^2).$$

The details of this relationship are explored in section 3.6 and, more specifically, in Example 3.6.7.

The cumulative distribution function of a normal does not have a closed form. For a standard normal, the symbol $\Phi$ is used to denote the cumulative distribution function. Thus, if $Z \sim \mathrm{N}(0, 1)$ then

$$\Phi(z) = F_Z(z) = \mathrm{P}(Z \le z).$$

In the past, tables of values of $\Phi$ were used to evaluate probabilities associated with normal distributions. Cheap computing power has rendered tables more or less obsolete. The only place you are likely to need to use a table is in an examination.

**Example 3.3.23** (Gamma distribution)
The final distribution that we consider in this section is the **Gamma distribution**. A Gamma-distributed random variable can only take positive values. The distribution is characterised by two positive parameters, often referred to as shape, $\alpha$, and rate, $\lambda$. If $X \sim \mathrm{Gamma}(\alpha, \lambda)$, the density of $X$ is

$$f_X(x) = \begin{cases} \frac{1}{\Gamma(\alpha)}\lambda^\alpha x^{\alpha-1}e^{-\lambda x} & \text{for } 0 < x < \infty, \\ 0 & \text{otherwise.} \end{cases}$$

Again, $\Gamma$ is the Gamma function defined by equation (3.9). The Gamma has strong associations with several other distributions. These associations are explored in examples in subsequent sections. We can observe immediately that the exponential is a special case of the Gamma. If $Y \sim \mathrm{Exp}(\lambda)$, then $Y \sim \mathrm{Gamma}(1, \lambda)$. Similarly to the exponential, the Gamma distribution is often parameterised in terms of its shape, $\alpha$, and scale, $\theta = 1/\lambda$.

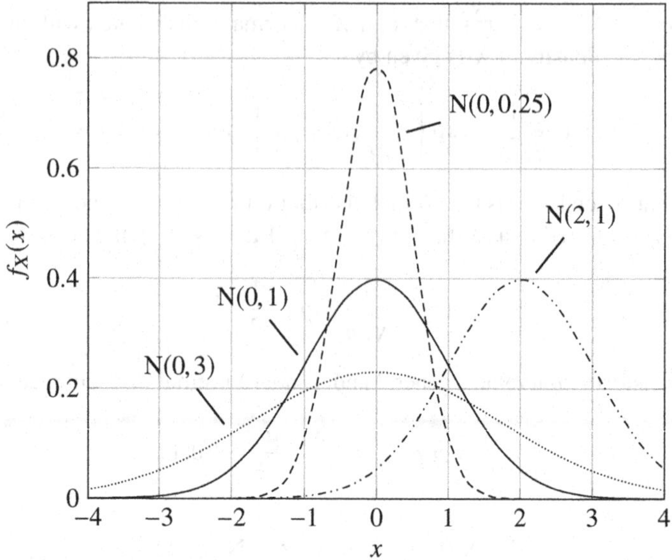


Figure 3.10 *The* $N(\mu, \sigma^2)$ *density function for various values of* $\mu$ *and* $\sigma^2$.

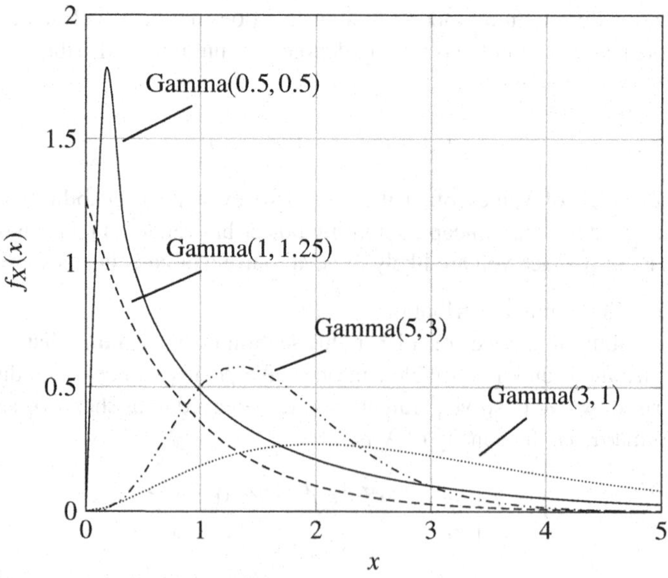


Figure 3.11 *The* Gamma$(\alpha, \lambda)$ *density function for various values of* $\alpha$ *and* $\lambda$.

**Example 3.3.24** (Exponential family)
We end the examples of this section with a representation of mass and density functions that will prove useful in later chapters. A distribution belongs to an **exponential**

**family** if its mass (in the discrete case) or density (in the continuous case) can be written in the form

$$f_Y(y) = h(y)a(\boldsymbol{\theta}) \exp\left[\sum_{i=1}^{k} \eta_i(\boldsymbol{\theta})t_i(y)\right], \tag{3.15}$$

where $\boldsymbol{\theta}$ is a vector of parameters, and $h(y) \geq 0$. The functions $h$ and $t_1, \ldots, t_k$ are functions of $y$ but not $\boldsymbol{\theta}$, and the functions $a$ and $\eta_1, \ldots, \eta_k$ are functions of $\boldsymbol{\theta}$ but not $y$. For any given family, the members of the family differ only in the values of the parameter vector $\boldsymbol{\theta}$. The special case where $k = 1$, $\eta_1(\theta) = \theta$, and $t_1(y) = y$ is known as a **natural exponential family** (of order 1).

Classes of discrete distribution that form exponential families include the binomial, Poisson, and negative binomial. For continuous distributions, the Beta, Gamma, and normal are all exponential families. The terms exponential family and exponential distribution should not be confused; exponential distributions (as a subclass of the Gamma distribution) form one of many possible exponential families. We illustrate by deriving the exponential family form of the binomial and normal distributions.

Consider a binomial distribution for which the parameter of interest is $p$, the probability of success at each trial. We can rewrite the binomial mass function in the form of equation (3.15) as

$$\begin{aligned} f_Y(y) &= \binom{n}{y} p^y (1-p)^{n-y} \\ &= \binom{n}{y} (1-p)^n \left(\frac{p}{1-p}\right)^y \\ &= \binom{n}{y} (1-p)^n \exp\left\{\log\left(\frac{p}{1-p}\right) y\right\}. \end{aligned}$$

In this representation, we have $h(y) = \binom{n}{y}$, $a(p) = (1-p)^n$, $\eta_1(p) = \log(\frac{p}{1-p})$, and $t_1(y) = y$. Notice that if we consider $\theta = \log(\frac{p}{1-p})$ to be the parameter of interest (instead of $p$), this is a natural exponential family,

$$f_Y(y) = \binom{n}{y} (1 + e^{\theta})^{-n} \exp(\theta y)$$

For the normal distribution, where the parameters of interest are $\mu$ and $\sigma^2$, we can rewrite the density as

$$\begin{aligned} f_Y(y) &= \frac{1}{\sqrt{2\pi\sigma^2}} \exp\left\{-\frac{1}{2\sigma^2}(y-\mu)^2\right\} \\ &= \frac{1}{\sqrt{2\pi\sigma^2}} \exp\left\{-\frac{1}{2\sigma^2}(y^2 - 2y\mu + \mu^2)\right\} \\ &= \frac{1}{\sqrt{2\pi\sigma^2}} \exp\left\{-\frac{\mu^2}{2\sigma^2}\right\} \exp\left\{\frac{\mu}{\sigma^2}y - \frac{1}{2\sigma^2}y^2\right\}. \end{aligned}$$

In this representation, we have $h(y) = 1$, $a(\mu, \sigma^2) = \frac{1}{\sqrt{2\pi\sigma^2}} \exp\left\{-\frac{\mu^2}{2\sigma^2}\right\}$, $\eta_1(\mu, \sigma^2) = \frac{\mu}{\sigma^2}$, $t_1(y) = y$, $\eta_2(\mu, \sigma^2) = -\frac{1}{2\sigma^2}$, and $t_2(y) = y^2$.

**Example 3.3.25** (Arguments of functions)
It is important to understand the role of the argument when dealing with distribution and density functions. Consider the density function value $f_X(x)$; in this statement the subscript $X$ (uppercase) identifies the random variable for which $f_X$ is the density, and the lowercase $x$ is the argument. The argument is just a place holder. Its only important feature is that it is a real number. We can give the argument any name we like; $f_X(y)$, $f_X(z)$ and $f_X(grapefruit)$ are all values of the density of $X$. When we integrate over a variable, the result is dependent only on the range over which we integrate, not the name we give the variable:

$$F_X(x) = \int_{-\infty}^{x} f_X(u)du = \int_{-\infty}^{x} f_X(y)dy = \int_{-\infty}^{x} f_X(oranges)doranges.$$

In the example above, note that we do not integrate over $x$ because we are already using $x$ as the argument for the distribution function $F_X$. This choice is sensible but arbitrary. We can use anything, so

$$F_X(y) = \int_{-\infty}^{y} f_X(u)du = \int_{-\infty}^{y} f_X(x)dx = \int_{-\infty}^{x} f_X(apples)dapples,$$

or even (don't try this at home)

$$F_X(u) = \int_{-\infty}^{u} f_X(z)dz,$$

$$F_Z(x) = \int_{-\infty}^{x} f_Z(y)dy,$$

$$F_{COCONUTS}(mangos) = \int_{-\infty}^{mangos} f_{COCONUTS}(pineapples)dpineapples.$$

Make sure you understand the role of *COCONUTS*, *mangos*, and *pineapples* in the equation above. Being clear about which parts identify the random variable and which parts are arguments becomes increasingly important as we move to considering multivariate distributions in Chapter 4.

### 3.3.3 *Parameters and families of distributions*

A parameter is a characteristic of a distribution that is of interest. Examples of parameters are the probability of success, $p$, for a binomial distribution and the mean, $\mu$, of a normal. Parameters often arise as terms in mass or density functions; the parameter, $\lambda$, of an exponential distribution determines the rate at which the density converges to zero. A distribution family is a set of distributions that differ only in the value of their parameters. For example, consider the number of heads when flipping a fair coin once (can be 0 or 1) and the number of sixes when throwing a fair die once (can also be 0 or 1).

**Exercise 3.3**

1. Show that $\Gamma(1/2) = \sqrt{\pi}$, and hence write down a formula for $\Gamma(k/2)$, where $k$ is a positive integer. [*Hint*: In the definition of $\Gamma(1/2)$, use the substitution $z = \sqrt{2u}$ and rearrange the integrand to obtain a standard normal density function.]
2. (**Geometric mass function**) Show that the function
$$f_X(x) = (1-p)^{x-1}p \text{ for } x = 1,2,\ldots,$$
is a valid mass function.
3. (**Geometric cumulative distribution function**) Suppose that $X \sim \text{Geometric}(p)$. Find the cumulative distribution function of $X$.
4. (**Negative binomial mass function**) Show that the function
$$f_X(x) = \binom{x-1}{r-1} p^r (1-p)^{x-r} \text{ for } x = r, r+1, \ldots$$
is a valid mass function.
5. (**Cumulative distribution function of a continuous uniform**) Suppose that $X \sim \text{Unif}[a,b]$. Derive the cumulative distribution function of $X$.
6. Show that the function
$$f_X(x) = \begin{cases} \frac{3}{2}x^2 + x & \text{for } 0 \le x \le 1, \\ 0 & \text{otherwise,} \end{cases}$$
is a valid density function.

## 3.4 Expectation, variance, and higher moments

### 3.4.1 *Mean of a random variable*

**Central tendency** is among the first concepts taught on any course in descriptive statistics. The hope is that calculating central tendency will provide us with some sense of the usual or average values taken by an observed variable. Among sample statistics commonly considered are the **mode** (most commonly occurring value), the **median** (middle value when observations are ordered) and the **arithmetic mean**. If we have a massless ruler with points of equal mass placed at locations corresponding to the observed values, the arithmetic mean is the point where we should place a fulcrum in order for the ruler to balance. We will follow the usual convention and refer to the arithmetic mean as just the **mean**.

These ideas transfer neatly to describing features of distributions. The measures of central tendency that are applied to describe data can also be applied to our models. For example, suppose that $X$ is a continuous random variable with density $f_X$ and cumulative distribution function $F_X$. We define $\text{mode}(X) = \arg\max_x f_X(x)$ and $\text{median}(X) = m$, where $m$ is the value satisfying $F_X(m) = 0.5$. We will now focus our attention on the mean.

**Definition 3.4.1** (Mean)
The **mean** of a random variable $X$, denoted $\mathbb{E}(X)$, is given by

$$\mathbb{E}(X) = \begin{cases} \sum_x x f_X(x) & \text{if } X \text{ discrete,} \\ \int_{-\infty}^{\infty} x f_X(x) dx & \text{if } X \text{ continuous,} \end{cases}$$

where, to guarantee that $\mathbb{E}(X)$ is well defined, we usually insist that $\sum_x |x| f_X(x) < \infty$ in the discrete case and $\int_{-\infty}^{\infty} |x| f_X(x) dx < \infty$ in the continuous case.

The Greek letter $\mu$ is often used to denote the mean. In this context it is clear that the mean may be viewed as a parameter of the distribution. The mean of $X$ is often also referred to as the **expected value** of $X$ or the **expectation** of $X$. During our discussion of the properties of random variables we will also use $\mathbb{E}(X)$ to denote the mean. The expectation operator $\mathbb{E}(.)$ defines a process of averaging; this process is widely applied and is the subject of the next subsection.

When talking about expectation, we will not repeatedly say "whenever this expectation exists". In many propositions about the mean, there is an implicit assumption that the proposition only holds when the mean is well defined. The same applies in later sections to other quantities defined as an expectation, such as the variance, higher moments, and moment-generating functions (more on these later).

We give two examples of computation of the mean using Definition 3.4.1, one a discrete random variable and the other continuous.

**Example 3.4.2** (Mean of a binomial)
Suppose that $X \sim \text{Bin}(n, p)$. We know that the mass function of $X$ is

$$f_X(x) = \binom{n}{x} p^x (1-p)^{n-x}, \quad \text{for } x = 0, \dots, n.$$

From Definition 3.4.1, the mean of $X$ is

$$\mathbb{E}(X) = \sum_x x f_X(x) = \sum_{x=0}^{n} x \binom{n}{x} p^x (1-p)^{n-x}.$$

The remaining parts of the calculation involve some tricks in series summation. In particular, we will get an expression involving the sum of an easily recognised mass function over its support. We know that this sum will be equal to one. Details of each step are given below.

$$
\begin{aligned}
\mathbb{E}(X) &= \sum_{x=1}^{n} x\binom{n}{x}p^x(1-p)^{n-x} && \text{the } x = 0 \text{ term is } 0 \\
&= \sum_{x=1}^{n} x\frac{n!}{x!(n-x)!}p^x(1-p)^{n-x} && \text{definition of } \binom{n}{x} \\
&= \sum_{x=1}^{n} \frac{n!}{(x-1)!(n-x)!}p^x(1-p)^{n-x} && \frac{x}{x!} = \frac{1}{(x-1)!} \text{ for } x > 0 \\
&= np\sum_{x=1}^{n} \frac{(n-1)!}{(x-1)!(n-x)!}p^{x-1}(1-p)^{n-x} && \text{taking out } np \\
&= np\sum_{r=0}^{n-1} \frac{(n-1)!}{r!(n-1-r)!}p^{r}(1-p)^{n-1-r} && \text{setting } r = x-1 \\
&= np\sum_{r=0}^{n-1} \binom{n-1}{r}p^{r}(1-p)^{n-1-r} && \text{definition of } \binom{n-1}{r} \\
&= np,
\end{aligned}
$$

since the final summand is a $\text{Bin}(n-1, p)$ mass. Thus if $X \sim \text{Bin}(n, p)$ then $\mathbb{E}(X) = np$. It should be clear that the letters $n$ and $p$ are just labels for the parameters; if $X \sim \text{Bin}(m, q)$ then $\mathbb{E}(X) = mq$, if $Y \sim \text{Bin}(k, \theta)$ then $\mathbb{E}(Y) = k\theta$, and so on.

**Example 3.4.3** (Mean of an exponential distribution)
Suppose that $X \sim \text{Exp}(\lambda)$. From Definition 3.4.1 the mean of $X$ is given by

$$\mathbb{E}(X) = \int_{-\infty}^{\infty} x f_X(x)dx = \int_{0}^{\infty} x\lambda e^{-\lambda x}dx.$$

Note that

$$x\lambda e^{-\lambda x} = -\lambda\frac{d}{d\lambda}e^{-\lambda x}.$$

Assuming that we can swap the order of integration with respect to $x$ and differentiation with respect to $\lambda$, we have

$$
\begin{aligned}
\mathbb{E}(X) &= -\lambda\frac{d}{d\lambda}\int_{0}^{\infty} e^{-\lambda x}dx = -\lambda\frac{d}{d\lambda}\left[-\frac{1}{\lambda}e^{-\lambda x}\right]_0^{\infty} \\
&= -\lambda\frac{d}{d\lambda}\frac{1}{\lambda} = \lambda\frac{1}{\lambda^2} = \frac{1}{\lambda}.
\end{aligned}
$$

Alternatively, you can use integration by parts, in which case you will need to use the fact that $xe^{-\lambda x} \to 0$ as $x \to \infty$ (see Exercise 3.4).

### 3.4.2 *Expectation operator*

The process of finding the mean is so commonly used that we adopt a special notation for it. The expectation operator, $\mathbb{E}$, when applied to a random variable yields its mean;

if $\mu$ is the mean of $X$ then $\mathbb{E}(X) = \mu$. We will often want to evaluate the mean of a function of a random variable, $g(X)$. We could work out the mass or density of $g(X)$ then use Definition 3.4.1 to evaluate $\mathbb{E}(g(X))$. However, this is not always a straightforward process. The following theorem provides a simple mechanism for calculating the expectation of a random variable without requiring us to derive its mass/density. We will discuss methods to find the mass/density of a function of a random variable in section 3.6.

**Theorem 3.4.4** (Expectation of a function of a random variable)
*For any well-behaved function $g : \mathbb{R} \longrightarrow \mathbb{R}$, the expectation of $g(X)$ is defined as*

$$\mathbb{E}[g(X)] = \begin{cases} \sum_x g(x) f_X(x) & \text{if } X \text{ discrete,} \\ \int_{-\infty}^{\infty} g(x) f_X(x) dx & \text{if } X \text{ continuous,} \end{cases}$$

*where, to guarantee that expectation is well defined, we usually insist that $\sum_x |g(x)| f_X(x) < \infty$ in the discrete case and $\int_{-\infty}^{\infty} |g(x)| f_X(x) dx < \infty$ in the continuous case.*

In the discrete case this is sometimes referred to (pejoratively by mathematicians) as the law of the unconscious statistician. We will prove the continuous case in section 3.6.

One of the key properties of expectation is its linearity.

**Proposition 3.4.5** (Linearity of expectation operator)
*For a random variable $X$ and real constants $a_0, a_1, \ldots, a_r$,*

$$\mathbb{E}\left(\sum_{i=0}^{r} a_i X^i\right) = \sum_{i=0}^{r} a_i \, \mathbb{E}(X^i).$$

The proof of this result is trivial as linearity is inherited directly from the definition of expectation in terms of a sum or integral. Another statement of the proposition is

$$\mathbb{E}(a_0 + a_1 X + \ldots + a_r X^r) = a_0 + a_1 \, \mathbb{E}(X) + \ldots + a_r \, \mathbb{E}(X^r).$$

From which it is clear that, for any real constants $a_0$ and $a_1$,

$$\mathbb{E}(a_0) = a_0 \quad \text{and} \quad \mathbb{E}(a_0 + a_1 X) = a_0 + a_1 \, \mathbb{E}(X).$$

As we might expect, the mean of a positive random variable is always non-negative.

**Claim 3.4.6** (Expectation of a positive random variable)
*If $X$ is a positive random variable then $\mathbb{E}(X) \geq 0$.*

This statement is readily verified using the definition of expectation. As we will see in the following subsection, an immediate consequence of Claim 3.4.6 is that variance must always be non-negative. Calculation of the variance illustrates the use of Theorem 3.4.4, so the related examples will appear in the next subsection.

### 3.4.3 Variance of a random variable

If measures of central tendency are the first thing taught in a course about descriptive statistics, then measures of spread are probably the second. One possible measure of spread is the interquartile range; this is the distance between the point that has a quarter of the probability below it and the point that has a quarter of the probability above it, $\text{IQR}(X) = F_X^{-1}(0.75) - F_X^{-1}(0.25)$. We will focus on the variance. The variance measures the average squared distance from the mean.

**Definition 3.4.7** (Variance and standard deviation)
If $X$ is a random variable, the **variance** of $X$ is defined as

$$\sigma^2 = \text{Var}(X) = \mathbb{E}[(X - \mathbb{E}(X))^2]$$
$$= \begin{cases} \sum_x (x - \mathbb{E}(X))^2 f_X(x) & \text{if } X \text{ discrete,} \\ \int_{-\infty}^{\infty} (x - \mathbb{E}(X))^2 f_X(x) dx & \text{if } X \text{ continuous,} \end{cases}$$

whenever this sum/integral is finite. The standard deviation is defined as $\sigma = \sqrt{\text{Var}(X)}$.

Some properties of the variance operator are given by the following proposition.

**Proposition 3.4.8** (Properties of variance)
*For a random variable $X$ and real constants $a_0$ and $a_1$, the variance has the following properties:*

*i.* $\text{Var}(X) \geq 0$,

*ii.* $\text{Var}(a_0 + a_1 X) = a_1^2 \text{Var}(X)$.

*Proof.*

Both properties are inherited from the definition of variance as an expectation.

i. By definition, $(X - \mathbb{E}(X))^2$ is a positive random variable, so $\text{Var}(X) = \mathbb{E}[(X - \mathbb{E}(X))^2] \geq 0$ by Claim 3.4.6.

ii. If we define $Y = a_0 + a_1 X$, then $\mathbb{E}(Y) = a_0 + a_1 \mathbb{E}(X)$, by linearity of expectation. Thus $Y - \mathbb{E}(Y) = a_1(X - \mathbb{E}(X))$ and so

$$\text{Var}(a_0 + a_1 X) = \text{Var}(Y) = \mathbb{E}[(Y - \mathbb{E}(Y))^2] = \mathbb{E}[a_1^2 (X - \mathbb{E}(X))^2]$$
$$= a_1^2 \text{Var}(X).$$

□

Note that the second of these properties says that the variance is not affected by translation (adding $a_0$). This is a desirable property for a measure of spread; when a distribution moves left or right along the $x$ axis, the dispersion about the mean is unaltered. Variance can be calculated directly using Definition 3.4.7. However, it is often more convenient to use one of the alternative forms provided by the following proposition.

**Proposition 3.4.9**
*For a random variable X the variance is given by*

*i.* $\text{Var}(X) = \mathbb{E}(X^2) - \mathbb{E}(X)^2$,
*ii.* $\text{Var}(X) = \mathbb{E}[X(X-1)] - \mathbb{E}(X)\mathbb{E}(X-1)$.

These results follow directly from the linearity of the expectation operator. Filling in the details is part of Exercise 3.4.

**Example 3.4.10** (Variance of a binomial)
Suppose that $X \sim \text{Bin}(n, p)$. We know that $\mathbb{E}(X) = np$. Thus, in order to calculate the variance we simply need to calculate $\mathbb{E}(X^2)$. In fact, in this case it is easier to calculate $\mathbb{E}[X(X-1)]$.

$$\begin{aligned}
\mathbb{E}[X(X-1)] &= \sum_x x(x-1)f_X(x) \\
&= \sum_{x=0}^{n} x(x-1)\binom{n}{x}p^x(1-p)^{n-x} \\
&= n(n-1)p^2 \sum_{x=2}^{n} \frac{(n-2)!}{(x-2)!(n-x)!}p^{x-2}(1-p)^{n-x} \\
&= n(n-1)p^2 \sum_{r=0}^{n-2} \frac{(n-2)!}{r!(n-2-r)!}p^r(1-p)^{n-2-r} \\
&= n(n-1)p^2,
\end{aligned}$$

thus,

$$\text{Var}(X) = n(n-1)p^2 - np(np-1) = np[(n-1)p - (np-1)] = np(1-p).$$

This is often written as $\text{Var}(X) = npq$, where $q = 1 - p$.

**Example 3.4.11** (Variance of a random variable with polynomial density)
Recall Example 3.3.21 in which we consider a random variable $X$ with polynomial density

$$f_X(x) = \begin{cases} \frac{3}{2}x^2 + x & \text{for } 0 \le x \le 1, \\ 0 & \text{otherwise.} \end{cases}$$

We can calculate the mean and variance using brute force.

$$\begin{aligned}
\mathbb{E}(X) &= \int_{-\infty}^{\infty} x f_X(x)dx = \int_0^1 \left(\frac{3}{2}x^3 + x^2\right) dx = \left[\frac{3}{8}x^4 + \frac{1}{3}x^3\right]_0^1 = \frac{17}{24} \\
\mathbb{E}(X^2) &= \int_{-\infty}^{\infty} x^2 f_X(x)dx = \int_0^1 \left(\frac{3}{2}x^4 + x^3\right) dx = \left[\frac{3}{10}x^5 + \frac{1}{4}x^4\right]_0^1 = \frac{11}{20} \\
\Rightarrow \text{Var}(X) &= \frac{11}{20} - \left(\frac{17}{24}\right)^2 = 0.048 \text{ (3dp)}.
\end{aligned}$$

Note that variances must be positive, so if we get a negative variance value we know that something has gone wrong in our calculation.

### 3.4.4 Inequalities involving expectation

In proving convergence results it is often useful to be able to provide bounds for probabilities and expectations. The propositions below provide bounds that are often rather loose. The appeal of these results is their generality.

**Proposition 3.4.12** (Markov inequality)
*If $Y$ is a positive random variable with $\mathbb{E}(Y) < \infty$, then $\text{P}(Y \geq a) \leq \mathbb{E}(Y)/a$ for any constant $a > 0$.*

*Proof.*

We prove this for the continuous case. A similar argument holds in the discrete case.

$$\begin{aligned} \text{P}(Y \geq a) &= \int_a^\infty f_Y(y)dy && \text{by definition} \\ &\leq \int_a^\infty \frac{y}{a} f_Y(y)dy && \text{since } 1 \leq \frac{y}{a} \text{ for } y \in [a, \infty) \\ &\leq \frac{1}{a}\int_0^\infty y f_Y(y)dy && \text{for positive } g, \int_a^\infty g(y)dy \leq \int_0^\infty g(y)dy \\ &\leq \frac{1}{a}\mathbb{E}(Y) && \text{by definition of } \mathbb{E}(Y). \end{aligned}$$

□

The Markov inequality provides us with a bound on the amount of probability in the upper tail of the distribution of a positive random variable. As advertised, this bound is fairly loose. Consider the following illustration. Let us suppose that $Y$ is the length of life of a British man. Life expectancy in Britain is not great, in fact, male life expectancy is around 79 years (it's all the lager and pies). If we take $\mathbb{E}(Y) = 79$, then we can calculate a bound on the probability that a British man lives to be over 158. Using the Markov inequality,

$$\text{P}(Y \geq 158) \leq \mathbb{E}(Y)/158 = 79/158 = 1/2.$$

Clearly, this is a loose bound. We would expect this probability to be pretty close to zero. The beauty of the Markov inequality lies not in tightness of the bounds but generality of application; no distributional assumptions are required.

The Markov inequality can be extended to random variables that are not necessarily positive.

**Proposition 3.4.13** (Chebyshev inequality)
*If $X$ is a random variable with $\text{Var}(X) < \infty$, then $\text{P}(|X - \mathbb{E}(X)| \geq a) \leq \text{Var}(X)/a^2$ for any constant $a > 0$.*

*Proof.*

This is an immediate consequence of the Markov inequality applied to the positive random variable $Y = (X - \mathbb{E}(X))^2$. Filling in the details is part of Exercise 3.4. □

**Corollary 3.4.14** (Chebyshev for a standardised distribution)
*IF $X$ is a random variable with $\mu = \mathbb{E}(X)$ and $\sigma^2 = \text{Var}(X)$, and $\lambda > 0$ is a real constant, then*

$$\text{P}\left(\frac{|X-\mu|}{\sigma} \geq \lambda\right) \leq \frac{1}{\lambda^2}.$$

*Proof.*

This is straight from the Chebyshev inequality taking $a = \lambda\sigma$. □

Consider the following illustration of the implications of Corollary 3.4.14. Suppose that we want a bound on the probability that a point from a distribution lies more than two standard deviations away from the mean. From Corollary 3.4.14

$$\text{P}(|X-\mu| \geq 2\sigma) \leq \frac{1}{4}.$$

For a specific distribution this bound may be rather inaccurate. For example, you may recall that for a normal distribution the probability a point lies more than two standard deviations from the mean is roughly 0.05. However, the fact that this result holds for all distributions with finite variance is remarkable.

The final proposition in this section deals with the expectation of a function of a random variable when the function falls into a particular class.

**Definition 3.4.15** (Convex functions)
A function $g : \mathbb{R} \to \mathbb{R}$ is **convex** if for any real constant $a$ we can find $\lambda$ such that

$$g(x) \geq g(a) + \lambda(x-a),$$

for all $x \in \mathbb{R}$.

The intuition here is that for any point on the curve $y = g(x)$ we can find a line which touches the curve (this is the $y = g(a) + \lambda(x-a)$ part) but always lies on or below it (Figure 3.12). Functions whose curves open upwards, such as $g(x) = x^2$, are convex. We can also define a concave function. A concave function is one for which we can always draw tangents that lie on or above the curve, such as $g(x) = -x^2$. Straight lines are both concave and convex. Many functions are neither concave nor convex, for example, $g(x) = x^3$.

**Proposition 3.4.16** (Jensen's inequality)
*If $X$ is a random variable with $\mathbb{E}(X)$ defined, and $g$ is a convex function with $\mathbb{E}(g(X)) < \infty$, then*

$$\mathbb{E}(g(X)) \geq g(\mathbb{E}(X)).$$

*Proof.*

We prove this inequality for continuous random variables; the discrete case is similar.

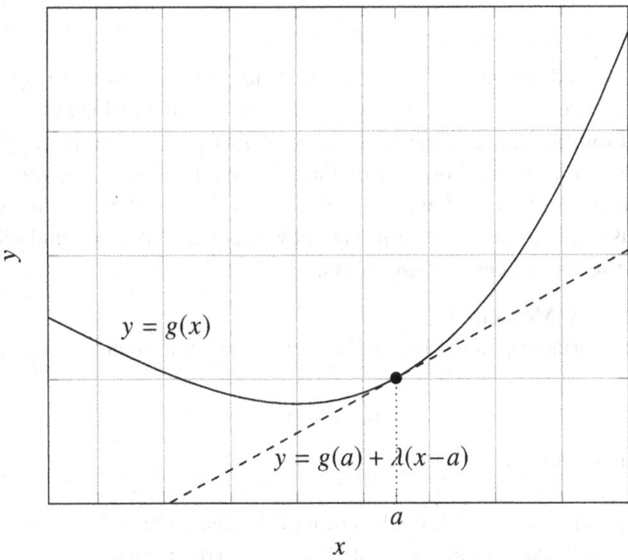


Figure 3.12 *Example of a convex function.*

Setting $a = \mathbb{E}(X)$ in the definition of a convex function, we can find $\lambda$ such that $g(x) \geq g(\mathbb{E}(X)) + \lambda(x - \mathbb{E}(X))$ for all $x \in \mathbb{R}$. We then have

$$\begin{aligned}
\mathbb{E}(g(X)) &= \int_{-\infty}^{\infty} g(x) f_X(x) dx \\
&\geq \int_{-\infty}^{\infty} [g(\mathbb{E}(X)) + \lambda(x - \mathbb{E}(X))] f_X(x) dx \\
&= \mathbb{E}[g(\mathbb{E}(X)) + \lambda(X - \mathbb{E}(X))] \\
&= g(\mathbb{E}(X)) + \lambda(\mathbb{E}(X) - \mathbb{E}(X)) \\
&= g(\mathbb{E}(X)) .
\end{aligned}$$

□

**Example 3.4.17** (Applications of Jensen's inequality)
Jensen's inequality allows us to derive many interesting relationships involving expectation. A few examples follow.

1. Straight line: if $g(x) = a_0 + a_1 x$ then $g$ is convex. Jensen's inequality gives us $\mathbb{E}(a_0 + a_1 X) \geq a_0 + a_1 \mathbb{E}(X)$. In this case we know, by linearity of expectation, that $\mathbb{E}(a_0 + a_1 X) = a_0 + a_1 \mathbb{E}(X)$.
2. Quadratic: if $g(x) = x^2$ then $g$ is convex. Jensen's inequality gives us $\mathbb{E}(X^2) \geq \mathbb{E}(X)^2$. Subtracting $\mathbb{E}(X)^2$ from both sides leads to $\text{Var}(X) \geq 0$, which is reassuring.
3. Reciprocal: if $g(x) = 1/x$ then $g$ is convex for $x \geq 0$. So if $X$ is a positive random variable then $\mathbb{E}(1/X) \geq 1/\mathbb{E}(X)$.

### 3.4.5 *Moments*

We have discussed measures of central tendency and measures of spread. As the names suggest, central tendency gives an indication of the location of the centre of a distribution, and spread measures how widely probability is dispersed. Other characteristics of a distribution that might be of interest include symmetry and the extent to which we find probability in the tails (fatness of tails). We can express commonly used measures of central tendency, spread, symmetry, and tail fatness in terms of **moments** and **central moments**.

**Definition 3.4.18** (Moments)
For a random variable $X$ and positive integer $r$, the $r^{\text{th}}$ moment of $X$ is denoted $\mu'_r$, where

$$\mu'_r = \mathbb{E}(X^r),$$

whenever this is well defined.

Moments depend on the horizontal location of the distribution. When we are measuring a characteristic like spread, we would like to use a quantity that remains unchanged when the distribution is moved left or right along the horizontal axis. This motivates the definition of central moments, in which we perform a translation to account for the value of the mean.

**Definition 3.4.19** (Central moments)
For a random variable $X$ and positive integer $r$, the $r^{\text{th}}$ central moment of $X$ is denoted $\mu_r$, where

$$\mu_r = \mathbb{E}[(X - \mathbb{E}(X))^r].$$

whenever this is well defined.

We may use $\mu'_{X,r}$ and $\mu_{X,r}$ for the $r^{\text{th}}$ moment and $r^{\text{th}}$ central moment when we need to identify the distribution with which the moments are associated. The notation here becomes rather cumbersome; we have a number of different ways of referring to the same thing. For example,

$$\mu'_1 = \mathbb{E}(X) = \mu,$$

all refer to the mean, and

$$\mu_2 = \mathbb{E}[(X - \mathbb{E}(X))^2] = \text{Var}(X) = \sigma^2,$$

all refer to the variance. Note that

$$\mu_1 = \mathbb{E}(X - \mu'_1) = 0,$$

that is, the first central moment is always zero. We can express central moments in terms of (non-central) moments: from Definition 3.4.19 and the binomial expansion we have

$$\begin{aligned}\mu_r &= \mathbb{E}\left(\sum_{j=0}^{r}\binom{r}{j}X^j(-\mathbb{E}(X))^{r-j}\right)\\ &= \sum_{j=0}^{r}\binom{r}{j}\mathbb{E}(X^j)(-\mathbb{E}(X))^{r-j}\\ &= \sum_{j=0}^{r}\binom{r}{j}\mu'_j(-\mu'_1)^{r-j}. \end{aligned} \tag{3.16}$$

An immediate consequence is that the second and third central moments can be written as

$$\begin{aligned}\mu_2 &= \mu'_2 - {\mu'_1}^2,\\ \mu_3 &= \mu'_3 - 3\mu'_1\mu'_2 + 2{\mu'_1}^3. \end{aligned} \tag{3.17}$$

Notice that the first of these is just another way of saying $\text{Var}(X) = \mathbb{E}(X^2) - \mathbb{E}(X)^2$.

We have seen that the first moment (the mean) is a commonly used measure of central tendency and the second central moment (the variance) is a commonly used measure of spread. The following definition provides moment-based measures of symmetry and fatness of tails.

**Definition 3.4.20** (Coefficient of skewness and coefficient of kurtosis)
For a random variable $X$ with $\text{Var}(X) = \sigma^2 < \infty$, the coefficient of **skewness** is given by

$$\gamma_1 = \mathbb{E}[(X-\mu)^3]/\sigma^3 = \mu_3/\mu_2^{\frac{3}{2}},$$

and the coefficient of **kurtosis** is given by

$$\gamma_2 = (\mathbb{E}[(X-\mu)^4]/\sigma^4) - 3 = (\mu_4/\mu_2^2) - 3.$$

The sense in which the third central moment is associated with symmetry is not immediately apparent. Note that, if $g(x) = x^3$, then $g$ is an odd function, that is, $g(-x) = -g(x)$. Thus, the third central moment is,

$$\begin{aligned}\mathbb{E}[(X-\mu)^3] &= \int_{-\infty}^{\infty}(x-\mu)^3 f_X(x)dx\\ &= \int_{-\infty}^{\infty} z^3 f_X(\mu+z)dz && \text{where } z = x-\mu\\ &= \int_{0}^{\infty} z^3[f_X(\mu+z) - f_X(\mu-z)]dz && \text{since } z^3 \text{ odd.}\end{aligned}$$

The term $[f_X(\mu+z) - f_X(\mu-z)]$ compares the density at a point of distance $z$ above the mean with the density at a point of distance $z$ below the mean. Any difference

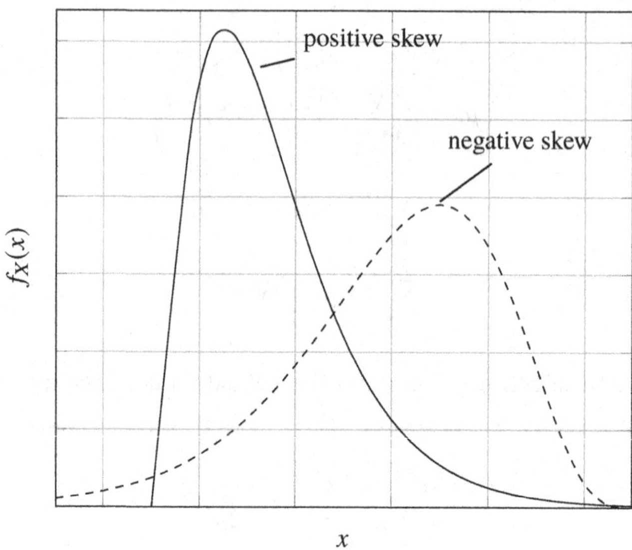


Figure 3.13 *Positive and negative skew.*

in these quantities is indicative of asymmetry. If there are sources of asymmetry far from the mean, multiplying by $z^3$ ensures that this will result in a large value of the third central moment.

In the coefficient of kurtosis, a $-3$ term appears rather mysteriously. By convention we measure kurtosis relative to a normal distribution. The fourth central moment of a normal distribution, $\mathrm{N}(\mu, \sigma^2)$, is $3\sigma^4$. By subtracting 3 we ensure that the coefficient of kurtosis is zero for a normal distribution, that is, $\gamma_2 = 0$. The coefficient of kurtosis is sometimes referred to as **excess kurtosis**, that is, kurtosis in excess of that displayed by a normal distribution.

To understand what kurtosis is measuring, consider two random variables with the same variance: $X$ almost always takes values moderately far from its mean, while $Y$ usually takes values very close to its mean, but can occasionally be very far away (i.e. it comes from a **heavy-tailed** distribution). Even though $\mathbb{E}(X - \mathbb{E}(X))^2$ is equal to $\mathbb{E}(Y - \mathbb{E}(Y))^2$ (they have the same variance), the extreme values mean that $\mathbb{E}(Y - \mathbb{E}(Y))^4$ will be higher than $\mathbb{E}(X - \mathbb{E}(X))^4$.

**Example 3.4.21** (Skewness for an exponential distribution)
Suppose that $X \sim \mathrm{Exp}(\lambda)$. In order to calculate the coefficient of skewness for $X$ we need the second and third central moments. We will calculate these from the first three (non-central) moments. In fact, we can work out an expression for the general

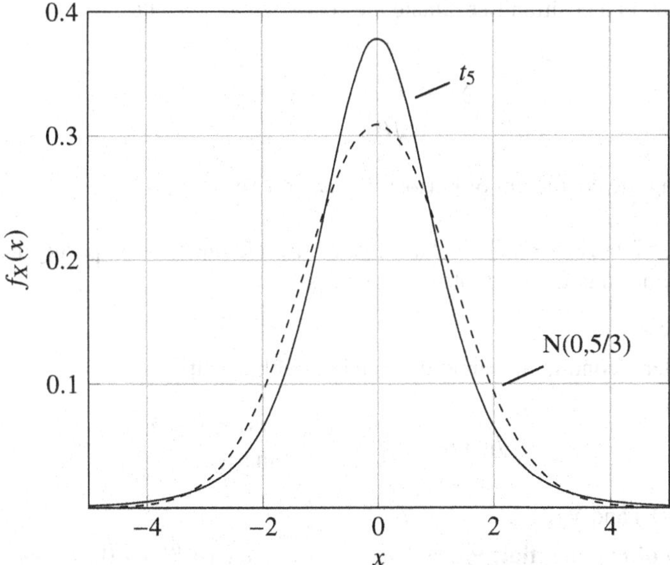


Figure 3.14 *The* $t_5$ *distribution (see Definition 8.1.3) has the same variance as the* N(0, 5/3), *but higher kurtosis. Notice that* $t_5$ *has higher density near 0, but heavier tails.*

($r^{\text{th}}$) moment at no extra cost:

$$\begin{aligned}
\mu'_r = \mathbb{E}(X^r) &= \int_{-\infty}^{\infty} x^r f_X(x)dx \\
&= \int_0^{\infty} x^r \lambda e^{-\lambda x} dx && \text{substituting exponential density} \\
&= \left[-x^r e^{-\lambda x}\right]_0^{\infty} + r\int_0^{\infty} x^{r-1} e^{-\lambda x} dx && \text{integrating by parts} \\
&= \frac{r}{\lambda}\mu'_{r-1}.
\end{aligned}$$

We now have a recursion for generating $\mu'_r$. The starting condition is readily evaluated as $\mu'_0 = 1$. Thus,

$$\mu'_r = \frac{r}{\lambda}\frac{r-1}{\lambda}\cdots\frac{2}{\lambda}\frac{1}{\lambda} = \frac{r!}{\lambda^r},$$

and in particular,

$$\mu'_1 = \frac{1}{\lambda}, \ \mu'_2 = \frac{2}{\lambda^2} \text{ and } \mu'_3 = \frac{6}{\lambda^3}.$$

Substituting in equation (3.17) gives

$$\mu_2 = \frac{1}{\lambda^2} \text{ and } \mu_3 = \frac{2}{\lambda^3}.$$

We are now in a position to evaluate the coefficient of skewness,

$$\gamma_1 = \frac{\mu_3}{\sqrt{\mu_2^3}} = \frac{2}{\lambda^3}\sqrt{\frac{\lambda^6}{1}} = 2.$$

Calculating the coefficient of kurtosis is part of Exercise 3.4.

In Chapters 7 and 8, we will see that we can draw inference by comparing the moments of a distribution to their sample counterparts.

**Exercise 3.4**

1. Consider a continuous random variable $Y$ with density

$$f_Y(y) = \begin{cases} y - \frac{1}{4}y^3 & \text{for } 0 \leq y \leq 2, \\ 0 & \text{otherwise.} \end{cases}$$

   Find $\mathbb{E}(Y)$ and $\text{Var}(Y)$.
2. (**Mean of exponential**) Prove that, if $\theta > 0$, then $xe^{-\theta x} \longrightarrow 0$ as $x \longrightarrow \infty$. Hence, use integration by parts to show that, if $X \sim \text{Exp}(\theta)$, then $\mathbb{E}(X) = \frac{1}{\theta}$.
3. (**Alternative representations of variance**) Prove that, for a random variable $X$,
   i. $\text{Var}(X) = \mathbb{E}(X^2) - \mathbb{E}(X)^2$,
   ii. $\text{Var}(X) = \mathbb{E}[X(X-1)] - \mathbb{E}(X)\,\mathbb{E}(X-1)$.
4. (**Variance of exponential**) Suppose that $X \sim \text{Exp}(\lambda)$. Find $\text{Var}(X)$.
5. (**Chebyshev inequality**) Prove the Chebyshev inequality, that is, prove that if $X$ is a random variable with $\text{Var}(X) < \infty$, then $\text{P}(|X - \mathbb{E}(X)| \geq a) \leq \text{Var}(X)/a^2$ for any constant $a > 0$.
6. Suppose that $X$ is a random variable. Show that $\mathbb{E}[\mathbf{1}_{(-\infty,x]}(X)] = F_X(x)$.
7. (**Coefficient of kurtosis for exponential**) Suppose that $X \sim \text{Exp}(\lambda)$. Find the coefficient of kurtosis of $X$.

## 3.5 Generating functions

### *3.5.1 Moment-generating functions*

For many distributions, all the moments $\mathbb{E}(X), \mathbb{E}(X^2), \ldots$ can be encapsulated in a single function. This function is referred to as the **moment-generating function**, and it exists for many commonly used distributions. It often provides the most efficient method for calculating moments. Moment-generating functions are also useful in establishing distributional results, such as the properties of sums of random variables, and in proving asymptotic results.

**Definition 3.5.1** (Moment-generating function)
The moment-generating function of a random variable $X$ is a function $M_X : \mathbb{R} \longrightarrow [0, \infty)$ given by

$$M_X(t) = \mathbb{E}(e^{tX}) = \begin{cases} \sum_x e^{tx} f_X(x) & \text{if } X \text{ discrete,} \\ \int_{-\infty}^{\infty} e^{tx} f_X(x)dx & \text{if } X \text{ continuous.} \end{cases}$$

where, for the function to be well defined, we require that $M_X(t) < \infty$ for all $t \in [-h, h]$ for some $h > 0$.

A few things to note about moment-generating functions.

1. Problems involving moment-generating functions almost always use the definition in terms of expectation as a starting point.
2. The moment-generating function $M_X(t) = \mathbb{E}(e^{tX})$ is a function of $t$. The $t$ is just a label, so $M_X(s) = \mathbb{E}(e^{sX})$, $M_X(\theta) = \mathbb{E}(e^{\theta X})$, $M_Y(p) = \mathbb{E}(e^{pY})$, and so on.
3. We need the moment-generating function to be defined in an interval around the origin. Later on we will be taking derivatives of the moment-generating function at zero, $M'_X(0)$, $M''_X(0)$, and so on.

The moment-generating function of $X$ is the expected value of an exponential function of $X$. Useful properties of moment-generating functions are inherited from the exponential function, $e^x$. The Taylor series expansion around zero, provides an expression for $e^x$ as a polynomial in $x$,

$$e^x = 1 + x + \frac{1}{2!}x^2 + \ldots + \frac{1}{r!}x^r + \ldots = \sum_{j=0}^{\infty} \frac{1}{j!}x^j.$$

This expansion (and any Taylor series expansion around zero) is often referred to as the **Maclaurin series** expansion. All the derivatives of $e^x$ are equal to $e^x$,

$$\frac{d^r}{dx^r}e^x = e^x \text{ for } r = 1, 2, \ldots$$

This can be checked by differentiation of the Maclaurin series expansion. These observations lead to two propositions that are of direct use in evaluating moments.

**Proposition 3.5.2** (Expressing the moment-generating function as a polynomial)
*The moment-generating function, $M_X(t)$, can be expressed as a polynomial in $t$,*

$$M_X(t) = 1 + t\,\mathbb{E}(X) + \frac{t^2}{2!}\,\mathbb{E}(X^2) + \ldots + \frac{t^r}{r!}\,\mathbb{E}(X^r) + \ldots = \sum_{j=0}^{\infty} \frac{\mathbb{E}(X^j)}{j!}t^j.$$

*Proof.*

This stems directly from the Maclaurin series for $e^x$ and the linearity of expectation.

As such it depends on us being able to swap the summation from the Maclaurin series and the summation/integration from taking expectations.

$$M_X(t) = \mathbb{E}(e^{tX}) = \mathbb{E}\left(\sum_{j=0}^{\infty} \frac{1}{j!}(tX)^j\right) = \sum_{j=0}^{\infty} \frac{\mathbb{E}(X^j)}{j!} t^j.$$

□

**Proposition 3.5.3** (Derivatives of a moment-generating function at zero)
*The $r^{th}$ derivative of the moment-generating function evaluated at zero is the $r^{th}$ moment,*

$$M_X^{(r)}(0) = \frac{d^r}{dt^r} M_X(t)\bigg|_{t=0} = \mathbb{E}(X^r).$$

*Proof.*
The proof uses the expansion given by Proposition 3.5.2 and assumes that we can swap the order of differentiation and summation.

$$\frac{d^r}{dt^r} M_X(t) = \mathbb{E}(X^r) + \mathbb{E}(X^{r+1})t + \frac{\mathbb{E}(X^{r+2})}{2!} t^2 + \ldots = \sum_{j=r}^{\infty} \frac{\mathbb{E}(X^j)}{(j-r)!} t^{j-r}.$$

When this is evaluated at $t = 0$, all but the first term vanish, proving the result. □

Propositions 3.5.2 and 3.5.3 provide us with two mechanisms for evaluating moments from the moment-generating function.

1. The coefficient of $t^r$ in the expansion of the moment-generating function is the $r^{\text{th}}$ moment divided by $r!$. We can calculate the $r^{\text{th}}$ moment by comparing coefficients, that is,

$$\text{if } M_X(t) = \sum_{j=0}^{\infty} a_j t^j \text{ then } \mathbb{E}(X^r) = r! a_r.$$

2. We can evaluate the $r^{\text{th}}$ moment directly using Proposition 3.5.3, that is, differentiate the moment-generating function $r$ times with respect to $t$ and evaluate the derivative at $t = 0$,

$$\mathbb{E}(X^r) = \frac{d^r}{dt^r} M_X(t)\bigg|_{t=0}.$$

In general, comparing coefficients tends to be quicker. However, for some moment-generating functions, the expansion as a polynomial in $t$ is not easy to derive. In these cases differentiation provides an alternative method for calculating moments.

As we suggested earlier in this section, moment-generating functions are used to establish properties of sums of random variables (section 4.7), and in proving convergence results. The key to their usefulness is that the moment-generating function uniquely characterises a distribution; if random variables have the same moment-generating function, then they have the same distribution. The following proposition, which we will not prove, summarises this property.

**Proposition 3.5.4** (Uniqueness of moment-generating function)
*If $X$ and $Y$ are random variables, and we can find $h > 0$ such that $M_X(t) = M_Y(t)$ for all $t \in [-h, h]$, then $F_X(x) = F_Y(x)$ for all $x \in \mathbb{R}$.*

In the following examples we derive the moment-generating functions for two distributions and use them to evaluate moments.

**Example 3.5.5** (Moment-generating function of Poisson distribution)
Suppose $X \sim \text{Pois}(\lambda)$. The moment-generating function of $X$ is

$$M_X(t) = \exp\{\lambda(e^t - 1)\}. \tag{3.18}$$

As with many moment-generating function calculations, we show that equation (3.18) holds by direct application of the definition,

$$\begin{aligned}
M_X(t) = \mathbb{E}(e^{tX}) &= \sum_x e^{tx} f_X(x) \\
&= \sum_{x=0}^{\infty} e^{tx} \frac{\lambda^x e^{-\lambda}}{x!} && \text{substituting Poisson mass} \\
&= e^{-\lambda} \sum_{x=0}^{\infty} \frac{(\lambda e^t)^x}{x!} && \text{by rearranging} \\
&= e^{-\lambda} e^{(\lambda e^t)} && \text{Maclaurin series} \\
&= \exp\{\lambda(e^t - 1)\}.
\end{aligned}$$

We can generate moments by differentiation. For example, the mean,

$$\begin{aligned}
&M'_X(t) = \lambda e^t \exp\{\lambda(e^t - 1)\} \\
\Rightarrow\ &M'_X(0) = \lambda \ \Rightarrow\ \mathbb{E}(X) = \lambda,
\end{aligned}$$

and the variance,

$$\begin{aligned}
&M''_X(t) = \lambda e^t \exp\{\lambda(e^t - 1)\} + \lambda^2 e^{2t} \exp\{\lambda(e^t - 1)\} \\
\Rightarrow\ &M''_X(0) = \lambda + \lambda^2 \ \Rightarrow\ \mathbb{E}(X^2) = \lambda + \lambda^2 \ \Rightarrow\ \text{Var}(X) = \lambda.
\end{aligned}$$

You might like to try using this method to calculate higher moments. This rapidly becomes quite tedious.

**Example 3.5.6** (Moment-generating function of Gamma distribution)
In order to calculate the moment-generating function of a Gamma-distributed random variable we use a favourite integration trick. We will manipulate the integrand so that it looks like a Gamma density. Suppose $X \sim \text{Gamma}(\alpha, \lambda)$,

$$\begin{aligned}
M_X(t) = \mathbb{E}(e^{tX}) &= \int_{-\infty}^{\infty} e^{tx} f_X(x) dx \\
&= \int_0^{\infty} e^{tx} \frac{1}{\Gamma(\alpha)} \lambda^\alpha x^{\alpha-1} e^{-\lambda x} dx && \text{substituting Gamma density} \\
&= \frac{1}{\Gamma(\alpha)} \lambda^\alpha \int_0^{\infty} x^{\alpha-1} e^{-(\lambda - t)x} dx && \text{by rearranging.}
\end{aligned}$$

At this stage, we have taken everything that does not involve $x$ (the variable that we are integrating with respect to) outside the integral. We look at the integrand and ask which Gamma distribution would give us the terms $x^{\alpha-1}e^{-(\lambda-t)x}$. It is clear that a Gamma$(\alpha, \lambda - t)$ would give us these terms. The full density would be

$$\frac{1}{\Gamma(\alpha)}(\lambda - t)^{\alpha} x^{\alpha-1} e^{-(\lambda-t)x},$$

but we do not have a term in $(\lambda - t)^{\alpha}$, so we introduce it by multiplying by $(\lambda - t)^{\alpha}$. In order for the equation to balance, we also have to divide by $(\lambda - t)^{\alpha}$. As $(\lambda - t)^{\alpha}$ does not involve $x$, we can place one term inside and one term outside the integral. The result is

$$M_X(t) = \frac{\lambda^{\alpha}}{(\lambda - t)^{\alpha}} \int_0^{\infty} \frac{1}{\Gamma(\alpha)}(\lambda - t)^{\alpha} x^{\alpha-1} e^{-(\lambda-t)x} dx = \left(\frac{\lambda}{\lambda - t}\right)^{\alpha},$$

where the integral is equal to 1 because we are integrating a density over the entire support. In order to ensure that we are not dividing by zero, we put a constraint on $t$. Insisting that $|t| < \lambda$ is sufficient to ensure that the moment-generating function is well defined. We can divide by $\lambda$ to give

$$M_X(t) = \left(1 - \frac{t}{\lambda}\right)^{-\alpha}.$$

We know that $|t/\lambda| < 1$. Applying the negative binomial expansion given by equation (3.8) yields

$$M_X(t) = \sum_{j=0}^{\infty} \binom{j + \alpha - 1}{\alpha - 1} \left(\frac{t}{\lambda}\right)^j = \sum_{j=0}^{\infty} \frac{(j + \alpha - 1)!}{\lambda^j(\alpha - 1)!} \frac{t^j}{j!}.$$

From Proposition (3.5.2), we know that the $r^{\text{th}}$ moment is the coefficient of $t^r/r!$ in the polynomial expansion of the moment-generating function. We conclude that, if $X \sim \text{Gamma}(\alpha, \lambda)$, the $r^{\text{th}}$ moment of $X$ is

$$\mathbb{E}(X^r) = \frac{(r + \alpha - 1)!}{\lambda^r(\alpha - 1)!}.$$

We mentioned in Example 3.3.23 that if $Y \sim \text{Exp}(\lambda)$ then $Y \sim \text{Gamma}(1, \lambda)$. Thus, the moment-generating function for an exponential with parameter $\lambda$ is

$$M_Y(t) = \frac{\lambda}{\lambda - t}.$$

Calculating the moment-generating function of a Gamma gives us the moment-generating function of an exponential for free.

### 3.5.2 Cumulant-generating functions and cumulants

It is often convenient to work with the log of the moment-generating function. It turns out that the coefficients of the polynomial expansion of the log of the moment-generating function have convenient interpretations in terms of moments and central moments.

**Definition 3.5.7** (Cumulant-generating function and cumulants)
The **cumulant-generating function** of a random variable $X$ with moment-generating function $M_X(t)$, is defined as

$$K_X(t) = \log M_X(t).$$

The $r^{\text{th}}$ **cumulant**, $\kappa_r$, is the coefficient of $t^r/r!$ in the expansion of the cumulant-generating function $K_X(t)$ so

$$K_X(t) = \kappa_1 t + \kappa_2 \frac{t^2}{2!} + \ldots + \kappa_r \frac{t^r}{r!} + \ldots = \sum_{j=1}^{\infty} \kappa_j \frac{t^j}{j!}.$$

It is clear from this definition that the relationship between cumulant-generating function and cumulants is the same as the relationship between moment-generating function and moments. Thus, to calculate cumulants we can either compare coefficients or differentiate.

1. Calculating the $r^{\text{th}}$ cumulant, $\kappa_r$, by comparing coefficients:

$$\text{if } K_X(t) = \sum_{j=0}^{\infty} b_j t^j \text{ then } \kappa_r = r! b_r.$$

2. Calculating the $r^{\text{th}}$ cumulant, $\kappa_r$, by differentiation:

$$\kappa_r = K_X^{(r)}(0) = \left. \frac{d^r}{dt^r} K_X(t) \right|_{t=0}.$$

Cumulants can be expressed in terms of moments and central moments. Particularly useful are the facts that the first cumulant is the mean and the second cumulant is the variance. In order to prove these results we will use the expansion, for $|x| < 1$,

$$\log(1 + x) = x - \frac{1}{2}x^2 + \frac{1}{3}x^3 - \ldots + \frac{(-1)^{j+1}}{j} x^j + \ldots$$

**Proposition 3.5.8** (Relationship between cumulants and moments)
*If $X$ is a random variable with moments $\{\mu'_r\}$, central moments $\{\mu_r\}$, and cumulants $\{\kappa_r\}$, then*

*i. the first cumulant is the mean,* $\kappa_1 = \mathbb{E}(X) = \mu'_1 = \mu$.

*ii. the second cumulant is the variance,* $\kappa_2 = \text{Var}(X) = \mu_2 = \sigma^2$.

iii. *the third cumulant is the third central moment,* $\kappa_3 = \mu_3$.

iv. *a function of the fourth and second cumulants yields the fourth central moment,* $\kappa_4 + 3\kappa_2^2 = \mu_4$.

*Proof.*

We proceed using the fact that $K_X(t) = \log M_X(t)$, differentiating and evaluating the derivative at zero. We will also use the fact that $M_X(0) = 1$, and the relationship between central and non-central moments given by equation (3.16). Finally, we need some properties of derivatives, in particular,

$$\begin{aligned}
&\frac{d}{dx}\log x = \frac{1}{x},\\
&\frac{d}{dx}(f(g(x)) = g'(x)f'(g(x)),\\
&\frac{d}{dx}(f(x)g(x)) = f'(x)g(x) + f(x)g'(x).
\end{aligned}$$

i. $K_X'(t) = M_X'(t)[M_X(t)]^{-1} \Rightarrow K_X'(0) = M_X'(0) = \mu_1' = \mu \Rightarrow \kappa_1 = \mu.$

ii. $K_X''(t) = M_X''(t)[M_X(t)]^{-1} - (M_X'(t))^2[M_X(t)]^{-2} \Rightarrow$
$K_X''(0) = \mu_2' - (\mu_1')^2 \Rightarrow \kappa_2 = \sigma^2.$

iii. Proof is part of Exercise 3.5.

iv. Proof is an optional exercise – tedious but not difficult.

□

**Example 3.5.9** (Cumulant-generating function of Poisson distribution)
The Poisson is an example of a distribution for which the cumulant-generating function has a more convenient form than the moment-generating function. If $X \sim \text{Pois}(\lambda)$ then, using equation (3.18) and taking logs, we have

$$K_X(t) = \log M_X(t) = \lambda(e^t - 1).$$

The Maclaurin series expansion of $e^t$ then gives

$$K_X(t) = \lambda(t + t^2/2! + \dots) = \sum_{j=1}^{\infty} \lambda \frac{t^j}{j!}.$$

Comparing coefficients we see that, for any positive integer $r$, the $r^{\text{th}}$ cumulant is $\lambda$, that is, $\kappa_r = \lambda$ for all $r$.

**Example 3.5.10** (Cumulant-generating function of standard normal)
In order to derive the cumulant-generating function, we first derive the moment-generating function then take logs. Suppose $Z \sim N(0,1)$, the moment-generating function of Z is

$$M_Z(t) = e^{t^2/2}.$$

In order to show this we will rearrange the integrand into the form of a normal density. This will require us to complete a square. Starting from the definition we have

$$
\begin{aligned}
M_Z(t) = \mathbb{E}(e^{tZ}) &= \int_{-\infty}^{\infty} e^{tz} f_Z(z) dz \\
&= \int_{-\infty}^{\infty} e^{tz} \frac{1}{\sqrt{2\pi}} e^{-z^2/2} dz && \text{N(0, 1) density} \\
&= \int_{-\infty}^{\infty} \frac{1}{\sqrt{2\pi}} \exp\left\{-\frac{1}{2}(z^2 - 2tz)\right\} dz && \text{by rearranging} \\
&= \int_{-\infty}^{\infty} \frac{1}{\sqrt{2\pi}} \exp\left\{-\frac{1}{2}(z^2 - 2tz + t^2) + \frac{1}{2}t^2\right\} dz && \text{completing the square} \\
&= e^{t^2/2} \int_{-\infty}^{\infty} \frac{1}{\sqrt{2\pi}} \exp\left\{-\frac{1}{2}(z - t)^2\right\} dz && \text{by rearranging} \\
&= e^{t^2/2},
\end{aligned}
$$

because the final integrand is a N($t$, 1) density. Taking logs yields the cumulant-generating function,

$$K_Z(t) = \frac{t^2}{2}.$$

We conclude that for a standard normal, $\kappa_2 = 1$, and all the other cumulants are zero. We will establish the corresponding result for the general case of the normal distribution in Example 3.6.7.

**Exercise 3.5**

1. Suppose that $X \sim \text{Exp}(\lambda)$. Derive the moment-generating function of $X$ and hence show $\mathbb{E}(X^r) = r!/\lambda^r$.
2. For each of the following, derive the moment-generating function and cumulant-generating function:
   (a) Bernoulli($p$),
   (b) Bin($n, p$),
   (c) Geometric($p$),
   (d) NegBin($r, p$).

   Comment on any associations.
3. Show that the third cumulant of a random variable $X$ is equal to its third central moment, that is, $\kappa_3 = \mu_3$.
4. Use cumulants to calculate the coefficient of skewness for a Poisson distribution.

## 3.6 Functions of random variables

We have encountered functions of random variables in sections 3.1 and 3.4. Definition 3.1.3 states that, if $g : \mathbb{R} \to \mathbb{R}$ is a well-behaved function, $X$ is a random variable and

$Y = g(X)$, then $Y$ is also a random variable. Using Theorem 3.4.4 we can evaluate $\mathbb{E}[g(X)]$ without reference to the distribution of $g(X)$. However, in many applications of statistical inference, information about the distribution of a function of a random variable is required. Methods for deriving the distribution and the associated mass or density is the subject of this section.

### 3.6.1 *Distribution and mass/density for $g(X)$*

Suppose that $X$ is a random variable defined on $(\Omega, \mathcal{F}, \text{P})$ and $g : \mathbb{R} \to \mathbb{R}$ is a well-behaved function. We would like to derive an expression for the cumulative distribution function of $Y$, where $Y = g(X)$. Some care is required here. We define $g$ to be a function, so for every real number input there is a single real number output. However, $g^{-1}$ is not necessarily a function, so a single input may have multiple outputs. To illustrate, let $g(x) = x^2$, then $g^{-1}$ corresponds to taking the square root, an operation that typically has two real outputs; for example, $g^{-1}(4) = \{-2, 2\}$. So, in general,

$$\text{P}(Y \le y) = \text{P}(g(X) \le y) \ne \text{P}(X \le g^{-1}(y)).$$

Our first step in deriving an expression for the distribution function of $Y$ is to consider the probability that $Y$ takes values in a subset of $\mathbb{R}$. We will use the idea of the inverse image of a set.

**Definition 3.6.1** (Inverse image)
If $g : \mathbb{R} \to \mathbb{R}$ is a function and $B$ is a subset of real numbers, then the **inverse image** of $B$ under $g$ is the set of real numbers whose images under $g$ lie in $B$, that is, for all $B \subseteq \mathbb{R}$ we define the inverse image of $B$ under $g$ as

$$g^{-1}(B) = \{x \in \mathbb{R} : g(x) \in B\}.$$

Then for any well-behaved $B \subseteq \mathbb{R}$,

$$\begin{aligned} \text{P}(Y \in B) &= \text{P}(g(X) \in B) = \text{P}(\{\omega \in \Omega : g(X(\omega)) \in B\}) \\ &= \text{P}(\{\omega \in \Omega : X(\omega) \in g^{-1}(B)\}) = \text{P}(X \in g^{-1}(B)). \end{aligned}$$

Stated loosely, the probability that $g(X)$ is in $B$ is equal to the probability that $X$ is in the inverse image of $B$. The cumulative distribution function of $Y$ is then

$$\begin{aligned} F_Y(y) &= \text{P}(Y \le y) = \text{P}(Y \in (-\infty, y]) = \text{P}(g(X) \in (-\infty, y]) \qquad (3.19) \\ &= \text{P}(X \in g^{-1}((-\infty, y])) \\ &= \begin{cases} \sum_{\{x:g(x)\le y\}} f_X(x) & \text{if } X \text{ is discrete,} \\ \int_{\{x:g(x)\le y\}} f_X(x)dx & \text{if } X \text{ is continuous.} \end{cases} \qquad (3.20) \end{aligned}$$

In the discrete case, we can use similar reasoning to provide an expression for the mass function,

$$f_Y(y) = \text{P}(Y = y) = \text{P}(g(X) = y) = \text{P}(X \in g^{-1}(y)) = \sum_{\{x:g(x)=y\}} f_X(x).$$

For the continuous case, the density is found by differentiating the cumulative distribution function. Although these forms for the distribution, mass, and density functions are rather cumbersome, they are often reasonably straightforward to implement. The following examples illustrate.

**Example 3.6.2** (Number of failures in repeated Bernoulli trials)
Consider a sequence of $n$ independent Bernoulli($p$) trials. We know that, if $X$ is the number of successes, then $X$ has a binomial distribution, $X \sim \text{Bin}(n, p)$ (see Example 3.3.8). By symmetry, we would expect that, if $Y$ is the number of failures, then $Y$ would also have a binomial distribution, $Y \sim \text{Bin}(n, 1-p)$. This is indeed the case. Notice that we can write $Y = n - X$, in other words, $Y = g(X)$ where $g(x) = n - x$. Using equation (3.20) yields

$$\begin{aligned}
F_Y(y) &= \sum_{\{x:g(x)\leq y\}} f_X(x) \\
&= \sum_{n-y}^{n} \binom{n}{x} p^x (1-p)^{n-x} && \text{substituting binomial mass} \\
&= \sum_{j=0}^{y} \binom{n}{n-j} p^{n-j} (1-p)^j && \text{letting } j = n - x \\
&= \sum_{j=0}^{y} \binom{n}{j} p^{n-j} (1-p)^j && \text{since } \binom{n}{n-j} = \binom{n}{j}.
\end{aligned}$$

This is the cumulative distribution function of a binomial, so we conclude that $Y \sim \text{Bin}(n, 1-p)$.

**Example 3.6.3** (Square of a continuous random variable)
Consider a continuous random variable $X$ and let $Y = X^2$. We can work directly to derive the cumulative distribution function of $Y$,

$$\begin{aligned}
F_Y(y) &= \text{P}(Y \leq y) = \text{P}(X^2 \leq y) = \text{P}(-\sqrt{y} \leq X \leq \sqrt{y}) \\
&= F_X(\sqrt{y}) - F_X(-\sqrt{y}).
\end{aligned}$$

The support of $Y$ must be a subset of the positive real line, so the cumulative distribution function in full is

$$F_Y(y) = \begin{cases} F_X(\sqrt{y}) - F_X(-\sqrt{y}) & \text{for } y \geq 0, \\ 0 & \text{otherwise.} \end{cases}$$

Differentiating with respect to $y$ gives the density function

$$f_Y(y) = \begin{cases} \frac{1}{2\sqrt{y}}[f_X(\sqrt{y}) + f_X(-\sqrt{y})] & \text{for } y \geq 0, \\ 0 & \text{otherwise.} \end{cases} \tag{3.21}$$

To illustrate, consider $X \sim \text{N}(0, 1)$ and let $Y = X^2$. The support of $X$ is the whole of the real line, so the support of $Y$ is the positive real line. Using the equation (3.21) and the density of a standard normal yields

$$f_Y(y) = \frac{1}{2\sqrt{y}}\left(\frac{1}{\sqrt{2\pi}}e^{-y/2} + \frac{1}{\sqrt{2\pi}}e^{-y/2}\right) = \frac{1}{\sqrt{2\pi y}}e^{-y/2}.$$

We know that $\Gamma(1/2) = \sqrt{\pi}$ (see Exercise 3.3), so we can write this density as

$$f_Y(y) = \begin{cases} \frac{(1/2)^{1/2}}{\Gamma(1/2)} y^{-1/2} e^{-y/2} & \text{for } y \geq 0, \\ 0 & \text{otherwise.} \end{cases}$$

This is the density of a Gamma distribution. In fact, if $X \sim \mathrm{N}(0,1)$ and $Y = X^2$, then $Y \sim \mathrm{Gamma}(\frac{1}{2}, \frac{1}{2})$.

### *3.6.2 Monotone functions of random variables*

Much of the discussion of the previous section is simplified if we assume the function of interest, $g$, is monotone.

**Definition 3.6.4** (Monotone function)
Suppose $g : \mathbb{R} \to \mathbb{R}$ is a function. We say that $g$ is:

i. **monotone increasing** if $g(x_1) \leq g(x_2)$ for all $x_1 < x_2$,
ii. **monotone decreasing** if $g(x_1) \geq g(x_2)$ for all $x_1 < x_2$.

We say that $g$ is monotone if it is either monotone increasing or monotone decreasing. We say that $g$ is strictly monotone if the inequalities above are replaced by strict inequalities.

For a strictly monotone function, the inverse image of an interval is also an interval. Of particular interest in deriving the cumulative distribution function is $g^{-1}((-\infty, y])$. If $g : \mathbb{R} \to \mathbb{R}$ is strictly monotone and $y$ is in the range of $g$, then

$$\begin{aligned} g^{-1}((-\infty, y]) &= \{x \in \mathbb{R} : g(x) \in (-\infty, y]\} \\ &= \begin{cases} (-\infty, g^{-1}(y)] & \text{for } g \text{ increasing,} \\ [g^{-1}(y), \infty) & \text{for } g \text{ decreasing.} \end{cases} \end{aligned}$$

This leads us to the following claim.

**Claim 3.6.5** (Distribution for a monotone function of a random variable)
*If $X$ is a random variable, $g : \mathbb{R} \to \mathbb{R}$ is strictly monotone on the support of $X$, and $Y = g(X)$, then the cumulative distribution function of $Y$ is*

$$\begin{aligned} F_Y(y) = \mathrm{P}(X \in g^{-1}(-\infty, y]) &= \begin{cases} \mathrm{P}(X \leq g^{-1}(y)) & \textit{for } g \textit{ increasing,} \\ \mathrm{P}(X \geq g^{-1}(y)) & \textit{for } g \textit{ decreasing,} \end{cases} \\ &= \begin{cases} F_X(g^{-1}(y)) & \textit{for } g \textit{ increasing,} \\ 1 - F_X(g^{-1}(y)-) & \textit{for } g \textit{ decreasing,} \end{cases} \end{aligned}$$

*where $y$ is in the range of $g$.*

Note that

$$F_X(g^{-1}(y)-) = \begin{cases} F_X(g^{-1}(y)) & \text{if } X \text{ is continuous,} \\ F_X(g^{-1}(y)) - \mathrm{P}(X = g^{-1}(y)) & \text{if } X \text{ is discrete.} \end{cases}$$

In the continuous case, we can differentiate to get an expression for the density of $Y$ in terms of the density of $X$.

**Proposition 3.6.6** (Density of a monotone function of a continuous random variable)
*If $X$ is a continuous random variable, $g : \mathbb{R} \to \mathbb{R}$ is a well-behaved, monotone function and $Y = g(X)$, then the density function of $Y$ is*

$$f_Y(y) = \begin{cases} f_X(g^{-1}(y)) \left| \frac{d}{dy} g^{-1}(y) \right| & \textit{for } y \textit{ in the range of } g, \\ 0 & \textit{otherwise.} \end{cases}$$

*Proof.*
By differentiating the expression given in Claim 3.6.5 we get

$$f_Y(y) = \begin{cases} f_X(g^{-1}(y)) \frac{d}{dy} g^{-1}(y) & \text{for } g \text{ increasing,} \\ -f_X(g^{-1}(y)) \frac{d}{dy} g^{-1}(y) & \text{for } g \text{ decreasing.} \end{cases}$$

The result follows by noting that if $g$ is increasing then $g^{-1}$ is also increasing, so $\frac{d}{dy} g^{-1}(y) \geq 0$. Similarly, if $g$ is decreasing then $\frac{d}{dy} g^{-1}(y) \leq 0$. □

Suppose that we define $h : \mathbb{R} \to \mathbb{R}$ as $h(y) = g^{-1}(y)$. Proposition 3.6.6 then has the following equivalent statements:

$$\begin{aligned} f_Y(y) &= f_X(h(y)) |h'(y)| \\ &= f_X(h(y)) \left| \frac{1}{g'(h(y))} \right|. \end{aligned}$$

The second line comes directly from the chain rule,

$$\frac{d}{dy} g(h(y)) = h'(y) g'(h(y)).$$

In our case, $h(y) = g^{-1}(y)$ so $g(h(y)) = y$. We conclude that $h'(y)g'(h(y)) = 1$ and hence $h'(y) = [g'(h(y))]^{-1}$.

**Example 3.6.7** (Location-scale transformation)
In our discussion of the normal distribution (Example 3.3.22) we mention the relationship between a standard and general normal distribution. In particular, if $Z \sim \mathrm{N}(0, 1)$ and $X = \mu + \sigma Z$ then $X \sim \mathrm{N}(\mu, \sigma^2)$. This is an example of a location-scale transformation.

In general, we will not assume normality. Suppose $Z$ is any random variable and $g$ is a linear function,

$$g(z) = \mu + \sigma z,$$

for some real values $\mu$ and $\sigma > 0$. If $X = g(Z)$, that is, if $X = \mu + \sigma Z$, then $X$ and $Z$ are related by a location-scale transformation. A consequence of the linearity of this association is that we can readily infer the properties of $X$ from those of $Z$. We start by considering the distribution function,

$$\begin{aligned} F_X(x) &= \mathrm{P}(X \leq x) \\ &= \mathrm{P}(\mu + \sigma Z \leq x) \\ &= \mathrm{P}\left(Z \leq \frac{x-\mu}{\sigma}\right) \\ &= F_Z\left(\frac{x-\mu}{\sigma}\right). \end{aligned}$$

If $Z$ is continuous, by differentiating with respect to $x$ we can derive an expression for the density of $X$ in terms of the density of $Z$,

$$f_X(x) = \frac{1}{\sigma} f_Z\left(\frac{x-\mu}{\sigma}\right). \tag{3.22}$$

There is also a close association between the moment-generating functions of $X$ and $Z$,

$$\begin{aligned} M_X(t) &= \mathbb{E}(e^{tX}) \\ &= \mathbb{E}(e^{t(\mu+\sigma Z)}) \\ &= e^{t\mu}\,\mathbb{E}(e^{t\sigma Z}) \\ &= e^{t\mu} M_Z(t\sigma). \end{aligned}$$

Taking logs yields

$$K_X(t) = \mu t + K_Z(t\sigma). \tag{3.23}$$

Comparing coefficients, we can establish the association between the cumulants of $X$ and cumulants of $Z$,

$$\kappa_{X,r} = \begin{cases} \mu + \sigma \kappa_{Z,1} & \text{for } r = 1, \\ \sigma^r \kappa_{Z,r} & \text{for } r = 2, 3, \ldots \end{cases}$$

One immediate consequence is that $\mathbb{E}(X) = \mu + \sigma\,\mathbb{E}(Z)$ and $\mathrm{Var}(X) = \sigma^2\,\mathrm{Var}(Z)$. This result is also readily obtained from the fact that $X = \mu + \sigma Z$ by direct computation.

To illustrate the use of the location-scale transformation, we return to the case where $Z \sim \mathrm{N}(0, 1)$ and $X = \mu + \sigma Z$. From equation (3.22) and the density of a standard normal, we have

$$\begin{aligned} f_X(x) &= \frac{1}{\sigma} f_Z\left(\frac{x-\mu}{\sigma}\right) \\ &= \frac{1}{\sigma}\frac{1}{\sqrt{2\pi}} \exp\left\{-\frac{1}{2}\left(\frac{x-\mu}{\sigma}\right)^2\right\} \\ &= \frac{1}{\sqrt{2\pi\sigma^2}} \exp\left\{-\frac{1}{2\sigma^2}(x-\mu)^2\right\}. \end{aligned}$$

This confirms that $X \sim N(\mu, \sigma^2)$. Now consider the cumulant-generating function of $X$. From equation (3.23) and the cumulant-generating function of a standard normal we have

$$K_X(t) = \mu t + K_Z(\sigma t) = \mu t + \frac{\sigma^2 t^2}{2}.$$

Thus, if $X \sim N(\mu, \sigma^2)$ then $\kappa_1 = \mu$ and $\kappa_2 = \sigma^2$ (as is obvious, since the first two cumulants are always the mean and the variance). Perhaps more interestingly, we have also established that $\kappa_r = 0$ if $r > 2$. By the uniqueness of cumulant-generating functions, any distribution with $\kappa_r = 0$ for $r > 2$ is a normal distribution.

It is generally not the case that $X = \mu + \sigma Z$ has the same distribution as $Z$. For example, if $Z \sim \text{Exp}(\lambda)$, then $X$ has support $(\mu, \infty)$ and is clearly not exponentially distributed. Special cases such as the normal where $X$ and $Z$ have the same distribution (but with different parameter values) are known as **location-scale families**.

**Example 3.6.8** (Probability integral transformation)
Consider a cumulative distribution function, $F$, of a continuous random variable. For all the cases we consider, $F$ is strictly increasing on the support of the random variable, say $D \subseteq \mathbb{R}$. The inverse of this cumulative distribution function, $F^{-1}$, is strictly increasing on $[0, 1]$. In fact, we can think of this function as $F^{-1} : [0, 1] \longrightarrow D$. Let $U$ be a continuous uniform random variable on the interval $[0, 1]$, that is, $U \sim \text{Unif}[0, 1]$. The cumulative distribution function of $U$ is

$$F_U(u) = u \text{ for } 0 \leq u \leq 1.$$

Suppose that we define $X = F^{-1}(U)$, so $X$ is the random variable generated by applying $F^{-1}$ to $U$. By Claim 3.6.5,

$$F_X(x) = F_U((F^{-1})^{-1}(x)) = F(x) \text{ for } x \in D.$$

In other words, $X$ is a random variable with cumulative distribution function $F$. For cases where $F^{-1}$ is readily computed, this provides a convenient mechanism for generating random numbers drawn from a particular distribution. If we sample $u$ from $\text{Unif}[0, 1]$, we can treat $F_X^{-1}(u)$ as a draw from the distribution of $X$.

To illustrate, consider the exponential distribution. If $X \sim \text{Exp}(\lambda)$, the cumulative distribution function of $X$ is

$$F_X(x) = 1 - e^{-\lambda x} \text{ for } x \geq 0.$$

The inverse is given by

$$F_X^{-1}(u) = \frac{1}{\lambda} \log\left(\frac{1}{1-u}\right) \text{ for } 0 \leq u \leq 1.$$

We conclude that, if $U \sim \text{Unif}[0, 1]$, then

$$\frac{1}{\lambda} \log\left(\frac{1}{1-U}\right) \sim \text{Exp}(\lambda).$$

If you have experience with a statistics package or programming language (or even a spreadsheet application), try writing a simple program that takes the output from a standard random-number generator and returns draws from an exponential distribution.

**Exercise 3.6**

1. Let $X$ be a continuous random variable whose support is the whole real line. Find an expression for the density of $|X|$ in terms of the density of $X$.
2. If $X \sim \mathrm{N}(\mu, \sigma^2)$, find the density of $|X - \mu|$.
3. Suppose that $X$ is a positive, continuous random variable. Find an expression for the density of $1/X$ in terms of the density of $X$.
4. If $X \sim \mathrm{Exp}(\lambda)$, find the density of $1/X$.
5. **(Lognormal distribution)** Suppose that $X \sim \mathrm{N}(\mu, \sigma^2)$. If we define $Y = \exp(X)$, then $Y$ has a **lognormal distribution**, popular as a right-skewed distribution for positive variables. Find the expectation, variance, and PDF of $Y$.

## 3.7 Sequences of random variables and convergence

Suppose that $x_1, x_2, \ldots$ is a sequence of real numbers. We denote this sequence $\{x_n\}$. The definition of convergence for a sequence of real numbers is well established.

**Definition 3.7.1** (Convergence of a real sequence)
Let $\{x_n\}$ be a sequence of real numbers and let $x$ be a real number. We say that $x_n$ **converges** to $x$ if and only if, for every $\varepsilon > 0$, we can find an integer $N$ such that $|x_n - x| < \varepsilon$ for all $n > N$. Under these conditions, we write $x_n \to x$ as $n \to \infty$.

This definition is based on an intuitively appealing idea (although in the formal statement given above, this might not be obvious). If we take any interval around $x$, say $[x - \varepsilon, x + \varepsilon]$, we can find a point, say $N$, beyond which all elements of the sequence fall in the interval. This is true for an arbitrarily small interval.

Now consider a sequence of random variables $\{X_n\}$ and a random variable $X$. We want to know what it means for $\{X_n\}$ to converge to $X$. Using Definition 3.7.1 is not possible; since $|X_n - X|$ is a random variable, direct comparison with the real number $\varepsilon$ is not meaningful. In fact, for a random variable there are many different forms of convergence. We define four distinct modes of convergence for a sequence of random variables.

**Definition 3.7.2** (Types of convergence)
Let $\{X_n\}$ be a sequence of random variables and let $X$ be a random variable.

i. **Convergence in distribution**: $\{X_n\}$ converges in distribution to $X$ if

$$\mathrm{P}(X_n \le x) \to \mathrm{P}(X \le x) \text{ as } n \to \infty,$$

for all $x$ at which the cumulative distribution function is continuous. This is denoted by $X_n \xrightarrow{d} X$. This could also be written as $F_{X_n}(x) \longrightarrow F_X(x)$. Convergence in distribution is sometimes referred to as convergence in law.

ii. **Convergence in probability**: $\{X_n\}$ converges in probability to $X$ if, for any $\varepsilon > 0$,

$$\mathrm{P}(|X_n - X| < \varepsilon) \longrightarrow 1 \text{ as } n \longrightarrow \infty.$$

This is denoted $X_n \xrightarrow{p} X$. An alternative statement of convergence in probability is that $\mathrm{P}(|X_n - X| > \varepsilon) \longrightarrow 0$ as $n \longrightarrow \infty$. Convergence in probability is sometimes referred to as convergence in measure.

iii. **Convergence almost surely**: $\{X_n\}$ converges to $X$ almost surely if, for any $\varepsilon > 0$,

$$\mathrm{P}\left(\lim_{n\to\infty} |X_n - X| < \varepsilon\right) = 1.$$

This is denoted $X_n \xrightarrow{a.s.} X$. An alternative statement of almost-sure convergence is that, if we define $A = \{\omega \in \Omega : X_n(\omega) \longrightarrow X(\omega) \text{ as } n \longrightarrow \infty\}$, then $\mathrm{P}(A) = 1$. Almost-sure convergence is sometimes referred to as convergence with probability 1.

iv. **Convergence in mean square**: $\{X_n\}$ converges to $X$ in mean square if

$$\mathbb{E}[(X_n - X)^2] \longrightarrow 0 \text{ as } n \longrightarrow \infty.$$

This is denoted $X_n \xrightarrow{m.s.} X$.

Each of the types of convergence given above involves an operation using random variables $X_n$ and $X$ to yield a sequence of real numbers. Convergence of the random variables is then defined in terms of convergence of this sequence of real numbers. The distinction between types of convergence is in the nature of the operation we use to go from random variables to real numbers.

Some of the types of convergence given in Definition 3.7.2 are stronger than others. If $\{X_n\}$ converges almost surely or if $\{X_n\}$ converges in mean square, then $\{X_n\}$ converges in probability. In turn, if $\{X_n\}$ converges in probability, then $\{X_n\}$ converges in distribution. Put another way, the set of all sequences that converge in distribution contains the set of all sequences that converge in probability which, in turn, contains both the set of all sequences that converge almost surely and the set of all sequences that converge in mean square.

We can summarise these relationships as follows:

$$\begin{array}{lcl} X_n \xrightarrow{a.s.} X & \searrow & \\ & & X_n \xrightarrow{p} X \Rightarrow X_n \xrightarrow{d} X. \\ X_n \xrightarrow{m.s.} X & \nearrow & \end{array} \tag{3.24}$$

We prove one of these relationships that will be useful later when we discuss consistency of estimators.

**Proposition 3.7.3** (Mean-square convergence implies convergence in probability)
*Suppose* $\{X_n\}$ *is a sequence of random variables. If* $X_n \xrightarrow{m.s.} X$ *then* $X_n \xrightarrow{p} X$.

*Proof.*

Consider $\mathrm{P}(|X_n - X| > \varepsilon)$ for $\varepsilon > 0$. Applying the Chebyshev inequality (see Proposition 3.4.13) we have

$$\mathrm{P}(|X_n - X| > \varepsilon) \leq \frac{\mathbb{E}[(X_n - X)^2]}{\varepsilon^2}.$$

If $X_n \xrightarrow{m.s.} X$ then $\mathbb{E}[(X_n - X)^2] \to 0$, so $\{\mathrm{P}(|X_n - X| > \varepsilon)\}$ is a sequence of positive real numbers bounded above by a sequence that converges to zero. We conclude that $\mathrm{P}(|X_n - X| > \varepsilon) \to 0$ as $n \to \infty$, and thus $X_n \xrightarrow{p} X$. □

For some of the statements given in equation (3.24) we can specify conditions under which the reverse implication holds. One example is convergence in distribution to a constant. We can view a constant, $c$ say, as a degenerate random variable, that is, a random variable with all of its mass at one point. The distribution function of this degenerate random variable is the step function,

$$F_c(x) = \begin{cases} 0 & \text{for } x < c, \\ 1 & \text{for } x \geq c. \end{cases} \tag{3.25}$$

We exploit this cumulative distribution function in the proof of the following proposition.

**Proposition 3.7.4** (Convergence to a degenerate distribution)
*Suppose* $\{X_n\}$ *is a sequence of random variables. If* $X_n \xrightarrow{d} c$*, where* $c$ *is a constant, then* $X_n \xrightarrow{p} c$.

*Proof.*

We start by considering $\mathrm{P}(|X_n - c| > \varepsilon)$,

$$\begin{aligned} \mathrm{P}(|X_n - c| > \varepsilon) &= \mathrm{P}(X_n - c > \varepsilon) + \mathrm{P}(X_n - c < -\varepsilon) \\ &= \mathrm{P}(X_n > c + \varepsilon) + \mathrm{P}(X_n < c - \varepsilon) \\ &\leq \mathrm{P}(X_n > c + \varepsilon) + \mathrm{P}(X_n \leq c - \varepsilon) \\ &= [1 - F_{X_n}(c + \varepsilon)] + F_{X_n}(c - \varepsilon). \end{aligned}$$

If $X_n \xrightarrow{d} c$ then $F_{X_n} \to F_c$, and thus $F_{X_n}(c + \varepsilon) \to 1$ and $F_{X_n}(c - \varepsilon) \to 0$. Thus, $\{\mathrm{P}(|X_n - c| > \varepsilon)\}$ is a sequence of positive real numbers bounded above by a sequence that converges to zero. We conclude that $\mathrm{P}(|X_n - c| > \varepsilon) \to 0$ as $n \to \infty$, and thus $X_n \xrightarrow{p} c$. □

There is a remarkably easy way to determine whether or not a sequence converges almost surely. In the notation of Definition 3.7.2 iii., consider $A^c$, the event consisting of all the outcomes $\omega$ where $X_n(\omega)$ does not converge to $X(\omega)$ as $n \to \infty$. If this is the case, there must exist some $\varepsilon > 0$ such that for every $n$ we can find $m \geq n$ with $|X_m(\omega) - X(\omega)| > \varepsilon$. In other words, there must be *infinitely many* $m$ where $|X_m(\omega) - X(\omega)| > \varepsilon$. If there were only finitely many, we could find the largest one ($M$, say), then pick $n = M + 1$.

Now define the sequence of events $E_1, E_2, \dots$ where $E_n = |X_n - X| > \varepsilon$. The **limit superior** (lim sup) of this sequence is defined as

$$\limsup_{n\to\infty} E_n = \bigcap_{n=1}^{\infty}\left(\bigcup_{m=n}^{\infty} E_m\right). \tag{3.26}$$

Consider the union in the round brackets: for a given $n$, this event occurs when at least one $E_m$ (with $m \geq n$) occurs. For the lim sup to occur, we need *all* the events in the intersection to occur, that is, we must be able to find such an $m \geq n$ for every $n$. We conclude that the following definitions are equivalent:

$$\begin{aligned} A^c &= \{\omega \in \Omega : X_n(\omega) \nrightarrow X(\omega) \text{ as } n \to \infty\} \\ &= \{\omega \in \Omega : |X_m(\omega) - X(\omega)| > \varepsilon \text{ for infinitely many } m\} \\ &= \limsup_{n\to\infty} E_n \end{aligned}$$

We can now prove the key results.

**Lemma 3.7.5** (First Borel-Cantelli lemma)
*If $(\Omega, \mathcal{F}, \mathrm{P})$ is a probability space and $E_1, E_2, \dots \in \mathcal{F}$ is a sequence of events with $\sum_{n=1}^{\infty} \mathrm{P}(E_n) < \infty$, then $\mathrm{P}(\limsup_{n\to\infty} E_n) = 0$.*

*Proof.*

Define the partial sum $S_n = \sum_{m=1}^{n} \mathrm{P}(E_n)$, and let $S_\infty = \lim_{n\to\infty} S_n$. In the definition of lim sup, the events $\bigcup_{m=n}^{\infty} E_m$ form a decreasing sequence of sets for $n = 1, 2, \dots$, so we have

$$\begin{aligned} \mathrm{P}(\limsup_{n\to\infty} E_n) &= \mathrm{P}\left(\bigcap_{n=1}^{\infty}\left(\bigcup_{m=n}^{\infty} E_m\right)\right) \\ &= \lim_{n\to\infty} \mathrm{P}\left(\bigcup_{m=n}^{\infty} E_m\right) && \text{Proposition 2.2.12} \\ &\leq \lim_{n\to\infty} \sum_{m=n}^{\infty} \mathrm{P}(E_m) && \text{Proposition 2.2.11} \\ &= \lim_{n\to\infty} (S_\infty - S_{n-1}) \\ &= S_\infty - S_\infty = 0 \end{aligned}$$

(In the last line of the proof, we need $S_\infty$ to be finite or the expression would be of the form $\infty - \infty$, which is indeterminate.) □

A direct consequence of the first Borel-Cantelli lemma is that we can establish $X_n \xrightarrow{a.s.} X$ by showing that the sum $\sum_{n=1}^{\infty} \mathrm{P}(|X_n - X| > \varepsilon)$ converges. If the sum diverges, we can prove that the sequence does not converge almost surely, but we need to make an additional assumption.

**Lemma 3.7.6** (Second Borel-Cantelli lemma)
*If $(\Omega, \mathcal{F}, \mathrm{P})$ is a probability space and $E_1, E_2, \ldots \in \mathcal{F}$ is a sequence of mutually independent events with $\sum_{n=1}^{\infty} \mathrm{P}(E_n) = \infty$, then $\mathrm{P}(\limsup_{n\to\infty} E_n) = 1$.*

*Proof.*

Consider the complement of the lim sup,

$$\left(\limsup_{n\to\infty} E_n\right)^c = \left[\bigcap_{n=1}^{\infty}\left(\bigcup_{m=n}^{\infty} E_m\right)\right]^c = \bigcup_{n=1}^{\infty}\left(\bigcup_{m=n}^{\infty} E_m\right)^c = \bigcup_{n=1}^{\infty}\left(\bigcap_{m=n}^{\infty} E_m^c\right),$$

which is known as the **limit inferior** (lim inf). The events $\bigcap_{m=n}^{\infty} E_m^c$ form an increasing sequence of sets for $n = 1, 2, \ldots$, so we have

$$\begin{aligned}
\mathrm{P}\left(\limsup_{n\to\infty} E_n\right) &= 1 - \mathrm{P}\left(\bigcup_{n=1}^{\infty}\left(\bigcap_{m=n}^{\infty} E_m^c\right)\right) \\
&= 1 - \lim_{n\to\infty} \mathrm{P}\left(\bigcap_{m=n}^{\infty} E_m^c\right) && \text{Proposition 2.2.12} \\
&= 1 - \lim_{n\to\infty} \prod_{m=n}^{\infty} \mathrm{P}(E_m^c) && \text{By independence} \\
&= 1 - \lim_{n\to\infty} \prod_{m=n}^{\infty} (1 - \mathrm{P}(E_m)) \\
&\geq 1 - \lim_{n\to\infty} \prod_{m=n}^{\infty} \exp\{-\mathrm{P}(E_m)\} && 1 - x \leq e^{-x} \text{ for } x \geq 0 \\
&= 1 - \lim_{n\to\infty} \exp\Bigg\{-\underbrace{\sum_{m=n}^{\infty} \mathrm{P}(E_m)}_{=\infty}\Bigg\} = 1.
\end{aligned}$$

The sum in the final line diverges because it is equal to the (infinite) sum $S_\infty$ minus the (finite) sum $S_{n-1}$. □

In other words, if the random variables in the sequence $\{X_n\}$ are independent and the sum $\sum_{n=1}^{\infty} \mathrm{P}(|X_n - X| > \varepsilon)$ diverges, we can conclude that the sequence does not converge to $X$ almost surely.

**Exercise 3.7**

1. Consider a sequence of random variables $X_1, X_2, \dots$ with cumulant-generating functions $K_{X_1}, K_{X_2}, \dots$, and a random variable $X$ with cumulant-generating function $K_X(t)$. Suppose, in addition, that all these cumulant-generating functions are well defined for $|t| < c$. If $K_{X_n}(t) \to K_X(t)$ as $n \to \infty$, for all $t$ such that $|t| < c$, what can we conclude?
2. Consider a sequence of random variables $X_1, X_2, \dots$ and constant $c$. Prove that $X_n \xrightarrow{p} c$ implies $X_n \xrightarrow{d} c$.
3. Let $X_1, X_2, \dots$ be a sequence of independent random variables. For each of the following cases, explain whether or not $\{X_n\}$ converges to 0 almost surely, in mean square, and/or in probability.
   (a) $X_n = 1$ with probability $n^{-1}$, and 0 otherwise.
   (b) $X_n = n$ with probability $n^{-2}$, and 0 otherwise.
   (c) $X_n = 2^n$ with probability $5^{-n}$, and 0 otherwise.

## 3.8 A more thorough treatment of random variables

In earlier sections of this chapter we refer rather vaguely to conditions on a set $B$ for $\text{P}(X \in B)$ to be well defined and conditions on a function $g$ for $g(X)$ to be a random variable. We also suggest that we are not really interested in random variables as maps and that, for many situations, the notion of an underlying sample space is not particularly useful. In this section, we attempt to provide some justification for these assertions. The material here is technical and may be excluded without affecting understanding of other parts of the text. We start by providing an alternative definition of a random variable. This is equivalent to Definition 3.1.2 but uses more abstract concepts; key among them is the Borel $\sigma$-algebra.

**Definition 3.8.1** (Borel $\sigma$-algebra on $\mathbb{R}$)
Let $C$ be the collection of all open intervals of $\mathbb{R}$. The **Borel $\sigma$-algebra** on $\mathbb{R}$ is the (unique) smallest $\sigma$-algebra that contains $C$. We denote the Borel $\sigma$-algebra on $\mathbb{R}$ by $\mathcal{B}$. An element of $\mathcal{B}$ is referred to as a Borel set.

From the definition it is clear that any open interval $(x, y)$ is a Borel set. It is also the case that closed intervals $[x, y]$, half-open intervals $(x, y]$ and $[x, y)$, and finite unions of interval are all Borel sets in $\mathbb{R}$. In fact, sets that are not Borel sets are hard to construct; any subset of $\mathbb{R}$ that you come across in a practical problem is likely to be a Borel set. Clearly, since $\mathcal{B}$ is a $\sigma$-algebra, $(\mathbb{R}, \mathcal{B})$ is a measurable space. The term measurable can also be applied to functions.

**Definition 3.8.2** (Measurable function)
Consider measurable spaces $(\Omega, \mathcal{F})$ and $(\mathbb{R}, \mathcal{B})$. We say that a function $h : \Omega \to \mathbb{R}$ is **$\mathcal{F}$-measurable** if $h^{-1}(B) \in \mathcal{F}$ for all $B \in \mathcal{B}$.

We can now give an alternative definition of a random variable.

**Definition 3.8.3** (Random variable)
For a probability space $(\Omega, \mathcal{F}, \mathrm{P})$ and measurable space $(\mathbb{R}, \mathcal{B})$, a random variable, $X$, is a measurable function $X : \Omega \to \mathbb{R}$.

For every random variable, $X$, we can define a measure on $\mathbb{R}$ that completely characterises the distribution of probability associated with $X$. This measure is sometimes referred to as the **law** of $X$; a non-rigorous account follows. Suppose that we have a probability space $(\Omega, \mathcal{F}, \mathrm{P})$ and that $B$ is a Borel set, that is, $B \subseteq \mathbb{R}$ and $B \in \mathcal{B}$. Our definition of a random variable ensures that the inverse image of $B$ under $X$ is an event, that is, $X^{-1}(B) \in \mathcal{F}$. As such, it makes sense to talk about the probability of the inverse image,

$$\mathrm{P}(X^{-1}(B)) = \mathrm{P}(\{\omega \in \Omega : X(\omega) \in B\} = \mathrm{P}(X \in B).$$

If we define $Q_X(B) = \mathrm{P}(X \in B)$, then $Q_X : \mathcal{B} \to [0, 1]$ and $Q_X$ inherits the properties of a probability measure from P. Thus, $(\mathbb{R}, \mathcal{B}, Q_X)$ is a probability space for any random variable $X$. The probability measure $Q_X$ is referred to as the **law** of $X$. The following definition summarises.

**Definition 3.8.4** (The law of a random variable)
Suppose we have a probability space $(\Omega, \mathcal{F}, \mathrm{P})$ and a random variable $X : \Omega \to \mathbb{R}$. Let $\mathcal{B}$ be the Borel $\sigma$-algebra on $\mathbb{R}$. If we define $Q_X : \mathcal{B} \to \mathbb{R}$ by $Q_X(B) = \mathrm{P}(X \in B)$, then $Q_X$ is a measure on $(\mathbb{R}, \mathcal{B})$. The measure $Q_X$ is referred to as the law of $X$.

We are usually interested in the distribution of probability associated with a random variable, rather than with the random variable as a map. In other words, we are interested in the way in which the cumulative distribution function associates probabilities with sets of real numbers, rather than the way in which random variables associate real numbers with outcomes. The mapping defined by a random variable is often rather trivial. In fact, for any random variable $X$ we can find a probability space in which the identity function is a random variable with the same cumulative distribution function as $X$.

**Proposition 3.8.5** (Identity function as random variable)
*Let $X$ be a random variable on $(\Omega, \mathcal{F}, P)$ with distribution function $F_X$ and law $Q_X$. If we define the identity function $J : \mathbb{R} \to \mathbb{R}$ by $J(x) = x$ for all $x$, then $J$ is a random variable on $(\mathbb{R}, \mathcal{B}, Q_X)$ with cumulative distribution function identical to that of $X$, that is, $F_J(x) = F_Y(x)$ for all $x \in \mathbb{R}$.*

*Proof.*

$J$ is clearly $\mathcal{B}$-measurable, so $J$ is a random variable on $(\mathbb{R}, \mathcal{B}, Q_X)$, and

$$\begin{aligned} F_J(x) &= Q_X(\{y \in \mathbb{R} : J(y) \leq x\}) = Q_X(\{y \in \mathbb{R} : y \leq x\}) = Q_X((-\infty, x]) \\ &= \mathrm{P}(X \in (-\infty, x]) = \mathrm{P}(X \leq x) = F_X(x). \end{aligned}$$

□

In subsection 3.3.2 we state that, for a continuous random variable $X$ with density function $f_X$,

$$\mathrm{P}(X \in B) = \int_B f_X(x)dx,$$

for reasonably well-behaved sets $B$. In fact, this relationship holds for any Borel set $B$. The reasoning here involves careful definition of what we mean by integration and is beyond the scope of this text. We will often check whether a function is a valid density by checking that it is positive and real-valued, and that the integral over the whole real line is equal to 1. The following proposition justifies this approach.

**Proposition 3.8.6** (Characterisation of a valid density)
*If $f : \mathbb{R} \to [0, \infty)$ is integrable and $\int_{\infty}^{\infty} f(x)dx = 1$, then we can find a random variable for which $f$ is the density function.*

*Proof.*
Consider the measurable space $(\mathbb{R}, B)$ and define the measure $Q : \mathcal{B} \to [0, 1]$ by $Q(B) = \int_B f(x)dx$. Then $(\mathbb{R}, B, Q)$ is a probability space and $f$ is the density of the identity random variable, $J : \mathbb{R} \to \mathbb{R}$ where $J(x) = x$ for all $x \in \mathbb{R}$. □

For a Borel set $B$ we can write $\mathrm{P}(X \in B)$ in terms of the cumulative distribution function,

$$\mathrm{P}(X \in B) = \int_B f_X(x)dx = \int_B \frac{dF_X(x)}{dx}dx = \int_B dF_X(x). \tag{3.27}$$

The advantage of this notation is that is can be extended to the discrete case. If $X$ is a discrete random variable taking values of $\{x_1, x_2, \ldots\}$, then

$$\mathrm{P}(X \in B) = \sum_{x:x \in B} f_X(x)$$

Recall that, for a discrete random variable,

$$F_X(x) - F_X(x-) = \begin{cases} f_X(x) & \text{for } x \in \{x_1, x_2, \ldots\}, \\ 0 & \text{otherwise.} \end{cases}$$

Thus, in the discrete case, we can write $\mathrm{P}(B) = \int_B dF_X(x)$ where we define $dF(x) = F_X(x) - F_X(x-)$, and this integration reduces to summation since the integrand is only non-zero on a countable subset of $B$. Notice that the proof of Proposition 3.8.5 provides a characterisation of law in terms of the cumulative distribution function, namely

$$Q_X((-\infty, x]) = F_X(x) \text{ for all } x \in \mathbb{R}.$$

This holds regardless of whether $X$ is discrete, continuous, or neither. Among the implications of this characterisation are

$$\begin{aligned} Q_X((x, \infty)) &= 1 - F_X(x), \\ Q_X((-\infty, x)) &= F_X(x-), \\ Q_X((x, y]) &= F_X(y) - F_X(x). \end{aligned}$$

This can be extended to any Borel set $B$ by taking $Q(B) = \text{P}(X \in B) = \int_B dF_X(x)$. Once again, a precise explanation requires more careful treatment of the fundamentals of integration.

We can extend the notation given in equation (3.27) to provide an expression for expectation. For a random variable $X$ with cumulative distribution function $F_X$, and a well-behaved, real-valued function $g$, the expected value of $g(X)$ is

$$\mathbb{E}[g(X)] = \int_{\mathbb{R}} g(x) dF_X(x).$$

Here, $g$ being well behaved ensures that $g(X)$ is a random variable. We are now in a position to specify precisely what we mean by well behaved.

**Proposition 3.8.7** (Conditions for $g(X)$ to be a random variable)
*Consider a probability space $(\Omega, \mathcal{F}, \text{P})$ and a random variable $X : \Omega \to \mathbb{R}$. If $g : \mathbb{R} \to \mathbb{R}$ is a function for which the inverse image of any Borel set is a Borel set, that is, if $g$ is $\mathcal{B}$-measurable, then $g(X)$ is a random variable.*

*Proof.*

Suppose we define $Z : \Omega \to \mathbb{R}$ by $Z(\omega) = g(X(\omega))$ for all $\omega \in \Omega$. The inverse of $Z$ is given by $Z^{-1}(x) = X^{-1}(g^{-1}(x))$ for $x \in \mathbb{R}$. For $Z$ to be a random variable we require that $Z^{-1}(B) \in \mathcal{F}$ for all $B \in \mathcal{B}$, that is, we require $X^{-1}(g^{-1}(B)) \in \mathcal{F}$ for all $B \in \mathcal{B}$. We know that $X$ is a random variable, so this holds provided $g^{-1}(B) \in \mathcal{B}$ for all $B \in \mathcal{B}$. □

For many functions (including all those we will work with) $\mathcal{B}$-measurability holds by virtue of a regularity property such as continuity or monotonicity.

## 3.9 Further exercises

1. Let $Y$ be a random variable that has a binomial distribution with $n$ trials and probability $p$ of success for each trial, that is, $Y \sim \text{Bin}(n, p)$. *Without using generating functions*:
   (a) show that $\mathbb{E}(Y) = np$,
   (b) work out $\text{Var}(Y)$,
   (c) explain why $\mathbb{E}(Y^r) = p$ for $r = 1, 2, \ldots$,
   (d) find the third central moment of $Y$.
2. Let $Y$ be a random variable that has a Poisson distribution with parameter $\lambda$, that is, $Y \sim \text{Pois}(\lambda)$. *Without using generating functions*:
   (a) show that $\mathbb{E}(Y) = \lambda$,
   (b) find $\mathbb{E}(Y^3)$.
3. Let $Y$ be a random variable that has an exponential distribution with parameter $\theta$. *Without using generating functions* show that

(a) $\mathbb{E}(Y) = \frac{1}{\theta}$,

(b) $\text{Var}(Y) = \frac{1}{\theta^2}$.

4. Find the cumulative distribution functions corresponding to the following density functions:

(a) Cauchy: $f_X(x) = 1/[\pi(1 + x^2)]$ for $x \in \mathbb{R}$

(b) Logistic: $f_X(x) = e^{-x}/(1 + e^{-x})^2$ for $x \in \mathbb{R}$

(c) Pareto: $f_X(x) = (a - 1)/(1 + x)^a$ for $x > 0$, where $a > 1$

(d) Weibull: $f_X(x) = c\tau x^{\tau-1}e^{-cx^\tau}$ for $x > 0$, where $c, \tau > 0$

5. If $X$ is a positive continuous random variable with density function $f_X(x)$ and mean $\mu$, show that

$$g(y) = \begin{cases} yf_X(y)/\mu & y \geq 0 \\ 0 & y < 0 \end{cases}$$

is a valid density function, and hence show that

$$\mathbb{E}(X^3)\,\mathbb{E}(X) \geq \{\mathbb{E}(X^2)\}^2.$$

6. Find the mean and the variance for the following distributions:

(a) Gamma: $f_X(x) = \frac{\lambda^\alpha}{\Gamma(\alpha)}e^{-\lambda x}x^{\alpha-1}$ for $x > 0$, where $\alpha, \lambda > 0$

(b) Poisson: $f_X(x) = e^{-\lambda}\lambda^x/x!$ for $x = 0, 1, 2, \ldots$, where $\lambda > 0$

(c) Pareto: $f_X(x) = (a - 1)/(1 + x)^a$ for $x > 0$, where $a > 1$

7. Let $Z$ be a random variable with density

$$f_Z(z) = \frac{1}{k}, \quad \text{for} \quad -b < z < b.$$

(a) Find $k$.

(b) Find the moment-generating function of $Z$.

(c) Is the moment-generating function you have specified well defined in an open interval around the origin? If not, how might you resolve this problem? Explain your answer clearly.

8. Let $Z$ be a random variable with mass function

$$f_Z(z) = k(1 - p)^{z-1}, \quad \text{for} \quad z = 1, 2, \ldots \quad \text{and} \quad 0 < p < 1.$$

(a) Find $k$.

(b) Find the moment-generating function of $Z$.

(c) Give an open interval around the origin in which the moment-generating function is well defined.

9. Let $Z$ be a random variable with density

$$f_Z(z) = \frac{1}{2}e^{-|z|}, \quad \text{for} \quad -\infty < z < \infty.$$

(a) Show that $f_Z$ is a valid density.

(b) Find the moment-generating function of $Z$ and give an open interval around the origin in which the moment-generating function is well defined.

(c) By considering the cumulant-generating function or otherwise, evaluate $\mathbb{E}(Z)$ and $\text{Var}(Z)$.

10. Derive the moment-generating function of
    (a) $X$ where $X \sim \text{Bernoulli}(p)$,
    (b) $N$ where $N \sim \text{Pois}(\lambda)$.

11. Find the moment-generating function of the **Laplace distribution**, also known as the **double exponential**, which has density function

$$f_X(x) = \frac{1}{2}e^{-|x|} \quad -\infty < x < \infty.$$

Use this expression to obtain its first four cumulants.

12. Consider the following lottery game involving two urns:
    - Urn A contains five balls, labelled 1 to 5,
    - Urn B contains five balls, four green and one red.

    Tickets cost £1 and players select two numbers between 1 and 5 (repetition not permitted). The lottery draw consists of two balls taken without replacement from Urn A, and one ball taken from Urn B. For a given ticket, there are three possible outcomes:
    - numbers on ticket do not match draw from Urn A $\Rightarrow$ win £0,
    - numbers on ticket match draw from Urn A and a green ball is drawn from Urn B $\Rightarrow$ win £1,
    - numbers on ticket match draw from Urn A and a red ball is drawn from Urn B $\Rightarrow$ win £$(z + 1)$.

    (a) Evaluate the probability of each of these three outcomes.
    (b) Define the random variable $R$ to be the return from playing the game with a single ticket, so $R$ is equal to our winnings minus the £1 cost of the ticket. Write down the probability mass function of $R$ and evaluate the expected return $\mathbb{E}(R)$. What value must $z$ take for the expected return to be zero?

13. Suppose that $X \sim F_X$ is a continuous random variable taking values between $-\infty$ and $+\infty$. Sometimes we want to *fold* the distribution of $X$ about the value $x = a$, that is, we want the distribution of the random variable $Y = |X - a|$.
    (a) Work out the density function of $Y$ in terms of $f_X$.
    [*Hint*: start with the CDF, $F_Y(y)$.]
    (b) Apply the result to the case where $X \sim \text{N}(\mu, \sigma^2)$ and $a = \mu$.

## 3.10 Chapter summary

On completion of this chapter you should be able to:

- give both formal and informal definitions of a random variable,
- formulate problems in terms of random variables,
- understand the characteristics of cumulative distribution functions,
- describe the distinction between discrete and continuous random variables,
- give the mass function and support for the Bernoulli, binomial, geometric, Poisson, and discrete uniform distributions,
- give the density function and support for the continuous uniform, exponential, and normal distributions,
- determine whether a function defines a valid mass or density,
- calculate moments from mass or density,
- prove and use inequalities involving expectation,
- find moment-generating functions for discrete and continuous distributions,
- calculate moments from a moment-generating function,
- calculate cumulants from a cumulant-generating function,
- find the distribution of a function of a random variable, and
- describe location-scale and probability integral transformations.

Chapter 4

# Multivariate distributions

So far we have encountered only univariate distributions, that is, distributions that arise when we consider a single random variable. In this chapter we introduce the concept of multivariate distributions. Random variables provide us with models of real situations. For most practical problems, we want to model the associations between several variables; in other words, we want to consider a multivariate distribution.

For illustration, suppose that we measure 10 traits of an individual, for example, age, gender, height, weight, educational level, income, and so on. These numbers are denoted $x_1, \dots, x_{10}$ and can be modelled as instances of random variables $X_1, \dots, X_{10}$. In this context, it would be unreasonable to assume independence between the variables. A multivariate model attempts to capture the nature of any dependencies between variables, whether these are viewed as predictors and outcomes (for example, using an individual's age, gender, and educational level to predict their income), or on equal footing (for example, finding a suitable bivariate model for height and weight).

As with the univariate case, the starting point for modelling is a function that gives probability below a point. In the multivariate context with $n$ random variables, the point will be an element of $\mathbb{R}^n$ and the function is referred to as the joint cumulative distribution function.

## 4.1 Joint and marginal distributions

The cumulative distribution function for a collection of random variables is referred to as the **joint cumulative distribution function**. This is a function of several variables.

**Definition 4.1.1** (General joint cumulative distribution function)
If $X_1, \dots, X_n$ are random variables, the joint cumulative distribution function is a function $F_{X_1,\dots,X_n} : \mathbb{R}^n \to [0, 1]$ given by

$$F_{X_1,\dots,X_n}(x_1, x_2, \dots, x_n) = \mathrm{P}(X_1 \le x_1, X_2 \le x_2, \dots, X_n \le x_n).$$

The notation associated with the general case of $n$ variables rapidly becomes rather cumbersome. Most of the ideas associated with multivariate distributions are entirely explained by looking at the two-dimensional case, that is, the bivariate distribution. The generalisations to $n$ dimensions are usually obvious algebraically, although $n$-dimensional distributions are considerably more difficult to visualise. The definition of a bivariate cumulative distribution function is an immediate consequence of Definition 4.1.1.

**Definition 4.1.2** (Bivariate joint cumulative distribution function)
For two random variables $X$ and $Y$, the joint cumulative distribution function is a function $F_{X,Y} : \mathbb{R}^2 \to [0,1]$ given by

$$F_{X,Y}(x,y) = \mathrm{P}(X \leq x, Y \leq y).$$

Notice that there is an implicit $\cap$ in the statement $\mathrm{P}(X \leq x, Y \leq y)$, so $F_{X,Y}(x,y)$ should be interpreted as the probability that $X \leq x$ and $Y \leq y$. The elementary properties of bivariate distributions are given by Claim 4.1.3 below. Part of Exercise 4.1 is to generalise these to the $n$-dimensional case.

**Claim 4.1.3** (Elementary properties of joint cumulative distribution functions)
*Suppose that $X$ and $Y$ are random variables. If $F_{X,Y}$ is the joint cumulative distribution function of $X$ and $Y$, then $F_{X,Y}$ has the following properties:*

*i.* $F_{X,Y}(-\infty, y) = \lim_{x\to-\infty} F_{X,Y}(x,y) = 0$,
$F_{X,Y}(x, -\infty) = \lim_{y\to-\infty} F_{X,Y}(x,y) = 0$,
$F_{X,Y}(\infty, \infty) = \lim_{x\to\infty, y\to\infty} F_{X,Y}(x,y) = 1$.

*ii.* *Right-continuous in $x$:* $\lim_{h\downarrow 0} F_{X,Y}(x+h, y) = F_{X,Y}(x,y)$,
*Right-continuous in $y$:* $\lim_{h\downarrow 0} F_{X,Y}(x, y+h) = F_{X,Y}(x,y)$.

In the univariate case, we use the cumulative distribution function to calculate the probability that the value taken by a random variable will be in a given subset of $\mathbb{R}$. In particular, we show as part of Proposition 3.2.4 that $\mathrm{P}(x_1 < X \leq x_2) = F_X(x_2) - F_X(x_1)$. In the bivariate case, we will often be interested in the probability that $X$ and $Y$ take values in a particular region of $\mathbb{R}^2$. The simplest example is a rectangular region $A = \{(x,y) : x_1 < x \leq x_2 \text{ and } y_1 < Y \leq y_2\}$ where $\mathrm{P}(A) = \mathrm{P}(x_1 < X \leq x_2, y_1 < Y \leq y_2)$.

**Lemma 4.1.4**
*If $X$ and $Y$ are random variables with joint cumulative distribution function $F_{X,Y}$, then*

$$\begin{aligned}&\mathrm{P}(x_1 < X \leq x_2, y_1 < Y \leq y_2)\\&= [F_{X,Y}(x_2,y_2) - F_{X,Y}(x_1,y_2)] - [F_{X,Y}(x_2,y_1) - F_{X,Y}(x_1,y_1)].\end{aligned}$$

*Proof.*

The proof of this lemma is a useful illustration of a divide-and-conquer approach. We start by considering $Y$.

$$\begin{aligned}
&\mathrm{P}(x_1 < X \leq x_2, y_1 < Y \leq y_2) \\
&= \mathrm{P}(x_1 < X \leq x_2, Y \leq y_2) - \mathrm{P}(x_1 < X \leq x_2, Y \leq y_1) \\
&= [\mathrm{P}(X \leq x_2, Y \leq y_2) - \mathrm{P}(X \leq x_1, Y \leq y_2)] \\
&\quad - [\mathrm{P}(X \leq x_2, Y \leq y_1) - \mathrm{P}(X \leq x_1, Y \leq y_1)] \\
&= [F_{X,Y}(x_2, y_2) - F_{X,Y}(x_1, y_2)] - [F_{X,Y}(x_2, y_1) - F_{X,Y}(x_1, y_1)].
\end{aligned}$$

□

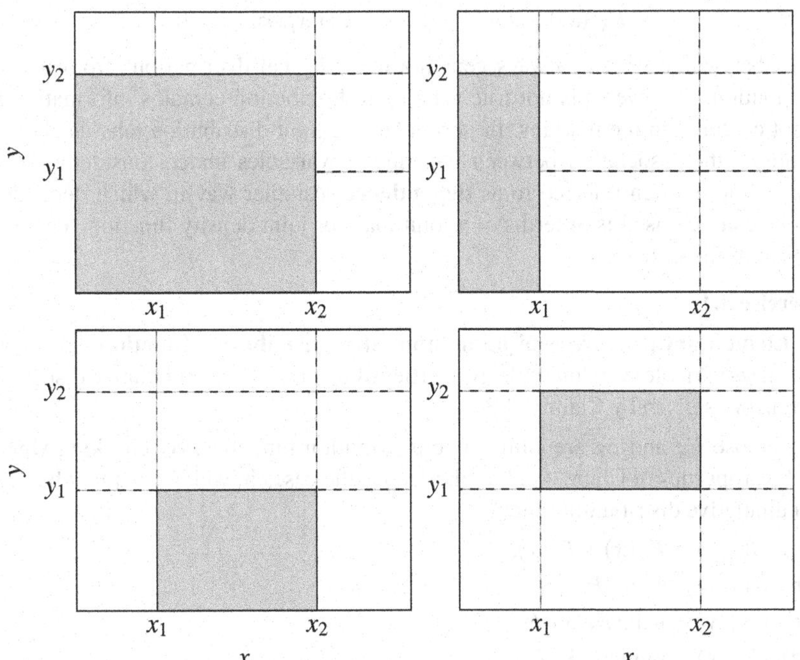


Figure 4.1 *To compute* $\mathrm{P}(x_1 < X \leq x_2, y_1 < Y \leq y_2)$*, we begin with* $F_{X,Y}(x_2, y_2)$ *(top left), then subtract* $F_{X,Y}(x_1, y_2)$ *(top right) and* $F_{X,Y}(x_2, y_1) - F_{X,Y}(x_1, y_1)$ *(bottom left) to obtain the final answer (bottom right).*

Just as the term joint refers to many variables, we use the term **marginal** to refer to the distributions of individual random variables. Thus, a marginal cumulative distribution function is just the usual cumulative distribution function of a single random variable. It is straightforward to generate marginal cumulative distribution functions from the joint distribution. The following claim is clear from the definition of the joint cumulative distribution function.

**Claim 4.1.5** (Marginal distribution from joint distribution)
*If* $F_{X,Y}$ *is the joint distribution function of* $X$ *and* $Y$*, then the marginal cumulative*

*distribution functions are given by*

$$F_X(x) = \lim_{y \to \infty} F_{X,Y}(x, y) = F_{X,Y}(x, \infty),$$
$$F_Y(y) = \lim_{x \to \infty} F_{X,Y}(x, y) = F_{X,Y}(\infty, y).$$

The bivariate case, where we generate the marginal cumulative distribution function for one variable by letting the other tend to infinity, readily generalises to the $n$-dimensional case. In general, if $X_1, \ldots, X_n$ are random variables with joint cumulative distribution function $F_{X_1,\ldots,X_n}$, the marginal cumulative distribution function of $X_j$ is given by

$$F_{X_j}(x_j) = F_{X_1,\ldots,X_n}(\infty, \ldots, \infty, x_j, \infty, \ldots, \infty).$$

Note that, while we can always generate the marginal distributions from the joint distribution, the reverse is not true. The joint distribution contains information that is not captured in the margins. In particular, the joint distribution tells us about the nature of the association between the random variables under consideration; this association is often referred to as **dependence**. Another way in which dependence can be characterised is in terms of a joint mass or joint density function. This is the subject of the next section.

**Exercise 4.1**

1. (**Elementary properties of joint cumulative distribution function**) Give a general ($n$-variable) version of the properties of bivariate joint cumulative distribution functions given by Claim 4.1.3.
2. Suppose $F_X$ and $F_Y$ are cumulative distribution functions. By checking whether the properties of Claim 4.1.3 hold, identify the cases in which $G$ *cannot* be a joint cumulative distribution function:
   (a) $G(x, y) = F_X(x) + F_Y(y)$,
   (b) $G(x, y) = F_X(x)F_Y(y)$,
   (c) $G(x, y) = \max[F_X(x), F_Y(y)]$,
   (d) $G(x, y) = \min[F_X(x), F_X(y)]$.

## 4.2 Joint mass and joint density

### *4.2.1 Mass for discrete distributions*

We will use the following example to illustrate ideas associated with discrete bivariate mass.

**Example 4.2.1** (Card drawing experiment)
We have a standard deck of 52 cards made up of 13 denominations from 4 suits. We select two cards at random without replacement and note the number of kings and the number of aces. We could use $X$ to denote the number of kings and $Y$ to denote the number of aces.

A natural way to characterise a discrete bivariate distribution is in terms of the probability that the variables of interest take particular values.

**Definition 4.2.2** (Joint mass function)
Suppose that $X$ and $Y$ are discrete random variables. The **joint mass function** of $X$ and $Y$ is the function $f_{X,Y} : \mathbb{R}^2 \to [0,1]$ given by

$$f_{X,Y}(x,y) = \mathrm{P}(X = x, Y = y).$$

Once again there is an implicit $\cap$ here, so $\mathrm{P}(X = x, Y = y)$ is the probability that $X = x$ <u>and</u> $Y = y$. For discrete distributions, events of the form $\{X = x_1, Y = y_1\}$ and $\{X = x_2, Y = y_2\}$ are disjoint for $x_1 \neq x_2$ and/or $y_1 \neq y_2$. As such, adding the probabilities of these events will give us the probability of the union. This idea can be extended to intervals, leading to the following claim.

**Claim 4.2.3**
*For discrete random variables $X$ and $Y$ with joint mass function $f_{X,Y}$, and real numbers $x_1 < x_2$ and $y_1 < y_2$, we have*

$$\mathrm{P}(x_1 < X \leq x_2,\ y_1 < Y \leq y_2) = \sum_{x_1 < x \leq x_2} \sum_{y_1 < y \leq y_2} f_{X,Y}(x,y).$$

A simple corollary of this claim is that

$$\mathrm{P}(X = x,\ y_1 < Y \leq y_2) = \sum_{y_1 < y \leq y_2} f_{X,Y}(x,y). \tag{4.1}$$

The principle of adding mutually exclusive outcomes to remove one variable from consideration can be applied generally.

**Claim 4.2.4** (Marginal mass from joint mass)
*For discrete random variables $X$ and $Y$ with joint mass function $f_{X,Y}$, the marginal mass functions are given by*

$$f_X(x) = \sum_y f_{X,Y}(x,y),$$
$$f_Y(y) = \sum_x f_{X,Y}(x,y).$$

This claim is clearly an application of the idea given by equation (4.1), with the summation ranging over every possible value of $y$. We return to our example to illustrate these ideas.

**Example 4.2.1 (Revisited)** Recall that we are drawing two cards at random without replacement from a standard deck. If $X$ is the number of kings and $Y$ is the number of aces, we can readily construct Table 4.1 (probabilities to 5 decimal places). Constructing this table is part of Exercise 4.2.

| $f_{X,Y}(x,y)$ | | $x$: 0 | 1 | 2 |
|---|---|---|---|---|
| | 0 | 0.71342 | 0.13273 | 0.00452 |
| $y$ | 1 | 0.13273 | 0.01207 | 0.00000 |
| | 2 | 0.00452 | 0.00000 | 0.00000 |

Table 4.1 *Joint mass function for a card drawing experiment.*

If we want to calculate the probability that $X \leq 1$ and $Y \leq 1$, we could do this by adding probabilities. In this example $X$ and $Y$ take values in $\{0, 1, 2\}$, so

$$\begin{aligned}&\text{P}(X \leq 1, Y \leq 1)\\&\quad= \text{P}(X = 0, Y = 0) + \text{P}(X = 0, Y = 1) + \text{P}(X = 1, Y = 0) + \text{P}(X = 1, Y = 1)\\&\quad= 0.713 + 0.133 + 0.133 + 0.012 = 0.991 \ \ (3 \text{ d.p.}).\end{aligned}$$

Suppose that we are not interested in the number of aces, $Y$. We could return to first principles to derive the marginal mass function for the number of kings, $X$. However, this information about marginal mass is available from the joint mass function. For example, in order to calculate the probability $\text{P}(X = 0)$, we would consider every case for which the outcome of the experiment is zero kings. Since these outcomes are mutually exclusive, we can add probabilities:

$$\begin{aligned}\text{P}(X=0) &= \text{P}(X=0, Y=0) + \text{P}(X=0, Y=1) + \text{P}(X=0, Y=2)\\&= 0.713 + 0.133 + 0.005 = 0.85 \ \ (2 \text{ decimal places}).\end{aligned}$$

Applying our results on marginal distributions (Claim 4.2.4) to Table 4.1, we sum the rows to get the marginal distribution of $X$, and sum the columns to get the marginal distribution of $Y$. Note that these figures appear in the margins of our table, hence the term *marginal* mass function.

| $f_{X,Y}(x,y)$ | | X: 0 | 1 | 2 | $f_Y(y)$ |
|---|---|---|---|---|---|
| | 0 | 0.71342 | 0.13273 | 0.00452 | 0.85067 |
| $y$ | 1 | 0.13273 | 0.01207 | 0.00000 | 0.14480 |
| | 2 | 0.00452 | 0.00000 | 0.00000 | 0.00452 |
| $f_X(x)$ | | 0.85067 | 0.14480 | 0.00452 | 1.00000 |

Table 4.2 *Joint and marginal mass functions for a card drawing experiment.*

*Multivariate case*

The multivariate case is a simple extension. Suppose that we have discrete random variables $X_1, \ldots, X_n$. The joint mass function is a function $f_{X_1,\ldots,X_n} : \mathbb{R}^n \to \mathbb{R}$ such that, for any real numbers $x_1, \ldots, x_n$,

$$f_{X_1,\ldots,X_n}(x_1, \ldots, x_n) = \text{P}(X_1 = x_1, \ldots, X_n = x_n).$$

The marginal mass function for $X_j$ is generated by summing over all the other variables,

$$f_{X_j}(x_j) = \sum_{x_1} \cdots \sum_{x_{j-1}} \sum_{x_{j+1}} \cdots \sum_{x_n} f_{X_1,\ldots,X_n}(x_1,\ldots,x_n).$$

### 4.2.2 *Density for continuous distributions*

Jointly continuous random variables can be characterised by an extension of the concept of density to more than one dimension.

**Definition 4.2.5** (Joint density function)
For jointly continuous random variables $X$ and $Y$ with joint cumulative distribution function $F_{X,Y}$, the **joint density function** is a function $f_{X,Y} : \mathbb{R}^2 \to [0,\infty)$ such that

$$F_{X,Y}(x,y) = \int_{-\infty}^{y} \int_{-\infty}^{x} f_{X,Y}(u,v)dudv \quad \text{for all} \quad x, y \in \mathbb{R}.$$

The term **jointly** continuous refers to cases in which the joint density function is well defined. It is possible to find pairs of continuous random variables that are not jointly continuous. All of the continuous multivariate examples that we deal with will be jointly continuous. As such, we will often just refer to the "discrete case" and the "continuous case".

As in the univariate case, the density function can be obtained from the cumulative distribution function by differentiation.

**Claim 4.2.6** (Joint density from joint distribution)
*For jointly continuous random variables X and Y with joint cumulative distribution function $F_{X,Y}$, the joint density function is given by*

$$f_{X,Y}(x,y) = \left. \frac{\partial^2}{\partial u \partial v} F_{X,Y}(u,v) \right|_{u=x,v=y}.$$

From the definition of joint density and the properties of cumulative distribution functions, it is clear that the joint density must integrate to 1 over the real plane $\mathbb{R}^2$. This is a key property that is useful in establishing whether a function is a valid joint density.

**Claim 4.2.7**
*If $f_{X,Y}$ is a joint density function then $f_{X,Y}$ is a positive, real-valued function with the property*

$$\int_{-\infty}^{\infty} \int_{-\infty}^{\infty} f_{X,Y}(x,y)dxdy = 1.$$

For a bivariate distribution, we can find the probability associated with a subset of $\mathbb{R}^2$ in much the same way that we did for a subset of $\mathbb{R}$ in the univariate case. For any reasonably well-behaved region $B \subseteq \mathbb{R}^2$, the probability that $(X,Y)$ takes a value in $B$ is the integral of the density function over $B$, that is,

$$\mathrm{P}((X,Y) \in B) = \int \int_B f_{X,Y}(x,y)dxdy. \tag{4.2}$$

In the univariate case, probability is associated with the area beneath the density curve. In the bivariate case, it is the volume beneath the density surface (Figure 4.2).

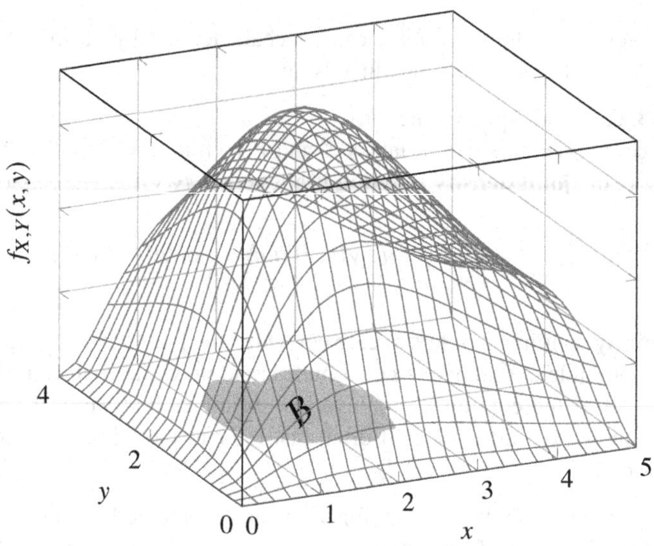


Figure 4.2 *The probability that $(X,Y)$ takes a value in $B$ is equal to the volume between the region $B$ and the surface $f_{X,Y}$.*

We will often be interested in rectangular regions.

**Claim 4.2.8** (Probability for a rectangular region)
*If $X$ and $Y$ are random variables with joint density function $f_{X,Y}(x,y)$, then*

$$\mathrm{P}(x_1 < X \leq x_2, y_1 < Y \leq y_2) = \int_{y_1}^{y_2} \int_{x_1}^{x_2} f_{X,Y}(x,y)dxdy.$$

When performing calculations with joint density functions we have to take extra care with the support. Careless application of results such as that given by Claim 4.2.8 can lead us to integrate over regions where the density is zero.

The marginal density for $X$ can be derived from the joint density. Consider the event $-\infty < Y < \infty$. This occurs with probability 1, thus

$$\mathrm{P}(-\infty < X \leq x, -\infty < Y < \infty) = \mathrm{P}(-\infty < X \leq x) = F_X(x).$$

By Claim 4.2.8 we then have

$$F_X(x) = \int_{-\infty}^{x} \int_{-\infty}^{\infty} f_{X,Y}(u,y)dydu.$$

Comparing with Definition 4.2.5 of the density, we conclude that

$$f_X(x) = \int_{-\infty}^{\infty} f_{X,Y}(u, y) dy,$$

thus the marginal density of $X$ is given by integrating the joint density $f_{X,Y}(x, y)$ with respect to $y$ over all possible values of $y$. This is clearly analogous to the discrete case, where we find the marginal of $X$ by summing the joint mass function $f_{X,Y}(x, y)$ over all possible values of $y$. The relationship between joint and marginal densities is summarised by the following claim.

**Claim 4.2.9** (Marginal density from joint density)
*If $X$ and $Y$ are jointly continuous random variables then*

$$f_X(x) = \int_{-\infty}^{\infty} f_{X,Y}(x, y) dy,$$
$$f_Y(y) = \int_{-\infty}^{\infty} f_{X,Y}(x, y) dx.$$

**Example 4.2.10** (Two polynomial joint density functions)
In this example we consider two polynomial density functions. While these examples are superficially similar, the support is rectangular in the first case and triangular in the second. This distinction is very important. The examples are set out as questions and solutions.

1. (**Polynomial density with rectangular support**) Consider the function

$$f_{X,Y}(x, y) = \begin{cases} x + y & \text{for } 0 < x < 1 \text{ and } 0 < y < 1, \\ 0 & \text{otherwise.} \end{cases}$$

   (a) Show that $f_{X,Y}$ is a valid density.
   (b) Derive the joint cumulative distribution function of $X$ and $Y$.
   (c) Evaluate $\mathrm{P}(0.2 < X < 0.5, 0 < Y < 0.3)$.
   (d) Find the marginal density of $X$.
   (e) Evaluate $\mathrm{P}(2X < Y)$.

   **Solutions**

   (a) To show that this is a valid density function we need to establish that it is a positive real-valued function that integrates to 1 over $\mathbb{R}^2$. The fact that it is positive and real-valued is clear from the definition. Evaluating the integral, we obtain

$$\int_{-\infty}^{\infty} \int_{-\infty}^{\infty} f_{X,Y}(x, y) dx dy$$
$$= \int_0^1 \int_0^1 (x + y) dx dy = \int_0^1 [x^2/2 + xy]_{x=0}^{x=1} dy$$
$$= \int_0^1 (\frac{1}{2} + y) dy = [y/2 + y^2/2]_0^1 = 1.$$

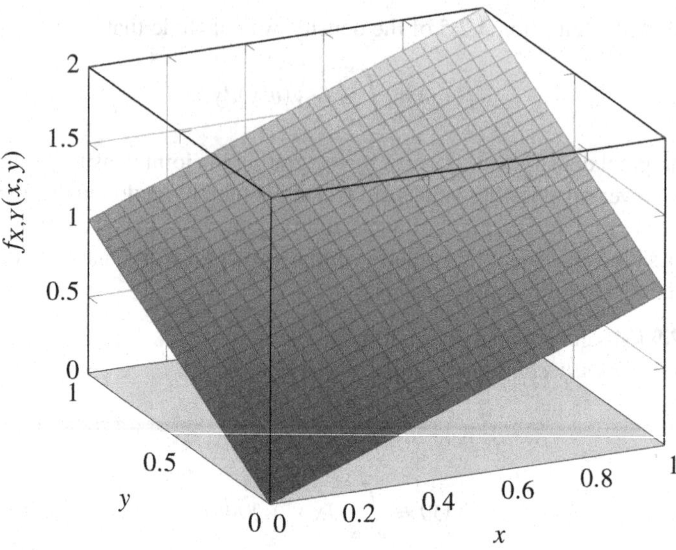


Figure 4.3 *Polynomial density with rectangular support. The shaded region on the xy-plane is the support.*

(b) We can derive the cumulative distribution function from the definition of joint density:

$$\begin{aligned}
F_{X,Y}(x,y) &= \int_{-\infty}^{y}\int_{-\infty}^{x} f(u,v)dudv \\
&= \int_{0}^{y}\int_{0}^{x}(u+v)dudv = \int_{0}^{y}[u^2/2+uv]_{u=0}^{u=x}dv \\
&= \int_{0}^{y}(x^2/2+xv)dv = [x^2v/2+xv^2/2]_{v=0}^{v=y} \\
&= \frac{1}{2}xy(x+y), \quad \text{for } 0<x<1, 0<y<1. \qquad (4.3)
\end{aligned}$$

The form of this function illustrates the importance of specifying the region over which a function is defined. From Claim 4.1.5, we know that the marginal distribution function of $X$ is given by $F_X(x) = \lim_{y\to\infty} F_{X,Y}(x,y)$. If we were to naively apply this to equation (4.3), we would conclude that the marginal cumulative distribution function for $X$ is nowhere finite; this is clearly incorrect. The point here is that the expression given by equation (4.3) for the cumulative distribution function is only valid for $0 < y < 1$. Values of $y$ that are greater than one have no impact on the cumulative distribution function since they do not have any density associated with them so, for $y \geq 1$ and $0 < x < 1$, we have $f_{X,Y}(x,y) = f_{X,Y}(x,1) = \frac{1}{2}x(x+1)$. The full specification of the cumulative distribution function is then rather cumbersome:

$$F_{X,Y}(x,y) = \begin{cases} \frac{1}{2}xy(x+y) & \text{for } 0 < x < 1 \text{ and } 0 < y < 1, \\ \frac{1}{2}x(x+1) & \text{for } 0 < x < 1 \text{ and } y \geq 1, \\ \frac{1}{2}y(y+1) & \text{for } x \geq 1 \text{ and } 0 < y < 1, \\ 1 & \text{for } x \geq 1 \text{ and } y \geq 1, \\ 0 & \text{otherwise.} \end{cases}$$

(c) We would like to know the probability that $X \in (0.2, 0.5)$ and $Y \in (0, 0.3)$. By Claim 4.2.8

$$\begin{aligned} &\mathrm{P}(0.2 < X < 0.5, 0 < Y < 0.3) \\ &= \int_0^{0.3} \int_{0.2}^{0.5} f_{X,Y}(x,y)dxdy \\ &= \int_0^{0.3} \int_{0.2}^{0.5} (x+y)dxdy = \int_0^{0.3} \left[x^2/2 + xy\right]_{x=0.2}^{x=0.5} dy \\ &= \int_0^{0.3} (0.105 + 0.3y)dy = \left[0.105y + 0.3y^2/2\right]_0^{0.3} \\ &= 0.45. \end{aligned}$$

Check that you get the same answer by performing the $y$ integral first.

(d) By Claim 4.2.9, the density of $X$ is given by integrating $f_{X,Y}(x,y)$ with respect to $y$, thus,

$$\begin{aligned} f_X(x) &= \int_{-\infty}^{\infty} f_{X,Y}(x,y)dy \\ &= \int_0^1 (x+y)dy = \left[xy + y^2/2\right]_{y=0}^{y=1} = x + \frac{1}{2}. \end{aligned}$$

In order for this to be a valid density, we have to take care in specifying the support. A complete answer is

$$f_X(x) = \begin{cases} x + \frac{1}{2} & \text{for } 0 < x < 1, \\ 0 & \text{otherwise.} \end{cases}$$

You should check that this is a valid density. It is clear, by symmetry, that the density for $Y$ will have an identical functional form.

(e) We can write $\mathrm{P}(2X < Y) = \mathrm{P}((X,Y) \in B)$ where $B$ is the triangular region $B = \{(x,y) : 0 < x < \frac{y}{2}, 0 < y < 1\}$. Thus, integrating with respect to $x$ first, the $x$ integral is over $(0, \frac{y}{2})$ and the $y$ integral is over $(0, 1)$,

$$\begin{aligned}
P(2X < Y) &= \int \int_B f_{X,Y}(x, y) dx dy \\
&= \int_0^1 \int_0^{y/2} (x + y) dx dy = \int_0^1 [x^2/2 + xy]_{x=0}^{x=y/2} dy \\
&= \int_0^1 (y^2/8 + y^2/2) dy = [y^3/24 + y^3/6]_0^1 \\
&= \frac{5}{24}.
\end{aligned}$$

If we integrate with respect to $y$ first, the $y$ integral is over $(2x, 1)$ and the $x$ integral is over $(0, \frac{1}{2})$. You should check that performing the integration in this order gives the same result.

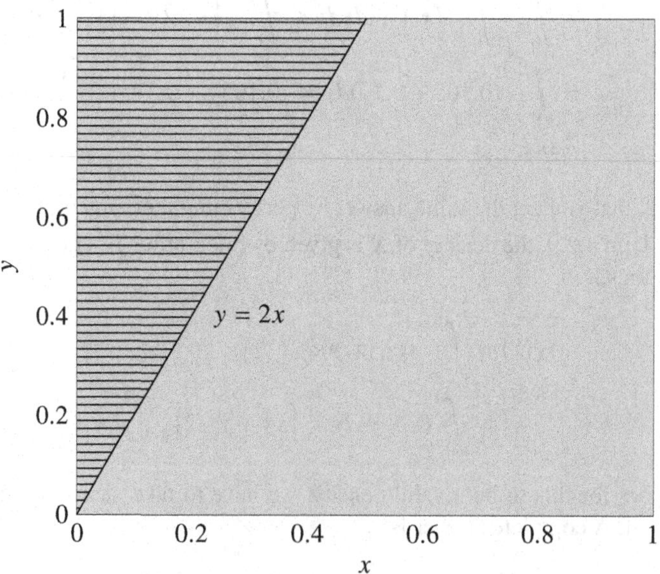


Figure 4.4 *We want to integrate $f_{X,Y}(x, y)$ where the support (light grey) overlaps with the region B (black lines). As y ranges from 0 to 1, the horizontal line segments extend from $x = 0$ to $x = y/2$. Integrating with respect to y first corresponds to shading the same region with vertical line segments instead.*

2. (**Polynomial density with triangular support**) Consider the function

$$f_{X,Y}(x, y) = \begin{cases} 8xy & \text{for } 0 < x < y < 1, \\ 0 & \text{otherwise.} \end{cases}$$

   (a) Show that $f_{X,Y}$ is a valid density.

(b) Find the marginal density of $X$.
(c) Evaluate $\mathrm{P}(Y < 2X)$.

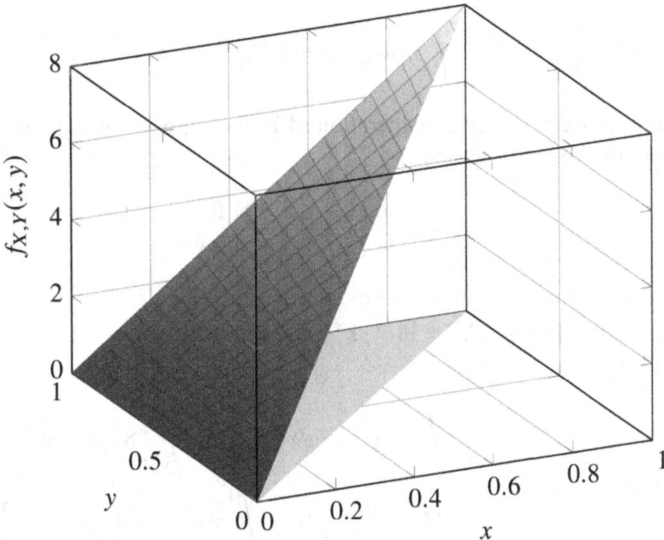


Figure 4.5 *Polynomial density with triangular support. The shaded region on the xy-plane is the support.*

**Solutions**

(a) It is clear from the form of $f_{X,Y}$ that it is a positive real-valued function. In order to prove that it is a valid density, we need to show that it integrates to 1 over the real plane,

$$\int_{-\infty}^{\infty}\int_{-\infty}^{\infty} f_{X,Y}(x,y) = \int_0^1\int_0^y 8xydxdy = \int_0^1 [4x^2y]_{x=0}^{x=y}dy$$
$$= \int_0^1 4y^3dy = [y^4]_0^1 = 1.$$

Note that the support is the triangular region $\{(x,y) : 0 < x < y, 0 < y < 1\}$. Integrating with respect to $x$ first, the $x$ integral is over $(0,y)$ and the $y$ integral is over $(0,1)$. An equivalent way in which the support can be described is $\{(x,y) : x < y < 1, 0 < x < 1\}$. Thus, if we integrate with respect to $y$ first, the $y$ integral is over $(x,1)$ and the $x$ integral is over $(0,1)$. You should check that performing the $y$ integral first gives the same answer.

(b) In order to evaluate the marginal density of $X$, we integrate with respect to $y$.

For fixed $x$ we have $x < y < 1$, thus,

$$\begin{aligned} f_X(x) &= \int_{-\infty}^{\infty} f_{X,Y}(x,y)dy \\ &= \int_x^1 8xydy = [4xy^2]_{y=x}^{y=1} = 4x(1-x^2). \end{aligned}$$

Since we have removed consideration of $Y$, the support for $X$ is $(0, 1)$. In full, the marginal density for $X$ is

$$f_X(x) = \begin{cases} 4x(1-x^2) & \text{for } 0 < x < 1, \\ 0 & \text{otherwise.} \end{cases}$$

You should check that this is a valid density function.

(c) We can write $P(Y < 2X) = P((X,Y) \in B)$ where $B = \{(x,y) : \frac{y}{2} < x < y, 0 < y < 1\}$, thus,

$$\begin{aligned} P(Y < 2X) &= \int\int_B f_{X,Y}(x,y)dxdy = \int_0^1 \int_{y/2}^y 8xydxdy \\ &= \int_0^1 [4x^2y]_{x=y/2}^{x=y} dy = \int_0^1 3y^3 dy = \left[\frac{3}{4}y^4\right]_0^1 = \frac{3}{4}. \end{aligned}$$

Performing the $y$ integral first is more difficult in this case. If $x > \frac{1}{2}$ then the upper limit for $y$ is 1 rather than $2x$ (Figure 4.6). Thus, another way to describe the region over which we need to integrate is

$$B = \left\{(x,y) : x < y < 2x, 0 < x < \frac{1}{2}\right\} \cup \left\{(x,y) : x < y < 1, \frac{1}{2} < x < 1\right\}.$$

The integral over $B$ is done in two parts, one corresponding to $0 < x < \frac{1}{2}$ and the other corresponding to $\frac{1}{2} < x < 1$. You should try this to check that you get the same answer as above.

*Multivariate case*

The generalisation to the $n$-dimensional multivariate case does not introduce any new ideas. Suppose that $X_1, \ldots, X_n$ are jointly continuous random variables with cumulative distribution function $F_{X_1,\ldots,X_n}$. The joint density is the function $f_{X_1,\ldots,X_n}$ satisfying

$$F_{X_1,\ldots,X_n}(x_1,\ldots,x_n) = \int_{-\infty}^{x_n} \cdots \int_{-\infty}^{x_1} f_{X_1,\ldots,X_n}(u_1,\ldots,u_n)du_1 \ldots du_n.$$

In order to generate the marginal density of $X_j$, we integrate the joint density with respect to all the other variables,

$$f_{X_j}(x_j) = \int_{-\infty}^{\infty} \cdots \int_{-\infty}^{\infty} f_{X_1,\ldots,X_n}(x_1,\ldots,x_n)dx_1 \ldots dx_{j-1}dx_{j+1} \ldots dx_n.$$

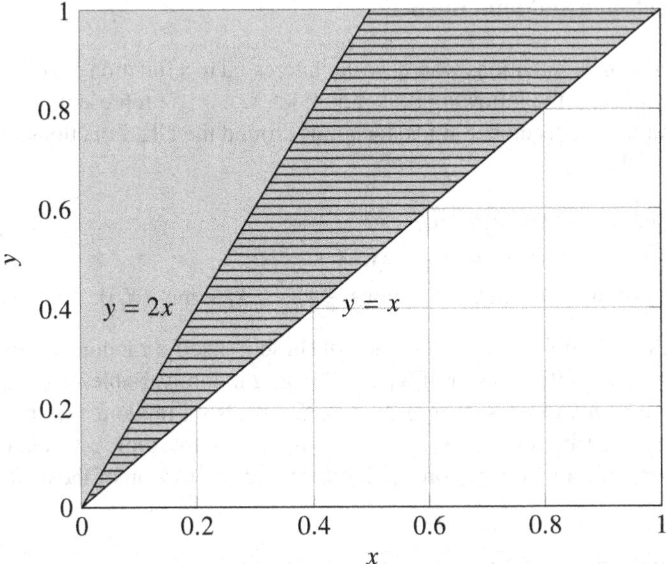


Figure 4.6 *We want to integrate $f_{X,Y}(x, y)$ where the support (light grey) overlaps with the region B (black lines). As y ranges from 0 to 1, the horizontal line segments extend from $x = y/2$ to $x = y$. Notice that it is more difficult to shade this region with vertical line segments, as we would need to consider the cases $0 < x < \frac{1}{2}$ and $\frac{1}{2} < x < 1$ separately.*

**Exercise 4.2**

1. I toss a coin three times. Let $X$ be the number of heads and let $Y$ be a random variable that takes the value 1 if I get a head on the first and last throw, and 0 otherwise.
   (a) Write down a table summarising the joint mass function of $X$ and $Y$.
   (b) Calculate the marginal mass functions. Are they what you expected?
2. Consider random variables $X$ and $Y$ with joint density

$$f_{X,Y}(x, y) = \begin{cases} kxy & \text{for } 0 < x < 1 \text{ and } 0 < y < 2, \\ 0 & \text{otherwise.} \end{cases}$$

   (a) Evaluate $k$.
   (b) Find $f_X(x)$ and $f_Y(y)$.
   (c) Evaluate $\mathrm{P}(Y > X)$.
3. Consider the function

$$f_{X,Y}(x, y) = \begin{cases} 2 & \text{for } 0 < x < y < 1, \\ 0 & \text{otherwise.} \end{cases}$$

   (a) Show that $f_{X,Y}$ is a valid density.
   (b) Find $f_X(x)$ and $f_Y(y)$.
   (c) Evaluate $\mathrm{P}(Y < X + \frac{1}{2})$.

## 4.3 Expectation and joint moments

We often encounter situations where we are interested in a function of several random variables. Consider the following illustration: let $X_1, \ldots, X_5$ represent our models for the total rainfall in December at five locations around the UK. Functions that may be of interest include:

- the mean across locations, $\frac{1}{5}\sum_{i=1}^{5} X_i$.
- the maximum across locations, $\max_i(X_i)$.
- the mean of the four rainiest locations, $\frac{1}{4}\left[\sum_{i=1}^{5} X_i - \min_i(X_i)\right]$.

As a function of random variables, each of these is itself a random variable. In the situations that we will consider, if $X_1, \ldots, X_n$ are random variables and $g : \mathbb{R}^n \to \mathbb{R}$ is a function of $n$ variables, then $g(X_1, \ldots, X_n)$ is also a random variable. In many instances, the distribution of $g(X_1, \ldots, X_n)$ is of interest; this topic is tackled in section 4.6. We start with something more straightforward: calculation of the mean.

### *4.3.1 Expectation of a function of several variables*

The expected value of a function of two random variables is given by an extension of Theorem 3.4.4.

**Proposition 4.3.1** (Expectation of a function of two random variables)
*If $g$ is a well-behaved, real-valued function of two variables, $g : \mathbb{R}^2 \longrightarrow \mathbb{R}$, and $X$ and $Y$ are random variables with joint mass/density function $f_{X,Y}$, then*

$$\mathbb{E}[g(X,Y)] = \begin{cases} \sum_y \sum_x g(x,y) f_{X,Y}(x,y) & \textit{(discrete case)} \\ \int_{-\infty}^{\infty}\int_{-\infty}^{\infty} g(x,y) f_{X,Y}(x,y)dxdy & \textit{(continuous case).} \end{cases} \tag{4.4}$$

An example using the simple polynomial density function of 4.2.10 follows.

**Example 4.3.2** (Expectation of a product for simple polynomial density)
Consider random variables $X$ and $Y$ with joint density function $f_{X,Y}(x,y) = x + y$ for $0 < x < 1$ and $0 < y < 1$. Suppose we are interested in calculating the expectation of the product of $X$ and $Y$, that is, $\mathbb{E}(XY)$. Using Proposition 4.3.1 with $g(X,Y) = XY$ we have

$$\begin{aligned} \mathbb{E}(XY) &= \int_{-\infty}^{\infty}\int_{-\infty}^{\infty} xy f_{X,Y}(x,y)dxdy \\ &= \int_0^1\int_0^1 xy(x+y)dxdy \\ &= \int_0^1 \left[\frac{x^3}{3}y + \frac{x^2}{2}y^2\right]_{x=0}^{x=1} dy = \int_0^1 \left(\frac{1}{3}y + \frac{1}{2}y^2\right) dy \\ &= \left[\frac{1}{6}y^2 + \frac{1}{6}y^3\right]_0^1 = \frac{1}{3}. \end{aligned}$$

Note that Proposition 4.3.1 is consistent with previous definitions, that is, we can deduce that $\mathbb{E}(X) = \int_{-\infty}^{\infty} x f_X(x) dx$ from the proposition. Showing that this is the case is part of Exercise 4.3. Another immediate consequence of Proposition 4.3.1 is that the expected value of a sum of random variables is the sum of the individual expectations.

**Corollary 4.3.3** (Expected value of a sum of random variables)
*If $X$ and $Y$ are random variables, then $\mathbb{E}(X+Y) = \mathbb{E}(X) + \mathbb{E}(Y)$.*

The proof of this corollary is part of Exercise 4.3. The generalisation of Proposition 4.3.1 to $n$ dimensions follows.

**Proposition 4.3.4** (Expectation of a function of $n$ variables)
*If $g$ is a well-behaved, real-valued function of $n$ variables, $g : \mathbb{R}^n \to \mathbb{R}$, and $X_1, \dots, X_n$ are random variables, then*

$$\mathbb{E}[g(X_1,\dots,X_n)]$$
$$= \begin{cases} \sum_{x_1} \cdots \sum_{x_n} g(x_1,\dots,x_n) f_{X_1,\dots,X_n}(x_1,\dots,x_n), & \textit{(discrete case)} \\ \int_{-\infty}^{\infty} \cdots \int_{-\infty}^{\infty} g(x_1,\dots,x_n) f_{X_1,\dots,X_n}(x_1,\dots,x_n) dx_1 \dots dx_n, & \textit{(continuous case)}. \end{cases}$$

### *4.3.2 Covariance and correlation*

In the univariate case, we discussed the use of single-number summaries for the features of a distribution. For example, we might use the mean as a measure of central tendency and the variance as a measure of spread. In the multivariate case we might, in addition, be interested in summarising the dependence between random variables. For a pair of random variables, a commonly used quantity for measuring the degree of (linear) association is correlation. The starting point for the definition of correlation is the notion of covariance.

**Definition 4.3.5** (Covariance)
For random variables $X$ and $Y$, the **covariance** between $X$ and $Y$ is defined as

$$\text{Cov}(X,Y) = \mathbb{E}[(X - \mathbb{E}(X))(Y - \mathbb{E}(Y))].$$

An alternative form for the covariance is

$$\text{Cov}(X,Y) = \mathbb{E}(XY) - \mathbb{E}(X)\,\mathbb{E}(Y).$$

Proving the equivalence of these two forms is part of Exercise 4.3. Covariance has a number of properties that are immediate consequences of its definition as an expectation.

**Claim 4.3.6** (Properties of covariance)
*For random variables $X$, $Y$, $U$, and $V$ the covariance has the following properties.*

i. *Symmetry:* $\text{Cov}(X,Y) = \text{Cov}(Y,X)$.

ii. *For constants a and b :* $\text{Cov}(aX, bY) = ab\text{Cov}(X,Y)$.

iii. *Relationship with variance:*

$$\begin{aligned}\text{Var}(X) &= \text{Cov}(X,X),\\ \text{Var}(X+Y) &= \text{Var}(X) + \text{Var}(Y) + 2\text{Cov}(X,Y),\\ \text{Var}(X-Y) &= \text{Var}(X) + \text{Var}(Y) - 2\text{Cov}(X,Y).\end{aligned}$$

iv. *Bilinearity:*

$\text{Cov}(U+V, X+Y) = \text{Cov}(U,X) + \text{Cov}(U,Y) + \text{Cov}(V,X) + \text{Cov}(V,Y)$.

Showing that these properties hold is part of Exercise 4.3. We are now in a position to define correlation.

**Definition 4.3.7** (Correlation)
For random variables $X$ and $Y$, the **correlation** between $X$ and $Y$ is defined as

$$\text{Corr}(X,Y) = \frac{\text{Cov}(X,Y)}{\sqrt{\text{Var}(X)\,\text{Var}(Y)}}.$$

Correlation measures the degree of linear association between two variables. While the magnitude of the covariance between $X$ and $Y$ depends on the variances of $X$ and $Y$, correlation is scaled so that $|\text{Corr}(X,Y)| \leq 1$ for any $X$ and $Y$. In the following theorem, we prove this result and shed some light on the sense in which correlation measures linear dependence.

**Theorem 4.3.8** (Values taken by correlation)
*For random variables $X$ and $Y$,*

$$-1 \leq \text{Corr}(X,Y) \leq 1\,.$$

*Furthermore, we have $|\text{Corr}(X,Y)| = 1$ if and only if we can find constants $\alpha$ and $\beta \neq 0$ such that $Y = \alpha + \beta X$, with*

$$\begin{aligned}\text{Corr}(X,Y) &= 1 \text{ if } \beta > 0,\\ \text{Corr}(X,Y) &= -1 \text{ if } \beta < 0.\end{aligned}$$

*Proof.*

Define the random variable $Z$ as $Z = Y - rX$, where $r$ is a real number. We can write

$$\text{Var}(Z) = \text{Var}(Y) - 2r\text{Cov}(X,Y) + r^2\,\text{Var}(X) = h(r).$$

Viewing the variance of $Z$ as a function of $r$, we see that $h(r)$ is a quadratic with roots when

$$r = \frac{\text{Cov}(X,Y) \pm \sqrt{\text{Cov}(X,Y)^2 - \text{Var}(X)\,\text{Var}(Y)}}{\text{Var}(X)}. \tag{4.5}$$

As $h(r)$ is defined as a variance, $h(r) \geq 0$. This means that $h(r)$ has at most one real root. This implies that the term under the square root in expression (4.5) (the discriminant of the quadratic) must be at most zero,

$$\begin{aligned}\mathrm{Cov}(X,Y)^2 - \mathrm{Var}(X)\,\mathrm{Var}(Y) \le 0 &\Rightarrow \frac{\mathrm{Cov}(X,Y)^2}{\mathrm{Var}(X)\,\mathrm{Var}(Y)} \le 1 \\ &\Rightarrow |\mathrm{Corr}(X,Y)| \le 1.\end{aligned}$$

This completes the proof of the first part of the proposition. For the second part, we begin by observing that, if $Y = \alpha + \beta X$, then

$$\mathrm{Var}(Y) = \beta^2\,\mathrm{Var}(X) \quad \text{and} \quad \mathrm{Cov}(X,Y) = \beta\,\mathrm{Var}(X),$$

and thus,

$$\mathrm{Corr}(X,Y) = \frac{\beta\,\mathrm{Var}(X)}{\sqrt{\beta^2\,\mathrm{Var}(X)\,\mathrm{Var}(X)}} = \frac{\beta}{|\beta|} = \mathrm{sign}(\beta),$$

that is, $\mathrm{Corr}(X,Y) = 1$ if $\beta > 0$ and $\mathrm{Corr}(X,Y) = -1$ if $\beta < 0$. For the "only if" part, notice that $|\mathrm{Corr}(X,Y)| = 1$ implies that the polynomial $h(r)$ will have a (double) root, which is

$$r_0 = \frac{\mathrm{Cov}(X,Y)}{\mathrm{Var}(X)} \neq 0\,.$$

as $r_0 = 0$ would imply $\mathrm{Corr}(X,Y) = 0$ (in fact, $r_0$, must have the same sign as $\mathrm{Corr}(X,Y)$). Now recall that $h(r)$ was defined as the variance of random variable $Z = Y - rX$. If this has variance zero, it must always take the same value,

$$h(r_0) = \mathrm{Var}(Y - r_0X) = 0 \Leftrightarrow Y - r_0X = k\,,$$

for some constant $k$. We deduce that $Y = k + r_0X$, which completes the proof. □

If the correlation between $X$ and $Y$ is zero, we say that $X$ and $Y$ are **uncorrelated**. It is clear that $X$ and $Y$ are uncorrelated if and only if $\mathbb{E}(XY) = \mathbb{E}(X)\,\mathbb{E}(Y)$. In saying that $X$ and $Y$ are uncorrelated, we are simply making a statement about the lack of linear association between $X$ and $Y$. It is possible to have random variables that are uncorrelated but still dependent. The following example illustrates covariance and correlation calculations.

**Example 4.3.9** (Correlation and polynomial densities)
Consider random variables $X$ and $Y$ with joint density function $f_{X,Y}(x,y) = x + y$ for $0 < x < 1$ and $0 < y < 1$. We have shown in Example 4.2.10 that $f_X(x) = x + \frac{1}{2}$ for $0 < x < 1$. Thus,

$$\begin{aligned}\mathbb{E}(X) &= \int_{-\infty}^{\infty} x f_X(x)dx = \int_0^1 (x^2 + x/2)dx = [x^3/3 + x^2/4]_0^1 = 7/12 \\ \mathbb{E}(X^2) &= \int_{-\infty}^{\infty} x^2 f_X(x)dx = \int_0^1 (x^3 + x^2/2)dx = [x^4/4 + x^3/6]_0^1 = 5/12 \\ \mathrm{Var}(X) &= \mathbb{E}(X^2) - \mathbb{E}(X)^2 = 11/144.\end{aligned}$$

By symmetry, $\mathbb{E}(Y) = \frac{7}{12}$ and $\text{Var}(Y) = \frac{11}{144}$. In Example 4.3.2 we show that $\mathbb{E}(XY) = \frac{1}{3}$. Thus,

$$\text{Cov}(X,Y) = \mathbb{E}(XY) - \mathbb{E}(X)\,\mathbb{E}(Y) = -\frac{1}{144},$$
$$\text{Corr}(X,Y) = \frac{\text{Cov}(X,Y)}{\sqrt{\text{Var}(X)\,\text{Var}(Y)}} = -\frac{1}{11}.$$

### 4.3.3 *Joint moments*

Joint moments provide information about the dependence structure between two random variables. For most practical purposes we only consider joint moments of low order. An example of such a joint moment is the covariance.

**Definition 4.3.10** (Joint moments and joint central moments)
If $X$ and $Y$ are random variables, then the $(r,s)^{\text{th}}$ **joint moment** of $X$ and $Y$ is

$$\mu'_{r,s} = \mathbb{E}(X^r Y^s).$$

The $(r,s)^{\text{th}}$ **joint central moment** of $X$ and $Y$ is

$$\mu_{r,s} = \mathbb{E}[(X - \mathbb{E}(X))^r (Y - \mathbb{E}(Y))^s].$$

Many familiar quantities can be expressed as joint moments.

i. $r^{\text{th}}$ moment for $X$: $\mathbb{E}(X^r) = \mu'_{r,0}$.
ii. $r^{\text{th}}$ central moment for $X$: $\mathbb{E}[(X - \mu_X)^r] = \mu_{r,0}$.
iii. Covariance: $\text{Cov}(X,Y) = \mu_{1,1}$.
iv. Correlation: $\text{Corr}(X,Y) = \mu_{1,1}/\sqrt{\mu_{2,0}\mu_{0,2}}$.

Joint moments are evaluated using Proposition 4.3.1. We return to the simple polynomial density of Example 4.2.10 to illustrate.

**Example 4.3.11**
Consider random variables $X$ and $Y$ with joint density function $f_{X,Y}(x,y) = x + y$ for $0 < x \leq 1$ and $0 < y \leq 1$. Suppose we are interested in calculating the $(r,s)^{\text{th}}$ joint moment of $X$ and $Y$, that is, $\mathbb{E}(X^r Y^s)$. Using Proposition 4.3.1 with $g(X,Y) = X^r Y^s$

we have

$$
\begin{aligned}
\mathbb{E}(X^rY^s) &= \int_{-\infty}^{\infty}\int_{-\infty}^{\infty} x^r y^s f_{X,Y}(x,y)dxdy \\
&= \int_0^1\int_0^1 x^r y^s (x+y)dxdy \\
&= \int_0^1 \left[\frac{x^{r+2}}{r+2}y^s + \frac{x^{r+1}}{r+1}y^{s+1}\right]_{x=0}^{x=1} dy \\
&= \int_0^1 \left(\frac{1}{r+2}y^s + \frac{1}{r+1}y^{s+1}\right) dy \\
&= \left[\frac{1}{(r+2)(s+1)}y^{s+1} + \frac{1}{(r+1)(s+2)}y^{s+2}\right]_0^1 \\
&= \frac{1}{(r+2)(s+1)} + \frac{1}{(r+1)(s+2)}.
\end{aligned}
$$

### 4.3.4 *Joint moment-generating functions*

The joint moments of a distribution are encapsulated in the joint moment-generating function.

**Definition 4.3.12** (Joint moment-generating function)
For random variables $X$ and $Y$, the **joint moment-generating function** is defined as

$$M_{X,Y}(t,u) = \mathbb{E}(e^{tX+uY}).$$

The argument of the expectation operator in the definition of the joint moment-generating function can be written as the product of two series. Assuming we can swap the order of summations and expectations, we have

$$M_{X,Y}(t,u) = \mathbb{E}\left(e^{tX}e^{uY}\right) = \mathbb{E}\left(\sum_{i=0}^{\infty}\frac{(tX)^i}{i!}\sum_{j=0}^{\infty}\frac{(uY)^j}{j!}\right) = \sum_{i=0}^{\infty}\sum_{j=0}^{\infty}\mathbb{E}\left(X^iY^j\right)\frac{t^iu^j}{i!j!}.$$

Thus, the joint moment-generating function is a polynomial in $t$ and $u$. The $(r,s)^{\text{th}}$ joint moment is the coefficient of $(t^ru^s)/(r!s!)$ in the polynomial expansion of the joint moment-generating function. One consequence is that the joint moments can be evaluated by differentiation:

$$M_{X,Y}^{(r,s)}(0,0) = \left.\frac{d^{r+s}}{dt^r du^s}M_{X,Y}(t,u)\right|_{t=0,u=0} = \mathbb{E}(X^rY^s) = \mu'_{r,s}.$$

The marginal moment-generating functions can be recovered from the joint,

$$
\begin{aligned}
M_X(t) &= \mathbb{E}(e^{tX}) = M_{X,Y}(t,0), \\
M_Y(t) &= \mathbb{E}(e^{uY}) = M_{X,Y}(0,u).
\end{aligned}
$$

The joint cumulant-generating function is evaluated by taking the log of the joint moment-generating function.

**Definition 4.3.13** (Joint cumulant-generating function and joint cumulants)
For random variables $X$ and $Y$, the **joint cumulant-generating function** is

$$K_{X,Y}(t,u) = \log M_{X,Y}(t,u).$$

We define the $(r,s)^{\text{th}}$ joint cumulant $\kappa_{r,s}$ as the coefficient of $(t^r u^s)/(r!s!)$ in the expansion of $K_{X,Y}(t,u)$.

The joint cumulant-generating function can be written as a polynomial in $t$ and $u$,

$$K_{X,Y}(t,u) = \sum_{i=0}^{\infty}\sum_{j=0}^{\infty} \kappa_{i,j} \frac{t^i u^j}{i!j!}.$$

The cumulants can be evaluated by comparing coefficients or by evaluating derivatives at zero. Joint cumulants provide another means of calculating correlation.

**Proposition 4.3.14** (Correlation in terms of cumulants)
*For random variables $X$ and $Y$ with $(r,s)^{th}$ joint cumulant $\kappa_{r,s}$, the correlation between $X$ and $Y$ is given by*

$$\text{Corr}(X,Y) = \frac{\kappa_{1,1}}{\sqrt{\kappa_{2,0}\kappa_{0,2}}}.$$

*Proof.*

First we establish that $\kappa_{2,0} = \text{Var}(X)$. If we set $u = 0$ in $K_{X,Y}(t,u)$ we get

$$K_{X,Y}(t,0) = \log M_{X,Y}(t,0) = \log M_X(t) = K_X(t).$$

Thus,

$$\left.\frac{\partial^r}{\partial t^r} K_{X,Y}(t,u)\right|_{t=0,u=0} = \left.\frac{\partial^r}{\partial t^r} K_{X,Y}(t,0)\right|_{t=0} = \left.\frac{\partial^r}{\partial t^r} K_X(t)\right|_{t=0}.$$

We conclude that $\kappa_{r,0}$ is the $r^{\text{th}}$ cumulant of $X$ and, in particular $\kappa_{2,0} = \text{Var}(X)$. To complete the proof we need to show that $\kappa_{1,1} = \text{Cov}(X,Y)$. We know that

$$M_{X,Y}(t,u) = 1 + \mu'_{1,0}t + \mu'_{0,1}u + \mu'_{1,1}tu + \ldots$$

Taking partial derivatives yields

$$\frac{\partial}{\partial u} K_{X,Y}(t,u) = (\mu'_{0,1} + \mu'_{1,1}t + \ldots)/M_{X,Y}(t,u),$$

$$\frac{\partial^2}{\partial t \partial u} K_{X,Y}(t,u) = \frac{\mu'_{1,1} + \ldots}{M_{X,Y}(t,u)} - \frac{(\mu'_{0,1} + \mu'_{1,1}t + \ldots)(\mu'_{1,0} + \mu'_{1,1}u + \ldots)}{[M_{X,Y}(t,u)]^2}.$$

Evaluating the derivative at $t = 0$ and $u = 0$ gives

$$\kappa_{1,1} = \left.\frac{\partial^2}{\partial t \partial u} K_{X,Y}(t,u)\right|_{t=0,u=0} = \mu'_{1,1} - \mu'_{0,1}\mu'_{1,0}.$$

We conclude that

$$\kappa_{1,1} = \mathbb{E}(XY) - \mathbb{E}(X)\,\mathbb{E}(Y) = \mathrm{Cov}(X,Y).$$

□

We will see applications of joint moment- and cumulant-generating functions in sections 4.7 and 4.8.

**Exercise 4.3**

1. (**Marginal expectation**) Show that Proposition 4.3.1 implies $\mathbb{E}(X) = \int_{-\infty}^{\infty} x f_X(x)dx$. [You may just consider the continuous case.]
2. (**Expectation of a sum of random variables**) Show that Proposition 4.3.1 implies $\mathbb{E}(X+Y) = \mathbb{E}(X) + \mathbb{E}(Y)$. [You may consider just the continuous case.]
3. (**Forms of covariance**) Using the definition of covariance, that is,
$$\mathrm{Cov}(X,Y) = \mathbb{E}[(X - \mathbb{E}(X))(Y - \mathbb{E}(Y))],$$
show that $\mathrm{Cov}(X,Y) = \mathbb{E}(XY) - \mathbb{E}(X)\,\mathbb{E}(Y)$.
4. (**Properties of covariance**) Using the definition of covariance show that the following hold:
   i. Symmetry: $\mathrm{Cov}(X,Y) = \mathrm{Cov}(Y,X)$.
   ii. For constants $a$ and $b$ : $\mathrm{Cov}(aX,bY) = ab\mathrm{Cov}(X,Y)$.
   iii. Relationship with variance:
   $\mathrm{Var}(X) = \mathrm{Cov}(X,X)$,
   $\mathrm{Var}(X+Y) = \mathrm{Var}(X) + \mathrm{Var}(Y) + 2\mathrm{Cov}(X,Y)$,
   $\mathrm{Var}(X-Y) = \mathrm{Var}(X) + \mathrm{Var}(Y) - 2\mathrm{Cov}(X,Y)$.
   iv. Bilinearity:
   $\mathrm{Cov}(U+V,X+Y) = \mathrm{Cov}(U,X) + \mathrm{Cov}(U,Y) + \mathrm{Cov}(V,X) + \mathrm{Cov}(V,Y)$.
5. Consider the following game. We pay a stake of £5 to play. A coin is tossed three times. If the first and last throws are heads, we get £10 for each head thrown. What is the expected return from playing the game? Derive the moment-generating function of the return.
6. Consider random variables $X$ and $Y$ with joint density
$$f_{X,Y}(x,y) = \begin{cases} 8xy & \text{for } 0 < x < y < 1, \\ 0 & \text{otherwise.} \end{cases}$$
Evaluate $\mathrm{Corr}(X,Y)$.

## 4.4 Independent random variables

The term independent and identically distributed (IID) is one that is used with great frequency in statistics. One of the key assumptions that is often made in inference is that we have a random sample. Assuming a sample is random is equivalent to stating that a reasonable model for the process that generates the data is a sequence of independent and identically distributed random variables. We start by defining what it means for a pair of random variables to be independent.

### *4.4.1 Independence for pairs of random variables*

**Definition 4.4.1** (Independent random variables)
The random variables $X$ and $Y$ are **independent** if and only if the events $\{X \leq x\}$ and $\{Y \leq y\}$ are independent for all $x$ and $y$.

One immediate consequence of this definition is that, for independent random variables, it is possible to generate the joint distribution from the marginal distributions.

**Claim 4.4.2** (Joint distribution of independent random variables)
*Random variables $X$ and $Y$ are independent if and only if the joint cumulative distribution function of $X$ and $Y$ is the product of the marginal cumulative distribution functions, that is, if and only if*

$$F_{X,Y}(x, y) = F_X(x)F_Y(y) \quad \textit{for all } x, y \in \mathbb{R}.$$

The claim holds since, by Definition 4.4.1, the events $\{X \leq x\}$ and $\{Y \leq y\}$ are independent if and only if the probability of their intersection is the product of the individual probabilities. Claim 4.4.2 states that, for independent random variables, knowledge of the margins is equivalent to knowledge of the joint distribution; this is an attractive property. The claim can be restated in terms of mass or density.

**Proposition 4.4.3** (Mass/density of independent random variables)
*The random variables $X$ and $Y$ are independent if and only if their joint mass/density is the product of the marginal mass/density functions, that is, if and only if*

$$f_{X,Y}(x, y) = f_X(x)f_Y(y) \quad \textit{for all } x, y \in \mathbb{R}.$$

*Proof.*

In the continuous case, this result is an immediate consequence of differentiating the result from Claim 4.4.2. For the discrete case, this comes about since independence of $\{X \leq x\}$ and $\{Y \leq y\}$ implies independence of $\{X = x\}$ and $\{Y = y\}$. □

An immediate corollary is that for independent random variables the expectation of the product is the product of the expectations.

**Lemma 4.4.4** (Expectation of the product of independent random variables)
*If $X$ and $Y$ are independent random variables, then*

$$\mathbb{E}(XY) = \mathbb{E}(X)\,\mathbb{E}(Y).$$

Another way to state this lemma is

$$X \text{ and } Y \text{ independent} \Rightarrow X \text{ and } Y \text{ uncorrelated.}$$

Note that the implication here is one way. Uncorrelated random variables are not necessarily independent. There is one notable exception: if $X, Y$ are jointly normally distributed (more on this later) and uncorrelated, then they are also independent.

We will often be interested in the properties of functions of random variables that we know are independent. Provided these functions are reasonably well behaved, we can show that the random variables generated by functions operating on an independent collection of random variables are also independent. This property is summarised by the following lemma that we will not prove.

**Lemma 4.4.5** (Independence of functions of independent random variables)
*If $X$ and $Y$ are independent random variables, and $g$ and $h$ are well-behaved, real-valued functions, then $g(X)$ and $h(Y)$ are independent random variables.*

An immediate consequence of this lemma is that the joint moment-generating function for independent random variables can be factorised as the product of the marginal moment-generating functions. If $X$ and $Y$ are independent then $e^{tX}$ and $e^{uY}$ are also independent, so

$$M_{X,Y}(t,u) = \mathbb{E}\left(e^{tX+uY}\right) = \mathbb{E}\left(e^{tX}e^{uY}\right) = \mathbb{E}\left(e^{tX}\right)\mathbb{E}\left(e^{uY}\right) = M_X(t)M_Y(u). \quad (4.6)$$

We return once again to the simple polynomial density of Example 4.2.10.

**Example 4.4.6** (Independence for random variables with polynomial joint density)
Consider random variables $X$ and $Y$ with joint density given by $f_{X,Y}(x,y) = x + y$ for $0 < x < 1$ and $0 < y < 1$. We know, from above, that $f_X(x) = x + \frac{1}{2}$ for $0 < x < 1$ and $f_Y(y) = y + \frac{1}{2}$ for $0 < y \leq 1$. Thus, the product of the marginals is

$$f_X(x)f_Y(y) = xy + \frac{x+y}{2} + \frac{1}{4} \quad \text{for } 0 < x < 1,\ 0 < y < 1.$$

Clearly, $f_X(x)f_Y(y) \neq f_{X,Y}(x,y)$ so in this case $X$ and $Y$ are not independent. We could also have established that $X$ and $Y$ are not independent by noting that $\text{Corr}(X,Y) = -\frac{1}{11}$ (see Example 4.3.9).

### 4.4.2 *Mutual independence*

We can readily extend the ideas of this section to a sequence of $n$ random variables. When considering many random variables, the terms **pairwise independent** and **mutually independent** are sometimes used. Pairwise independent, as the name suggests, means that every pair is independent in the sense of Definition 4.4.1.

**Definition 4.4.7** (Mutually independent random variables)
The random variables $X_1, X_2, \ldots, X_n$ are **mutually independent** if and only if the events $\{X_1 \leq x_1\}, \{X_2 \leq x_2\}, \ldots, \{X_n \leq x_n\}$ are mutually independent for all choices of $x_1, x_2, \ldots, x_n$.

When $X_1, X_2, \ldots, X_n$ are mutually independent the term "mutually" is often dropped and we just say $X_1, X_2, \ldots, X_n$ are independent or $\{X_i\}$ is a sequence of independent random variables. Note that this is a stronger property than pairwise independence; mutually independent implies pairwise independent but the reverse implication does not hold.

Any one of the equivalent statements summarised in the following claim could be taken to be a definition of independence.

**Claim 4.4.8** (Equivalent statements of mutual independence)
*If $X_1, \ldots, X_n$ are random variables, the following statements are equivalent:*

i. *The events $\{X_1 \leq x_1\}, \{X_2 \leq x_2\}, \ldots, \{X_n \leq x_n\}$ are independent for all $x_1, \ldots, x_n$.*
ii. $F_{X_1,\ldots,X_n}(x_1, \ldots, x_n) = F_{X_1}(x_1)F_{X_2}(x_2)\ldots F_{X_n}(x_n)$ *for all* $x_1, \ldots, x_n$.
iii. $f_{X_1,\ldots,X_n}(x_1, \ldots, x_n) = f_{X_1}(x_1)f_{X_2}(x_2)\ldots f_{X_n}(x_n)$ *for all* $x_1, \ldots, x_n$.

The implications of mutual independence may be summarised as follows.

**Claim 4.4.9** (Implications of mutual independence)
*If $X_1, \ldots, X_n$ are mutually independent random variables, then*

i. $\mathbb{E}(X_1 X_2 \ldots X_n) = \mathbb{E}(X_1)\,\mathbb{E}(X_2)\ldots\mathbb{E}(X_n)$,
ii. *if, in addition, $g_1, \ldots, g_n$ are well-behaved, real-valued functions, then the random variables $g_1(X_1), \ldots, g_n(X_n)$ are also mutually independent.*

### 4.4.3 *Identical distributions*

Another useful simplifying assumption is that of identical distributions.

**Definition 4.4.10** (Identically distributed random variables)
The random variables $X_1, X_2, \ldots, X_n$ are **identically distributed** if and only if their cumulative distribution functions are identical, that is

$$F_{X_1}(x) = F_{X_2}(x) = \ldots = F_{X_n}(x) \quad \text{for all } x \in \mathbb{R}.$$

If $X_1, X_2, \ldots, X_n$ are identically distributed we will often just use the letter $X$ to denote a random variable that has the distribution common to all of them. So the cumulative distribution function of $X$ is $\mathrm{P}(X \leq x) = F_X(x) = F_{X_1}(x) = \ldots = F_{X_n}(x)$. If $X_1, X_2, \ldots, X_n$ are independent and identically distributed, we may sometimes denote this as $\{X_i\} \sim \mathrm{IID}$.

**Exercise 4.4**

1. Suppose $X_1, \ldots, X_n$ is a sequence of $n$ independent and identically distributed standard normal random variables. Find an expression for the joint density of $X_1, \ldots, X_n$. [We denote this by $\{X_i\} \sim \text{NID}(0, 1)$, where NID stands for "normal and independently distributed".]
2. Let $X_1, \ldots, X_n$ be a sequence of $n$ independent random variables with cumulant-generating functions $K_{X_1}, \ldots, K_{X_n}$. Find an expression for the joint cumulant-generating function $K_{X_1,\ldots,X_n}$ in terms of the individual cumulant-generating functions.

## 4.5 Random vectors and random matrices

In this section we introduce notation that is used in the discussion of transformations of random variables (see section 4.6). We will continue to use this notation in the remainder of the text.

**Definition 4.5.1** (Random vector and random matrix)
An $n \times 1$ **random vector** $\boldsymbol{X}$ is an $n \times 1$ vector whose entries are random variables,

$$\boldsymbol{X} = \begin{pmatrix} X_1 \\ X_2 \\ \vdots \\ X_n \end{pmatrix}. \tag{4.7}$$

An $n \times n$ **random matrix** $\boldsymbol{W}$ is an $n \times n$ matrix whose entries are random variables,

$$\boldsymbol{W} = \begin{pmatrix} W_{1,1} & W_{1,2} & \cdots & W_{1,n} \\ W_{2,1} & W_{2,2} & & \vdots \\ \vdots & & \ddots & \vdots \\ W_{n,1} & \cdots & \cdots & W_{n,n} \end{pmatrix}. \tag{4.8}$$

We will regularly use random vectors as a means of simplifying notation. The cumulative distribution function of a random vector $\boldsymbol{X}$ is simply the joint cumulative distribution function of the random variables that make up $\boldsymbol{X}$. The same is true for the mass/density and moment-generating function. If $\boldsymbol{x} = (x_1, \ldots, x_n)^T$ and $\boldsymbol{t} = (t_1, \ldots, t_n)^T$ then

$$\begin{aligned} F_{\boldsymbol{X}}(\boldsymbol{x}) &= F_{X_1,\ldots,X_n}(x_1, \ldots, x_n) \\ f_{\boldsymbol{X}}(\boldsymbol{x}) &= f_{X_1,\ldots,X_n}(x_1, \ldots, x_n) \\ M_{\boldsymbol{X}}(\boldsymbol{t}) &= \mathbb{E}(e^{\boldsymbol{t}.\boldsymbol{X}}) = M_{X_1,\ldots,X_n}(t_1, \ldots, t_n) \end{aligned}$$

The expectation operator acts on non-scalar arguments element by element so, if $\boldsymbol{X}$

and $W$ are the random vector and random matrix given by equations (4.7) and (4.8), respectively,

$$\mathbb{E}(X) = \begin{pmatrix} \mathbb{E}(X_1) \\ \mathbb{E}(X_2) \\ \vdots \\ \mathbb{E}(X_n) \end{pmatrix} \quad \text{and} \quad \mathbb{E}(W) = \begin{pmatrix} \mathbb{E}(W_{1,1}) & \mathbb{E}(W_{1,2}) & \cdots & \mathbb{E}(W_{1,n}) \\ \mathbb{E}(W_{2,1}) & \mathbb{E}(W_{2,2}) & & \vdots \\ \vdots & & \ddots & \vdots \\ \mathbb{E}(W_{n,1}) & \cdots & \cdots & \mathbb{E}(W_{n,n}) \end{pmatrix}.$$

In general, if $g$ is a well-behaved function on $\mathbb{R}^n$, then $\mathbb{E}(g(X))$ is of the same dimension as $g(X)$ with

$$\mathbb{E}[g(X)] = \int_{\mathbb{R}^n} g(x) f_X(x) dx.$$

Note here that $\int_{\mathbb{R}^n} \ldots dx$ is used to denote $n$ integrals,

$$\int_{\mathbb{R}^n} g(x) f_X(x) dx = \int_{-\infty}^{\infty} \cdots \int_{-\infty}^{\infty} g(x_1, \ldots, x_n) f_{X_1,\ldots,X_n}(x_1, \ldots, x_n) dx_1 \ldots dx_n.$$

An important distinction between the scalar and the vector cases is that the variance of a random vector is a matrix.

**Definition 4.5.2** (Variance of a random vector)
Suppose that $X$ is an $n \times 1$ random vector. The variance of $X$ is given by

$$\text{Var}(X) = \mathbb{E}\left[(X - \mathbb{E}(X))(X - \mathbb{E}(X))^T\right],$$

an $n \times n$ matrix of real numbers.

Taking element-by-element expectations in the matrix $(X - \mathbb{E}(X))(X - \mathbb{E}(X))^T$ we can show that

$$\text{Var}(X) = \begin{pmatrix} \text{Var}(X_1) & \text{Cov}(X_2, X_1) & \cdots & \text{Cov}(X_n, X_1) \\ \text{Cov}(X_1, X_2) & \text{Var}(X_2) & & \vdots \\ \vdots & & \ddots & \vdots \\ \text{Cov}(X_1, X_n) & \cdots & \cdots & \text{Var}(X_n) \end{pmatrix}.$$

For this reason the variance of a random vector is usually referred to as the **covariance matrix**. Filling in the details of the derivation is part of Exercise 4.5. The covariance matrix is always symmetric. If the entries in $X$ are uncorrelated, then $\text{Var}(X)$ is a diagonal matrix. If the entries are uncorrelated and identically distributed, we can write $\text{Var}(X) = \sigma^2 I_n$ where $\sigma^2 = \text{Var}(X_1)$ is the common variance and $I_n$ is the $n \times n$ identity matrix.

For a scalar variable the variance is non-negative, that is, $\text{Var}(X) \geq 0$. In order to make a similar statement about the covariance matrix, we must first define a matrix version of what it means for a quantity to be non-negative.

**Definition 4.5.3** (Positive-definite and non-negative-definite matrices)
Let $\boldsymbol{A}$ be an $n \times n$ matrix.

i. $\boldsymbol{A}$ is **positive definite** if $\boldsymbol{b}^T \boldsymbol{A} \boldsymbol{b} > 0$ for all $\boldsymbol{b} \in \mathbb{R}^n$.
ii. $\boldsymbol{A}$ is **non-negative definite** (also called **positive semidefinite**) if $\boldsymbol{b}^T \boldsymbol{A} \boldsymbol{b} \geq 0$ for all $\boldsymbol{b} \in \mathbb{R}^n$.

We use this definition to describe an important property of covariance matrices.

**Proposition 4.5.4** (Covariance matrix is non-negative definite)
*If $\boldsymbol{X}$ is a random vector and $\boldsymbol{\Sigma} = \text{Var}(\boldsymbol{X})$ then $\boldsymbol{\Sigma}$ is a non-negative-definite matrix.*

*Proof.*

Suppose $\boldsymbol{X}$ is $n \times 1$ so $\boldsymbol{\Sigma}$ is $n \times n$. For any $\boldsymbol{b} \in \mathbb{R}^n$ we can construct a scalar random variable $Y = \boldsymbol{b}^T \boldsymbol{X}$. We know from Proposition 3.4.8 that $\text{Var}(Y) \geq 0$. From the definition of $Y$ and $\text{Var}(\boldsymbol{X})$ we have

$$\begin{aligned}
\text{Var}(Y) &= \mathbb{E}[(Y - \mathbb{E}(Y))^2] = \mathbb{E}[(\boldsymbol{b}^T \boldsymbol{X} - \mathbb{E}(\boldsymbol{b}^T \boldsymbol{X}))^2] \\
&= \mathbb{E}[(\boldsymbol{b}^T \boldsymbol{X} - \mathbb{E}(\boldsymbol{b}^T \boldsymbol{X}))(\boldsymbol{b}^T \boldsymbol{X} - \mathbb{E}(\boldsymbol{b}^T \boldsymbol{X}))^T] \\
&= \mathbb{E}[\boldsymbol{b}^T (\boldsymbol{X} - \mathbb{E}(\boldsymbol{X}))(\boldsymbol{X} - \mathbb{E}(\boldsymbol{X}))^T \boldsymbol{b}] \\
&= \boldsymbol{b}^T\, \mathbb{E}[(\boldsymbol{X} - \mathbb{E}(\boldsymbol{X}))(\boldsymbol{X} - \mathbb{E}(\boldsymbol{X}))^T] \boldsymbol{b} \\
&= \boldsymbol{b}^T\, \text{Var}(\boldsymbol{X}) \boldsymbol{b} = \boldsymbol{b}^T \boldsymbol{\Sigma} \boldsymbol{b}.
\end{aligned}$$

Thus $\boldsymbol{b}^T \boldsymbol{\Sigma} \boldsymbol{b} \geq 0$ for all $\boldsymbol{b} \in \mathbb{R}^n$. □

For random vectors $\boldsymbol{X}$ and $\boldsymbol{Y}$ the covariance matrix is defined as

$$\text{Cov}(\boldsymbol{X}, \boldsymbol{Y}) = \mathbb{E}[(\boldsymbol{X} - \mathbb{E}(\boldsymbol{X}))(\boldsymbol{Y} - \mathbb{E}(\boldsymbol{Y}))^T].$$

The matrix $\text{Cov}(\boldsymbol{X}, \boldsymbol{Y})$ is not generally symmetric. In fact, it need not be a square matrix; if $\boldsymbol{X}$ is $n \times 1$ and $\boldsymbol{Y}$ is $m \times 1$, then $\text{Cov}(\boldsymbol{X}, \boldsymbol{Y})$ is $n \times m$. Using the fact that expectation works element by element on matrices, we can show that

$$\text{Cov}(\boldsymbol{Y}, \boldsymbol{X}) = \text{Cov}(\boldsymbol{X}, \boldsymbol{Y})^T.$$

Proving this result is part of Exercise 4.5

**Notation 4.5.5** (Scalar- and vector-valued functions)
We use bold face to denote functions whose return value is a vector. For example, if $\boldsymbol{g} : \mathbb{R}^n \to \mathbb{R}^m$, then $\boldsymbol{g}(\boldsymbol{x}) \in \mathbb{R}^m$ for all $\boldsymbol{x} \in \mathbb{R}^n$. We do not extend this convention to $\mathbb{E}$, Var, and Cov.

**Exercise 4.5**

1. (**Covariance matrix**) Suppose that $\boldsymbol{X}$ is a random vector. Using the definition of the covariance matrix $\text{Var}(\boldsymbol{X}) = \mathbb{E}\left[(\boldsymbol{X} - \mathbb{E}(\boldsymbol{X}))(\boldsymbol{X} - \mathbb{E}(\boldsymbol{X}))^T\right]$, show that

$$\text{Var}(\boldsymbol{X}) = \begin{pmatrix} \text{Var}(X_1) & \text{Cov}(X_2, X_1) & \cdots & \text{Cov}(X_n, X_1) \\ \text{Cov}(X_1, X_2) & \text{Var}(X_2) & & \vdots \\ \vdots & & \ddots & \vdots \\ \text{Cov}(X_1, X_n) & \cdots & \cdots & \text{Var}(X_n) \end{pmatrix}.$$

2. (**Covariance matrix of a random vector**) Suppose that we define random vectors $\boldsymbol{X} = (X_1, \dots, X_m)^T$ and $\boldsymbol{Y} = (Y_1, \dots, Y_n)^T$.
   (a) Give an expression for the entries in $\text{Cov}(\boldsymbol{X}, \boldsymbol{Y})$.
   (b) Show that $\text{Cov}(\boldsymbol{Y}, \boldsymbol{X}) = \text{Cov}(\boldsymbol{X}, \boldsymbol{Y})^T$.

## 4.6 Transformations of continuous random variables

Recall from subsection 3.6.2 that if $Y$ is given by a strictly monotone function of a continuous random variable $X$, we can derive an expression for the density of $Y$ in terms of the density of $X$. We extend these ideas to functions of several continuous random variables, starting as usual with the bivariate case. We require a standard definition first.

**Definition 4.6.1** (One-to-one and onto)
Consider a function $\boldsymbol{g}$ with domain $D$ and range $R$. We say that $\boldsymbol{g}$ is:

i. **one-to-one** if $\boldsymbol{g}(\boldsymbol{x}_1) = \boldsymbol{g}(\boldsymbol{x}_2) \Rightarrow \boldsymbol{x}_1 = \boldsymbol{x}_2$ for all $\boldsymbol{x}_1, \boldsymbol{x}_2 \in D$,
ii. **onto** if for all $\boldsymbol{y} \in R$ we can find $\boldsymbol{x} \in D$ such that $\boldsymbol{y} = \boldsymbol{g}(\boldsymbol{x})$.

### 4.6.1 *Bivariate transformations*

We are interested in transforming one pair of random variables into another pair of random variables. Consider pairs of random variables $(U, V)$ and $(X, Y)$. Suppose that $X$ and $Y$ are both functions of $U$ and $V$, say

$$\begin{aligned} X &= g_1(U, V), \\ Y &= g_2(U, V). \end{aligned} \tag{4.9}$$

We will make extensive use of the inverse transformation,

$$\begin{aligned} U &= h_1(X, Y), \\ V &= h_2(X, Y). \end{aligned}$$

We label the overall transformation as $\boldsymbol{g}$, so $(X, Y) = \boldsymbol{g}(U, V)$, and the inverse as $\boldsymbol{h}$, so $(U, V) = \boldsymbol{g}^{-1}(X, Y) = \boldsymbol{h}(X, Y)$. Suppose $\boldsymbol{g}$ is a well-behaved function $\boldsymbol{g} : \mathbb{R}^2 \to \mathbb{R}^2$.

More specifically, suppose that, for domain $D \subseteq \mathbb{R}^2$, the function $\boldsymbol{g}$ is a one-to-one map onto range $R \subseteq \mathbb{R}^2$. Then, if $(U,V)$ is a pair of continuous random variables with support $D$ and $(X,Y) = \boldsymbol{g}(U,V)$, the joint density of $X$ and $Y$ is

$$f_{X,Y}(x,y) = \begin{cases} f_{U,V}(\boldsymbol{h}(x,y))|J_{\boldsymbol{h}}(x,y)| & \text{for } (x,y) \in R, \\ 0 & \text{otherwise.} \end{cases} \tag{4.10}$$

This is often referred to as the **change-of-variables formula**. Here, $J_{\boldsymbol{h}}(x,y)$ is the **Jacobian** of the inverse transformation, defined as

$$\begin{aligned} J_{\boldsymbol{h}}(x,y) &= \begin{vmatrix} \frac{\partial}{\partial x} h_1(x,y) & \frac{\partial}{\partial x} h_2(x,y) \\ \frac{\partial}{\partial y} h_1(x,y) & \frac{\partial}{\partial y} h_2(x,y) \end{vmatrix} \\ &= \frac{\partial}{\partial x} h_1(x,y) \frac{\partial}{\partial y} h_2(x,y) - \frac{\partial}{\partial x} h_2(x,y) \frac{\partial}{\partial y} h_1(x,y). \end{aligned}$$

The Jacobian can be expressed in terms of the Jacobian of the original transformation, which is

$$J_{\boldsymbol{g}}(u,v) = \begin{vmatrix} \frac{\partial}{\partial u} g_1(u,v) & \frac{\partial}{\partial u} g_2(u,v) \\ \frac{\partial}{\partial v} g_1(u,v) & \frac{\partial}{\partial v} g_2(u,v) \end{vmatrix}.$$

It holds that

$$J_{\boldsymbol{h}}(x,y) = [J_{\boldsymbol{g}}(h_1(x,y), h_2(x,y))]^{-1}.$$

There are a number of equivalent statements for the change-of-variables formula; for $(x,y)$ in the range of $\boldsymbol{g}$,

$$\begin{aligned} f_{X,Y}(x,y) &= f_{U,V}(h_1(x,y), h_2(x,y))|J_{\boldsymbol{h}}(x,y)| \\ &= f_{U,V}(\boldsymbol{g}^{-1}(x,y)) \left|[J_{\boldsymbol{g}}(\boldsymbol{g}^{-1}(x,y))]^{-1}\right| \\ &= f_{U,V}(g_1^{-1}(x,y), g_2^{-1}(x,y)) \left|[J_{\boldsymbol{g}}(g_1^{-1}(x,y), g_2^{-1}(x,y))]^{-1}\right|. \end{aligned}$$

The change-of-variables formula may seem a little bewildering at first glance. However, questions involving a change of variables can be answered in a mechanical fashion.

1. Determine the transformation: find $g_1$ and $g_2$ such that

$$\begin{aligned} x &= g_1(u,v), \\ y &= g_2(u,v). \end{aligned}$$

   This is usually clear from the question. Make sure you specify the domain and range of this transformation.

2. Invert the transformation: find $h_1$ and $h_2$ such that

$$\begin{aligned} u &= h_1(x,y), \\ v &= h_2(x,y). \end{aligned}$$

3. Compute the Jacobian: find the partial derivatives of $h_1$ and $h_2$ with respect to $x$ and $y$ and hence calculate

$$J_{\boldsymbol{h}}(x,y) = \begin{vmatrix} \frac{\partial}{\partial x} h_1(x,y) & \frac{\partial}{\partial x} h_2(x,y) \\ \frac{\partial}{\partial y} h_1(x,y) & \frac{\partial}{\partial y} h_2(x,y) \end{vmatrix}.$$

   Remember that it is the absolute value of the Jacobian, $|J_{\boldsymbol{h}}(x,y)|$, that is used in the formula.

4. Construct the joint density of $X$ and $Y$,

$$f_{X,Y}(x,y) = f_{U,V}(h_1(x,y), h_2(x,y))|J_h(x,y)|.$$

   Make sure that you specify the support for this joint density.

We illustrate this procedure with a case that arises frequently in practice.

**Example 4.6.2** (Product of two continuous random variables)
Suppose that $U$ and $V$ are continuous random variables with joint density $f_{U,V}$. We would like an expression for the density of their product, $UV$, in terms of $f_{U,V}$. We define a bivariate transformation $\boldsymbol{g} : \mathbb{R}^2 \to \mathbb{R}^2$ by

$$g_1(u,v) = uv,$$
$$g_2(u,v) = v.$$

Let $X = g_1(U,V) = UV$ and $Y = g_2(U,V) = V$. We will go through the steps described above to derive the joint density of $X$ and $Y$.

1. The transformation can be written

$$x = uv,$$
$$y = v.$$

   If the domain of this transformation is the whole of $\mathbb{R}^2$, then the range is also the whole of $\mathbb{R}^2$.

2. The inverse transformation $\boldsymbol{h}$ is defined by

$$u = x/y,$$
$$v = y.$$

3. The Jacobian is

$$J_{\boldsymbol{h}}(x,y) = \begin{vmatrix} \frac{1}{y} & 0 \\ -\frac{x}{y^2} & 1 \end{vmatrix} = \frac{1}{y}.$$

   The absolute value of the Jacobian is $\frac{1}{|y|}$.

4. The joint density of $X$ and $Y$ is

$$f_{X,Y}(x,y) = f_{U,V}(x/y, y)\frac{1}{|y|}.$$

We are actually interested in the density of the product, that is, the density of $X$. We can calculate this by integrating with respect to $y$ in the usual way,

$$\begin{aligned} f_X(x) &= \int_{-\infty}^{\infty} f_{X,Y}(x,y)dy \\ &= \int_{-\infty}^{\infty} f_{U,V}(x/y,y)\frac{1}{|y|}dy. \end{aligned}$$

**Notation 4.6.3**

We use $\boldsymbol{g}$ and $\boldsymbol{h}$ to denote a transformation and its inverse. These functions are often treated as implicit; given the inverse transformation $u = h_1(x,y)$ and $v = h_2(x,y)$, the Jacobian may be written as,

$$J_{\boldsymbol{h}}(x,y) = \begin{vmatrix} \partial u/\partial x & \partial v/\partial x \\ \partial u/\partial y & \partial v/\partial y \end{vmatrix}.$$

This provides a useful way of remembering the Jacobian. In this context, we are viewing $u$ and $v$ as functions of $x$ and $y$.

### 4.6.2 *Multivariate transformations*

We now consider transformations of $n$ random variables. We will use the random-vector notation established in section 4.5. Let $\boldsymbol{X} = (X_1,\dots,X_n)^T$ be a continuous random vector and let $\boldsymbol{g} : \mathbb{R}^n \to \mathbb{R}^n$ be a well-behaved function. In fact, we will assume that, if $D \subseteq \mathbb{R}^n$ is the support of $\boldsymbol{X}$, then $\boldsymbol{g}$ is a one-to-one mapping from $D$ onto the range $R \subseteq \mathbb{R}^n$. As before, we will make extensive use of the inverse transformation $\boldsymbol{h}(\boldsymbol{y}) = \boldsymbol{g}^{-1}(\boldsymbol{y})$ and, on occasion, consider individual components of vectors,

$$\begin{aligned} \boldsymbol{x} &= (x_1,\dots,x_n)^T, \\ \boldsymbol{g}(\boldsymbol{x}) &= (g_1(\boldsymbol{x}),\dots,g_n(\boldsymbol{x}))^T, \end{aligned}$$

and so on. Note here that, for $j = 1,\dots,n$, each $g_j$ is a function of $n$ variables, $g_j : \mathbb{R}^n \to \mathbb{R}$, so we could write

$$\boldsymbol{g}(\boldsymbol{x}) = (g_1(x_1,\dots,x_n),\dots,g_n(x_1,\dots,x_n))^T.$$

Now define a random vector $\boldsymbol{Y}$ by $\boldsymbol{Y} = \boldsymbol{g}(\boldsymbol{X})$. The density of $\boldsymbol{Y}$ is given by

$$f_{\boldsymbol{Y}}(\boldsymbol{y}) = \begin{cases} f_{\boldsymbol{X}}(\boldsymbol{h}(\boldsymbol{Y}))|J_{\boldsymbol{h}}(\boldsymbol{y})| & \text{for } \boldsymbol{y} \in R, \\ 0 & \text{otherwise.} \end{cases} \tag{4.11}$$

The Jacobian is defined as

$$J_{\boldsymbol{h}}(\boldsymbol{y}) = \left|\frac{\partial}{\partial \boldsymbol{y}}\boldsymbol{h}(\boldsymbol{y})\right| = \begin{vmatrix} \frac{\partial}{\partial y_1}h_1(\boldsymbol{y}) & \frac{\partial}{\partial y_1}h_2(\boldsymbol{y}) & \cdots & \frac{\partial}{\partial y_1}h_n(\boldsymbol{y}) \\ \frac{\partial}{\partial y_2}h_1(\boldsymbol{y}) & \frac{\partial}{\partial y_2}h_2(\boldsymbol{y}) & & \vdots \\ \vdots & & \ddots & \vdots \\ \frac{\partial}{\partial y_n}h_1(\boldsymbol{y}) & \cdots & \cdots & \frac{\partial}{\partial y_n}h_n(\boldsymbol{y}) \end{vmatrix}.$$

The Jacobian can also be written as

$$J_h(y) = [J_g(h(y))]^{-1},$$

where $J_g(x) = \left|\frac{\partial}{\partial x} g(x)\right|$. We illustrate with an example.

**Example 4.6.4** (Linear transformation)
A simple function of a random vector that arises frequently in practice is the linear transformation,

$$Y = AX.$$

We usually insist that $A$ is non-singular, that is, $|A| \neq 0$. This condition ensures that the inverse, $A^{-1}$, is well defined. We can readily verify that $J_g(x) = |A|$ and thus $J_h(y) = |A|^{-1}$. We conclude that

$$f_Y(y) = \frac{1}{||A||} f_X(A^{-1}y),$$

where $||A||$ is the absolute value of the determinant of $A$.

**Exercise 4.6**

1. (**Ratio of two random variables**) Suppose that $U$ and $V$ are continuous random variables with joint density $f_{U,V}$. Show that the density of the ratio $X = U/V$ is given by $f_X(x) = \int_{-\infty}^{\infty} f_{U,V}(xy, y)|y|dy$.
2. Let $U$ and $V$ be independent and identically distributed exponential random variables. Find the density of $U/V$.
3. Give an alternative expression for the density of a product of two random variables.
4. Suppose that $X = (X_1, \dots, X_n)^T$ is a vector of independent and identically distributed standard normal random variables, and that $A$ is an $n \times n$ non-singular matrix. Give an expression for the density of $Y$, where $Y = AX$.

## 4.7 Sums of random variables

In many practical situations, the natural model for a quantity of interest is a sum of random variables. Consider the following illustrations.

1. Suppose that in a given season $n$ hurricanes develop in the Atlantic Basin. Each has probability $p$ of making landfall, independent of all other hurricanes. If $Y$ is the total number of hurricanes making landfall in the season, we could write $Y = \sum_{j=1}^{n} X_j$ where $\{X_j\}$ is a sequence of independent Bernoulli($p$) random variables.
2. We take measurements of total December rainfall at 5 sites across the UK. If the random variable $X_i$ represents our model for the total December rainfall at site $i$, then the mean total rainfall across locations is $\bar{X} = \frac{1}{5}\sum_{j=1}^{5} X_j$. A key component of the calculation of this mean is a sum of random variables.

We start by considering the sum of a pair of random variables.

### 4.7.1 Sum of two random variables

We already have two results for the sum of a pair of random variables from Corollary 4.3.3 and Claim 4.3.6. If $X$ and $Y$ are random variables then

$$\begin{aligned}\mathbb{E}(X+Y) &= \mathbb{E}(X) + \mathbb{E}(Y),\\ \text{Var}(X+Y) &= \text{Var}(X) + 2\text{Cov}(X,Y) + \text{Var}(Y).\end{aligned}$$

In fact, using the linearity of expectation and the binomial expansion, the $r^{\text{th}}$ moment of the sum of two random variables is

$$\mathbb{E}[(X+Y)^r] = \sum_{j=0}^{r} \binom{r}{j} \mathbb{E}(X^j Y^{r-j}).$$

We can readily derive the mass or density function for a sum of two random variables.

**Proposition 4.7.1** (Mass/density for the sum of two random variables)
*Let $X$ and $Y$ be random variables with joint mass/density given by $f_{X,Y}$. If $Z = X+Y$ then the mass/density of $Z$ is*

$$f_Z(z) = \begin{cases} \sum_u f_{X,Y}(u, z-u) & \textit{(discrete case)}\\ \int_{-\infty}^{\infty} f_{X,Y}(u, z-u)du & \textit{(continuous case).}\end{cases}$$

*Proof.*

In the continuous case the result is a direct consequence of the change-of-variables formula (4.10). Working out the details is part of Exercise 4.7. In the discrete case, consider the event $\{Z = z\}$. By definition, this is identical to $\{X+Y = z\}$. If $X$ takes any value on its support, say $X = u$, then we must have $Y = z - u$. Thus,

$$\{X+Y = z\} = \bigcup_u \{X = u, Y = z-u\}.$$

Since $X$ and $Y$ are discrete this is a countable union and, by construction, events of the form $\{X = u, Y = z-u\}$ are disjoint for different values of $u$. We conclude that

$$f_Z(z) = \text{P}(Z = z) = \text{P}(X+Y = z) = \sum_u \text{P}(X = u, Y = z-u) = \sum_u f_{X,Y}(u, z-u).$$

□

When $X$ and $Y$ are independent, we can exploit the factorisation of the mass or density function.

**Corollary 4.7.2** (Mass/density for the sum of two independent random variables)
*If $X$ and $Y$ are independent random variables and $Z = X+Y$, then the mass or density of $Z$ is given by*

$$f_Z(z) = \begin{cases} \sum_u f_X(u) f_Y(z-u) & \textit{(discrete case)}\\ \int_{-\infty}^{\infty} f_X(u) f_Y(z-u)du & \textit{(continuous case).}\end{cases}$$

The operation of combining two functions defined by this corollary is referred to as **convolution**. The convolution of two functions is denoted using $*$, thus,

$$f_Z = f_X * f_Y \Leftrightarrow f_Z(z) = \int_{-\infty}^{\infty} f_X(u) f_Y(z-u) du.$$

Convolution is commutative, so $f_X * f_Y = f_Y * f_X$.

Many problems involving sums of random variables are greatly simplified by the use of moment-generating functions. If $X$ and $Y$ are independent, the mass or density function of the sum is given by a convolution integral, which may be difficult to evaluate. However, the moment-generating function is simply the product of the marginal moment-generating functions.

**Proposition 4.7.3** (MGF of the sum of two independent random variables)
*If $X$ and $Y$ are independent random variables and $Z = X + Y$, then the moment- and cumulant-generating function of $Z$ are given by*

$$\begin{aligned} M_Z(t) &= M_X(t) M_Y(t), \\ K_Z(t) &= K_X(t) + K_Y(t). \end{aligned}$$

*Proof.*

As with many moment-generating function results, we proceed directly from the definition and exploit the independence of $X$ and $Y$. We have

$$\begin{aligned} M_Z(t) &= \mathbb{E}(e^{tZ}) = \mathbb{E}(e^{t(X+Y)}) = \mathbb{E}(e^{tX} e^{tY}) \\ &= \mathbb{E}(e^{tX})\,\mathbb{E}(e^{tY}) && \text{by independence} \\ &= M_X(t) M_Y(t). \end{aligned}$$

The result for the cumulant-generating function follows directly by taking logs. □

**Example 4.7.4** (Sum of independent exponentials)
Suppose that $X$ and $Y$ are independent exponentially distributed random variables, $X \sim \text{Exp}(\lambda)$ and $Y \sim \text{Exp}(\theta)$ with $\lambda \neq \theta$. Let $Z = X + Y$. We can work out the density of $Z$ using Corollary 4.7.2,

$$\begin{aligned} f_Z(z) &= \int_{-\infty}^{\infty} f_X(u) f_Y(z-u) du \\ &= \int_0^z \lambda e^{-\lambda u} \theta e^{-\theta(z-u)} du && \text{substituting exponential density} \\ &= \lambda\theta e^{-\theta z} \int_0^z e^{-(\lambda-\theta)u} du && \text{rearranging} \\ &= \lambda\theta e^{-\theta z} \left[ -\frac{1}{\lambda-\theta} e^{-(\lambda-\theta)u} \right]_0^z \\ &= \frac{\lambda\theta}{\lambda-\theta}(e^{-\theta z} - e^{-\lambda z}), && \text{for } 0 \leq z < \infty. \end{aligned}$$

**Example 4.7.5** (Sum of independent normals)
Suppose that $X$ and $Y$ are independent normal random variables with $X \sim \text{N}(\mu_1, \sigma_1^2)$ and $Y \sim \text{N}(\mu_2, \sigma_2^2)$. If $Z = X + Y$ then we can show that $Z \sim \text{N}(\mu_1 + \mu_2, \sigma_1^2 + \sigma_2^2)$. From Example 3.6.7, we know that the cumulant-generating functions of $X$ and $Y$ are

$$K_X(t) = \mu_1 t + \sigma_1^2 t^2/2,$$
$$K_Y(t) = \mu_2 t + \sigma_2^2 t^2/2.$$

By Proposition 4.7.3,

$$K_Z(t) = K_X(t) + K_Y(t) = (\mu_1 + \mu_2)t + (\sigma_1^2 + \sigma_2^2)t^2/2.$$

Thus, by uniqueness of cumulant-generating functions, $Z \sim \text{N}(\mu_1 + \mu_2, \sigma_1^2 + \sigma_2^2)$.

### 4.7.2 *Sum of n independent random variables*

Consider a sequence of $n$ independent random variables $X_1, \dots, X_n$. Let $S$ be the sum, that is,

$$S = X_1 + X_2 + \ldots + X_n = \sum_{j=1}^{n} X_j.$$

The mass or density function of $S$ is the $n$-fold convolution of the marginal mass or density functions,

$$f_s = f_{X_1} * f_{X_2} * \ldots * f_{X_n}.$$

In general, evaluating this convolution requires repeated summation or integration. Moment-generating functions provide a useful alternative. We can readily show that the moment-generating function of the sum is the product of the marginal moment-generating functions,

$$M_S(t) = M_{X_1}(t) M_{X_2}(t) \ldots M_{X_n}(t).$$

Working out the details is part of Exercise 4.7. Note that this result requires independence of $X_1, \dots, X_n$. If, in addition, $X_1, \dots, X_n$ are identically distributed, we can derive an expression for the moment-generating function of the sum in terms of the common moment-generating function of the $X_j$s.

**Proposition 4.7.6** (MGF of a sum of $n$ independent random variables)
*Suppose $X_1, \dots, X_n$ is a sequence of independent and identically distributed random variables, and let $S = \sum_{j=1}^{n} X_j$. The moment- and cumulant-generating function of $S$ are given by*

$$M_S(t) = [M_X(t)]^n,$$
$$K_S(t) = n K_X(t),$$

*where $M_X$ and $K_X$ are respectively the common moment- and cumulant-generating functions of the $X_j$s.*

Proving this result is part of Exercise 4.7.

**Example 4.7.7** (Sum of $n$ independent Bernoullis)
We can use Proposition 4.7.6 to confirm that the sum of $n$ independent Bernoulli random variables has a binomial distribution. Let $X_1, \ldots, X_n$ be a sequence of independent Bernoulli($p$) random variables and let $S = X_1 + \ldots + X_n$. We know from Exercise 3.5 that

$$M_X(t) = 1 - p + pe^t.$$

Thus, by Proposition 4.7.6,

$$M_S(t) = (1 - p + pe^t)^n.$$

By uniqueness of moment-generating functions, this implies that $S \sim \text{Bin}(n, p)$.

**Exercise 4.7**

1. Suppose $X$ and $Y$ are random variables with joint density

$$f_{X,Y}(x, y) = \begin{cases} 2 & \text{for } 0 < x < y < 1, \\ 0 & \text{otherwise.} \end{cases}$$

   Find the density function of $Z$ where $Z = X + Y$.

2. (**Sum of positive random variables**) Suppose that $X$ and $Y$ are positive continuous random variables and let $Z = X + Y$. Using the result given by Proposition 4.7.1 as a starting point, show that

$$f_Z(z) = \begin{cases} \int_0^z f_{X,Y}(u, z - u)du & \text{for } z \geq 0, \\ 0 & \text{otherwise.} \end{cases}$$

3. Let $X$ and $Y$ be independent random variables with $X \sim \text{Gamma}(\alpha_1, \lambda_1)$ and $Y \sim \text{Gamma}(\alpha_2, \lambda_2)$. Give an expression for the moment-generating function of $Z$ where $Z = X + Y$. Under what conditions does $Z$ have a Gamma distribution?

4. Show from first principles that if $X_1, \ldots, X_n$ is a sequence of independent and identically distributed random variables, and $S = \sum_{i=1}^n X_i$ then

$$M_S(t) = [M_X(t)]^n,$$
$$K_S(t) = nK_X(t).$$

5. Let $S$ be a sum of $n$ independent binomially distributed random variables. Under what conditions does $S$ also have a binomial distribution?

## 4.8 Multivariate normal distribution

In Example 3.3.22 we discuss some of the properties of normal distributions. In particular, it is clear that a normal distribution is uniquely specified by its mean and variance. In the multivariate case, we can show that the relationship between normal distributions is completely characterised by their correlation. Thus, if random variables are (jointly) normally distributed and uncorrelated, they are also independent.

### 4.8.1 Bivariate case

Our starting point is to consider a pair of independent standard normal random variables. If $U$ and $V$ are independent $\mathrm{N}(0,1)$ random variables, then their joint density and joint moment-generating functions are, respectively,

$$\begin{aligned} f_{U,V}(u,v) &= \frac{1}{2\pi} e^{-(u^2+v^2)/2}, \quad \text{for } u,v \in \mathbb{R} \\ M_{U,V}(s,t) &= e^{(s^2+t^2)/2}, \quad \text{for } s,t \in \mathbb{R}. \end{aligned}$$

This is a simple consequence of independence; joint density is a product of the marginal densities, and the joint moment-generating function is the product of the marginal moment-generating functions (see Exercise 4.8). The assumption of independence is rather restrictive. In cases of practical interest, the variables under consideration are correlated. What we would like is a bivariate version of the standard normal distribution. The following proposition indicates how a standard bivariate normal distribution may be constructed.

**Proposition 4.8.1** (Construction of a standard bivariate normal distribution)
*Suppose $U$ and $V$ are independent $\mathrm{N}(0,1)$ random variables. If we let $X = U$ and $Y = \rho U + \sqrt{1-\rho^2}V$, then*

*i.* $X \sim \mathrm{N}(0,1)$ *and* $Y \sim \mathrm{N}(0,1)$,

*ii.* $\mathrm{Corr}(X,Y) = \rho$,

*iii.* $f_{X,Y}(x,y) = \frac{1}{2\pi\sqrt{1-\rho^2}} \exp\left[-(x^2 - 2\rho xy + y^2)/(2(1-\rho^2))\right]$, *for* $x,y \in \mathbb{R}$,

*iv.* $M_{X,Y}(s,t) = \exp[\frac{1}{2}(s^2 + 2\rho st + t^2)]$, *for* $s,t \in \mathbb{R}$.

*Proof.*

i. We know that $X \sim \mathrm{N}(0,1)$ since $X = U$ and $U \sim \mathrm{N}(0,1)$. From Example 4.7.5, a sum of independent normal random variables is normal, so $Y$ is normal. By definition of $Y$ and independence of $U$ and $V$ we have

$$\mathbb{E}(Y) = 0 \text{ and } \mathrm{Var}(Y) = \rho^2 \mathrm{Var}(U) + (1-\rho^2)\mathrm{Var}(V) = 1.$$

Thus $Y \sim \mathrm{N}(0,1)$.

ii. To work out the correlation we exploit the independence of $U$ and $V$, and the fact that all the variables involved have mean 0 and variance 1. We have

$$\begin{aligned} \mathrm{Cov}(X,Y) &= \mathrm{Cov}\left(U, \rho U + \sqrt{1-\rho^2}V\right) \\ &= \mathrm{Cov}(U,\rho U) + \mathrm{Cov}\left(U, \sqrt{1-\rho^2}V\right) \\ &= \rho\,\mathrm{Var}(U) = \rho\,. \end{aligned}$$

Thus, since $\mathrm{Var}(X) = \mathrm{Var}(Y) = 1$, we have

$$\mathrm{Corr}(X,Y) = \rho.$$

iii. The transformation here is $x = u$ and $y = \rho u + \sqrt{1-\rho^2}v$, so the inverse transformation is $u = x$ and $v = (1-\rho^2)^{-1/2}(y - \rho x)$. We can readily see that the Jacobian of the inverse is $(1-\rho^2)^{-1/2}$. Thus, by change-of-variables formula (4.10),

$$\begin{aligned} f_{X,Y}(x,y) &= f_{U,V}(x,(1-\rho^2)^{-1/2}(y-\rho x))(1-\rho^2)^{-1/2} \\ &= \frac{1}{2\pi\sqrt{1-\rho^2}} \exp\left[-\frac{1}{2}\left(x^2 + \frac{(y-\rho x)^2}{(1-\rho^2)}\right)\right] \\ &= \frac{1}{2\pi\sqrt{1-\rho^2}} \exp\left[-(x^2 - 2\rho xy + y^2)/(2(1-\rho^2))\right]. \end{aligned}$$

iv. This is part of Exercise 4.9.

□

A general bivariate normal can be constructed from a standard bivariate normal using location-scale transformations. By Example 3.6.7, if we define

$$X^* = \mu_X + \sigma_X X \text{ and } Y^* = \mu_Y + \sigma_Y Y,$$

then $X^* \sim \mathrm{N}(\mu_X, \sigma_X^2)$ and $Y^* \sim \mathrm{N}(\mu_Y, \sigma_Y^2)$. It is straightforward to show that $\mathrm{Corr}(X^*, Y^*) = \rho$ and, by change-of-variables formula (4.10),

$$f_{X^*,Y^*}(x,y) = \frac{1}{\sigma_X\sigma_Y} f_{X,Y}\left(\frac{x-\mu_X}{\sigma_X}, \frac{y-\mu_Y}{\sigma_Y}\right).$$

Note that it is possible to construct a bivariate distribution which has normal margins but for which the joint density is not that of a bivariate normal distribution. For example, take $X \sim \mathrm{N}(0,1)$ and define $Y$ as

$$Y = \begin{cases} X & \text{with probability } 1/2, \\ -X & \text{with probability } 1/2. \end{cases}$$

We can readily show that $Y \sim \mathrm{N}(0,1)$, but the joint distribution of $X, Y$ is not bivariate normal; if we fix $X$, then $Y$ can only take two possible values.

### 4.8.2 *n-dimensional multivariate case*

We can construct a general $n$-dimensional multivariate normal distribution from a vector of independent standard normal random variables. Suppose that $\boldsymbol{U}$ is an $n \times 1$ random vector, such that $\boldsymbol{U} \sim \mathrm{N}(\boldsymbol{0}, \boldsymbol{I}_n)$. By independence, the density of $\boldsymbol{U}$ is

$$f_{\boldsymbol{U}}(\boldsymbol{u}) = \prod_{i=1}^{n} (2\pi)^{-1/2} \exp(-u_i^2/2) = (2\pi)^{-n/2} \exp\left[-\boldsymbol{u}^T\boldsymbol{u}/2\right].$$

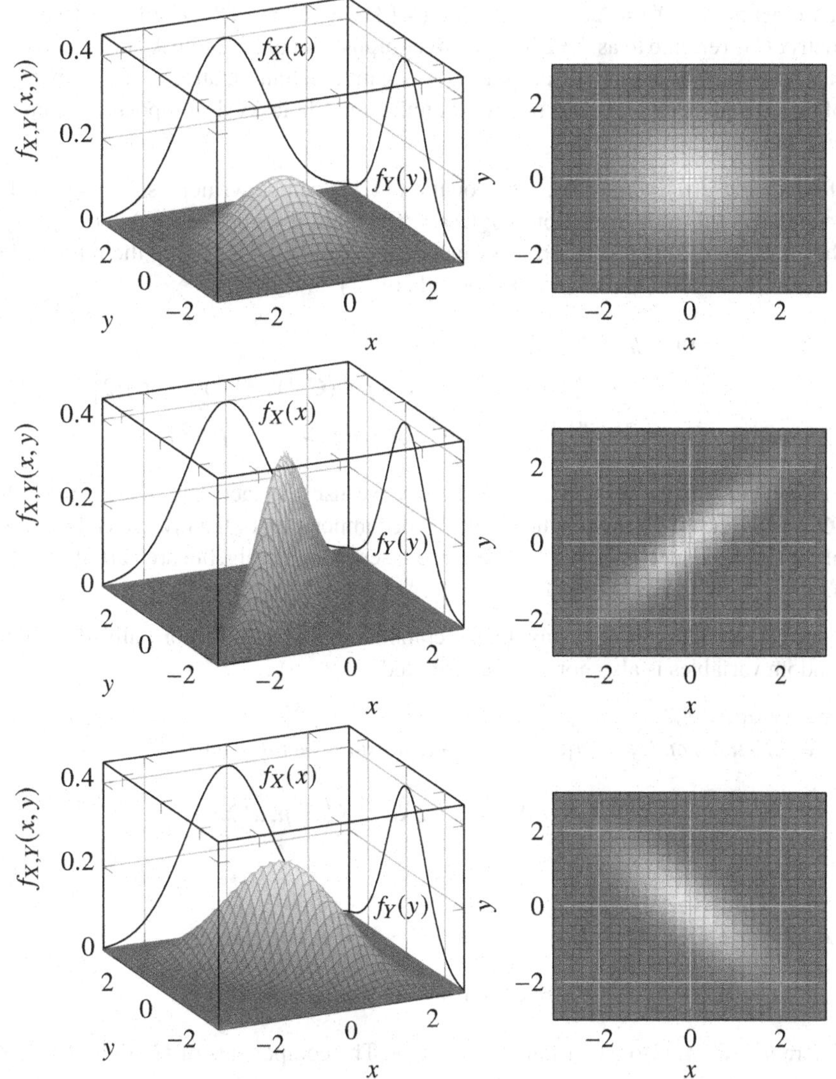


Figure 4.7 *The standard bivariate normal with* $\rho = 0$ *(top),* $\rho = 0.9$ *(middle), and* $\rho = -0.8$ *(bottom). Notice that the marginal densities are always the same.*

Consider the transformation

$$X = \mu + CU, \tag{4.12}$$

where $\mu$ is an $n \times 1$ vector and $C$ is an $n \times n$ non-singular, lower-triangular matrix. This is a multivariate version of the scale-location transformation. The reason for specifying that $C$ be lower-triangular will become apparent shortly. Under this transformation,

$$\mathbb{E}(X) = \mu \quad \text{and} \quad \text{Var}(X) = CC^T. \tag{4.13}$$

If we define $\text{Var}(\boldsymbol{X}) = \Sigma$, then equation (4.13) says $\Sigma = \boldsymbol{C}\boldsymbol{C}^T$. The lower-triangular matrix $\boldsymbol{C}$ is referred to as the **Cholesky decomposition** of $\Sigma$. Cholesky decomposition can be viewed as a generalisation to non-negative-definite matrices of the operation of taking square roots. If $\Sigma$ is positive definite, the Cholesky decomposition is unique and non-singular.

If we define $\boldsymbol{g}(\boldsymbol{u}) = \boldsymbol{\mu} + \boldsymbol{C}\boldsymbol{u}$, then equation (4.12) can be written as $\boldsymbol{X} = \boldsymbol{g}(\boldsymbol{U})$. The Jacobian of this transformation is $J_{\boldsymbol{g}}(\boldsymbol{u}) = |\boldsymbol{C}|$. Thus, if we define $\boldsymbol{h}(\boldsymbol{x}) = \boldsymbol{C}^{-1}(\boldsymbol{x} - \boldsymbol{\mu})$, the Jacobian of this inverse transformation is $J_{\boldsymbol{h}}(\boldsymbol{x}) = |\boldsymbol{C}|^{-1}$. Using the multivariate version of the change-of-variables formula (4.11) now yields

$$\begin{aligned} f_{\boldsymbol{X}}(\boldsymbol{x}) &= f_{\boldsymbol{U}}(\boldsymbol{h}(\boldsymbol{x}))|J_{\boldsymbol{h}}(\boldsymbol{x})| \\ &= (2\pi)^{-n/2}|\boldsymbol{C}|^{-1} \exp\left[-(\boldsymbol{x} - \boldsymbol{\mu})^T(\boldsymbol{C}^{-1})^T(\boldsymbol{C}^{-1})(\boldsymbol{x} - \boldsymbol{\mu})/2\right] \\ &= (2\pi)^{-n/2}|\Sigma|^{-1/2} \exp\left[-(\boldsymbol{x} - \boldsymbol{\mu})^T\Sigma^{-1}(\boldsymbol{x} - \boldsymbol{\mu})/2\right]. \end{aligned}$$

In deriving this expression for the density we use the fact that $|\boldsymbol{C}| = |\Sigma|^{1/2}$ and $(\boldsymbol{C}^{-1})^T(\boldsymbol{C}^{-1}) = \Sigma^{-1}$, that is, the Cholesky decomposition of the inverse is the inverse of the Cholesky decomposition. These exercises in matrix algebra are part of Exercise 4.8.

A direct corollary is that any linear combination of jointly normally distributed random variables is also normally distributed.

**Proposition 4.8.2**
*If $\boldsymbol{X} \sim \text{N}(\boldsymbol{\mu}, \Sigma)$ and $\boldsymbol{a} = (a_1, \ldots, a_n)^T$ is a constant vector, then*

$$\boldsymbol{a}^T\boldsymbol{X} = a_1X_1 + \ldots a_nX_n \sim \text{N}\left(\boldsymbol{a}^T\boldsymbol{\mu}, \boldsymbol{a}^T\Sigma\boldsymbol{a}\right)$$

*Proof.*

Let $\boldsymbol{X} = \boldsymbol{\mu} + \boldsymbol{C}\boldsymbol{U}$, where $\Sigma = \boldsymbol{C}\boldsymbol{C}^T$, and $\boldsymbol{U} \sim \text{N}(\boldsymbol{0}, \boldsymbol{I}_n)$. We have

$$\boldsymbol{a}^T\boldsymbol{X} = \boldsymbol{a}^T\boldsymbol{\mu} + \boldsymbol{a}^T\boldsymbol{C}\boldsymbol{U} = \boldsymbol{a}^T\boldsymbol{\mu} + \boldsymbol{b}^T\boldsymbol{U},$$

where $\boldsymbol{b} = \boldsymbol{C}^T\boldsymbol{a}$, also a constant $n \times 1$ vector. The components of $\boldsymbol{U}$ are independent standard normal, so

$$\boldsymbol{b}^T\boldsymbol{U} = \sum_{i=1}^{n} b_iU_i \sim \text{N}\left(0, \sum_{i=1}^{n} b_i^2\right).$$

The variance in this expression is

$$\boldsymbol{b}^T\boldsymbol{b} = \boldsymbol{a}^T\boldsymbol{C}\boldsymbol{C}^T\boldsymbol{a} = \boldsymbol{a}^T\Sigma\boldsymbol{a},$$

and the result follows. □

**Exercise 4.8**

1. Show that if $U$ and $V$ are independent standard normals then the joint density and joint moment-generating functions of $U$ and $V$ are given by

$$\begin{aligned} f_{U,V}(u,v) &= \tfrac{1}{2\pi} e^{-(u^2+v^2)/2}, \\ M_{U,V}(s,t) &= e^{(s^2+t^2)/2}. \end{aligned}$$

2. If $U$ and $V$ are independent standard normals and we define $X = U$ and $Y = \rho U + \sqrt{1-\rho^2}V$, show that $M_{X,Y}(s,t) = \exp[\frac{1}{2}(s^2 + 2\rho st + t^2)]$.
3. If $\boldsymbol{\Sigma}$ is a positive-definite matrix and $\boldsymbol{C}$ is the Cholesky decomposition of $\boldsymbol{\Sigma}$, show that $|\boldsymbol{C}| = |\boldsymbol{\Sigma}|^{-1/2}$ and $(\boldsymbol{C}^{-1})(\boldsymbol{C}^{-1})^T = \boldsymbol{\Sigma}^{-1}$.

## 4.9 Further exercises

1. Suppose that the joint distribution of $X$ and $Y$ is standard bivariate normal with correlation $\rho$.
   (a) Let $U = X$ and $V = (Y - \rho X)/\sqrt{1-\rho^2}$. Show that $U$ and $V$ are uncorrelated and that $\text{Var}(V) = 1$.
   (b) Now let $W = \sigma X$. Find the $r^{\text{th}}$ moment of W.
2. Let $X \sim \text{Pois}(\lambda)$, so

$$f_X(x) = \frac{\lambda^x e^{-\lambda}}{x!}, \quad x = 0, 1, \ldots$$

   Let $X_1, \ldots, X_n$ be a sequence of independent $\text{Pois}(\lambda)$ random variables and $\bar{X} = \frac{1}{n}\sum_{i=1}^n X_i$.
   (a) Derive the cumulant-generating function of $X$.
   (b) Find $K_{\bar{X}}(t)$, the cumulant-generating function of $\bar{X}$. Find the limit as $n \to \infty$ of $K_{\bar{X}}(t)$. Comment on the result.
3. Suppose that $U$ and $V$ are independent random variables with cumulant-generating functions $K_U$ and $K_V$. Define random variables $X$ and $Y$ by $X = aU + bV$ and $Y = cU + dV$.
   (a) Show that the joint cumulant-generating function of $X$ and $Y$ is given by

$$K_{X,Y}(s,t) = K_U(sa + tc) + K_V(sb + td).$$

   (b) Suppose $U \sim \text{N}(0,1)$ and $V \sim \text{N}(0,1)$. Let

$$X = U,$$

   and

$$Y = \rho U + \sqrt{1-\rho^2}V.$$

   Derive the joint cumulants of $X$ and $Y$, $\kappa_{i,j}$, for all $i$ and $j$. What properties of $X$ and/or $Y$ are given by $\kappa_{0,1}$, $\kappa_{1,1}$, and $\kappa_{2,0}$?

4. Let $\boldsymbol{X}$ be an $n \times 1$ random vector with mean $\boldsymbol{\mu}$ and covariance matrix $\boldsymbol{\Sigma}$. Let $\boldsymbol{a}$ and $\boldsymbol{b}$ be two $n \times 1$ constant vectors. Find $\mathbb{E}(\boldsymbol{a}^T \boldsymbol{X})$ and $\text{Cov}(\boldsymbol{a}^T \boldsymbol{X}, \boldsymbol{b}^T \boldsymbol{X})$.
5. Let $X$ be a random variable uniformly distributed on the interval $[-2, 1]$. Find the probability density function of $X^2$.
6. Suppose that the random variables $X$ and $Y$ have joint probability density function

$$f(x, y) = \begin{cases} 120y(x-y)(1-x) & \text{for } 0 \leq y \leq x,\ 0 < x \leq 1, \\ 0 & \text{otherwise.} \end{cases}$$

   Find the density function of $Z = Y/\sqrt{X}$. [*Hint*: Find the joint density function of $(U, Z)$ first with $U = \sqrt{X}$.]
7. Consider the following joint density for random variables $X$ and $Y$:

$$f_{X,Y}(x, y) = \begin{cases} xy, & 0 \leq x < 1 \text{ and } 0 \leq y < 2, \\ 0, & \text{otherwise.} \end{cases}$$

   (a) Verify that $f_{X,Y}$ is a valid density function.
   (b) Find the marginal densities and show that $X$ and $Y$ are independent.
   (c) Evaluate $\mathbb{E}(X)$ and $\mathbb{E}(Y)$. Hence, write down $\mathbb{E}(XY)$.
   (d) Evaluate $\text{P}(Y \leq 1 - X)$.
   (e) Suppose $Z = X + Y$. Find the density function of $Z$.
8. (a) Let $f_{X,Y}(x, y)$ be the joint density function of random variables $(X, Y)$. Show that, under the transformation $U = X/Y$ and $V = Y$, the joint density of $(U, V)$ is

$$f_{U,V}(u, v) = f_{X,Y}(uv, v)|v|,$$

   and hence write down a formula for the density function of $U$.
   (b) Let $X$ and $Y$ be two independent random variables, $X \sim \text{N}(0, \sigma_1^2)$, and $Y \sim \text{N}(0, \sigma_2^2)$. Find the density function of $U = X/Y$.
9. Consider the random variables $X$ and $Y$ with joint density

$$f_{X,Y}(x, y) = \begin{cases} \lambda^2 e^{-\lambda x} & \text{for } 0 < y < x < \infty, \\ 0 & \text{otherwise.} \end{cases}$$

   Show that their joint moment-generating function is

$$M_{X,Y}(t, u) = \frac{\lambda^2}{(\lambda - t)(\lambda - t - u)}.$$

   For which values of $t$ and $u$ is this well defined?
10. (a) Let $f_{X,Y}(x, y)$ be the joint density function for random variables $X$ and $Y$. Derive the marginal density function for

    i. $U = Y - X$,
    ii. $V = X + Y$.

(b) If the joint density function of $X$ and $Y$ is

$$f_{X,Y}(x,y) = \begin{cases} 2 & \text{for } 0 < x < y < 1, \\ 0 & \text{otherwise,} \end{cases}$$

find the marginal density for

i. $U = Y - X$,
ii. $V = X + Y$.

(c) Let $X_1, \cdots, X_k$ be independent Poisson random variables, and $\mathbb{E}(X_i) = \mu_i$ for $i = 1, \ldots, k$. Find the distribution of $Y = X_1 + \ldots + X_k$.

11. Let $X$ and $Y$ be random variables with joint density

$$f_{X,Y}(x,y) = \begin{cases} ke^{-\lambda x}, & 0 < y < x < \infty, \\ 0, & \text{otherwise.} \end{cases}$$

(a) Find $k$.
(b) Derive the marginal density for $Y$ and hence evaluate $\mathbb{E}(Y)$, $\mathbb{E}(Y^2)$, and $\text{Var}(Y)$.
(c) Find $\text{P}(X + 2Y \leq 3)$.

12. Suppose that we perform $n$ independent replications of an experiment that has $k$ possible outcomes. For $i = 1, \ldots, k$, let $p_i$ denote the probability corresponding to the $i^{\text{th}}$ outcome (so $\sum_{i=1}^{k} p_i = 1$), and $X_i$ the number of times the $i^{\text{th}}$ outcome occurs. We say that $\boldsymbol{X} = (X_1, \ldots, X_k)^T$ follows a **multinomial distribution**, $\boldsymbol{X} \sim \text{Multinom}(n, \boldsymbol{p})$, where $\boldsymbol{p} = (p_1, \ldots, p_k)^T$.

(a) Show that the mass function of $\boldsymbol{X}$ is

$$f_{\boldsymbol{X}}(\boldsymbol{x}) = \begin{cases} (x_1, \ldots, x_k)! \prod_{i=1}^{k} p_i^{x_i} & \text{for } x_1, \ldots, x_n \in \{0, 1, \ldots, n\}, \ \sum_{i=1}^{k} x_i = n, \\ 0 & \text{otherwise,} \end{cases}$$

where $(x_1, \ldots, x_k)!$ is a multinomial coefficient (see Proposition 2.3.8). What happens in the case $k = 2$?

(b) What is the marginal distribution of $X_i$?

## 4.10 Chapter summary

On completion of this chapter you should be able to:

- define the terms joint and marginal,
- generate the marginal cumulative distribution function of a random variable from a joint cumulative distribution function involving the variable,
- construct a table giving the joint and marginal mass functions for a discrete bivariate distribution,
- determine whether a function of two variables is a valid mass or density function,
- derive marginal mass or density from joint mass or density (even when the support is an awkward shape),

- calculate probabilities from bivariate mass or density,
- calculate expectations for functions of several variables,
- calculate joint moments from bivariate mass or density,
- describe the distinction between pairwise and mutual independence,
- describe the distinction between uncorrelated and independent,
- show that the covariance matrix of a random vector is non-negative definite,
- find the density of a transformation of continuous random variables,
- find the distribution of a sum of random variables,
- construct a standard bivariate normal distribution, and
- derive the density of a linear transformation of independent normal random variables.

Chapter 5

# Conditional distributions

This chapter extends the concept of conditional probability to random variables. We have seen how we can use tools such as the joint distribution, density, and moment-generating functions to capture the nature of dependencies between several variables. In practice, we often have information about the values taken by some of these random variables, and wish to determine how this information affects the distribution of other variables. For example, if we know a person's medical history, what can we say about the distribution of their lifespan?

As with all probability measures, in order to define conditional probability we need events. Recall that for events $A$ and $B$ we have $\text{P}(A|B) = \text{P}(A \cap B)/\text{P}(B)$, provided $\text{P}(B) > 0$ (section 2.4). In the discrete case, the definition of conditional probability can be exploited directly to define conditional mass. Similar ideas are used in the continuous case, although we have to be more careful about interpretation. Conditioning for discrete and continuous random variables form the first two sections of this chapter.

## 5.1 Discrete conditional distributions

Consider discrete random variables $X$ and $Y$. Suppose we know that $X$ takes some particular value, $x$. This knowledge will affect our assessment of the distribution of probability associated with $Y$. We return to an example from Chapter 4 to illustrate.

**Example 5.1.1** (Card drawing example again)
Recall in Example 4.2.1 we draw two cards at random without replacement from a deck of 52 and define $X$ to be the number of kings, and $Y$ to be the number of aces. Suppose we are told that exactly one king has been drawn, that is, $X = 1$. This will affect our view of the distribution of the number of aces. The most obvious immediate consequence is that we now know that there cannot be two aces. We can work out the other probabilities using our knowledge of conditional probability and the results

from Table 4.1.

$$\begin{aligned}
\text{P}(0 \text{ aces}|1 \text{ king}) = \text{P}(Y = 0|X = 1) &= \frac{\text{P}(X = 1, Y = 0)}{\text{P}(X = 1)} \\
&= \frac{f_{X,Y}(1,0)}{f_X(1)} = \frac{0.1327}{0.1448} = 0.916, \\
\text{P}(1 \text{ aces}|1 \text{ king}) = \text{P}(Y = 1|X = 1) &= \frac{\text{P}(X = 1, Y = 1)}{\text{P}(X = 1)} \\
&= \frac{f_{X,Y}(1,1)}{f_X(1)} = \frac{0.0121}{0.1448} = 0.084, \\
\text{P}(2 \text{ aces}|1 \text{ king}) = \text{P}(Y = 2|X = 1) &= \frac{\text{P}(X = 1, Y = 2)}{\text{P}(X = 1)} \\
&= \frac{f_{X,Y}(1,2)}{f_X(1)} = \frac{0.0000}{0.1448} = 0.000.
\end{aligned}$$

Note that $\text{P}(0 \text{ aces}|1 \text{ king}) + \text{P}(1 \text{ aces}|1 \text{ king}) + \text{P}(2 \text{ aces}|1 \text{ king}) = 1$.

For the discrete case, the basic principles of conditional probability hold. We treat $\{X = x\}$ as an event and, provided $\text{P}(X = x) > 0$, we can condition on $\{X = x\}$ in the usual way. Values of the conditional mass function can be interpreted as conditional probabilities.

**Definition 5.1.2** (Conditional mass function)
Suppose that $X$ and $Y$ are discrete random variables with joint mass function $f_{X,Y}$. The **conditional mass function** of $Y$ given $X = x$ is defined as

$$f_{Y|X}(y|x) = \text{P}(Y = y|X = x), \text{ where } \text{P}(X = x) > 0.$$

The conditional mass can be written in terms of the joint mass of $X$ and $Y$ and the marginal mass of $X$,

$$f_{Y|X}(y|x) = \begin{cases} \frac{f_{X,Y}(x,y)}{f_X(x)} & \text{for } f_X(x) > 0, \\ 0 & \text{otherwise.} \end{cases} \tag{5.1}$$

This is a simple consequence of the definition of the conditional probability 2.4.1.

We can readily show that $f_{Y|X}(y|x)$ as defined by (5.1) is a valid mass function. It is clearly positive and real-valued. We are, in effect, taking the row of the joint probability table corresponding to $X = x$, then dividing the probabilities by a common factor, $f_X(x)$, to ensure that they sum to 1. Summing over all possible values of $y$,

$$\sum_y f_{Y|X}(y|x) = \sum_y \frac{f_{X,Y}(x,y)}{f_X(x)} = \frac{\sum_y f_{X,Y}(x,y)}{f_X(x)} = \frac{f_X(x)}{f_X(x)} = 1.$$

The relationship between the conditional mass function and the conditional cumulative distribution function is the same as in the unconditional case. The conditional cumulative distribution function of $Y$ given $X = x$ is defined simply as a sum of values of the conditional mass function.

**Definition 5.1.3** (Conditional cumulative distribution function (discrete case))
Suppose that $X$ and $Y$ are discrete random variables and let $f_{Y|X}$ be the conditional mass function of $Y$ given $X = x$. The **conditional cumulative distribution function** of $Y$ given $X = x$ is defined as

$$F_{Y|X}(y|x) = \sum_{y_i \leq y} f_{Y|X}(y_i|x).$$

In the discrete case, values of the conditional distribution function can be interpreted as conditional probabilities,

$$F_{Y|X}(y|x) = \sum_{y_i \leq y} \frac{f_{X,Y}(x, y_i)}{f_X(x)} = \frac{\mathrm{P}(Y \leq y, X = x)}{\mathrm{P}(X = x)} = \mathrm{P}(Y \leq y|X = x).$$

We use $Y|X = x$ to denote the random variable with mass function $f_{Y|X}(y|x)$ and cumulative distribution function $F_{Y|X}(y|x)$. This is a random variable in the usual sense and, provided we are careful with notation, it can be treated like any other random variable. For example, we can define its expected value.

**Definition 5.1.4** (Expected value of the conditional distribution (discrete case))
Suppose that $X$ and $Y$ are discrete random variables, with $f_{Y|X}$ the conditional mass function of $Y$ given $X = x$. The expected value of the conditional distribution is given by

$$\mathbb{E}(Y|X = x) = \sum_{y} y f_{Y|X}(y|x).$$

Definition 5.1.4 introduces no new concepts. In order to calculate an expectation, we multiply a value from the distribution by the mass function and sum over all possible values. This expectation should not be confused with the conditional expectation which is defined in section 5.4.

**Notation 5.1.5** (Conditioning notation)
It is important to understand the notation here; we will usually think of $f_{Y|X}(y|x)$ as a function of $y$. The subscript $Y|X$ indicates that we are considering the mass of the random variable $Y$ conditional on the random variable $X$ taking some value. The notation $.|x$ indicates that the value of $X$ that we are conditioning on is $x$. As usual, $x$ and $y$ are arbitrary place holders. We could write

$$f_{Y|X}(a|b) = \mathrm{P}(Y = a|X = b) = \frac{f_{X,Y}(b, a)}{f_X(b)},$$

$$f_{Y|X}(peach|plum) = \mathrm{P}(Y = peach|X = plum) = \frac{f_{X,Y}(plum, peach)}{f_X(plum)}.$$

**Exercise 5.1**

1. Consider the card-drawing Example 5.1.1. Give expressions for the conditional mass functions $f_{Y|X}(y|0)$ and $f_{Y|X}(y|2)$.
2. Consider the game described in section 4.3, where $X$ is the number of heads from tossing a coin three times, while $Y$ takes the value 1 if the first throw and third throw are heads, and the value 0 otherwise. Suppose we know that the first throw is a head. Write down the mass functions for $X$ and $Y$ conditional on this event.

## 5.2 Continuous conditional distributions

For continuous random variables, we define the conditional density to have the same form as the conditional mass function.

**Definition 5.2.1** (Conditional density)
Suppose that $X$ and $Y$ are jointly continuous random variables with joint density $f_{X,Y}$ and that the marginal density of $X$ is $f_X$. The **conditional density** of $Y$ given $X = x$ is defined as

$$f_{Y|X}(y|x) = \begin{cases} \frac{f_{X,Y}(x,y)}{f_X(x)} & \text{for } f_X(x) > 0, \\ 0 & \text{otherwise.} \end{cases}$$

We cannot interpret conditional density as a conditional probability; recall that, if $X$ is continuous, then $\text{P}(X = x) = 0$ for any real number $x$. However, we can readily show that, viewed as a function of $y$, the function $f_{Y|X}(y|x)$ is a valid density function. This is part of Exercise 5.2.

We define $Y|X = x$ as the random variable with density function $f_{Y|X}(y|x)$. As we might expect, the cumulative distribution function and expected value of $Y|X = x$ are found by integration,

$$F_{Y|X}(y|x) = \int_{-\infty}^{y} f_{Y|X}(u|x)du, \text{ and}$$

$$\mathbb{E}(Y|X = x) = \int_{-\infty}^{\infty} y f_{Y|X}(y|x)dy.$$

We illustrate these ideas using some familiar polynomial densities.

**Example 5.2.2** (Two polynomial densities again)
We return to the polynomial densities first described in Example 4.2.10.

1. Consider random variables $X$ and $Y$ with joint density function $f_{X,Y}(x,y) = x + y$ for $0 < x < 1, 0 < y < 1$. We have shown that $f_X(x) = x + \frac{1}{2}$ for $0 < x < 1$, thus,

$$f_{Y|X}(y|x) = \frac{f_{X,Y}(x,y)}{f_X(x)} = \begin{cases} \frac{x+y}{x+\frac{1}{2}} & \text{for } 0 < y < 1, \\ 0 & \text{otherwise.} \end{cases}$$

   Check that this is a valid density. Figure 5.1 is a visual illustration of this result.

2. Consider random variables $X$ and $Y$ with joint density function $f_{X,Y}(x,y) = 8xy$ for $0 < x < y < 1$. We have shown that $f_X(x) = 4x(1 - x^2)$ for $0 < x < 1$, thus,

$$f_{Y|X}(y|x) = \frac{f_{X,Y}(x,y)}{f_X(x)} = \begin{cases} \frac{2y}{1-x^2} & \text{for } x < y < 1, \\ 0 & \text{otherwise,} \end{cases}$$

   where the support now depends on $x$ (Figure 5.2). Check that this is a valid density.

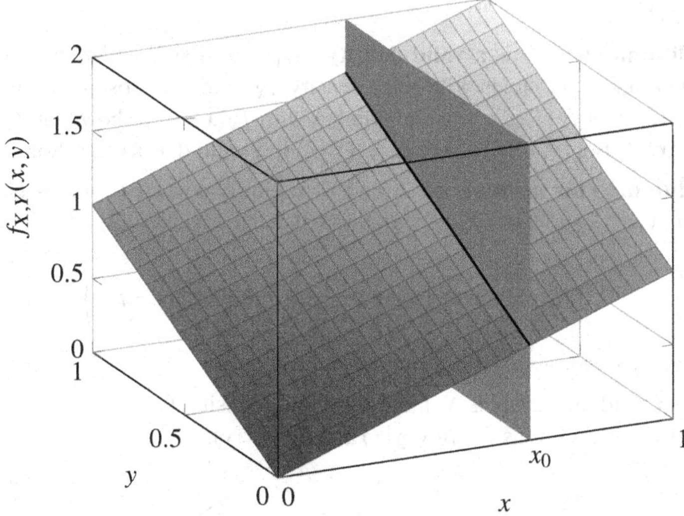


Figure 5.1 *The conditional density $f_{Y|X}(y|x_0)$ corresponds to the intersection of the density $f_{X,Y}(x, y) = x + y$ and the plane $x = x_0$ (dark grey), which is the line $g(y) = x_0 + y$ (black line). In other words, we take a slice of the joint density surface at $X = x_0$, then rescale it so that the resulting density function integrates to 1. This is in direct analogy to the discrete case, where we took a slice (row or column) of the joint probability table and rescaled it. Notice that the support of the conditional density is always $0 < y < 1$ in this example.*

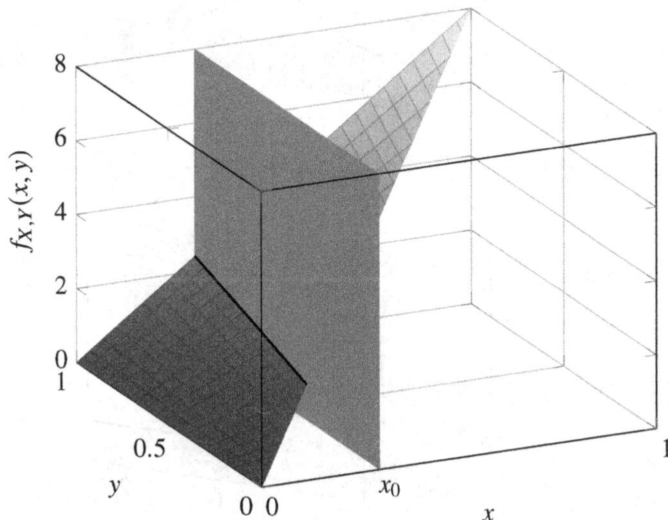


Figure 5.2 *The conditional density $f_{Y|X}(y|x_0)$ in the example where the support of the joint density is triangular. Notice that the support of the conditional density now depends on $x_0$.*

**Exercise 5.2**

1. (**Conditional density is a valid density**) Suppose that $X$ and $Y$ are jointly continuous random variables with joint density $f_{X,Y}$ and suppose that the marginal density for $X$ is $f_X$. Let $x$ be a real number with $f_X(x) > 0$. Show that if we define $f_{Y|X}(y|x) = f_{X,Y}(x,y)/f_X(x)$ then $f_{Y|X}(.|x)$ is a valid density function.
2. (**Conditioning for normal**) Let $X$ and $Y$ be jointly normal with $X \sim \text{N}(\mu_X, \sigma_X^2)$, $Y \sim \text{N}(\mu_Y, \sigma_Y^2)$, and $\text{Corr}(X,Y) = \rho$. Show that

$$Y|X = x \sim \text{N}\left(\mu_y + \rho\frac{\sigma_Y}{\sigma_X}(x - \mu_X),\ \sigma_Y^2(1-\rho^2)\right).$$

   See Figure 5.3 for a visual illustration of this result.
3. Consider random variables $X$ and $Y$ with joint density $f_{X,Y}(x,y) = 2$ for $0 < x < y < 1$. Find expressions for $f_{Y|X}(y|x)$ and $f_{X|Y}(x|y)$.

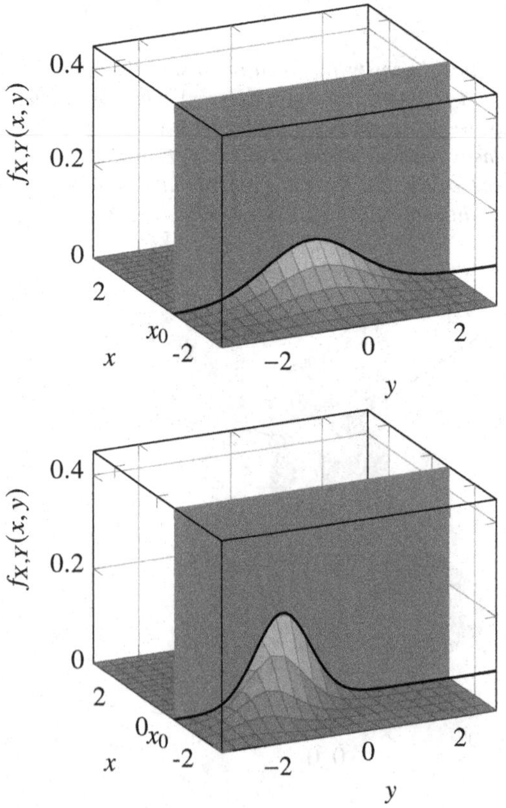


Figure 5.3 *Graphical representation of the conditional density $f_{Y|X}(y|x_0)$, where $X$ and $Y$ are jointly standard normal with correlation $\rho = 0$ (top) and $\rho = 0.8$ (bottom). The variance of the conditional distribution decreases as the correlation approaches 1 (or −1).*

## 5.3 Relationship between joint, marginal, and conditional

The similarity in the definitions of conditional mass and conditional density allows us to write an expression for the joint mass/density in terms of the conditional mass/density and the marginal mass/density of the variable that we are conditioning on,

$$f_{X,Y}(x,y) = f_{Y|X}(y|x)f_X(x). \tag{5.2}$$

Notice that the situation is symmetric so it is also true that

$$f_{X,Y}(x,y) = f_{X|Y}(x|y)f_Y(y).$$

The terminology here becomes rather cumbersome. To avoid repeatedly using the term mass/density we will just use joint, conditional and marginal when describing results that hold for both discrete and continuous cases. Thus, equation (5.2) states that the joint is the product of the conditional and the marginal.

It is useful to note that we can define joint distributions that combine discrete and continuous random variables. No problems arise, provided we sum with respect to the discrete variable and integrate with respect to the continuous variable. Consider, for illustration, $f_{X,Y}(x,y)$ where $X$ is discrete and $Y$ is continuous. We have

$$\int_{-\infty}^{\infty} \sum_x f_{X,Y}(x,y)dy = 1,$$

$$f_X(x) = \int_{-\infty}^{\infty} f_{X,Y}(x,y)dy \qquad f_Y(y) = \sum_x f_{X,Y}(x,y),$$

$$\mathbb{E}(X) = \sum_x \int_{-\infty}^{\infty} x f_{X,Y}(x,y)dy \qquad \mathbb{E}(Y) = \int_{-\infty}^{\infty} \sum_x y f_{X,Y}(x,y)dy,$$

and so on.

In many instances (see Example 5.3.1 and section 5.5) we have a quantity of interest $Y$ about which we initially only have conditional information. In these situations, we can use equation (5.2) and the usual method for going from joint to marginal to evaluate $f_Y(y)$,

$$f_Y(y) = \begin{cases} \sum_x f_{Y|X}(y|x)f_X(x) & (X \text{ discrete}), \\ \int_{-\infty}^{\infty} f_{Y|X}(y|x)f_X(x)dx & (X \text{ continuous}). \end{cases} \tag{5.3}$$

Note that if we define $f_{Y|X}(y|X)$ as a function of $X$ in the obvious way, then equation (5.3) says

$$f_Y(y) = \mathbb{E}(f_{Y|X}(y|X)).$$

To illustrate, consider the hurricanes example.

**Example 5.3.1** (Landfalling hurricanes)
Suppose that $X$ is the number of hurricanes that form in the Atlantic Basin in a given year and $Y$ is the number making landfall. We assume that each hurricane has

probability $p$ of making landfall independent of all other hurricanes. If we know the number of hurricanes that form in the basin, say $x$, we can view $Y$ as the number of successes in $x$ independent Bernoulli trials. Put another way, $(Y|X = x) \sim \text{Bin}(x, p)$. If, in addition, we know the distribution of $X$, say $X \sim \text{Pois}(\lambda)$, we can use equation (5.3) to work out the marginal mass of $Y$. In this calculation, we will exploit the fact that the number of landfalling hurricanes must be less than or equal to the total number of hurricanes that form, that is, $X \geq Y$. The mass function of $Y$ is then

$$
\begin{aligned}
f_Y(y) &= \sum_x f_{Y|X}(y|x) f_X(x) \\
&= \sum_{x=y}^{\infty} \binom{x}{y} p^y (1-p)^{x-y} \frac{\lambda^x e^{-\lambda}}{x!} && \text{substituting mass functions} \\
&= \frac{\lambda^y p^y e^{-\lambda}}{y!} \sum_{x=y}^{\infty} \frac{[\lambda(1-p)]^{x-y}}{(x-y)!} && \text{rearranging} \\
&= \frac{(\lambda p)^y e^{-\lambda}}{y!} \sum_{j=0}^{\infty} \frac{[\lambda(1-p)]^j}{j!} && \text{where } j = x - y \\
&= \frac{(\lambda p)^y e^{-\lambda}}{y!} e^{\lambda(1-p)} && \text{Maclaurin series expansion} \\
&= \frac{(\lambda p)^y e^{-\lambda p}}{y!}, \quad y = 0, 1, \dots
\end{aligned}
$$

This is the mass function of a Poisson. We conclude that $Y \sim \text{Pois}(\lambda p)$.

**Exercise 5.3**

1. Suppose that $Y$ is the number of insurance claims made by an individual policyholder in one year. We assume that $(Y|X = x) \sim \text{Pois}(x)$, where $X \sim \text{Gamma}(\alpha, \theta)$. What is the probability that the policyholder makes $y$ claims in a year?
2. Suppose that $Y$ is the number of minutes that a bus is late. The lateness of the bus depends on the volume of the traffic, $X$. In fact, $(Y|X = x) \sim \text{Exp}(x)$, where $X \sim \text{Gamma}(\alpha, \theta)$. Find the density of $Y$.

## 5.4 Conditional expectation and conditional moments

In this section, we deal with results that arise when we take expectations with respect to a conditional distribution. Although the notation becomes slightly more cumbersome, the definitions and results are analogous to the unconditional case. One key distinction is that, while unconditional expected values are just numbers, conditional expectation is a random variable.

### 5.4.1 Conditional expectation

We have specified that $Y|X = x$ denotes a random variable with density function $f_{Y|X}(y|x)$. The expected value of this random variable is a function of $x$, say $\psi(x)$. It turns out that the random variable $\psi(X)$ (generated by applying the function $\psi$ to the random variable $X$) has a number of useful applications. We refer to this random variable as the **conditional expectation** of $Y$ given $X$.

**Definition 5.4.1** (Conditional expectation)
Suppose $X$ and $Y$ are random variables. We define

$$\psi(x) = \mathbb{E}(Y|X = x) = \begin{cases} \sum_y y f_{Y|X}(y|x) & \text{(discrete case)}, \\ \int_{-\infty}^{\infty} y f_{Y|X}(y|x) dy & \text{(continuous case)}. \end{cases}$$

The conditional expectation of $Y$ given $X$ is $\mathbb{E}(Y|X) = \psi(X)$, a random variable.

We can use the conditional expectation $\mathbb{E}(Y|X)$ to find the expected value of $Y$. This result is often referred to as the **law of iterated expectations**.

**Proposition 5.4.2** (Law of iterated expectations)
*If $X$ and $Y$ are random variables with conditional expectation $\mathbb{E}(Y|X)$, then*

$$\mathbb{E}(Y) = \mathbb{E}[\mathbb{E}(Y|X)].$$

*Proof.*

We start with the right-hand side and use the definition of conditional expectation. We give the proof for the continuous case. The discrete case is similar.

$$\begin{aligned} \mathbb{E}[\mathbb{E}(Y|X)] = \mathbb{E}[\psi(X)] &= \int_{-\infty}^{\infty} \psi(x) f_X(x) dx \\ &= \int_{-\infty}^{\infty} \left[ \int_{-\infty}^{\infty} y f_{Y|X}(y|x) dy \right] f_X(x) dx \\ &= \int_{-\infty}^{\infty} \int_{-\infty}^{\infty} y \frac{f_{X,Y}(x,y)}{f_X(x)} f_X(x) dy dx \\ &= \int_{-\infty}^{\infty} \int_{-\infty}^{\infty} y f_{X,Y}(x,y) dx dy \\ &= \int_{-\infty}^{\infty} y f_Y(y) dy = \mathbb{E}(Y). \end{aligned}$$

□

The key consequence of the law of iterated expectations is that we can calculate the expected value of $Y$ without referring explicitly to the marginal mass or density of $Y$,

$$\mathbb{E}(Y) = \begin{cases} \sum_x \mathbb{E}(Y|X = x) f_X(x) & \text{(discrete case)}, \\ \int_{-\infty}^{\infty} \mathbb{E}(Y|X = x) f_X(x) dx & \text{(continuous case)}. \end{cases}$$

This is not surprising, since we establish in section 5.3 that the marginal for $Y$ can be generated from the conditional for $Y$ given $X$ and the marginal for $X$. We return to this theme in section 5.5.

**Example 5.4.3** (Conditional expectation from polynomial density)
Consider random variables $X$ and $Y$ with joint density function $f_{X,Y}(x, y) = x + y$, for $0 < x < 1$ and $0 < y < 1$. We have shown in Example 5.2.2 that the conditional density of $Y$ given $X$ is $f_{Y|X}(y|x) = 2y(1 - x^2)^{-1}$ for $x < y < 1$. In order to calculate the conditional expectation, we start by calculating $\mathbb{E}(Y|X = x)$,

$$\begin{aligned}\psi(x) = \mathbb{E}(Y|X = x) &= \int_{-\infty}^{\infty} y f_{Y|X}(y|x) dy \\ &= \int_x^1 \frac{2y^2}{1 - x^2} dy = \left[\frac{2y^3}{3(1 - x^2)}\right]_{y=x}^{y=1} = \frac{2(1 - x^3)}{3(1 - x^2)} = \frac{2(1 + x + x^2)}{3(1 + x)}.\end{aligned}$$

The conditional expectation is then given by

$$\mathbb{E}(Y|X) = \psi(X) = \frac{2(1 + X + X^2)}{3(1 + X)}.$$

**Example 5.4.4** (Recognising densities)
We can often avoid explicit calculation of $\mathbb{E}(Y|X = x)$ by identifying the distribution of $Y|X = x$, as the following example illustrates. Consider random variables $X$ and $Y$ with joint density

$$f_{X,Y}(x, y) = \begin{cases} xe^{-xy}e^{-x} & \text{for } x > 0 \text{ and } y > 0, \\ 0 & \text{otherwise.} \end{cases}$$

We want to find the conditional expectation, $\mathbb{E}(Y|X)$. We start by calculating the marginal density of $X$,

$$\begin{aligned} f_X(x) &= \int_{-\infty}^{\infty} f_{X,Y}(x, y) dy = \int_0^{\infty} xe^{-xy}e^{-x} dy \\ &= e^{-x}\left[-e^{-xy}\right]_0^{\infty} = e^{-x} \quad \text{for } x > 0 \\ \Rightarrow f_{Y|X}(y|x) &= \frac{f_{X,Y}(x, y)}{f_X(x)} = xe^{-xy} \quad \text{for } y > 0. \end{aligned}$$

Thus, $(Y|X = x) \sim \text{Exp}(x)$. We conclude that $\mathbb{E}(Y|X = x) = 1/x$, and so $\mathbb{E}(Y|X) = 1/X$.

**Example 5.4.5** (Using iterated expectations)
In the final example of this subsection we use a question-and-solution format. Consider random variables $X$ and $Y$ with joint density

$$f_{X,Y}(x, y) = \begin{cases} 6y & \text{for } 0 < y < x < 1, \\ 0 & \text{otherwise.} \end{cases}$$

1. Find $f_X$, $\mathbb{E}(X)$, $\mathbb{E}(X^2)$, and $\text{Var}(X)$.

2. Find $f_{Y|X}$ and $\mathbb{E}(Y|X)$
3. Hence evaluate $\mathbb{E}(Y)$ and $\mathrm{Cov}(X,Y)$

**Solution**

1. We start with the marginal density for $X$,

$$f_X(x) = \int_{-\infty}^{\infty} f_{X,Y}(x,y)dy = \int_0^x 6y dy = \left[3y^2\right]_0^x = 3x^2.$$

In full,

$$f_X(x) = \begin{cases} 3x^2 & \text{for } 0 < x < 1, \\ 0 & \text{otherwise.} \end{cases}$$

Using this density we can calculate $\mathbb{E}(X)$, $\mathbb{E}(X^2)$, and $\mathrm{Var}(X)$,

$$\begin{aligned}
\mathbb{E}(X) &= \int_{-\infty}^{\infty} x f_X(x)dx = \int_0^x 3x^3 dx = \left[\frac{3}{4}x^4\right]_0^1 = \frac{3}{4}, \\
\mathbb{E}(X^2) &= \int_{-\infty}^{\infty} x^2 f_X(x)dx = \int_0^1 3x^4 dx = \frac{3}{5}, \\
\mathrm{Var}(X) &= \mathbb{E}(X^2) - \mathbb{E}(X)^2 = \frac{3}{5} - \left(\frac{3}{4}\right)^2 = \frac{3}{80}.
\end{aligned}$$

2. The conditional density is given by

$$f_{Y|X}(y|x) = \frac{f_{X,Y}(x,y)}{f_X(x)} = \frac{2y}{x^2}.$$

In full,

$$f_{Y|X}(y|x) = \begin{cases} 2y/x^2 & \text{for } 0 < y < x, \\ 0 & \text{otherwise.} \end{cases}$$

Using this density we can calculate $\mathbb{E}(Y|X = x)$,

$$\mathbb{E}(Y|X = x) = \int_{-\infty}^{\infty} y f_{Y|X}(y|x)dy = \int_0^x \frac{2y^2}{x^2}dy = \left[\frac{2y^3}{3x^2}\right]_{y=0}^{y=x} = \frac{2}{3}x.$$

Thus, the conditional expectation is

$$\mathbb{E}(Y|X) = \frac{2}{3}X.$$

3. We could work out $\mathbb{E}(Y)$ by finding the marginal density $f_Y$ and then integrating. However, using iterated expectations avoids additional integration:

$$\mathbb{E}(Y) = \mathbb{E}[\mathbb{E}(Y|X)] = \mathbb{E}\left(\frac{2}{3}X\right) = \frac{2}{3}\mathbb{E}(X) = \frac{2}{3}\cdot\frac{3}{4} = \frac{1}{2}.$$

We can exploit iterated expectations in calculating $\text{Cov}(X, Y)$ by noting

$$\mathbb{E}(XY) = \mathbb{E}[\mathbb{E}(XY|X)] = \mathbb{E}[X\,\mathbb{E}(Y|X)] = \mathbb{E}\left(\frac{2}{3}X^2\right) = \frac{2}{3}\cdot\frac{3}{5} = \frac{2}{5}.$$

Thus,

$$\text{Cov}(X, Y) = \mathbb{E}(XY) - \mathbb{E}(X)\,\mathbb{E}(Y) = \frac{2}{5} - \frac{3}{4}\cdot\frac{1}{2} = \frac{1}{40}.$$

You can check these calculations by evaluating $\mathbb{E}(Y)$ and $\mathbb{E}(XY)$ directly.

### 5.4.2 *Conditional moments*

We start with a result that will prove crucial in evaluating conditional moments. We can evaluate the conditional expectation of a function of a random variable using the conditional version of Theorem 3.4.4.

**Claim 5.4.6** (Conditional expectation of a function of $Y$)
*Suppose $g : \mathbb{R} \to \mathbb{R}$ is a well-behaved function. Let $X$ and $Y$ be random variables with conditional mass/density $f_{Y|X}$. If we define*

$$h(x) = \mathbb{E}(g(Y)|X = x) = \begin{cases} \sum_y g(y) f_{Y|X}(y|x) & \textit{(discrete case)}, \\ \int_{-\infty}^{\infty} g(y) f_{Y|X}(y|x)dy & \textit{(continuous case)}, \end{cases}$$

*then the conditional expectation of $g(Y)$ given $X$ is $\mathbb{E}(g(Y)|X) = h(X)$, a random variable.*

An important feature of conditional expectation is that any function of $X$ can be treated as a constant with respect to expectations that are conditional on $X$. Thus,

$$\mathbb{E}(XY|X) = X\,\mathbb{E}(Y|X),$$

and, since $\mathbb{E}(Y|X)$ is a function of $X$,

$$\mathbb{E}[\mathbb{E}(Y|X)Y|X] = \mathbb{E}(Y|X)^2.$$

In general, for well-behaved functions $g$ and $h$,

$$\mathbb{E}[h(X)g(Y)|X] = h(X)\,\mathbb{E}[g(Y)|X].$$

You should always bear in mind that any conditional expectation is a function of the random variable we are conditioning on.

Conditional moments are defined as conditional expectations. Thus, each conditional moment is a function of the random variable we are conditioning on and, as such, is itself a random variable.

**Definition 5.4.7** (Conditional moments and conditional central moments)
For random variables $X$ and $Y$, the $r^{\text{th}}$ **conditional moment** of $Y$ given $X$ is $\mathbb{E}(Y^r|X)$ and the $r^{\text{th}}$ **conditional central moment** of $Y$ given $X$ is $\mathbb{E}[(Y - \mathbb{E}(Y|X))^r|X]$.

Of particular interest is the second conditional central moment, that is, the conditional variance.

**Definition 5.4.8** (Conditional variance)
For random variables $X$ and $Y$, the **conditional variance** of $Y$ given $X$ is

$$\text{Var}(Y|X) = \mathbb{E}[(Y - \mathbb{E}(Y|X))^2|X].$$

We can use Claim 5.4.6 to evaluate the conditional variance. Define

$$\omega(x) = \text{Var}(Y|X = x) = \begin{cases} \sum_y [y - \mathbb{E}(Y|X = x)]^2 f_{Y|X}(y|x) & \text{(discrete case),} \\ \int_{-\infty}^{\infty} [y - \mathbb{E}(Y|X = x)]^2 f_{Y|X}(y|x) dy & \text{(continuous case).} \end{cases}$$

The conditional variance of $Y$ given $X$ is $\text{Var}(Y|X) = \omega(X)$, a random variable. An alternative representation that often proves useful in evaluating the conditional variance is given by the following lemma. The proof is part of Exercise 5.4.

**Lemma 5.4.9** (Alternative representation of conditional variance)
*For random variables $X$ and $Y$, the conditional variance of $Y$ given $X$ can be written as*

$$\text{Var}(Y|X) = \mathbb{E}(Y^2|X) - \mathbb{E}(Y|X)^2.$$

The conditional variance is a random variable. It is natural to ask what the expected value of the conditional variance is. Drawing parallels with the law of iterated expectations, we might suppose that $\mathbb{E}[\text{Var}(Y|X)]$ would be the variance of $Y$. However, this is not quite the case.

**Proposition 5.4.10** (Decomposition of variance)
*For random variables $X$ and $Y$, the variance of $Y$ is given by*

$$\text{Var}(Y) = \mathbb{E}[\text{Var}(Y|X)] + \text{Var}[\mathbb{E}(Y|X)].$$

*Proof.*

The proof exploits iterated expectations.

$$\begin{aligned}
&\text{Var}(Y) \\
&= \mathbb{E}(Y^2) - \mathbb{E}(Y)^2 \\
&= \mathbb{E}[\mathbb{E}(Y^2|X)] - \mathbb{E}[\mathbb{E}(Y|X)]^2 && \text{iterated expectations} \\
&= \mathbb{E}\left[\text{Var}(Y|X) + \mathbb{E}(Y|X)^2\right] - \mathbb{E}[\mathbb{E}(Y|X)]^2 && \text{Lemma 5.4.9} \\
&= \mathbb{E}[\text{Var}(Y|X)] + \mathbb{E}\left[\mathbb{E}(Y|X)^2\right] - \mathbb{E}[\mathbb{E}(Y|X)]^2 && \text{rearranging} \\
&= \mathbb{E}[\text{Var}(Y|X)] + \text{Var}[\mathbb{E}(Y|X)] && \text{variance definition.}
\end{aligned}$$

□

We return to the hurricanes example to illustrate the use of conditional variance and conditional expectation.

**Example 5.4.11** (Moments in hurricanes problem)
In Example 5.3.1 we show that, if $(Y|X = x) \sim \text{Bin}(x,p)$ and $X \sim \text{Pois}(\lambda)$, then $Y \sim \text{Pois}(\lambda p)$. It follows that $\mathbb{E}(Y) = \lambda p$ and $\text{Var}(Y) = \lambda p$. We could have derived these results without working out the mass function of $Y$. Since $X \sim \text{Pois}(\lambda)$, we have

$$\mathbb{E}(X) = \lambda \text{ and } \text{Var}(X) = \lambda,$$

and, since $(Y|X = x) \sim \text{Bin}(x,p)$, we have

$$\mathbb{E}(Y|X) = Xp \text{ and } \text{Var}(Y|X) = Xp(1-p).$$

Using Propositions 5.4.2 and 5.4.10 now yields

$$\mathbb{E}(Y) = \mathbb{E}[\mathbb{E}(Y|X)] = \mathbb{E}(Xp) = p\,\mathbb{E}(X) = p\lambda,$$

and

$$\begin{aligned}\text{Var}(Y) &= \mathbb{E}[\text{Var}(Y|X)] + \text{Var}[\mathbb{E}(Y|X)]\\ &= \mathbb{E}(Xp(1-p)) + \text{Var}(Xp) = p(1-p)\lambda + p^2\lambda = p\lambda.\end{aligned}$$

**Example 5.4.12** (Exploiting variance formula)
Suppose $X$ and $Y$ are random variables with joint density $f_{X,Y}(x,y) = 6y$, for $0 < y < x < 1$.

1. Find the conditional variance $\text{Var}(Y|X)$.
2. Hence evaluate $\text{Var}(Y)$.

**Solution**

1. We start by calculating $\text{Var}(Y|X = x)$. We know that

   $$\text{Var}(Y|X = x) = \mathbb{E}(Y^2|X = x) - \mathbb{E}(Y|X = x)^2.$$

   In Example 5.4.5 we show $\mathbb{E}(Y|X = x) = \frac{2}{3}x$ and give an expression for the conditional density $f_{Y|X}(.|x)$. We can find $\mathbb{E}(Y^2|X = x)$ using this conditional density,

   $$\mathbb{E}(Y^2|X = x) = \int_{-\infty}^{\infty} y^2 f_{Y|X}(y|x)dy = \int_0^x \frac{2y^3}{x^2}dy = \left[\frac{2y^4}{4x^2}\right]_{y=0}^{y=x} = \frac{1}{2}x^2.$$

   Thus,

   $$\begin{aligned}\text{Var}(Y|X = x) &= \frac{1}{2}x^2 - \left(\frac{2}{3}x\right)^2 = \frac{1}{18}x^2\\ \Rightarrow \text{Var}(Y|X) &= \frac{1}{18}X^2.\end{aligned}$$

2. From Proposition 5.4.10 we have

$$\operatorname{Var}(Y) = \mathbb{E}[\operatorname{Var}(Y|X)] + \operatorname{Var}[\mathbb{E}(Y|X)].$$

Substituting previously calculated expressions yields

$$\operatorname{Var}(Y) = \mathbb{E}\left(\frac{1}{18}X^2\right) + \operatorname{Var}\left(\frac{2}{3}X\right) = \frac{1}{18}\cdot\frac{3}{5} + \left(\frac{2}{3}\right)^2\frac{3}{80} = \frac{1}{20}.$$

As before, this expression can be checked by direct evaluation.

### 5.4.3 *Conditional moment-generating functions*

The conditional moment-generating function is defined as a conditional expectation.

**Definition 5.4.13** (Conditional moment-generating function)
For random variables $X$ and $Y$, the **conditional moment-generating function** of $Y$ given $X$ is

$$M_{Y|X}(u|X) = \mathbb{E}(e^{uY}|X).$$

The conditional moment-generating function can be evaluated using Claim 5.4.6. If we define

$$v(x) = M_{Y|X}(u|x) = \mathbb{E}(e^{uY}|X = x) = \begin{cases} \sum_y e^{uy} f_{Y|X}(y|x) & \text{(discrete case)}, \\ \int_{-\infty}^{\infty} e^{uy} f_{Y|X}(y|x)dy & \text{(continuous case)}. \end{cases}$$

The conditional moment-generating function of $Y$ given $X$ is $M_{Y|X}(u|X) = v(X)$. This is a conditional expectation, so it is a random variable.

The conditional moment-generating function can be used to calculate the marginal moment-generating function of $Y$ and the joint moment-generating function of $X$ and $Y$.

**Proposition 5.4.14** (Properties of conditional moment-generating function)
*If $X$ and $Y$ are random variables with conditional moment-generating function $M_{Y|X}(u|X)$, then*

*i.* $M_Y(u) = \mathbb{E}[M_{Y|X}(u|X)]$,
*ii.* $M_{X,Y}(t,u) = \mathbb{E}[e^{tX} M_{Y|X}(u|X)]$,

*where $M_Y$ is the marginal moment-generating function of $Y$, and $M_{X,Y}$ is the joint moment-generating function of $X$ and $Y$.*

Both of these results are consequences of the law of iterated expectations. Proving them is part of Exercise 5.4. In the following example we show how the hurricanes problem can be tackled using moment-generating functions.

**Example 5.4.15** (Hurricanes problem using moment-generating functions)
The moment-generating function results of Proposition 5.4.14 provide another approach to the hurricane problem. Since $X \sim \text{Pois}(\lambda)$ and $(Y|X = x) \sim \text{Bin}(x, p)$, we have

$$M_X(t) = \exp[\lambda(e^t - 1)] \text{ and } M_{Y|X}(u|X) = (1 - p + pe^t)^X.$$

By Proposition 5.4.14,

$$\begin{aligned}
M_Y(u) &= \mathbb{E}[M_{Y|X}(u|X)] \\
&= \mathbb{E}[(1 - p + pe^t)^X] && \text{substituting conditional MGF} \\
&= \mathbb{E}[\exp\{X \log(1 - p + pe^t)\}] && \text{using } a^b = \exp(b \log a) \\
&= M_X(\log(1 - p + pe^t)) && \text{by definition of } M_X \\
&= \exp\{\lambda[(1 - p + pe^t) - 1]\} && \text{since } X \sim \text{Pois}(\lambda) \\
&= \exp[\lambda p(e^t - 1)].
\end{aligned}$$

This is the moment-generating function of a Poisson. As before, we conclude that $Y \sim \text{Pois}(\lambda p)$.

**Exercise 5.4**

1. Suppose that $X$ and $Y$ are random variables. Show that the conditional variance of $Y$ given $X$ can be written as
$$\text{Var}(Y|X) = \mathbb{E}(Y^2|X) - \mathbb{E}(Y|X)^2.$$
2. Suppose that $X$ and $Y$ are random variables with conditional moment-generating function $M_{Y|X}(u|X)$. Show that
   i. $M_Y(u) = \mathbb{E}[M_{Y|X}(u|X)]$,
   ii. $M_{X,Y}(t, u) = \mathbb{E}[e^{tX} M_{Y|X}(u|X)]$,

   where $M_Y$ is the marginal moment-generating function of $Y$, and $M_{X,Y}$ is the joint moment-generating function of $X$ and $Y$.
3. Consider random variables $X$ and $Y$ with joint density
$$f_{X,Y}(x, y) = \begin{cases} 2 & \text{for } 0 < y < x < 1, \\ 0 & \text{otherwise.} \end{cases}$$
   Find $\mathbb{E}(X)$, $\mathbb{E}(X^2)$, $f_{Y|X}$, and $\mathbb{E}(Y|X)$. Hence evaluate $\mathbb{E}(Y)$ and $\text{Cov}(X, Y)$.

## 5.5 Hierarchies and mixtures

In many situations it is natural to build a model as a hierarchy. In a **hierarchical model**, the random variable that we are interested in, say $Y$, has a distribution that depends on other random variables. In this situation we say that $Y$ has a **mixture distribution**. Consider the following illustrations. A two-level hierarchical model can be summarised as follows. If $Y$ is the quantity of interest, in the first instance we do not know the marginal distribution of $Y$ directly. Instead, the situation is most naturally described in terms of the conditional distribution of $Y$ given $X = x$ and the marginal distribution of $X$. We have encountered this situation in the hurricanes problem.

**Example 5.5.1** (Hurricanes problem as mixing)
In the hurricanes problem we are interested in the number of landfalling hurricanes $Y$. Conditional on the number of hurricanes that form in the basin, $X$, we know the distribution of $Y$. Thus, if the marginal distribution of $X$ is known, we can derive the marginal distribution of $Y$. This is precisely what we do in Example 5.3.1 (using mass functions) and in Example 5.4.15 (using moment-generating functions).

The theory developed in sections 5.3 and 5.4 is used to solve problems involving hierarchies. The key results are

$$\begin{aligned}\mathbb{E}(Y) &= \mathbb{E}[\mathbb{E}(Y|X)],\\ \text{Var}(Y) &= \mathbb{E}[\text{Var}(Y|X)] + \text{Var}[\mathbb{E}(Y|X)],\\ f_Y(y) &= \mathbb{E}[f_{Y|X}(y|X)],\\ M_Y(u) &= \mathbb{E}[M_{Y|X}(u|X)]. \end{aligned} \tag{5.4}$$

The first two results are useful if we are only interested in calculating $\mathbb{E}(Y)$ and $\text{Var}(Y)$. Either of the last two results allows us to characterise the entire distribution of $Y$.

**Example 5.5.2** (Poisson mixing)
Suppose that $(Y|\Lambda = \lambda) \sim \text{Pois}(\lambda)$ for some positive random variable $\Lambda$. We can immediately see that

$$\mathbb{E}(Y|\Lambda) = \Lambda \text{ and } \text{Var}(Y|\Lambda) = \Lambda.$$

Using the results summarised in equation (5.4) we have

$$\begin{aligned}\mathbb{E}(Y) &= \mathbb{E}(\Lambda),\\ \text{Var}(Y) &= \mathbb{E}(\Lambda) + \text{Var}(\Lambda).\end{aligned}$$

**Exercise 5.5**

1. Show that if $(Y|\Lambda = \lambda) \sim \text{Pois}(\lambda)$ then $M_Y(t) = M_\Lambda(e^t - 1)$.
2. Find the moment-generating function for $Y$ in the insurance claim example of Exercise 5.3.
3. We can choose one of two investment opportunities:

   (A) has return £$X$ with density $f_X$,

   (B) has return £$Y$ with density $f_Y$.

   We choose (A) with probability $p$ and (B) with probability $(1 - p)$. The return on our investment is $Z$. Formulate this setup as a mixture. Give the density, expected value, and variance of $Z$ in terms of the density, expected value, and variance of $X$ and $Y$.

## 5.6 Random sums

In section 4.7 we discuss sums of random variables. We assume that the number of random variables we are adding together is some fixed number, often denoted $n$. It is easy to describe situations in which the number of random variables that we are summing is not fixed but can be modelled using a given distribution. Consider the following illustration.

**Example 5.6.1** (Health insurance)
Each year the value of claims (in £000s) made by individual owners of a health insurance policy are exponentially distributed with mean $\alpha$, and independent of previous claims. At the end of each year, there is probability $p$ that an individual will cancel their policy, independent of previous years (and of the value of the claims). We are interested in the total cost of this policy to the insurer. This problem can be formulated as follows. Let $X_j$ be the amount claimed in year $j$ and let $N$ be the number of years for which the policy is held. From the question, we know that

$$X_j \sim \text{Exp}\left(\frac{1}{\alpha}\right) \quad \text{and } N \sim \text{Geometric}(p),$$

where the $X_j$s are independent of each other, and also independent of $N$. The total cost to the insurer is then

$$S = X_1 + \ldots + X_N = \sum_{j=1}^{N} X_j.$$

We will return to this example after we have developed the theory needed to find the moments and the distribution of $S$.

We refer to sums of random variables in which the number of variables included in the sum is itself a random variable as **random sums**. The situation is typically described as follows

$$Y = X_1 + \ldots + X_n = \sum_{j=1}^{N} X_j.$$

We will assume that $\{X_j\}$ is a sequence of independent identically distributed random variables. A random sum can be viewed as a mixture; conditional on $N$, the sum $Y$ is just a sum of independent identically and distributed random variables,

$$(Y|N = n) = \sum_{j=1}^{n} X_j.$$

The results that we derive in section 4.7 for sums of random variables can be applied to $Y|N = n$. Then, using our definitions of conditional expectation, conditional variance, and the conditional moment-generating function, we can prove the following lemma (the proof is part of Exercise 5.6).

**Lemma 5.6.2** (Conditional results for random sums)
*Suppose that $\{X_j\}$ is a sequence of independent and identically distributed random variables, and that $N$ is a random variable taking non-negative integer values (independently of the $\{X_j\}$). If $Y = \sum_{j=1}^{N} X_j$, then*

*i.* $\mathbb{E}(Y|N) = N\,\mathbb{E}(X)$,

*ii.* $\text{Var}(Y|N) = N\,\text{Var}(X)$,

*iii.* $M_{Y|N}(t|N) = [M_X(t)]^N$,

*iv.* $K_{Y|N}(t|N) = NK_X(t)$.

The marginal results for $Y$ follow by applications of results from sections 5.3 and 5.4.

**Proposition 5.6.3** (Marginal results for random sums)
*Suppose that $\{X_j\}$ is a sequence of independent and identically distributed random variables, and that $N$ is a random variable taking non-negative integer values (independently of the $\{X_j\}$). If $Y = \sum_{j=1}^{N} X_j$, then*

*i.* $\mathbb{E}(Y) = \mathbb{E}(N)\,\mathbb{E}(X)$,

*ii.* $\text{Var}(Y) = \mathbb{E}(N)\,\text{Var}(X) + \mathbb{E}(X)^2\,\text{Var}(N)$,

*iii.* $M_Y(t) = M_N(\log M_X(t))$,

*iv.* $K_Y(t) = K_N(K_X(t))$.

Proving this proposition is part of Exercise 5.6.

**Example 5.6.1 (Revisited)** We have established that $X_j \sim \text{Exp}(\frac{1}{\alpha})$ and $N \sim \text{Geometric}(p)$. Thus, $\mathbb{E}(X) = \alpha$ and $\mathbb{E}(N) = \frac{1}{p}$. If we are only interested in the average cost, we can use the result $\mathbb{E}(S) = \mathbb{E}(N)\,\mathbb{E}(X) = \frac{\alpha}{p}$. In order to get the distribution we use the cumulant-generating function result. The cumulant-generating functions of $X$ and $N$ are

$$\begin{aligned} K_X(t) &= -\log(1-\alpha t), \\ K_N(t) &= -\log\left(1 - \frac{1}{p} + \frac{1}{p}e^{-t}\right). \end{aligned}$$

Make sure you understand why this is the case. The cumulant-generating function result then gives

$$\begin{aligned} K_S(t) &= K_N(K_X(t)) \\ &= -\log\left(1 - \frac{1}{p} + \frac{1}{p}(1-\alpha t)\right) \\ &= -\log\left(1 - \frac{\alpha}{p}t\right). \end{aligned}$$

This is the moment-generating function of an exponential. We conclude that

$$S \sim \text{Exp}\left(\frac{p}{\alpha}\right).$$

**Exercise 5.6**

1. Prove Lemma 5.6.2 and Proposition 5.6.3.
2. Show that the landfalling hurricanes problem can be formulated as a random sum.
3. Suppose that we have a sequence of independent and identically distributed random variables $X_1, X_2, \ldots$. We define the random sum $S = \sum_{j=1}^{N} X_j$ where $N \sim \text{Pois}(\lambda)$, and $N$ is independent of the $X_j$. Show that
   (a) $\mathbb{E}(S) = \lambda\,\mathbb{E}(X)$,
   (b) $\text{Var}(S) = \lambda\,\mathbb{E}(X^2)$,
   (c) the $r^{\text{th}}$ cumulant of $S$ is $\lambda\,\mathbb{E}(X^r)$.

## 5.7 Conditioning for random vectors

Consider the random vectors $\boldsymbol{X} = (X_1, \ldots, X_m)^T$ and $\boldsymbol{Y} = (Y_1, \ldots, Y_n)^T$. The joint of $\boldsymbol{X}$ and $\boldsymbol{Y}$ is just the joint of all the variables in $\boldsymbol{X}$ and all the variables in $\boldsymbol{Y}$,

$$f_{\boldsymbol{X},\boldsymbol{Y}}(\boldsymbol{x}, \boldsymbol{y}) = f_{X_1,\ldots,X_m,Y_1,\ldots,Y_n}(x_1, \ldots, x_m, y_1, \ldots, y_n)$$

We can define conditional mass/density in exactly the same way as in the univariate case,

$$f_{\boldsymbol{Y}|\boldsymbol{X}}(\boldsymbol{y}|\boldsymbol{x}) = \frac{f_{\boldsymbol{X},\boldsymbol{Y}}(\boldsymbol{x}, \boldsymbol{y})}{f_{\boldsymbol{X}}(\boldsymbol{x})} \quad \text{where} \quad f_{\boldsymbol{X}}(\boldsymbol{x}) > 0.$$

We will make use of the expression

$$f_{\boldsymbol{X},\boldsymbol{Y}}(\boldsymbol{x}, \boldsymbol{y}) = f_{\boldsymbol{Y}|\boldsymbol{X}}(\boldsymbol{y}|\boldsymbol{x}) f_{\boldsymbol{X}}(\boldsymbol{x}). \tag{5.5}$$

Consider the case of three random variables $X_1$, $X_2$, and $X_3$. We can start by grouping these as $(x_3)$ and $(x_1, x_2)$. Equation (5.5) then implies

$$f_{X_1,X_2,X_3}(x_1, x_2, x_3) = f_{X_3|X_1,X_2}(x_3|x_1, x_2) f_{X_1,X_2}(x_1, x_2). \tag{5.6}$$

Applying (5.5) again yields $f_{X_1,X_2}(x_1, x_2) = f_{X_2|X_1}(x_2|x_1) f_{X_1}(x_1)$. Substituting this result in equation (5.6) gives us

$$f_{X_1,X_2,X_3}(x_1, x_2, x_3) = f_{X_3|X_1,X_2}(x_3|x_1, x_2) f_{X_2|X_1}(x_2|x_1) f_{X_1}(x_1).$$

In general, we can decompose the joint mass/density function of a random vector into a product of conditional mass/density functions. The following proposition provides the details.

**Proposition 5.7.1** (Decomposition of mass/density)
*If we define an n-dimensional random vector* $\boldsymbol{X}_n = (X_1, \ldots, X_n)^T$, *and* $\boldsymbol{x}_n = (x_1, \ldots, x_n)^T \in \mathbb{R}^n$, *then*

$$\begin{aligned} f_{\boldsymbol{X}_n}(\boldsymbol{x}_n) &= f_{X_n|\boldsymbol{X}_{n-1}}(x_n|\boldsymbol{x}_{n-1}) f_{X_{n-1}|\boldsymbol{X}_{n-2}}(x_{n-1}|\boldsymbol{x}_{n-2}) \ldots f_{X_2|X_1}(x_2|x_1) f_{X_1}(x_1) \\ &= \prod_{j=1}^{n} f_{X_j|\boldsymbol{X}_{j-1}}(x_j|\boldsymbol{x}_{j-1}), \end{aligned}$$

*where we define* $f_{X_1|X_0}(x_1|x_0) = f_{X_1}(x_1)$.

The proof is part of Exercise 5.7.

We now state the multivariate version of a result we encountered in Exercise 5.2.

**Proposition 5.7.2** (Conditional distribution for multivariate normal)
*Suppose that the random vectors* $\boldsymbol{X} = (X_1, \dots, X_n)^T$ *and* $\boldsymbol{Y} = (Y_1, \dots, Y_m)^T$ *(for some integers n and m) are jointly normal, with* $\boldsymbol{X} \sim \mathrm{N}(\boldsymbol{\mu}_X, \boldsymbol{\Sigma}_X)$ *and* $\boldsymbol{Y} \sim \mathrm{N}(\boldsymbol{\mu}_Y, \boldsymbol{\Sigma}_Y)$*. If, in addition,* $\mathrm{Cov}(\boldsymbol{X}, \boldsymbol{Y}) = \boldsymbol{\Sigma}_{XY} = \boldsymbol{\Sigma}_{YX}^T$*, then*

$$\begin{aligned}\mathbb{E}(\boldsymbol{Y}|\boldsymbol{X}) &= \boldsymbol{\mu}_Y + \boldsymbol{\Sigma}_{YX}\boldsymbol{\Sigma}_X^{-1}(\boldsymbol{X} - \boldsymbol{\mu}_X),\\ \mathrm{Var}(\boldsymbol{Y}|\boldsymbol{X}) &= \boldsymbol{\Sigma}_Y - \boldsymbol{\Sigma}_{YX}\boldsymbol{\Sigma}_X^{-1}\boldsymbol{\Sigma}_{XY},\end{aligned}$$

*and the conditional distribution of* $\boldsymbol{Y}$ *given* $\boldsymbol{X} = \boldsymbol{x}$ *is multivariate normal.*

A proof can be found in Appendix A. Notice that the conditional variance, $\mathrm{Var}(\boldsymbol{Y}|\boldsymbol{X})$, is independent of $\boldsymbol{X}$, hence constant.

**Exercise 5.7**

1. Prove Proposition 5.7.1.
2. Let $\boldsymbol{X}$, $\boldsymbol{Y}$, and $\boldsymbol{Z}$ be jointly normally distributed random vectors. If $\boldsymbol{Y}$ and $\boldsymbol{Z}$ are uncorrelated, show that

$$\mathbb{E}(\boldsymbol{X}|\boldsymbol{Y}, \boldsymbol{Z}) = \mathbb{E}(\boldsymbol{X}|\boldsymbol{Y}) + \mathrm{Cov}(\boldsymbol{X}, \boldsymbol{Z})\,\mathrm{Var}(\boldsymbol{Z})^{-1}[\boldsymbol{Z} - \mathbb{E}(\boldsymbol{Z})]$$

   and

$$\mathrm{Var}(\boldsymbol{X}|\boldsymbol{Y}, \boldsymbol{Z}) = \mathrm{Var}(\boldsymbol{X}|\boldsymbol{Y}) - \mathrm{Cov}(\boldsymbol{X}, \boldsymbol{Z})\,\mathrm{Var}(\boldsymbol{Z})^{-1}\mathrm{Cov}(\boldsymbol{Z}, \boldsymbol{X})\,.$$

## 5.8 Further exercises

1. Let $X_1, X_2, \dots$ be a sequence of independent and identically distributed random variables with moment-generating function $M_X$. Let $S = \sum_{i=1}^{N} X_i$ where $N \sim \mathrm{Pois}(\beta)$.
   (a) Show that $\mathbb{E}(S) = \beta\,\mathbb{E}(X)$ and $\mathrm{Var}(S) = \beta\,\mathbb{E}(X^2)$.
   (b) Show that the cumulant-generating function of $S$ is $K_S(t) = \beta(M_X(t) - 1)$.
2. The total number of offspring produced by an individual of a species of small rodent follows a $\mathrm{Pois}(4\lambda)$ distribution, where $\lambda$ is the length of its life in months. The distribution of length of life in months is $\Lambda \sim \mathrm{Exp}(0.05)$. If $R$ is the distribution of the total number of offspring, find $\mathbb{E}(R)$, $\mathrm{Var}(R)$, and $\mathrm{Cov}(R, \Lambda)$.
3. I buy $N$ raffle tickets, where $N \sim \mathrm{Pois}(\lambda)$. Each ticket costs £1. For each ticket, I have probability $\frac{1}{50}$ of winning £10, otherwise I win nothing. You may assume independence of amounts won on each ticket.
   (a) Let $X$ be my profit from a single ticket. Write down the moment-generating function of $X$.
   (b) Let $S$ be my total profit. Find an expression for the cumulant-generating function of $S$ and hence evaluate $\mathbb{E}(S)$.

4. The annual number of hurricanes forming in the Atlantic basin has a Poisson distribution with parameter $\lambda$. Each hurricane that forms has probability $p$ of making landfall, independent of all other hurricanes. Let $X$ be the number of hurricanes that form in the basin and $Y$ be the number that make landfall. Find $\mathbb{E}(X|Y)$, $\text{Var}(X|Y)$, $\text{Cov}(X,Y)$, and $\text{Corr}(X,Y)$.

5. Consider the following function of two variables:

$$f_{X,Y}(x,y) = \begin{cases} 4y^2, & 0 < x < y < 1, \\ 0, & \text{otherwise.} \end{cases}$$

   (a) Show that $f_{X,Y}$ is a valid density function.
   (b) Derive an expression for the marginal density $f_Y(y)$ and hence evaluate $\mathbb{E}(Y)$ and $\text{Var}(Y)$.
   (c) Derive the conditional density, $f_{X|Y}(x|y)$, and the conditional expectation, $\mathbb{E}(X|Y)$. Hence evaluate $\mathbb{E}(X)$.
   (d) Derive an expression for $\mathbb{E}(X^r Y^s)$ where $r$ and $s$ are positive integers. Hence evaluate $\text{Corr}(X,Y)$.
   (e) Evaluate $\text{P}(2X > Y \mid Y < \frac{1}{2})$.

6. Consider the following game involving two players and a bag containing 3 disks; 2 blue and 1 red. The players take turns. At each turn, the player puts £X into the kitty, removes a disk from the bag, looks at the colour, and replaces it in the bag. If the disk is blue, the game continues with the other player's turn. If the disk is red the game stops and the player who picked the red disk wins the money in the kitty. Suppose $X \sim \text{Exp}(\theta)$. Let $Y$ be the number of turns in a game, so $Y = 1$ if the red disk is chosen on the first turn. Let $Z$ be the amount of money in the kitty when the game ends.

   (a) Evaluate $\text{P}(Y = 1)$, $\text{P}(Y = 2)$ and $\text{P}(Y = 3)$. Write down the probability mass function of $Y$.
   (b) Derive the moment-generating function of $X$ and the moment-generating function of $Y$. Give intervals around the origin for which each function is well defined.
   (c) Derive the moment-generating function of $Z$, $M_Z(t)$. Express $M_Z(t)$ as a polynomial in $t$, and hence find $\mathbb{E}(Z)$ and $\text{Var}(Z)$.
   (d) Evaluate the probability that the person who has the first turn wins the game. Given the choice, would you choose to start or go second in the game? Explain your reasoning.

7. Suppose that $(Y|Z = z) \sim \text{N}(0, z)$ where $Z \sim \text{Exp}(\lambda)$. Find the variance and the fourth cumulant of $Y$.

8. Let $X$ and $Y$ be random variables with joint density

$$f_{X,Y}(x,y) = \begin{cases} ke^{-\lambda x}, & 0 < y < x < \infty, \\ 0, & \text{otherwise.} \end{cases}$$

   (a) Find $k$.

(b) Derive the conditional density, $f_{X|Y}(x|y)$, and the conditional expectation, $\mathbb{E}(X|Y)$.

(c) Evaluate $\mathbb{E}(X)$ and $\text{Cov}(X,Y)$.

9. For the multinomial distribution in Exercise 4.9 question 12, work out the conditional mass function of $X_i|X_j = x$ for $i \neq j$, and show that $\text{Cov}(X_i, X_j) = -np_i p_j$.

## 5.9 Chapter summary

On completion of this chapter you should be able to:

- find conditional mass/density functions from joint mass/density functions,
- write down the relationship between joint, marginal, and conditional, and exploit this relationship to solve problems,
- calculate conditional expectations of a function of a random variable,
- use the law of iterated expectations,
- prove that variance can be decomposed as the sum of the expected value of the conditional variance and the variance of the conditional expectation,
- find and use conditional moment-generating functions and conditional cumulant-generating functions, and
- solve problems involving hierarchies, mixtures, and random sums.

## Chapter 6

# Statistical models

In this chapter, we exploit the terminology of distribution theory to provide a brief account of statistical modelling. Statistical analysis requires data, which may be viewed as an observed sample from a population or, more generally, as the recorded output of a process. Models for processes that generate data are the fundamental cogs in our inferential machine.

The coverage here is broad rather than deep. We touch on linear modelling (and its generalised counterpart), survival analysis, time series models, and stochastic processes. For each of the approaches we consider, models are specified in terms of the properties of a sequence of random variables. We begin by establishing some basic terminology for dealing with these sequences.

## 6.1 Modelling terminology, conventions, and assumptions

In its simplest form, a data set is a collection of real numbers, $y_1, \ldots, y_n$. The data can be thought of as one instance of a collection of random variables, $Y_1, \ldots, Y_n$. The properties that we ascribe to these random variables form our model for the data.

### 6.1.1 *Sample, observed sample, and parameters*

Many statistical problems may be thought of in the context of a sample from a population. The terminology is useful even when there is no well-defined population to sample from. Although it may sometimes be rather cumbersome, we will attempt to maintain the distinction between the observed sample, which is a collection of numbers, and the sample, which is the corresponding collection of random variables. To clarify:

- The **observed sample** is a collection of numbers, also known as the observations, the data or the data set. These are represented using lowercase letters, for example, $y_1, \ldots, y_n$. In some instances it may be notationally convenient to treat the observed sample as a vector of real numbers, $\boldsymbol{y} = (y_1, \ldots, y_n)^T$. We use both $y_1, \ldots, y_n$ and $\boldsymbol{y}$ to refer to the observed sample.

- The **sample** is a collection of random variables whose properties form our model for the data. These are represented using uppercase letters, for example, $Y_1, \ldots, Y_n$. As with the observed sample, we will occasionally treat the sample as a vector of random variables and use the notation $\boldsymbol{Y} = (Y_1, \ldots, Y_n)^T$.

A common constituent of model definitions is a collection of parameters. In the first instance we will treat the parameters as fixed, unknown constants. The usual convention is to use lowercase Greek characters for parameters. For example, $\mu$ is often used to denote the mean, and $\sigma^2$ to denote the variance. We will use $\theta$ as a generic parameter. In situations where there is more than one parameter, we will refer to the parameter vector $\boldsymbol{\theta} = (\theta_1, \ldots, \theta_r)^T$. We may also treat $\boldsymbol{\theta}$ as a single (vector) parameter.

### 6.1.2 *Structural and distributional assumptions*

The properties of a sample are completely characterised by the joint distribution of the sample members. However, rather than referring explicitly to the joint distribution, we often specify models in terms of properties from which the joint distribution may be inferred. These assumptions are divided into two broad categories.

- Structural: describe the relationships between members of the sample.
- Distributional: describe the marginal distributions of the members of the sample.

A widely used structural assumption is that the members of the sample are independent and identically distributed (IID); these terms are introduced in section 4.4. Sometimes slightly weaker structural assumptions are used. For example, we might assume that the members of the sample are uncorrelated (rather than independent) or that they are identically distributed (without making any assumptions about the nature of the dependence between sample members). If the members of the sample, $\boldsymbol{Y} = (Y_1, \ldots, Y_n)^T$, are identically distributed, we may make a assumption about their shared distributional form and use $Y$ (without an index) to denote a generic sample member. For example, we may take all sample members to be normally distributed with mean $\mu$ and variance $\sigma^2$. In this case, a generic sample member is $Y \sim \mathrm{N}(\mu, \sigma^2)$.

**Exercise 6.1**

1. Describe the advantages and potential pitfalls of building a model in which we make structural assumptions, that is, assumptions about the relationship between variables.
2. Carefully explain why a sequence of identically distributed, uncorrelated (jointly) normal random variables is an IID sequence.

## 6.2 Independent and identically distributed sequences

Sequences of independent and identically distributed random variables arise in a number of different modelling contexts. In many situations an IID sequence is taken to be a reasonable model for the observations themselves. When we come to introduce models in which there is dependence between variables, IID sequences feature as **error** or **noise** terms. We start by describing circumstances under which an IID sequence is a reasonable model for the data.

### *6.2.1 Random sample*

We shall often assume that the sample consists of independent and identically distributed random variables drawn from the population distribution. This is referred to as a **random sample.**

**Definition 6.2.1** (Random sample)
A collection of random variables $Y_1, \ldots, Y_n$ is a random sample from a population with cumulative distribution function $F_Y$ if $Y_1, \ldots, Y_n$ are mutually independent and $Y_i \sim F_Y$ for $i = 1, \ldots, n$.

A random sample is used as a model for situations where we sample from a population and every member of the population has equal probability of being selected. In fact, if the population is a tangible, finite collection of individuals, we should sample the population with replacement. When sampling with replacement, each time we sample, we draw from the population distribution. Thus, the sample $Y_1, \ldots, Y_n$, consists of identically distributed random variables, as each of them has the population distribution, $Y_i \sim F_Y$ for $i = 1, \ldots, n$. In addition, $Y_1, \ldots, Y_n$ are mutually independent. To illustrate independence, suppose that we know that the first member of the observed sample is $y_1$. This knowledge does not change the distribution of $Y_2$, since

$$P(Y_2 \leq y_2 | Y_1 = y_1) = P(Y_2 \leq y_2) = F_Y(y_2).$$

If we sample without replacement, the first sample member is drawn from the population distribution but, as there is no replacement, each subsequent member is chosen from a smaller group. The composition of this smaller group depends on which members of the population have already been included in the sample. Thus, the random variables making up the sample are neither identically distributed nor independent. However, if the population is large, removing a few individuals will have little effect on the distribution of the remaining group. We conclude that, if the sample size is small relative to the size of the population, a random sample may be a good approximate model when sampling from the population without replacement.

### *6.2.2 Error sequences*

We often use model formulations in which IID variables enter as error terms. We use $\varepsilon$ to denote a generic error term. We will often adopt the convention that error terms have mean 0 and variance 1. A random sample from a population with mean $\mu$ and variance $\sigma^2$ can be represented using error sequence notation. Thus, if $\{Y_i\} \sim \text{IID}(\mu, \sigma^2)$, we can write the $i^{\text{th}}$ sample member as

$$Y_i = \mu + \sigma\varepsilon_i \text{ where } \{\varepsilon_i\} \sim \text{IID}(0, 1). \tag{6.1}$$

The real benefits of this type of representation become apparent when we consider models in which the variables are dependent. The first model of this class that we consider are linear models in which the variable of interest, $Y$, is dependent on another variable, $X$.

**Exercise 6.2**

1. A political polling company conducts a telephone survey based on dialling random digits. Is the resulting sample a random sample of the electorate?
2. We draw a random sample of size $n$ from a finite population of size $N$ (with replacement). What is the probability that at least one member of the population appears more than once in the sample?

## 6.3 Linear models

Consider a situation in which we have two variables $X$ and $Y$. We believe that changes in the value of $Y$ can be explained by changes in the value of $X$. In this context, $X$ is referred to as the explanatory variable and $Y$ is referred to as the response. Both of these terms have alternative names. Explanatory variables may be referred to (in slightly different contexts) as independent variables, covariates, factors, treatments or predictors; the response variable is sometimes referred to as the dependent variable or target variable. One way in which we could model the relationship between $Y$ and $X$ is as a straight line. This simple idea is the basis for a powerful and widely used class of models. We refer to this class as **linear models**. The starting point for a discussion of linear models is **linear regression**.

### 6.3.1 *Simple linear regression*

In a regression model the association between $Y$ and $X$ is represented by a straight line. Suppose we have a random sample of pairs $\{(X_1, Y_1), \dots, (X_n, Y_n)\}$. Using the notation of equation (6.1) we can write, for $i = 1, \dots, n$,

$$Y_i = \alpha + \beta X_i + \sigma\varepsilon_i \quad \text{where} \quad \{\varepsilon_i\} \sim \text{IID}(0, 1). \tag{6.2}$$

The usual practice is to conduct inference by conditioning on the observed values of the variable $X$. Thus, equation (6.2) becomes

$$(Y_i | X_i = x_i) = \alpha + \beta x_i + \sigma\varepsilon_i. \tag{6.3}$$

A convenient way to think about the model is in terms of the conditional mean and variance functions,

$$\mu(x_i) = \mathbb{E}(Y_i | X_i = x_i) = \alpha + \beta x_i,$$
$$\sigma^2(x_i) = \text{Var}(Y_i | X_i = x_i) = \sigma^2.$$

The model defined by (6.2) has a single explanatory variable and for this reason is often referred to as **simple** linear regression.

You may have encountered such models expressed in the more common notation

$$Y_i = \alpha + \beta X_i + \varepsilon_i \tag{6.4}$$

where each error term $\varepsilon_i$ has mean 0 and variance $\sigma^2$. This is equivalent to equation (6.2), but our notation has the advantage that it makes explicit the presence of the unknown parameter $\sigma^2$.

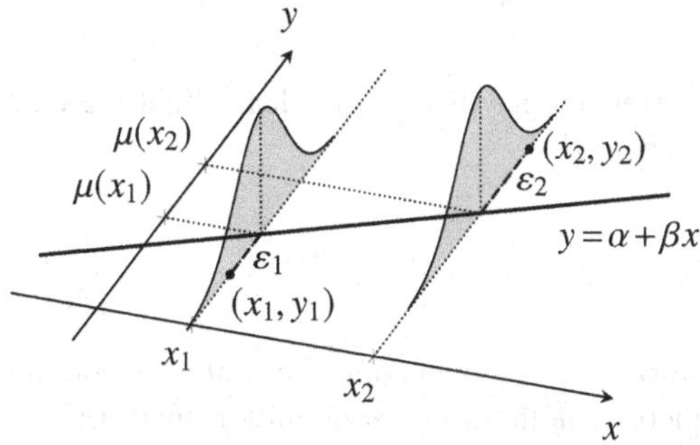


Figure 6.1 *A visual illustration of simple linear regression. For a given value of x, the value of y is drawn from a density with mean $\mu_x$ and constant variance $\sigma^2$. In the diagram, the grey shaded curves have different means but the same variance. The error term $\varepsilon_i$ corresponds to the (signed) distance between $\mu(x_i)$ and $y_i$, so $\varepsilon_1 < 0$ and $\varepsilon_2 > 0$.*

### 6.3.2 *Multiple linear regression*

Suppose now that we have $p$ explanatory variables, $X_1, \dots, X_p$. Note the change of notation from the simple regression case; $X_1, \dots, X_p$ is a collection of $p$ different variables, not a sample of $p$ variables from the same distribution. A sample of size $n$ consists of $n$ values of the response, each of which is associated with $p$ explanatory variable values. We could express this as

$$\{(X_{1,1}, \dots, X_{1,p}, Y_1), \dots, (X_{n,1}, \dots, X_{n,p}, Y_n)\},$$

where $X_{i,j}$ denotes the $i^{\text{th}}$ observation from the $j^{\text{th}}$ variable. The model for linear association between $Y$ and the $p$ explanatory variables is then

$$Y_i = \beta_0 + X_{i,1}\beta_1 + \ldots + X_{i,p}\beta_p + \sigma\varepsilon_i \quad \text{for } i = 1, \dots, n. \tag{6.5}$$

The $n$ equations defined by (6.5) can be neatly summarised in vector form as

$$\boldsymbol{Y} = \boldsymbol{X}\boldsymbol{\beta} + \sigma\boldsymbol{\varepsilon}, \tag{6.6}$$

where

$$\boldsymbol{Y} = \begin{pmatrix} Y_1 \\ Y_2 \\ \vdots \\ \vdots \\ Y_n \end{pmatrix}, \quad \boldsymbol{X} = \begin{pmatrix} 1 & X_{1,1} & \dots & \dots & X_{1,p} \\ 1 & X_{2,1} & \dots & \dots & X_{2,p} \\ \vdots & \vdots & & & \vdots \\ \vdots & \vdots & & & \vdots \\ 1 & X_{n,1} & \dots & \dots & X_{n,p} \end{pmatrix},$$

$$\boldsymbol{\beta} = \begin{pmatrix} \beta_0 \\ \beta_1 \\ \vdots \\ \beta_p \end{pmatrix} \quad \text{and} \quad \boldsymbol{\varepsilon} = \begin{pmatrix} \varepsilon_1 \\ \varepsilon_2 \\ \vdots \\ \vdots \\ \varepsilon_n \end{pmatrix}.$$

The benefits of this formulation will become clear when we come to talk about estimating the parameters in $\boldsymbol{\beta}$.

### 6.3.3 *Applications*

The regression framework is often thought of in the context of continuous explanatory variables. However, there is nothing in the definition that prevents us from using discrete explanatory variables. In fact, the linear model formulation given by equation (6.6) is remarkably flexible. Consider the model given by (6.2) and suppose that we take $x_i = 0$ for all $i$. The resulting model is

$$Y_i = \alpha + \sigma\varepsilon_i,$$

that is, $\{Y_i\} \sim \text{IID}(\alpha, \sigma^2)$. If we take $x_i = 1$ then

$$Y_i = (\alpha + \beta) + \sigma\varepsilon_i,$$

that is, $\{Y_i\} \sim \text{IID}(\alpha + \beta, \sigma^2)$. This simple observation provides a mechanism for representing two distinct populations in a single equation. Suppose that we have a sample from population $A$ with mean $\mu_A$ and a sample from population $B$ with mean $\mu_B$. Stacking these observations into a single vector, $\boldsymbol{Y}$, we have a common representation of the response for both samples,

$$Y_i = \alpha + \beta x_i + \sigma\varepsilon_i,$$

where

$$x_i = \begin{cases} 0 & \text{if } i^{\text{th}} \text{ element of } \boldsymbol{Y} \text{ is from population } A, \\ 1 & \text{if } i^{\text{th}} \text{ element of } \boldsymbol{Y} \text{ is from population } B. \end{cases}$$

Taking $\alpha = \mu_A$ and $\beta = \mu_B - \mu_A$ completes the model. Notice that in this context the explanatory variable is not random. The model also contains the tacit assumption that both populations have the same variance, $\sigma^2$.

We can extend this idea to any number of populations by introducing additional explanatory variables. Suppose that we have $k$ populations; for reasons that will become apparent shortly, we will label these populations $0, 1, \ldots, k-1$. The population means are, respectively, $\mu_0, \mu_1, \ldots, \mu_{k-1}$. Taking

$$Y_i = \beta_0 + x_{i,1}\beta_1 + \ldots + x_{i,k-1}\beta_{k-1} + \sigma\varepsilon_i. \tag{6.7}$$

where, for $j = 1, \ldots, k-1$, each $x_{i,j}$ is an indicator of membership of population $j$,

$$x_{i,j} = \begin{cases} 1 & \text{if } i^{\text{th}} \text{ element from population } j, \\ 0 & \text{if } i^{\text{th}} \text{ element not from population } j. \end{cases}$$

Then, taking $\beta_0 = \mu_0$ and $\beta_j = \mu_j - \mu_0$ for $j = 1, \ldots, k-1$, completes the model. Notice that, as before, the model assumes that the $k$ populations have equal variance. The model defined by (6.7) forms the basis of one-way **analysis of variance** (ANOVA); see the further exercises in Chapter 8.

**Exercise 6.3**

1. For the multiple linear regression model in section 6.3.2, let $\bar{X}_j$ and $s_j$ denote, respectively, the sample mean and standard deviation of $X_{1,j}, \ldots, X_{n,j}$. We can then define the standardised explanatory variables $X_1^*, \ldots, X_p^*$, where for each $j$ we set $X_{i,j}^* = (X_{i,j} - \bar{X}_j)/s_j$, for $i = 1, \ldots, n$. Write equation (6.5) in terms of the standardised explanatory variables and show that this is still a linear regression model.
2. Consider the model
$$\boldsymbol{Y} = \boldsymbol{X}\boldsymbol{\beta} + \boldsymbol{C}\boldsymbol{\varepsilon}, \tag{6.8}$$
where $\boldsymbol{Y}$, $\boldsymbol{X}$, $\boldsymbol{\beta}$, and $\boldsymbol{\varepsilon}$ are as defined in equation (6.6), but $\boldsymbol{C}$ is an $n \times n$ matrix.
   (a) What is the variance of the term $\boldsymbol{C}\boldsymbol{\varepsilon}$?
   (b) Show that the multiple linear regression model is a special case of this model.
   (c) Explain what assumption we are making about the variance of $Y_i | X_i = x_i$ if
      i. $\boldsymbol{C} = \text{diag}(\sigma_1, \ldots, \sigma_n)$, a diagonal matrix with entries $\sigma_1, \ldots, \sigma_n$.
      ii. $\boldsymbol{C}$ is non-diagonal.

## 6.4 Generalised linear models

### 6.4.1 *Motivation*

For the linear modelling framework described in section 6.3, the mean function is a linear function of the explanatory variables. In some instances, assuming linearity of the association between explanatory variables and mean may not be appropriate.

For example, consider a situation in which a lender is building a model for the risk associated with lending to individuals. The model is built on data from past customers. The explanatory variables are customer characteristics. The response is a binary variable taking a value of 0 for customers who pay back the loan, and 1 for customers who default. Recall that the mean function is the mean of the response conditional on the covariates. Thus, it is reasonable to insist that the mean function is constrained to take a value between 0 and 1. There is no way to impose this constraint in the framework of section 6.3.

### 6.4.2 *Link function*

The **generalised linear model** provides a more general framework that allows some of the assumptions associated with the linear model to be relaxed. The important implication for practicing statisticians is that the ideas of linear modelling may be more widely applied. For the linear model defined by equation (6.6), the mean function is a linear function of the explanatory variables,

$$\mu(\boldsymbol{x}_i) = \mathbb{E}(Y_i|\boldsymbol{X}_i = \boldsymbol{x}_i) = \boldsymbol{x}_i\boldsymbol{\beta}.$$

In the generalised linear model, we take a function of the mean function to be linear, that is,

$$g(\mu(\boldsymbol{x}_i)) = \boldsymbol{x}_i\boldsymbol{\beta}.$$

The function $g$ is a continuous monotone function referred to as the **link function**. As the link function is invertible, we can write an expression for the mean function in terms of its inverse,

$$\mu(\boldsymbol{x}_i) = g^{-1}(\boldsymbol{x}_i\boldsymbol{\beta}).$$

The quantity $\boldsymbol{x}_i\boldsymbol{\beta}$ is known as the **linear predictor** of $Y_i$.

Two questions immediately arise: how do we choose a link function, and what are the implications for the variance function? It turns out that these questions can be answered rather neatly if we restrict our attention to a particular group of distributions. If the conditional mass/density for a single response is of the form

$$f_{Y_i|\boldsymbol{X}_i}(y_i|\boldsymbol{x}_i) = h(y_i, \phi)\exp\left\{\frac{\theta_i y_i - b(\theta_i)}{\tau(\phi)}\right\}, \tag{6.9}$$

where $\theta_i$ depends on $\boldsymbol{x}_i$, we say that the distribution belongs to an **exponential dispersion family**. This is a generalisation of the natural exponential family we encountered in Chapter 3.

The parameters $\theta_i$ and $\phi$ are, in effect, location and scale parameters. In fact, we can show (see Exercise 6.4) that the mean and variance functions are

$$\begin{aligned} \mu_i(\boldsymbol{x}_i) &= \mathbb{E}(Y_i|\boldsymbol{X}_i = \boldsymbol{x}_i) = b'(\theta_i), \\ \sigma_i^2(\boldsymbol{x}_i) &= \mathrm{Var}(Y_i|\boldsymbol{X}_i = \boldsymbol{x}_i) = \tau(\phi)\, b''(\theta_i). \end{aligned} \tag{6.10}$$

If we take the location parameter to be a linear function of the explanatory variables, $\theta_i = \boldsymbol{x}_i\boldsymbol{\beta}$, equation (6.10) yields

$$\mu_i(\boldsymbol{x}_i) = b'(\boldsymbol{x}_i\boldsymbol{\beta}) \Rightarrow b'^{-1}(\mu(\boldsymbol{x}_i)) = \boldsymbol{x}_i\boldsymbol{\beta}.$$

Thus, $b'^{-1}$ is a natural choice of link function and is often referred to as the **canonical link function**.

**Example 6.4.1** (Bernoulli response and logistic regression)
We are often interested in response variables that relate to events that either happen or do not happen. Our earlier illustration relating to loan default falls into this category. An individual with a fixed-term loan either defaults on the loan or does not default, and these two are the only possible outcomes. In this type of situation our response is a Bernoulli random variable. In practice we would look at historical data on defaults and, in additional to the Bernoulli default/no default response variable, we would collect variables relating to customer characteristics at the time of application. The hope is that we can build a model that uses these characteristics to give an estimate of an individual's probability of default.

We model the response conditional on a collection of explanatory variables,

$$(Y_i|\boldsymbol{X}_i = \boldsymbol{x}_i) \sim \text{Bernoulli}(p(\boldsymbol{x}_i)).$$

Notice that, in the context of our earlier example, $p(\boldsymbol{x}_i)$ can be interpreted as the probability of default for an individual whose characteristics are summarised in the vector $\boldsymbol{x}_i$. By the properties of Bernoulli random variables,

$$p(\boldsymbol{x}_i) = \mathbb{E}(Y_i|\boldsymbol{X}_i = \boldsymbol{x}_i),$$

so $p(\boldsymbol{x}_i)$ is the mean function. Using the notationally convenient shorthand, $p_i = p(\boldsymbol{x}_i)$, the conditional mass function is

$$\begin{aligned} f_{Y_i|\boldsymbol{X}_i}(y_i|\boldsymbol{x}_i) &= p_i^{y_i}(1-p_i)^{(1-y_i)} \\ &= \exp\{y_i \log p_i + (1-y_i)\log(1-p_i)\} \\ &= \exp\{y_i \log(p_i/(1-p_i)) - \log(1/(1-p_i))\}. \end{aligned} \tag{6.11}$$

The rearrangement given above makes clear that we can express the conditional mass function as a member of an exponential dispersion family. In the notation of equation (6.9) we have

$$\theta_i = \log\left(\frac{p_i}{1-p_i}\right), \tag{6.12}$$

and the canonical link function is, thus,

$$\log\left(\frac{p_i}{1-p_i}\right) = \boldsymbol{x}_i\boldsymbol{\beta}.$$

This function is referred to as the **logit** (pronounced "LOW-jit") or log-odds, and generalised linear modelling with this link function is often called **logistic regression**.

In this example, we can readily confirm the results of (6.10). Solving equation (6.12) for $p_i$, we find

$$p_i = \frac{e^{\theta_i}}{1 + e^{\theta_i}},$$

so equation (6.11) gives

$$b(\theta_i) = \log\left(\frac{1}{1 - p_i}\right) = \log(1 + e^{\theta_i})$$

and $\tau(\phi) = 1$. Differentiating $b(\theta)$ with respect to $\theta$ now yields the mean and variance functions,

$$\begin{aligned} b'(\theta_i) &= e^{\theta_i}/(1 + e^{\theta_i}) = p_i, \\ b''(\theta_i) &= e^{\theta_i}/(1 + e^{\theta_i}) - e^{2\theta_i}/(1 + e^{\theta_i})^2 = p_i(1 - p_i). \end{aligned}$$

**Example 6.4.2** (Poisson response)
Consider now the problem of predicting the number of hurricanes that develop in the Atlantic Basin in a year. A reasonable starting point would be to model this as a Poisson random variable, where the parameter depends on various environmental factors (for example, the temperature of the ocean, or the level of tropical cyclone activity). We write

$$(Y_i | \boldsymbol{X}_i = \boldsymbol{x}_i) \sim \text{Pois}(\lambda(\boldsymbol{x}_i)),$$

so the mean function is

$$\mu(\boldsymbol{x}_i) = \mathbb{E}(Y_i | \boldsymbol{X}_i = \boldsymbol{x}_i) = \lambda(\boldsymbol{x}_i), \tag{6.13}$$

which we interpret as the expected number of hurricanes in a year whose environmental characteristics are summarised in the vector $\boldsymbol{x}_i$. Using the shorthand $\lambda_i = \lambda(\boldsymbol{x}_i)$ the conditional mass function is

$$f_{Y_i|\boldsymbol{X}_i}(y_i|\boldsymbol{x}_i) = \frac{e^{-\lambda_i}\lambda_i^{y_i}}{y_i!} = \exp\{y_i \log\lambda_i - \lambda_i - \log(y_i!)\}.$$

We can express this as a member of an exponential dispersion family by setting

$$\theta_i = \log\lambda_i, \quad b(\theta_i) = \lambda_i = e^{\theta_i}, \quad \tau(\phi) = 1,$$

and thus, the canonical link function is

$$b'^{-1}(\lambda_i) = \log\lambda_i = \theta_i = \boldsymbol{x}_i\boldsymbol{\beta}. \tag{6.14}$$

**Exercise 6.4**

1. Suppose that the distribution of $Y$ is a member of an exponential dispersion family with parameters $\theta$ and $\phi$, that is,

$$f_Y(y) = h(y, \phi)\exp\left\{\frac{\theta y - b(\theta)}{\tau(\phi)}\right\},$$

Show that $\mathbb{E}(Y) = b'(\theta)$ and $\text{Var}(Y) = \tau(\phi)b''(\theta)$.

2. Consider a situation where a bus operator wants to model passenger waiting times using explanatory variables such as the frequency/number of buses on the route, the time of day, current traffic conditions, etc. The response (waiting time) is a positive, continuous variable, so we could start by modelling it as an exponential random variable.
   (a) Show that we can express this as a member of an exponential dispersion family.
   (b) What is the canonical link function?
   (c) What practical problem does this introduce?
3. In a linear regression model, each coefficient $\beta_j$ has an obvious interpretation: it is the change in the mean of the response variable, $Y$, if the $j^{\text{th}}$ explanatory variable increases by one unit (holding everything else constant). In a logistic regression model, $\beta_j$ is the corresponding change in the log-odds, which is more difficult to interpret. Suppose that $\theta^* = \theta + \beta_j$ and consider the mean function $p(\theta) = e^{\theta}/(1 + e^{\theta})$, known as the **expit** function.
   (a) Show that the derivative of the expit function is maximised at $\theta = 0$. What is $p'(0)$?
   (b) Explain why the change in probability $|p(\theta^*) - p(\theta)|$ cannot be greater than $|\beta_j|/4$.
   (c) When is $p(\theta) + \beta_j/4$ a good approximation for $p(\theta^*)$?

   [This is known as the **divide-by-4 rule**. It is a useful rule of thumb for interpreting logistic regression coefficients.]

## 6.5 Time-to-event models

Consider a long-term clinical trial. Patients with a specific illness are given a treatment and the length of time that they live subsequent to treatment is recorded. The data take the form of times until an event; in this case the event is death. In this context, the analysis of time-to-event data is often referred to as **survival analysis**. The treatment is deemed effective if the survival times for the treatment group are longer than those for the control group. The problem of determining whether differences in survival times are significant is one of statistical inference. It is often the case that there are a number of covariates that have an impact on survival time. Obvious examples include the patient's age, the stage in the illness when treatment is administered, and the lifestyle of the patient (their eating habits, whether or not they smoke, etc.).

Time-to-event data arise in a number of other contexts. In measuring the reliability of a manufactured product, the variable of interest is the length of time until the product fails. Banks are interested in the length of time that customers hold accounts with them. Government agencies may analyse the length of time that individuals spend claiming a particular benefit. It is hard to get away from emotive terms in time-to-event analysis. For the sake of clarity, we will use terms that make sense in the reliability context. Thus, the variable of interest is the **survival time** or **length of life** of an entity. At the start, the entity is functioning. The event of interest is failure, and our data consist of a collection of failure times.

We start by defining functions that provide an alternative perspective on the distribution of probability associated with a random variable. These are particularly useful in situations where the random variable in question measures survival time. We will assume that time is measured from 0, so all survival time variables are non-negative; if $Y$ is the survival time, then $Y \geq 0$.

### 6.5.1 *Survival function and hazard function*

Suppose that $Y$ is a random variable modelling length of life. The event $\{Y > y\}$ can be viewed as the survival beyond time $y$.

**Definition 6.5.1** (Survival function)
If $Y$ is a random variable with cumulative distribution function $F_Y$, the **survival function** $S_Y$ (also known as the **reliability function**) is defined as

$$S_Y(y) = \mathrm{P}(Y > y) = 1 - F_Y(y).$$

The survival function inherits a number of properties directly from its definition in terms of the cumulative distribution function.

**Lemma 6.5.2** (Properties of survival functions)
*If $S_Y$ is a survival function, then*

*i* *$S_Y$ is non-increasing, $S_Y(x) \geq S_Y(y)$ for $x < y$,*
*ii* $\lim_{y \to -\infty} S_Y(y) = 1$ *and* $\lim_{y \to \infty} S_Y(y) = 0$,
*iii* *$S_Y$ is right-continuous, $S_Y(y+) = S_Y(y)$.*

Proving these results is part of Exercise 6.5.

We will often make the distinction between discrete and continuous models for survival time. In the discrete case the support for survival time will be the non-negative integers, $\{0, 1, 2, \ldots\}$, and in the continuous case it will be the non-negative real line, $[0, \infty)$. In many instances, explanations are more straightforward in the discrete case as we can give simple interpretations in terms of probability. We start with a general definition of the hazard function before justifying this definition for both discrete and continuous cases.

**Definition 6.5.3** (Hazard function)
The **hazard function** for a random variable $Y$ is

$$\lambda_Y(y) = \frac{f_Y(y)}{S_Y(y-)} = \begin{cases} f_Y(y)/S_Y(y-1) & \text{for } Y \text{ discrete,} \\ f_Y(y)/S_Y(y) & \text{for } Y \text{ continuous.} \end{cases}$$

*Hazard in the discrete case*

In the discrete case, the hazard function gives us the probability of failure at time $y$

given that we know the entity is functioning at time $y - 1$. This is a result of applying elementary rules of conditional probability,

$$P(Y = y|Y > y - 1) = \frac{P(Y = y \cap Y > y - 1)}{P(Y > y - 1)} = \frac{P(Y = y)}{P(Y > y - 1)} = \frac{f_Y(y)}{S_Y(y - 1)}.$$

With some manipulation, we can express the survival function in terms of the hazard function,

$$S_Y(y) = \prod_{u=0}^{y} (1 - \lambda_Y(u)).$$

Establishing this result is part of Exercise 6.5.

*Hazard in the continuous case*

In the continuous case, the probability of failure at a particular time is zero so we must adopt a slightly different approach. We start by defining a random variable that is the survival time conditional on survival beyond time $y$ (that is, given that we know the entity is functioning at time $y$). We denote this variable $W = Y|Y > y$. The cumulative distribution function of $W$ is

$$F_W(w) = P(Y \leq w|Y > y) = \frac{P(Y \leq w \cap Y > y)}{P(Y > y)} = \frac{F_Y(w) - F_Y(y)}{S_Y(y)} \quad \text{for } w \geq y.$$

Differentiating with respect to $w$ then yields the density function

$$f_W(w) = \begin{cases} f_Y(w)/S_Y(y) & \text{for } w \geq y, \\ 0 & \text{otherwise.} \end{cases} \tag{6.15}$$

Checking that this is a valid density function is part of Exercise 6.5.

Comparing equation (6.15) with Definition 6.5.3, we see that the hazard function is the value at $y$ of the density conditional on survival beyond $y$. It is important to note that the hazard function value is a value from a conditional density; the hazard function itself is NOT a density function. We can interpret the hazard function as giving an indication of the chance of instantaneous failure at time $y$. Put loosely, if the hazard evaluated at $y$ is high, there is high probability of failure in the time shortly after $y$; if hazard is low, this probability is small.

Notice that we can obtain the hazard function directly from the survival function,

$$-\frac{d}{dy}\log S_Y(y) = -\frac{1}{S_Y(y)}\frac{d}{dy}S_Y(y) = -\frac{1}{S_Y(y)}\frac{d}{dy}(1 - F_Y(y)) = \frac{f_Y(y)}{S_Y(y)} = \lambda_Y(y). \tag{6.16}$$

The hazard function for a continuous random variable has a number of interesting characterisations. The **cumulative hazard function** is to hazard what the cumulative distribution function is to density. We can express cumulative hazard as

$$H_Y(y) = \int_0^y \lambda_Y(u)du = [-\log S_Y(u)]_0^y = -\log S_Y(y). \tag{6.17}$$

Notice here that, since $Y \geq 0$, the lower limit of the integrals is 0, and $S_Y(0) = 1$. Rearranging equation (6.17) yields

$$S_Y(y) = \exp(-H_Y(y)) = \exp\left(\int_0^y \lambda_Y(u)du\right),$$

and thus,

$$F_Y(y) = 1 - \exp\left(\int_0^y \lambda_Y(u)du\right). \tag{6.18}$$

From equation (6.18), we can see that the cumulative distribution function (and thus, everything we know about a distribution) can be characterised in terms of hazard.

**Example 6.5.4** (Geometric as a survival model)
Consider a discrete survival time model. A natural distribution to consider is the geometric; this is the distribution of the trial at which the first success occurs in repeated independent Bernoulli trials. If $Y \sim \text{Geometric}(p)$ then

$$f_Y(y) = (1-p)^{y-1}p \quad \text{for } y = 1, 2, \ldots$$

The cumulative distribution function is

$$F_Y(y) = \sum_{u=1}^{y} f_Y(u) = \sum_{u=1}^{y}(1-p)^{u-1}p = p\frac{1-(1-p)^y}{1-(1-p)} = 1-(1-p)^y.$$

Thus, the survival function is

$$S_Y(y) = (1-p)^y.$$

From Definition 6.5.3, the hazard function is

$$\lambda_Y(y) = \frac{f_Y(y)}{S_Y(y-1)} = \frac{(1-p)^{y-1}p}{(1-p)^{y-1}} = p.$$

Notice that the hazard is constant, that is, hazard does not depend on the value of $y$. We can view constant hazard from the perspective of the conditional survival function. Consider an entity that has survived until time $x$. The probability of survival for a further period of time that is at least $y$ is given by

$$\text{P}(Y > y + x | Y > x) = \frac{S_Y(y+x)}{S_Y(x)} = \frac{(1-p)^{y+x}}{(1-p)^x} = (1-p)^y = S_Y(y). \tag{6.19}$$

From equation (6.19), the probability of survival for time $y$ beyond $x$ is identical to the probability of a total life length greater than $y$. This property is known as the **memoryless property**, and the geometric is the only discrete distribution that has it.

**Example 6.5.5** (Exponential as a survival model)
If $X \sim \text{Exp}(\lambda)$, the survival function of $X$ is

$$S_X(x) = 1 - F_X(x) = \begin{cases} e^{-\lambda x} & \text{for } 0 \leq x < \infty, \\ 1 & \text{otherwise.} \end{cases}$$

The probability that $X$ takes a value larger than $x$ decreases exponentially as $x$ increases. We can readily show that the exponential has constant hazard,

$$\lambda_Y(y) = \frac{f_Y(y)}{S_Y(y)} = \frac{\lambda e^{-\lambda x}}{e^{-\lambda x}} = \lambda.$$

Demonstrating the equivalence between constant hazard and the memoryless property is part of Exercise 6.5.

**Example 6.5.6** (Weibull distribution)
Consider the distribution defined by the cumulative distribution function

$$F_X(x) = \begin{cases} 1 - \exp(-\lambda x^\alpha) & \text{for } 0 < x < \infty, \\ 0 & \text{otherwise,} \end{cases}$$

where $\alpha, \lambda$ are positive parameters. This is known as the **Weibull distribution**, and is commonly used for time-to-event analysis. The hazard function is

$$\lambda_X(x) = -\frac{d}{dx}\log S_X(x) = -\frac{d}{dx}\log(\exp(-\lambda x^\alpha)) = \alpha\lambda x^{\alpha-1}.$$

Thus, the form of the Weibull hazard function depends on the value of the parameter $\alpha$. If $\alpha > 1$, the hazard is increasing; if $\alpha = 1$, the hazard is constant (and the Weibull reduces to the exponential distribution); and if $\alpha < 1$, the hazard is decreasing.

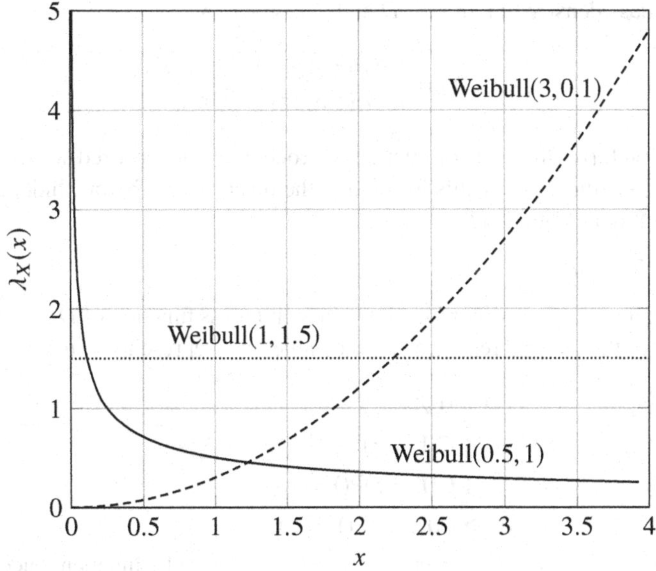


Figure 6.2 *The* Weibull$(\alpha, \lambda)$ *hazard function for various values of* $\alpha$ *and* $\lambda$.

### 6.5.2 *Censoring of time-to-event data*

It is often the case that we do not know the time at which failure occurred. The only information we have is that the entity was functioning at a particular time. If this is the case, we say that the data are **censored**. In the context of clinical trials, censoring arises when a patient is lost to follow-up or is still alive when the study ends. Suppose that $Y$ models survival time and $C$ models the time at which censoring occurs. We will assume that $Y$ and $C$ are independent. Let $Z$ denote a sample value in a study with censoring, then

$$Z = \begin{cases} Y & \text{if } Y \leq C, \\ C & \text{if } Y > C. \end{cases}$$

In other words, if failure occurs before the censoring time, we record the survival time, $Y$; if censoring occurs before failure, we record the time at which censoring occurred, $C$. In time-to-event data, we have information that allows us to distinguish between points that are censored and those that are true survival times. We represent this knowledge using an indicator for censoring,

$$D = \begin{cases} 0 & \text{if } Y \leq C, \\ 1 & \text{if } Y > C. \end{cases}$$

The data takes the form of instances of the pair $(D, Z)$. For example, the pair $(0, 21)$ would indicate an observed survival time of 21, while $(1, 11)$ is indicative of an entity that was functioning at time 11 but for whom no further information is available.

The joint mass/density function of $D$ and $Z$ is given by

$$f_{D,Z}(d, z) = \begin{cases} S_C(z-) f_Y(z) & \text{for } d = 0, \\ f_C(z) S_Y(z) & \text{for } d = 1. \end{cases}$$

This function forms the basis of inferential procedures for censored survival data. We give a brief justification for this formula in the discrete case below. Independence of $C$ and $Y$ is fundamental here.

*Censoring in the discrete case*

If $Y$ is discrete we can readily write down the joint mass function of $D$ and $Z$. If $D = 0$ then we have the uncensored case and the value of $Z$ represents a survival time,

$$\begin{aligned} f_{D,Z}(0, z) &= \mathrm{P}(D = 0, Z = z) \\ &= \mathrm{P}(Y \leq C, Y = z) \\ &= \mathrm{P}(Y \leq C | Y = z)\mathrm{P}(Y = z) \\ &= \mathrm{P}(C \geq z | Y = z)\mathrm{P}(Y = z) \\ &= \mathrm{P}(C \geq z)\mathrm{P}(Y = z) && \text{by independence} \\ &= S_C(z - 1) f_Y(z) \,. \end{aligned}$$

Using similar reasoning to establish the result for $D = 1$ is part of Exercise 6.5.

### 6.5.3 Covariates in time-to-event models

We identify two key mechanisms by which the impact of covariates on life length is modelled. These are referred to as **accelerated-life** and **proportional-hazard** models.

*Accelerated-life models*

In an accelerated-life model, the log life length is taken to be a linear function of the covariates,

$$\log(Y_i) = \boldsymbol{X}_i\boldsymbol{\beta} + \sigma\varepsilon_i. \tag{6.20}$$

Exponentiating both sides of (6.20) yields

$$Y_i = \exp(\boldsymbol{X}_i\boldsymbol{\beta})U_i,$$

where $U_i = \exp(\sigma\varepsilon_i)$. In order to get a clearer idea of the impact of the covariates in this model, consider the conditional survival function,

$$\begin{aligned} S_{Y_i|\boldsymbol{X}_i}(y|\boldsymbol{x}_i) &= P(Y_i > y|\boldsymbol{X}_i = \boldsymbol{x}_i) \\ &= P(\exp(\boldsymbol{x}_i\boldsymbol{\beta})U_i > y|\boldsymbol{X}_i = \boldsymbol{x}_i) = S_U(y/\exp(\boldsymbol{x}_i\boldsymbol{\beta})). \end{aligned} \tag{6.21}$$

From equation (6.21), the survival function has a fixed form, but its argument is scaled by the values of the explanatory variables. For illustration, consider the case with two populations $A$ and $B$, where $X_i$ is an indicator that the $i^{\text{th}}$ sample member is from population $B$, and we take $\beta = \log 2$. The survival function is then

$$S_{Y_i|X_i}(y|x_i) = \begin{cases} S_U(y) & \text{for population } A, \\ S_U(y/2) & \text{for population } B. \end{cases}$$

The probability that a member of population $A$ survives beyond 40 is the same as the probability that a member of population $B$ survives beyond 20. The impact of the covariate is to accelerate the lives of members of population $B$. In effect, members of population $B$ are aging twice as fast as those in population $A$.

*Proportional-hazard models*

In a proportional-hazard model the covariates act on a baseline hazard function via a multiplicative factor,

$$\lambda_{Y_i|\boldsymbol{X}_i}(y|\boldsymbol{x}_i) = \lambda_0(y)\exp(\boldsymbol{x}_i\boldsymbol{\beta}).$$

The baseline hazard function, $\lambda_0$, is the hazard when all covariates are set to zero, $\lambda_0(y) = \lambda_{Y_i|\boldsymbol{X}_i}(y|\boldsymbol{0})$. For other values of the covariates, the hazard is simply a scaled version of this underlying function. The log hazard function is linear in the covariates,

$$\log(\lambda_{Y_i|\boldsymbol{X}_i}(y|\boldsymbol{x}_i)) = \boldsymbol{x}_i\boldsymbol{\beta} + \log\lambda_0(y).$$

In this formulation, positive values of $\boldsymbol{x}_i\boldsymbol{\beta}$ correspond to increased hazard.

**Exercise 6.5**

1. Prove Lemma 6.5.2.
2. Suppose that $Y$ is a discrete random variable. Using the definitions of survival and hazard, show that
$$\lambda_Y(y) = 1 - \frac{S_Y(y)}{S_Y(y-1)}.$$
Hence or otherwise, show that
$$S_Y(y) = \prod_{u=0}^{y} (1 - \lambda_Y(u)).$$
3. Suppose that $Y$ is a continuous random variable. Show that the function
$$f_W(w) = \begin{cases} f_Y(w)/S_Y(y) & \text{for } w \geq y, \\ 0 & \text{otherwise,} \end{cases}$$
is a valid density.
4. From Definition 6.5.3 we can see that the density function can be written in terms of the hazard and survival functions,
$$f_Y(y) = \lambda_Y(y) S_Y(y).$$
Use differentiation to confirm that equation (6.18) is consistent with this result.
5. Suppose that $Y$ is survival time, $C$ is censoring time, and that $Y$ and $C$ are independent. Further, suppose that $Z$ is the observed time and $D$ is an indicator for censoring. Show that when $D = 1$ the joint mass/density of $D$ and $Z$ is given by
$$f_{D,Z}(1, z) = f_C(z) S_Y(z).$$
You should treat the cases where $Y$ is discrete and $Y$ is continuous separately.

## 6.6 Time series models

The simplest definition of a time series is a collection of observations taken over time. In time series analysis, we usually assume that observations are made at regular intervals, for example, every day or every month. Weekly maximum temperature readings at a weather station, daily closing prices for a stock exchange, and monthly passenger volume data at a particular airport are all examples of time series data (Figure 6.3).

In general, what happens today is dependent on what happened yesterday. Thus, a random sample is not usually a good model for time series data. The central importance of dependence between sample members is what characterises time series analysis. In this section we describe some simple time series models and use them to illustrate basic model properties that frequently arise in time series analysis.

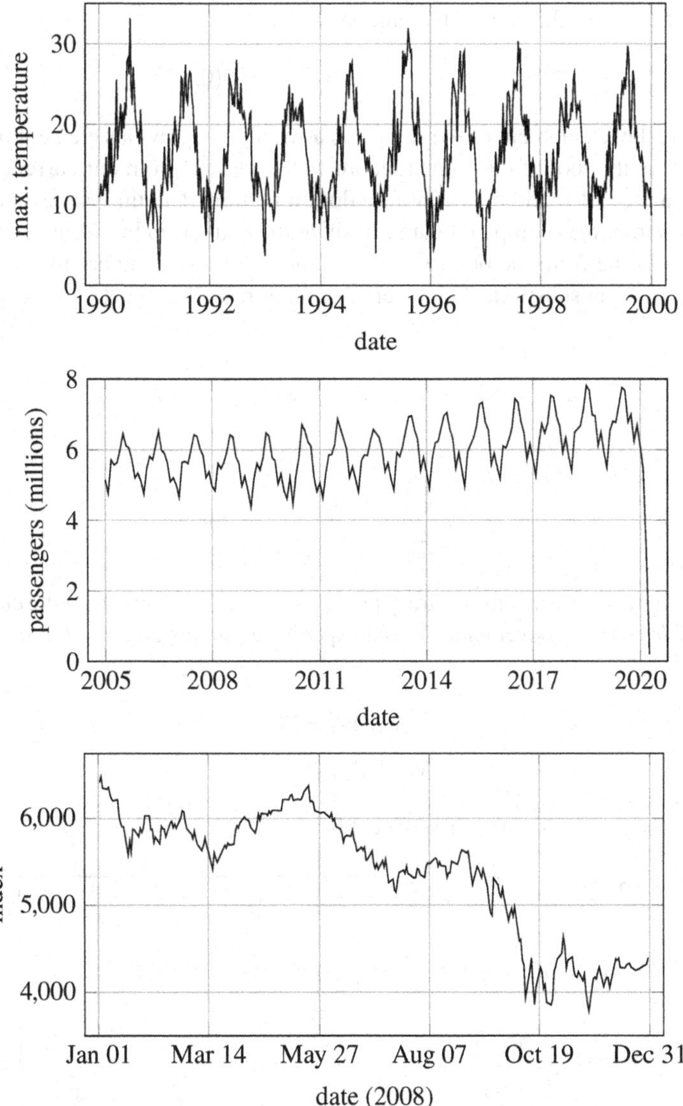


Figure 6.3 *Some examples of time series. Top: weekly maximum temperatures from the Central England Temperature data set, 1990–2000. Middle: monthly passenger volume at Heathrow airport, January 2005 – April 2020. Bottom: daily closing prices for the FTSE100 index, 2008.*

A good time series model should be able to generate accurate predictions of future values. Time series models formalise our intuition that the future is more like the recent past than the distant past. It is often reasonable to suppose that recent values are more relevant in making inferences about what will happen next. A very simple model that captures this idea is the random walk,

$$Y_t = Y_{t-1} + \varepsilon_t, \qquad \{\varepsilon_t\} \sim \text{IID}(0, \sigma_\varepsilon^2). \tag{6.22}$$

In this model each element of the sample is a simply the previous element plus some noise. This is the root of the name random walk; we start from our current position, then take a step of random size in a random direction. The time series plot of this process is sometimes compared humorously to the path taken by a drunken professor on his way home from the pub (this is probabilist humour; you are forgiven for not laughing). We can substitute recursively in equation (6.22), to get

$$\begin{aligned} Y_t &= Y_{t-2} + \varepsilon_{t-1} + \varepsilon_t \\ &= Y_{t-3} + \varepsilon_{t-2} + \varepsilon_{t-1} + \varepsilon_t \\ &\ \ \vdots \\ &= Y_{t-h} + \sum_{i=0}^{h-1} \varepsilon_{t-i} \qquad \text{for } h > 0. \end{aligned} \tag{6.23}$$

Thus, the route to $Y_t$ from any arbitrary previous point, $Y_{t-h}$, consists of the cumulative impact of $h$ random movements. We can specify moments conditional on a starting value, say $Y_0$,

$$\begin{aligned} \mathbb{E}(Y_t|Y_0) &= Y_0, \\ \text{Var}(Y_t|Y_0) &= t\sigma_\varepsilon^2. \end{aligned}$$

Proving these results is part of Exercise 6.6.

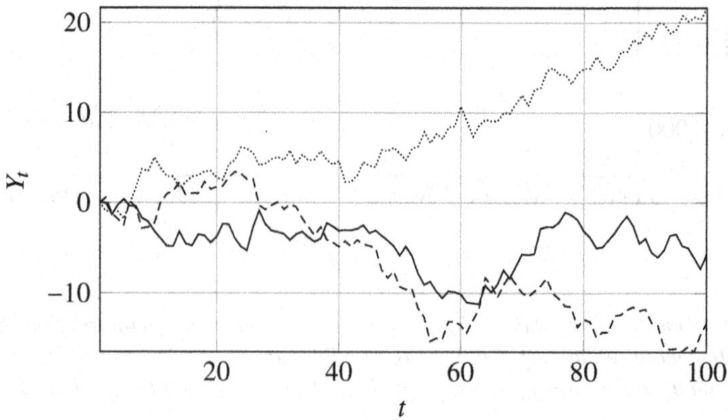


Figure 6.4 *Each line represents 100 observations from a random walk, with* $\varepsilon_t \sim \text{N}(0, 1)$. *Notice that the process tends to drift away from zero.*

### 6.6.1 *Autoregressive models*

The variance of a random walk grows over time. The potential to wander off indefinitely makes a random walk an unattractive model for many real-world phenomena. In practice, many phenomena are mean-reverting, so we would like a model whose values remain close to the mean. One possibility is rooted in the ideas of linear regression. In an **autoregressive** model, the explanatory variables are previous members of the time series. An autoregressive model of order 1, denoted AR(1), is

$$Y_t = \alpha + \phi Y_{t-1} + \varepsilon_t .$$

The relationship to the simple regression model given by equation (6.2) is clear. The mean-reverting properties become apparent if we reparameterise; taking $\alpha = (1-\phi)\mu$ yields

$$(Y_t - \mu) = \phi(Y_{t-1} - \mu) + \varepsilon_t . \tag{6.24}$$

Recursive substitution in (6.24) yields

$$(Y_t - \mu) = \sum_{i=0}^{\infty} \phi^i \varepsilon_{t-i} .$$

Thus, provided $|\phi| < 1$, we have

$$\begin{aligned} \mathbb{E}(Y_t) &= \mu , \\ \mathrm{Var}(Y_t) &= \sigma_\varepsilon^2/(1-\phi^2) . \end{aligned} \tag{6.25}$$

Notice that the variance of $Y_t$ does not grow over time. To establish the formula for the variance, we use the geometric series expansion,

$$(1-\phi)^{-1} = \sum_{i=0}^{\infty} \phi^i , \text{ provided } |\phi| < 1 .$$

This analysis is dependent on the condition $|\phi| < 1$. The importance of this condition will be reiterated in our discussion of the **stationarity** of time series models. Before moving on to this discussion, we introduce one more model.

### 6.6.2 *Moving-average models*

In the autoregressive model, there is explicit dependence on previous values of the series. Dependence can also be introduced by allowing explicit dependence on previous members of the error series. The resulting model is referred to as a **moving-average** model. A moving-average model of order 1, denoted MA(1), is

$$(Y_t - \mu) = \varepsilon_t + \theta \varepsilon_{t-1} .$$

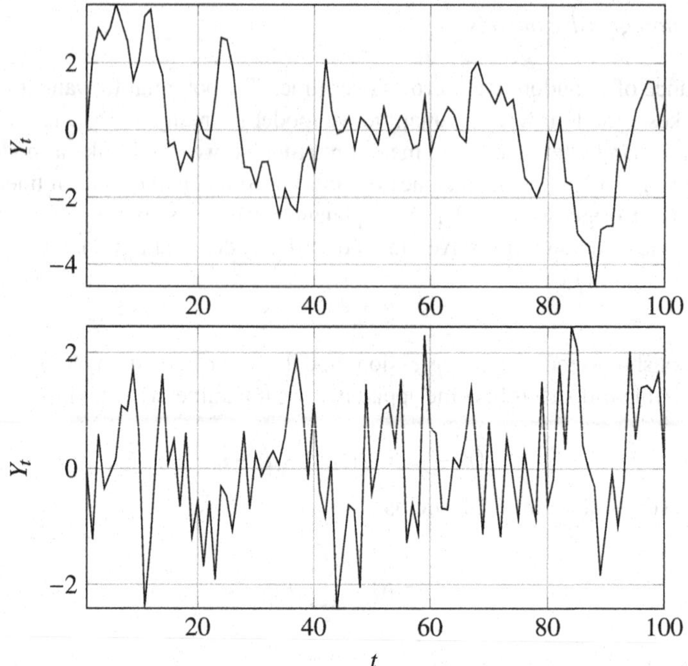


Figure 6.5 *Each plot shows 100 observations from an AR(1) model with $\mu = 0$ and $\sigma_\varepsilon = 1$. In the top plot we have $\phi = 0.9$ so the time series can drift quite far from zero, whereas in the bottom plot $\phi = 0.2$ and the time series tends to oscillate around zero. In both cases, the process is mean-reverting.*

The expectation and variance are given by

$$\mathbb{E}(Y_t) = \mu,$$
$$\mathrm{Var}(Y_t) = \sigma_\varepsilon^2(1 + \theta^2).$$

In the remainder of this section we will consider processes with zero mean, that is, $\mu = 0$. This is for notational convenience, as the generalisations to non-zero-mean processes are always straightforward.

### 6.6.3 *Autocovariance, autocorrelation, and stationarity*

Our key interest in time series analysis is the relationship between members of our sample. In practical terms we want to know how alike today is to yesterday, how alike today is to one week ago, and so on. A key tool in measuring dependence is correlation. In time series analysis, we are interested in the correlation between members of the sample, that is, we are interested in **autocorrelation**.

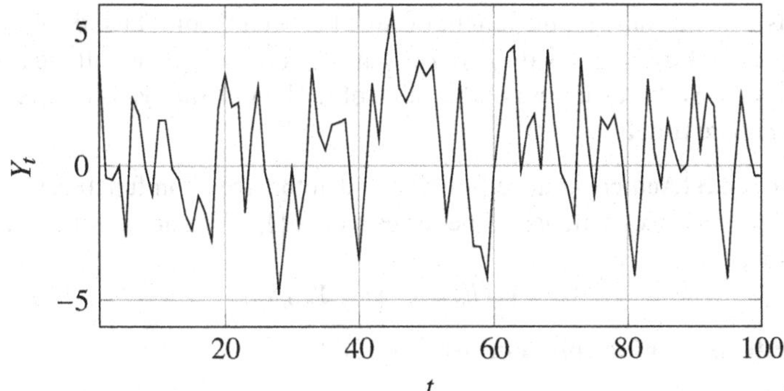


Figure 6.6 *100 observations from the MA(1) process* $Y_t = \varepsilon_t + 0.8\varepsilon_{t-1}$ *with* $\sigma_\varepsilon = 2$.

**Definition 6.6.1** (Autocovariance and autocorrelation)
Consider a time series model $\{Y_t\}$. The **autocovariance** at lag $h$ is

$$\text{Cov}(Y_t, Y_{t-h}) = \mathbb{E}[(Y_t - \mathbb{E}(Y_t))(Y_{t-h} - \mathbb{E}(Y_{t-h}))],$$

and the **autocorrelation** at lag $h$ is

$$\text{Corr}(Y_t, Y_{t-h}) = \frac{\text{Cov}(Y_t, Y_{t-h})}{\sqrt{\text{Var}(Y_t)\,\text{Var}(Y_{t-h})}}.$$

The autocovariance and autocorrelation measure the relationship between $Y_t$ and the point $h$ time steps earlier. We can also view autocovariance in the context of a vector representation of the series. If $\boldsymbol{Y} = (Y_1, \dots, Y_n)^T$, the covariances are the entries in the covariance matrix of $\boldsymbol{Y}$. In particular, if $\boldsymbol{\Sigma}_Y = \text{Var}(\boldsymbol{Y})$, the $(i, j)^{\text{th}}$ entry is given by $[\boldsymbol{\Sigma}_Y]_{i,j} = \text{Cov}(Y_i, Y_j)$, that is, the autocovariance at lag $i - j$.

The definition of autocovariance given above is very loose. As a starting point for a discussion of the properties of time series models, we place some restrictions on the form that the autocovariance can take. These restrictions are the basis of the definition of covariance stationarity.

**Definition 6.6.2** (Covariance stationarity)
A time series model $\{Y_t\}$ is **covariance stationary** if

i $\mathbb{E}(Y_t) = \mu_Y$, where $\mu_Y$ is constant.

ii $\text{Cov}(Y_t, Y_{t-h}) = \gamma_Y(h) < \infty$, for all integer $h$, where $\gamma_Y$ is not a function of $t$.

This definition can be summarised loosely as "a time series model is covariance stationary if the mean and covariance structure do not change over time". Recall that covariance measures the strength of the relationship between different points in the series; covariance stationarity means that the strength of the relationship between two points depends only on the length of time between them. Thus, if $\{Y_t\}$ is covariance stationary, $\text{Cov}(Y_6, Y_1)$, $\text{Cov}(Y_{105}, Y_{100})$ and $\text{Cov}(Y_{227}, Y_{222})$ are all the same since $\text{Cov}(Y_t, Y_{t-5}) = \gamma_Y(5)$ regardless of the value of $t$. The function $\gamma_Y$ is referred to as the **autocovariance function**.

**Definition 6.6.3** (Autocovariance function and autocorrelation function)
Consider a covariance-stationary time series model $\{Y_t\}$. The autocovariance function (ACVF) of $\{Y_t\}$ is

$$\gamma_Y(h) = \text{Cov}(Y_t, Y_{t-h}), \tag{6.26}$$

and the **autocorrelation function** (ACF) of $\{Y_t\}$ is

$$\rho_Y(h) = \gamma_Y(h)/\sigma^2, \tag{6.27}$$

where $\sigma^2 = \gamma_Y(0) = \text{Var}(Y_t)$. Both functions are defined for integer values of $h$.

It is important to note here that $\gamma_Y$ is a function of $h$ alone. The basic properties of autocorrelation are set out in the following claim.

**Claim 6.6.4** (Properties of autocorrelation)
*Suppose that $\rho_Y$ is the autocorrelation function of a covariance-stationary time series $\{Y_t\}$. The following properties hold:*

*i. Symmetry: $\rho_Y(-h) = \rho_Y(h)$,*
*ii. Correlation at lag 0: $\rho_Y(0) = 1$,*
*iii. Correlation at other lags: $|\rho_Y(h)| \leq 1$ for $h > 0$.*

**Example 6.6.5** (Autocorrelation of MA(1))
Consider the MA(1) process $\{Y_t\}$, where

$$Y_t = \varepsilon_t + \theta\varepsilon_{t-1} \qquad \{\varepsilon_t\} \sim \text{IID}(0, \sigma_\varepsilon^2).$$

The different terms of the error series $\{\varepsilon_t\}$ are independent, so we have

$$\begin{aligned}
&\gamma_Y(1) = \text{Cov}(Y_t, Y_{t-1}) = \text{Cov}(\varepsilon_t + \theta\varepsilon_{t-1}, \varepsilon_{t-1} + \theta\varepsilon_{t-2}) = \theta\text{Cov}(\varepsilon_{t-1}, \varepsilon_{t-1}) = \theta\sigma_\varepsilon^2 \\
&\gamma_Y(h) = 0 \ \text{ for } h = 2, 3, \ldots
\end{aligned}$$

Thus, the autocorrelation function is

$$\rho_Y(h) = \begin{cases} 1 & h = 0 \\ \theta/(1+\theta^2) & h = \pm 1 \\ 0 & h = \pm 2, \pm 3, \ldots \end{cases}$$

**Example 6.6.6** (Autocorrelation of AR(1))
Suppose that $\{Y_t\}$ is an AR(1) process, with model equation

$$Y_t = \phi Y_{t-1} + \varepsilon_t \qquad \{\varepsilon_t\} \sim \text{IID}(0, \sigma_\varepsilon^2),$$

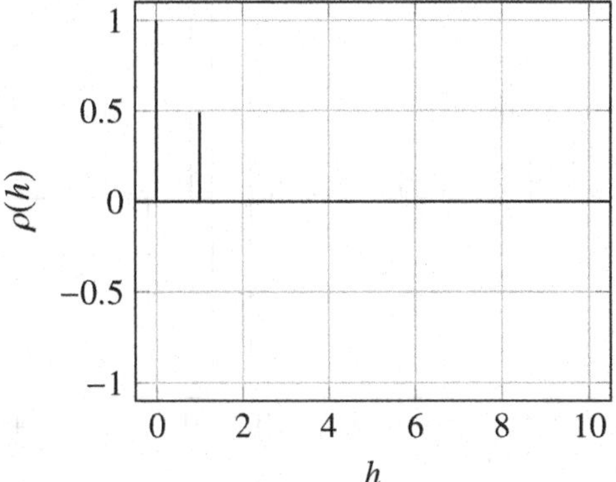


Figure 6.7 *Autocorrelation function of an MA(1) process with $\theta = 0.8$. Notice that it cuts off after lag 1.*

where $|\phi| < 1$. Recursive substitution yields

$$Y_t = \phi^h Y_{t-h} + \sum_{i=0}^{h-1} \phi^i \varepsilon_{t-i} .$$

We have already established that $\text{Var}(Y_t) = \sigma_\varepsilon^2/(1+\phi^2)$. The autocovariance at lag $h$ is, thus,

$$\begin{aligned}\text{Cov}(Y_t, Y_{t-h}) &= \text{Cov}\left(\phi^h Y_{t-h} + \sum_{i=0}^{h-1} \phi^i \varepsilon_{t-i}, Y_{t-h}\right) \\ &= \text{Cov}\left(\phi^h Y_{t-h}, Y_{t-h}\right) = \phi^h \frac{\sigma_\varepsilon^2}{1+\phi^2},\end{aligned}$$

as $Y_{t-h}$ is uncorrelated with the error terms $\varepsilon_{t-h+1}, \ldots, \varepsilon_t$ in the summation. The autocorrelation function is then

$$\rho_Y(h) = \frac{\phi^h \sigma_\varepsilon^2/(1+\phi^2)}{\sigma_\varepsilon^2/(1+\phi^2)} = \phi^h .$$

Notice that the ACF of the AR(1) process gradually decays to zero (Figure 6.8), unlike that of the MA(1) process, which cuts off after lag 1 (Figure 6.7).

Covariance stationarity refers only to the first and second moment of a series. Strict stationarity is a property that ensures temporal homogeneity of the joint distribution of any collection of elements of the series.

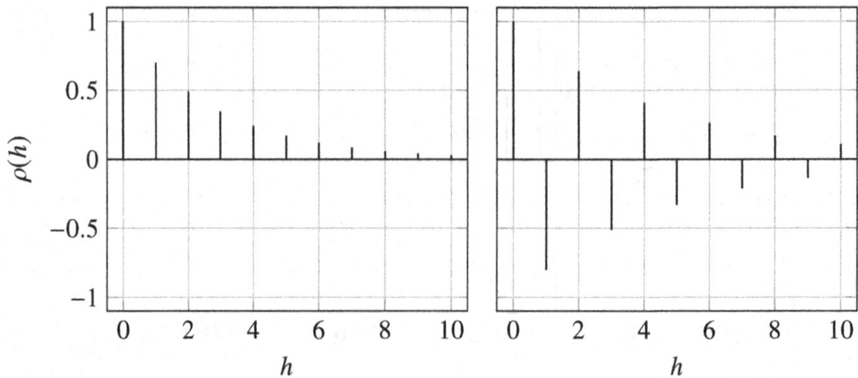


Figure 6.8 *Autocorrelation function of an AR(1) process with $\phi = 0.7$ (left) and $\phi = -0.8$ (right).*

**Definition 6.6.7** (Strict stationarity)
A time series $\{Y_t\}$ is **strictly stationary** if, for any collection of indices $t_1, \dots, t_m$, the joint distributions of $(Y_{t_1}, \dots, Y_{t_m})$ and $(Y_{t_1+h}, \dots, Y_{t_m+h})$ are identical for all integer $h$.

The ideas of covariance stationarity and strict stationarity are well established. One rather unfortunate side effect is that they have gathered several distinct labels. You may see covariance stationarity referred to as weak stationarity or second-order stationarity. Similarly, strict stationarity is sometimes referred to as strong stationarity or stationarity in the wider sense. These labels are misleading as they suggest that all strictly stationary series are covariance stationary; this is not the case (see Exercise 6.6). In what follows, we use the term stationary to refer to a covariance-stationary model. A non-stationary model is a model that is not covariance stationary. We have already encountered a non-stationary model; for the random walk, if we condition on some fixed starting value, the variance is a function of time. The unconditional variance is not finite.

**Exercise 6.6**

1. Give a brief justification for each of the following statements.
   (a) If $\{Y_t\}$ is strictly stationary then the variables $Y_1, Y_2, \dots$ are identically distributed.
   (b) Any IID sequence is strictly stationary.
   (c) A strictly stationary series with finite second moments is covariance stationary.
   (d) A covariance-stationary jointly normally distributed series is strictly stationary.
2. Show that if $\{Y_t\}$ is a random walk then $\mathbb{E}(Y_t|Y_0) = Y_0$ and $\text{Var}(Y_t|Y_0) = t\sigma_\varepsilon^2$.

3. Consider the processes $\{X_t\}$ and $\{Y_t\}$ where

$$X_t = \phi X_{t-1} + \varepsilon_t, \qquad \text{where } \{\varepsilon_t\} \sim \text{IID}(0, \sigma_\varepsilon^2),$$
$$Y_t = \eta_t + \theta\eta_{t-1}, \qquad \text{where } \{\eta_t\} \sim \text{IID}(0, \sigma_\eta^2),$$

Carefully show that

$$\text{Var}(X_t) = \sigma_\varepsilon^2/(1 - \phi^2),$$
$$\text{Var}(Y_t) = \sigma_\eta^2(1 + \theta^2).$$

4. If $\{X_t\}$ and $\{Y_t\}$ are mutually uncorrelated stationary sequences, that is, $X_r$ and $Y_s$ are uncorrelated for every $r$ and $s$, show that $\{X_t + Y_t\}$ is a stationary process with autocovariance function equal to the sum of the autocovariance functions of $\{X_t\}$ and $\{Y_t\}$.

## 6.7 Poisson processes

### *6.7.1 Stochastic processes and counting processes*

Informally, a **stochastic process** is a collection of random variables indexed by time. We consider both discrete-time stochastic processes, that is, collections of the form $\{Y_1, Y_2, Y_3, \dots\}$ or $\{\dots, Y_{-1}, Y_0, Y_1, \dots\}$, and continuous-time stochastic process, that is, collections of the form $\{Y_t : t \in [0, \infty)\}$ or $\{Y_t : t \in \mathbb{R}\}$. We can give a more formal definition, which closely mirrors Definition 3.8.3.

**Definition 6.7.1** (Stochastic process)
For a probability space $(\Omega, \mathcal{F})$ and measurable space $(S, \mathcal{B})$, a stochastic process is a collection of random variables,

$$\{Y_t : t \in T\},$$

where $Y_t : \Omega \to S$. The set $T$ is the indexing set of the process, and $S$ is the state space.

In the cases that we consider in this section, the indexing set, $T$, is the positive real numbers, so we have continuous-time stochastic processes. Continuous-time processes are used as models for situations where events can occur at any time.

Statistical models in finance are often continuous-time processes. For example, the price of a stock is determined by the price paid at the last time the stock was traded. As trades can occur at any time (when the exchange is open), price can be modelled using a continuous-time process. As always, the practicalities of data collection mean that the real situation is somewhat messier. Prices are actually treated as changing at discrete times. However, the discrete scale is very fine so continuity provides a useful model.

Prices on an exchange can move up or down. In this section, we focus on processes that can only move upwards, namely **counting processes**. Consider balls being thrown into a bucket, where the arrival of a ball in the bucket is an event of interest. At each arrival, the number of balls in the bucket increases by one. A counting process, $Y(t)$, arises as a model for the number of balls in the bucket at time $t$. In other words, a counting process models the number of arrivals that occur up to and including time $t$. In order for it to be a reasonable model for a count, a counting process must be

i. positive and integer-valued: $Y(t)$ takes values in $\{0, 1, 2, \dots\}$,
ii. non-decreasing: $Y(t + s) \geq Y(t)$ for $s > 0$,
iii. right-continuous: $\lim_{h \downarrow 0} Y(t + h) = Y(t)$.

Note that the property of right continuity comes from the fact that we define the value of the counting process at time $t$ as being the number of arrivals up to *and including* time $t$.

We will often be interested in the number of arrivals in an interval, say $(t, t + h]$; this is often referred to as an increment.

**Definition 6.7.2** (Increment)
The **increment** associated with an interval is the number of arrivals that occur in the interval. For the interval $(t, t + s]$ we define

$$N_{(t,t+s]} = Y(t + s) - Y(t).$$

By the properties of counting processes, increments are always non-negative, that is, $N_{(t,t+s]} \geq 0$, for all $t \geq 0$ and $s > 0$.

The notation $N_{(t,t+h]}$ is read as "the number of arrivals in the interval $(t, t + h]$". The properties of simple models for counts are conveniently expressed in terms of the properties of increments. In particular, we will refer to independent increments and stationary increments.

**Definition 6.7.3** (Independent increments)
A counting process, $\{Y(t)\}$, has **independent increments** if, for any collection of real values $0 \leq t_1 < t_2 < \dots < t_n$, the set of increments

$$\{Y(t_2) - Y(t_1), Y(t_3) - Y(t_2), \dots, Y(t_{n-1}) - Y(t_n)\}$$

is a set of independent random variables.

Using our previously defined notation for increments, Definition 6.7.3 could be rewritten as

$$N_{(t_1,t_2]}, N_{(t_2,t_3]}, \dots, N_{(t_{n-1},t_n]} \text{ are independent.}$$

An obvious consequence of this definition is that the increments associated with non-overlapping intervals are independent.

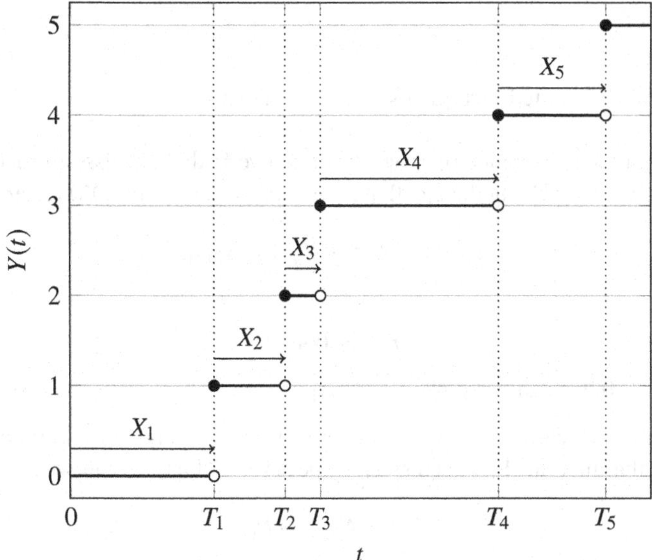


Figure 6.9 *Example of a counting process. The* $\{T_i\}$ *are the arrival times, while the* $\{X_i\}$ *are known as the* interarrival *times.*

**Definition 6.7.4** (Stationary increments)
A counting process, $\{Y(t)\}$, has **stationary increments** if, for any $t_1, t_2 \geq 0$ and $s > 0$, the increments

$$Y(t_1 + s) - Y(t_1) \text{ and } Y(t_2 + s) - Y(t_2)$$

have the same distribution.

Clearly, this definition could be rewritten as "$N_{(t_2,t_2+s]}$ and $N_{(t_1,t_1+s]}$ have the same distribution". If a counting process has stationary increments, the distribution of the increments depends on the width but not on the location of the interval. For example, the distribution of the number of arrivals for the interval between times 1005 and 1010 is identical to that for the number of arrivals times 5 and 10 (also identical to the number of arrivals between times 0.2 and 5.2, 17.36 and 22.36, and so on).

### *6.7.2 Definitions of the Poisson process*

A **Poisson process** is an example of a continuous-time counting process. The definition that we give initially is useful for demonstrating the key properties of Poisson processes. Later on we will show how this definition can be derived from more basic principles. A Poisson process has a single parameter, $\lambda$, often referred to as the **rate** or **intensity**.

**Definition 6.7.5** (Poisson process)
A Poisson process $\{Y(t)\}$ with rate $\lambda$ is a continuous-time counting process with

i. $Y(0) = 0$,

ii. independent increments,

iii. Poisson-distributed increments, $N_{(t,t+s]} \sim \text{Pois}(\lambda s)$.

An immediate consequence of property iii. above is that a Poisson process has stationary increments. Using the fact that $Y(0) = 0$, we can express $Y(t)$ as an increment,

$$Y(t) = Y(t) - Y(0) = N_{(0,t]},$$

and, thus,

$$Y(t) \sim \text{Pois}(\lambda t).$$

Since $Y(0) = 0$, we can write $N_{(0,t]} = N_{[0,t]}$.

For any value $s$ such that $0 < s < t$, we know that $N_{[0,s]}$ and $N_{(s,t]}$ are the number of arrivals in the intervals $[0, s]$ and $(s, t]$, respectively. Thus, we can write

$$Y(t) = N_{[0,s]} + N_{(s,t]}.$$

As the distributions of $Y(t)$, $N_{[0,s]}$ and $N_{(s,t]}$ are all Poisson, the consistency of Definition 6.7.5 depends on the properties of a sum of Poisson distributions established in Exercise 4.7.

**Proposition 6.7.6** (Elementary properties of Poisson processes)
*If $\{Y_t\}$ is a Poisson process with intensity $\lambda$, and $0 < s < t$, then*

*i.* $\mathbb{E}(Y(t)) = \lambda t$,

*ii.* $\text{Var}(Y(t)) = \lambda t$,

*iii.* $\text{Cov}(Y(s), Y(t)) = \lambda s$,

*iv.* $(N_{(s,t]} | Y(s) = m) \sim \text{Pois}(\lambda(t - s))$,

*v.* $Y(s)|Y(t) = n \sim \text{Bin}(n, s/t)$.

The choice of the Poisson distribution in the definition of this process may seem a bit restrictive, but a Poisson process can arise from much simpler conditions.

**Proposition 6.7.7** (Poisson process from first principles)
*If $Y(t)$ is a continuous-time counting process with*

*i.* $Y(0) = 0$,

*ii. independent increments,*

*iii.* $\text{P}(N_{(t,t+h]} = j) = \begin{cases} 1 - \lambda h + o(h) & \text{for } j = 0 \\ \lambda h + o(h) & \text{for } j = 1 \\ o(h) & \text{for } j > 1 \end{cases}$

*then $Y(t)$ is a Poisson process with intensity $\lambda$.*

Note: a function $g(x)$ is said to be $o(x)$ (read "little-oh of x") if $\frac{g(x)}{x} \to 0$ as $x \to 0$. The implication is that, for small values of $x$, $g(x)$ is much smaller than $x$. Thus, if $K$ is any quantity that does not depend on $x$, then $Kg(x)$ is also $o(x)$. Intuitively, the definition states that, in a small interval of time $h$:

- the probability of exactly one arrival is (approximately) proportional to the length of the interval, and
- the probability of more than one arrival is very small and can be ignored.

*Proof.*

It suffices to show that $Y(t)$ is Poisson-distributed with parameter $\lambda t$.

For notational convenience, we define $p_n(t) = \mathrm{P}(Y(t) = n)$. Our aim is to show that $p_n(t) = e^{-\lambda t}(\lambda t)^n/n!$, which we will prove with induction on $n$. For $n = 0$ we have

$$\begin{aligned} p_0(t+h) &= \mathrm{P}(Y(t+h) = 0) \\ &= \mathrm{P}(Y(t+h) - Y(t) = 0, Y(t) = 0) \\ &= \mathrm{P}(Y(t+h) - Y(t) = 0)\mathrm{P}(Y(t) = 0) \\ &= (1 - \lambda h + o(h))p_0(t), \end{aligned}$$

which we rearrange to obtain

$$\begin{aligned} p_0(t+h) - p_0(t) &= -\lambda h p_0(t) + o(h) \\ \Rightarrow \frac{p_0(t+h) - p_0(t)}{h} &= -\lambda p_0(t) + \frac{o(h)}{h}. \end{aligned}$$

Letting $h \downarrow 0$ leaves us with

$$\frac{d}{dt}p_0(t) = -\lambda p_0(t), \tag{6.28}$$

a differential equation with general solution $p_0(t) = C_1 e^{-\lambda t}$. The boundary condition is $p_0(0) = \mathrm{P}(Y(0) = 0) = 1 \Rightarrow C_1 = 1$, thus,

$$p_0(t) = e^{-\lambda t}, \quad t > 0,$$

and the formula holds for $n = 0$. Obtaining a general expression for $p_n(t)$ is slightly trickier, but the approach is similar. We start by conditioning on the number of arrivals at time $t$,

$$\begin{aligned}
p_n(t+h) &= \mathrm{P}(Y(t+h)=n) \\
&= \sum_{j=0}^{n} \mathrm{P}(Y(t+h)-Y(t)=j, Y(t)=n-j) \\
&= \sum_{j=0}^{n} \mathrm{P}(Y(t+h)-Y(t)=j)\mathrm{P}(Y(t)=n-j) \\
&= p_0(h)p_n(t) + p_1(h)p_{n-1}(t) + \sum_{j=2}^{n} p_j(h)p_{n-j}(t) \\
&= (1-\lambda h + o(h))p_n(t) + (\lambda h + o(h))p_{n-1}(t) + o(h).
\end{aligned}$$

Rearranging as before gives

$$\frac{p_n(t+h)-p_n(t)}{h} = -\lambda p_n(t) + \lambda p_{n-1}(t) + \frac{o(h)}{h},$$

and we let $h \downarrow 0$ to find

$$\frac{d}{dt}p_n(t) = -\lambda p_n(t) + \lambda p_{n-1}(t). \tag{6.29}$$

To simplify this expression we multiply both sides by $e^{\lambda t}$,

$$\begin{aligned}
e^{\lambda t}\frac{d}{dt}p_n(t) &= -\lambda e^{\lambda t}p_n(t) + \lambda e^{\lambda t}p_{n-1}(t) \\
\Rightarrow \frac{d}{dt}\left(e^{\lambda t}p_n(t)\right) &= \lambda e^{\lambda t}p_{n-1}(t). \qquad (6.30)
\end{aligned}$$

Now we assume that the formula holds for $n-1$ and, substituting $p_{n-1}(t)$ into the equation above, we obtain

$$\frac{d}{dt}\left(e^{\lambda t}p_n(t)\right) = \lambda e^{\lambda t}\frac{e^{-\lambda t}(\lambda t)^{n-1}}{(n-1)!} = \frac{\lambda^n t^{n-1}}{(n-1)!},$$

a differential equation with solution

$$\begin{aligned}
e^{\lambda t}p_n(t) &= \frac{\lambda^n t^n}{(n-1)!n} + C_2 \\
\Rightarrow p_n(t) &= \frac{e^{-\lambda t}(\lambda t)^n}{n!} + C_2 e^{-\lambda t}.
\end{aligned}$$

For $n > 0$ the boundary condition is $p_n(0) = \mathrm{P}(Y(0) = n) = 0$, so $C_2 = 0$, and we conclude that the formula holds for all non-negative integers $n$.

□

### 6.7.3 Thinning and superposition

**Proposition 6.7.8** (Thinning)
*Consider a Poisson process $\{Y(t)\}$ with intensity $\lambda$ where each arrival is classified as being of Type I (with probability $p$) or Type II (with probability $1-p$), independently of all other arrivals. Let $Y_1(t)$ and $Y_2(t)$ denote, respectively, the number of Type I and Type II arrivals by time $t$. Then*

*i. $\{Y_1(t)\}$ and $\{Y_2(t)\}$ are Poisson processes with intensities $\lambda p$ and $\lambda(1-p)$, respectively,*
*ii. $Y_1(t)$ and $Y_2(t)$ are independent.*

*Proof.*

i. Conditional on the total number of arrivals, both $Y_1(t)$ and $Y_2(t)$ are binomially distributed, so this is fundamentally the same setup as Example 5.3.1. Filling in the details of the proof is left as an exercise.
ii. Their joint PMF is

$$\begin{aligned} P(Y_1(t)=m, Y_2(t)=n) &= P(Y_1(t)=m, Y_2(t)=n|Y(t)=m+n)P(Y(t)=m+n) \\ &= \binom{m+n}{m} p^m (1-p)^n \frac{e^{-\lambda t}(\lambda t)^{m+n}}{(m+n)!} \\ &= \frac{e^{-\lambda p\, t}(\lambda p\, t)^m}{m!} \frac{e^{-\lambda(1-p)t}(\lambda(1-p)t)^n}{n!} \\ &= P(Y_1(t)=m)P(Y_2(t)=n), \end{aligned}$$

so they are independent. This result is perhaps a bit surprising, as it implies that the number of Type I arrivals tells us nothing about the number of Type II arrivals.

□

**Proposition 6.7.9** (Superposition)
*Let $\{Y_1(t)\}$ and $\{Y_2(t)\}$ be independent Poisson processes with intensities $\lambda$ and $\mu$, respectively. If we define $Y(t)=Y_1(t)+Y_2(t)$, then $\{Y(t)\}$ is a Poisson process with intensity $\lambda+\mu$.*

Proving this result is part of Exercise 6.7.

### 6.7.4 Arrival and interarrival times

Thus far, we have studied Poisson processes by looking at the number of arrivals in a set time interval, which is a discrete random variable. It is also useful to consider the process in terms of the arrival times, which are continuous random variables.

**Definition 6.7.10** (Arrival and interarrival times)
For a Poisson process $\{Y(t)\}$, the $n^{\text{th}}$ **arrival time** is

$$T_n = \begin{cases} 0 & n = 0 \\ \inf\{t : Y(t) = n\} & n > 0. \end{cases} \tag{6.31}$$

The $n^{\text{th}}$ **interarrival time** is then defined as $X_n = T_n - T_{n-1}$, that is, the time that elapses between the $(n-1)^{\text{th}}$ and $n^{\text{th}}$ arrivals (Figure 6.9).

**Proposition 6.7.11** (Properties of arrival times)
*If $Y(t)$ is a Poisson process with rate $\lambda$, arrival times $T_n$, and interarrival times $X_n$, then*

*i.* $X_1, X_2, X_3, \ldots$ *are IID* $\text{Exp}(\lambda)$,

*ii.* $T_n \sim \text{Gamma}(n, \lambda)$, *for* $n = 1, 2, 3, \ldots$,

*iii.* $\text{Cov}(T_m, T_n) \sim \text{Gamma}(m, \lambda)$, *for* $m, n \in \{1, 2, 3, \ldots\}$, $m < n$.

*Proof.*

We prove the first property; the rest are part of Exercise 6.7.

i. Since events in non-overlapping intervals are independent, the interarrival times are clearly independent random variables. Now suppose that we start observing the process at an arbitrary time point $s$, and let $W(s)$ be the waiting time to the next arrival. We have

$$P(W(s) > t) = P(Y(s+t) - Y(s) = 0) = \frac{e^{-\lambda t}(\lambda t)^0}{0!} = e^{-\lambda t},$$

which is the exponential survival function, so $W(s) \sim \text{Exp}(\lambda)$, independently of the choice of starting point $s$. In particular, if $s$ is the time of the $(n-1)^{\text{th}}$ arrival, then $W_s$ is the $n^{\text{th}}$ interarrival time, $X_n$. We deduce that the interarrival times are all distributed as $\text{Exp}(\lambda)$.

□

Suppose that the number of taxicabs that pass a particular point on a busy street follows a Poisson process with rate $\lambda = 12$ per hour. The expected time between consecutive taxicabs is $1/\lambda = 1/12$ hours, that is, 5 minutes. Due to the memoryless property of the exponential distribution, if you start looking for a taxicab at an arbitrary point in time, your expected waiting time is still 5 minutes, regardless of how much time has passed since the last arrival. Thus, you are likelier to be in a longer-than-average interarrival interval. This somewhat counterintuitive result is known as the **waiting-time paradox**.

### 6.7.5 *Compound Poisson process*

In a Poisson process, $Y(t)$ jumps by 1 every time there is an arrival. Suppose now that the magnitude of each jump is itself a random variable, for example, when the arrivals occur in groups.

**Definition 6.7.12** (Compound Poisson process)
Suppose that $Y(t)$ is a Poisson process with rate $\lambda$, and $U_1, U_2, \ldots \sim F_U$ are IID non-negative discrete random variables. We define a **compound Poisson process** with rate $\lambda$ and jump distribution $F_U$ as

$$X(t) = \begin{cases} \sum_{i=1}^{Y(t)} U_i & \text{if } Y(t) > 0 \\ 0 & \text{if } Y(t) = 0. \end{cases} \tag{6.32}$$

We can think of the usual Poisson process as a special case of the compound Poisson process, where the jumps $\{U_i\}$ are always equal to 1. Notice that a compound Poisson process is a random sum, so we can use the results from section 5.6.

**Proposition 6.7.13** (Properties of compound Poisson processes)
*If $X(t)$ is a compound Poisson process with rate $\lambda$ and jump distribution $F_U$, and $0 < s < t$, then*

*i.* $\mathbb{E}(X(t)) = \lambda t\, \mathbb{E}(U)$,

*ii.* $\mathrm{Var}(X(t)) = \lambda t\, \mathbb{E}(U^2)$,

*iii.* $\mathrm{Cov}(X(s), X(t)) = \lambda s\, \mathbb{E}(U^2)$,

*iv.* $\mathrm{Cov}(X(t), Y(t)) = \lambda t\, \mathbb{E}(U)$.

*v.* $K_{X(t)}(v) = \lambda t(M_U(v) - 1)$.

Proving these results is part of Exercise 6.7.

### 6.7.6 *Non-homogeneous Poisson process*

Thus far, we have assumed that the rate of arrivals, $\lambda$, remains constant over time. There are many practical applications where we may wish to relax this assumption; for example, if we are monitoring the number of calls to a call centre, we would expect more occurrences during peak times. We can model this behaviour with a **non-homogeneous Poisson process**.

**Definition 6.7.14** (Non-homogeneous Poisson process)
If $Y(t)$ is a continuous-time counting process with

i. $Y(0) = 0$,

ii. independent increments,

iii. $$\mathrm{P}(N_{(t,t+h]} = j) = \begin{cases} 1 - \lambda(t)h + o(h) & \text{for } j = 0 \\ \lambda(t)h + o(h) & \text{for } j = 1 \\ o(h) & \text{for } j > 1 \end{cases}$$

then $Y(t)$ is a non-homogeneous Poisson process with rate function $\lambda(t)$.

The rate function $\lambda(t)$ is a non-negative function of time. We will assume that it is deterministic, i.e. we know the value of $\lambda(t)$ for all values of $t$. There are important models in financial mathematics and biology where $\lambda(t)$ is itself a random variable, but these are beyond the scope of this book.

The **mean-value function**, $\mu(t)$, of a non-homogeneous Poisson process $Y(t)$ is defined as

$$\mu(t) = \int_0^t \lambda(s)ds\,. \tag{6.33}$$

Notice that, if $\lambda(t) = \lambda$ (constant) for all $t$, then $\mu(t) = \lambda t$, which we know is the mean of $Y(t)$.

**Proposition 6.7.15** (Properties of non-homogeneous Poisson processes)
*If $Y(t)$ is a non-homogeneous Poisson process with rate function $\lambda(t)$ and mean-value function $\mu(t)$, and $0 < s < t$, then*

*i.* $Y(t) \sim \text{Pois}(\mu(t))$,
*ii.* $\text{Cov}(Y(s), Y(t)) = \mu(s)$,
*iii.* $(N_{(s,t]}|Y(s) = n) \sim \text{Pois}(\mu(t) - \mu(s))$,
*iv.* $(Y(s)|Y(t) = n) \sim \text{Bin}\,(n, \mu(s)/\mu(t))$.

Proving these results is part of Exercise 6.7.

**Exercise 6.7**

1. Prove Propositions 6.7.13 and 6.7.15.
2. Let $Y(t)$ be a Poisson process with intensity $\lambda = 1$. Evaluate the following:
   (a) $\text{P}(Y(4) = j)$ for $j = 1, 2$,
   (b) $\text{P}(Y(4) = j|Y(3) = 1)$ for $j = 1, 2$,
   (c) $\text{Var}(Y(4)|Y(3) = 1)$,
   (d) $\text{Cov}(Y(3), Y(4))$,
   (e) $\text{P}(Y(1) = 0|Y(3) = 1)$,
   (f) $\mathbb{E}(Y(1)|Y(3) = j)$.
3. Let $W$ be a random variable with a $\text{Gamma}(\alpha, \theta)$ distribution and consider a counting process $N(t)$ which, conditional on $W = w$, is a Poisson process with intensity $w$. Thus, $(N(t)|W = w) \sim \text{Pois}(wt)$ for $t > 0$. Derive an expression for the marginal mass function of $N(t)$.
4. Let $\lambda(t)$ be a non-negative function of time, and suppose that there exists $\lambda^* > 0$ such that $\lambda(t) \leq \lambda^*$ for any $t \geq 0$. Now consider a Poisson process $\{Y^*(t)\}$ with rate $\lambda^*$, and let $t_1, t_2, t_3 \ldots$ be the arrival times. If we classify the $i^{\text{th}}$ arrival as being of Type I with probability $\lambda(t_i)/\lambda^*$, show that the Type I arrivals form a non-homogeneous Poisson process with rate function $\lambda(t)$.

## 6.8 Markov chains

Consider the following problem: the current score in a game of a tennis is deuce (40-40), that is, each player has won three points. To win the game, a player needs to be two points ahead. Starting from deuce, the player who wins the next point is said to have the advantage. If the player with the advantage wins the next point, that player wins the game; if the player loses the next point, the game returns to deuce.

Suppose that Player A, who is serving, wins each point with probability $p$. If different points are independent, what is the probability that Player A wins this game?

We can represent this problem with Figure 6.10.

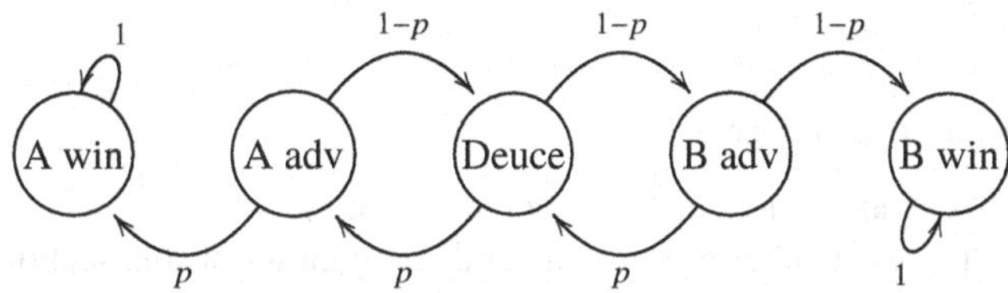


Figure 6.10 *State transition diagram for a game of tennis at deuce.*

The techniques we have encountered so far are not particularly well suited to solving this problem. One approach would be to consider every possible path from "Deuce" to "A win" that does not pass through "B win", but generating these paths is not trivial.

This situation is an example of a **Markov chain**. This section is intended as a concise introduction to Markov chains; proofs of the theorems are omitted. Markov chains are a subclass of stochastic processes, as the following definition makes clear.

**Definition 6.8.1** (Markov chain)
Consider a stochastic process $\{X_0, X_1, X_2, \ldots\}$ with state space $S = \{1, 2, \ldots, M\}$, where $M$ may be infinite. This process is a Markov chain if

$$P(X_n = x_n | X_{n-1} = x_{n-1}, \ldots, X_0 = x_0) = P(X_n = x_n | X_{n-1} = x_{n-1}),$$

for all $x_0, x_1, x_2, \ldots, x_n \in S$ and $n \in \mathbb{Z}^+$. This is known as the **Markov property**.

Definition 6.8.1 has an intuitive interpretation. If $X_{n-1}$ represents the present state of the chain, then $X_0, \ldots, X_{n-2}$ is the past, and $X_n$ is the future. The Markov property states that, if the present is known, any additional information about the past does not affect the future. In the context of the tennis game, any predictions about the future will be based only on the current score.

We are mostly interested in Markov chains that are **homogeneous**.

**Definition 6.8.2** (Homogeneity)
A Markov chain is homogeneous if, for all $t \in \mathbb{Z}^+$ and $i, j \in S$,

$$P(X_{t+1} = j | X_t = i) = P(X_1 = j | X_0 = i) = p_{i,j},$$

where the probability $p_{i,j}$ does not depend on $t$.

For a homogeneous Markov chain, we can define the $M \times M$ **transition matrix** $\boldsymbol{P}$ which has $(i, j)^{\text{th}}$ element $p_{i,j}$, that is,

$$[\boldsymbol{P}]_{i,j} = \mathrm{P}(X_1 = j | X_0 = i) = \mathrm{P}(X_{t+1} = j | X_t = i), \text{ for all } t \in \mathbb{Z}^+.$$

**Proposition 6.8.3** (Properties of transition matrices)
*The transition matrix, $\boldsymbol{P}$, of a homogeneous Markov chain has the following properties:*

1. $[\boldsymbol{P}]_{i,j} \geq 0$ *for all* $i, j \in S$.
2. $\sum_{j=1}^{M} [\boldsymbol{P}]_{i,j} = 1$, *for all* $j \in S$.

The proof is part of Exercise 6.8.

**Example 6.8.4** (Transition matrix for the tennis game)
We return to the motivating example from the beginning of this section. Labelling the states in the diagram $1, \ldots, 5$ from left to right, we can write down the transition matrix,

$$\boldsymbol{P} = \begin{pmatrix} 1 & 0 & 0 & 0 & 0 \\ p & 0 & 1-p & 0 & 0 \\ 0 & p & 0 & 1-p & 0 \\ 0 & 0 & p & 0 & 1-p \\ 0 & 0 & 0 & 0 & 1 \end{pmatrix}.$$

Notice that if the chain reaches state 1 (Player A wins) or state 5 (Player B wins) it cannot leave. Such states are known as **absorbing** states.

The transition matrix $\boldsymbol{P}$ describes the evolution of the Markov chain in a single time step. We are often interested in the evolution of the chain over a longer time period; for this, we need to obtain the ***n*-step transition matrix**.

**Theorem 6.8.5** ($n$-step transition matrix – Chapman-Kolmogorov equations)
*The $n$-step transition matrix of a homogeneous Markov chain is equal to its transition matrix, $\boldsymbol{P}$, raised to the $n^{th}$ power, that is,*

$$[\boldsymbol{P}^n]_{i,j} = p_{i,j}^{(n)} = \mathrm{P}(X_n = j | X_0 = i),$$

*for all $n \in \mathbb{Z}^+$ and $i, j \in S$.*

*Proof.*

The result clearly holds if $n = 1$. Assume that it is true for $n = n'$, and consider

$n = n' + 1$. Conditioning on $X_{n'}$, we have

$$
\begin{aligned}
&\mathrm{P}(X_{n'+1} = j | X_0 = i) \\
&= \sum_{k=1}^{M} \mathrm{P}(X_{n'+1} = j | X_{n'} = k, X_0 = i)\mathrm{P}(X_{n'} = k | X_0 = i) && \text{law of total probability} \\
&= \sum_{k=1}^{M} \mathrm{P}(X_{n'+1} = j | X_{n'} = k)\mathrm{P}(X_{n'} = k | X_0 = i|) && \text{Markov property} \\
&= \sum_{k=1}^{M} [\boldsymbol{P}]_{k,j} [\boldsymbol{P}^{n'}]_{i,k} && \text{assume true for } n = n' \\
&= [\boldsymbol{P}^{n'} \cdot \boldsymbol{P}]_{i,j} && \text{matrix multiplication} \\
&= [\boldsymbol{P}^{n'+1}]_{i,j} \,.
\end{aligned}
$$

We conclude that, by induction, the result holds for all $n \in \mathbb{Z}^+$. □

**Example 6.8.6** (Computing the $n$-step transition matrix)
We can find the general form for $\boldsymbol{P}^n$ using the technique of matrix diagonalisation, which you may have encountered in a linear algebra course. Consider the simple two-state Markov chain in Figure 6.11, where $p_1, p_2 \in (0, 1)$.

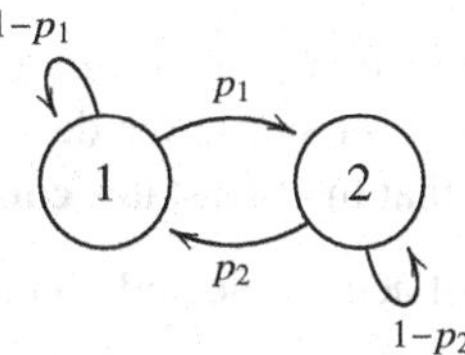


Figure 6.11 *State transition diagram for a simple Markov chain.*

The transition matrix is

$$\boldsymbol{P} = \begin{pmatrix} 1 - p_1 & p_1 \\ p_2 & 1 - p_2 \end{pmatrix}.$$

The eigenvectors, $\boldsymbol{v}_i$, and eigenvalues, $\lambda_i$, of $\boldsymbol{P}$ are the solutions to the equation $\boldsymbol{P}\boldsymbol{v} = \lambda \boldsymbol{v}$. We find the eigenvalues by setting

$$|\boldsymbol{P} - \lambda \boldsymbol{I}| = 0 \Rightarrow \begin{vmatrix} 1 - p_1 - \lambda & p_1 \\ p_2 & 1 - p_2 - \lambda \end{vmatrix} = (1 - p_1 - \lambda)(1 - p_2 - \lambda) - p_1 p_2 = 0\,.$$

This has solutions $\lambda_1 = 1$ and $\lambda_2 = 1 - p_1 - p_2$, and corresponding eigenvectors $\boldsymbol{v}_1 = (1, -1)^T$ and $\boldsymbol{v}_2 = (p_2, p_1)^T$. If we define the matrices $\boldsymbol{\Lambda} = \operatorname{diag}(\lambda_1, \lambda_2)$ and

$U = (v_1, v_2)$, we can write

$$
\begin{aligned}
P &= U\Lambda U^{-1} \Rightarrow \\
P^n &= U\Lambda^n U^{-1} = \begin{pmatrix} 1 & p_2 \\ -1 & p_1 \end{pmatrix} \begin{pmatrix} 1 & 0 \\ 0 & 1 - p_1 - p_2 \end{pmatrix}^n \begin{pmatrix} 1 & p_2 \\ -1 & p_1 \end{pmatrix}^{-1} \\
&= \begin{pmatrix} 1 & p_2 \\ -1 & p_1 \end{pmatrix} \begin{pmatrix} 1 & 0 \\ 0 & (1 - p_1 - p_2)^n \end{pmatrix} \begin{pmatrix} p_1 & -p_2 \\ 1 & 1 \end{pmatrix} |U|^{-1} \\
&= \begin{pmatrix} p_2 + p_1(1 - p_1 - p_2)^n & p_1 - p_1(1 - p_1 - p_2)^n \\ p_2 - p_2(1 - p_1 - p_2)^n & p_1 + p_2(1 - p_1 - p_2)^n \end{pmatrix} (p_1 + p_2)^{-1}.
\end{aligned}
$$

Notice that $(1 - p_1 - p_2)^n \to 0$ (as long as $|1 - p_1 - p_2| \neq 1$), so

$$
P^{\infty} = \lim_{n \to \infty} P^n = \begin{pmatrix} p_2 & p_1 \\ p_2 & p_1 \end{pmatrix} (p_1 + p_2)^{-1},
$$

that is, the probability that the chain is in state 1 at some point in the distant future is equal to $p_2/(p_1 + p_2)$, regardless of the initial state. We will return to this concept in the context of limiting distributions.

### 6.8.1 *Classification of states and chains*

State $j$ is **accessible** from state $i$ if $p_{i,j}^{(n)} > 0$ for some $n \in \mathbb{Z}^+$, that is, if $j$ can be reached from $i$ in a finite number of steps. We denote this by $i \to j$. If $i \to j$ and $j \to i$, we write $i \leftrightarrow j$ and say that the two states **communicate**.

**Proposition 6.8.7** (Properties of accessible and communicating states)
*For states $i, j, k \in S$,*

*i. If $i \to j$ and $j \to k$, then $i \to k$.*
*ii. If $i \leftrightarrow j$ and $j \leftrightarrow k$, then $i \leftrightarrow k$.*

The proof is part of Exercise 6.8.

A Markov chain is **irreducible** if every state is accessible from any other state. Furthermore, if $S$ is finite and there exists some value $N \in \mathbb{Z}^+$ for which every element of $P^N$ is positive, the chain is said to be **regular**. A regular chain is irreducible (it is possible to reach any $j \in S$ from any $i \in S$ in exactly $N$ steps), but the converse is not true; we provide a counterexample in subsection 6.8.3.

In general, we can partition the state space of any Markov chain into **communicating classes**.

**Definition 6.8.8** (Communicating class)
A set of states $C \subset S$ is a communicating class if

i. $i \leftrightarrow j$ for any $i, j \in C$,

ii. $i \nleftrightarrow k$ for any $i \in C$ and $k \in S \backslash C$.

This means that we can write $S = C_1 \cup C_2 \cup \ldots \cup C_K$, where $K$ may be infinite, and $C_i \cap C_j = \emptyset$. This partition is unique – the proof is left as an (easy) exercise. If the chain is irreducible, the state space forms a single communicating class. In the tennis example, there are three communicating classes – $\{1\}$, $\{2, 3, 4\}$, and $\{5\}$ – so the chain is reducible.

Communicating classes are either **open** or **closed**. $C$ is a **closed class** if, for all $i \in C$ and $j \in S \backslash C$, we have $i \nrightarrow j$. In other words, if the chain enters a closed class, it cannot leave. If $i$ is an absorbing state, the class $\{i\}$ is a closed class.

### 6.8.2 *Absorption*

Consider again the simple two-state chain from Example 6.8.6. If we set $p_2 = 0$, state 2 is an absorbing state. What is the probability that the chain is eventually absorbed at state 2? If we define $a_i^{(j)}$ to be the probability of absorption at state $j$ given that the chain begins at state $i$, we have

$$a_1^{(2)} = 1 - \mathrm{P}(\text{never leave state 1}) = 1 - \lim_{n \to \infty} p_1^n = \begin{cases} 0 & \text{if } p_1 < 1\,, \\ 1 & \text{if } p_1 = 1\,, \end{cases}$$

and, obviously, $a_2^{(2)} = 1$. We conclude that the probability of eventual absorption at state 2 is certain, except in the degenerate case where state 1 is also an absorbing state.

If a Markov chain has an absorbing state that is accessible from any other state, we call it an **absorbing chain**. For an absorbing state $j$, there is a simple technique that allows us to find $a_i^{(j)}$ as the solution to a system of linear equations. Suppose that the chain begins at state $i$ and consider the first time step. If $A_j$ is the event that the chain is eventually absorbed at $j$, we can write

$$\mathrm{P}(A_j | X_0 = i) = \sum_{k=1}^{M} \mathrm{P}(A_j | X_0 = i, X_1 = k) \mathrm{P}(X_1 = k | X_0 = i)\,.$$

By the Markov property, conditional on $X_1 = k$, $A_j$ is independent of $X_0 = i$, and we have

$$a_i^{(j)} = \sum_{k=1}^{M} p_{i,k} a_k^{(j)} \Leftrightarrow \boldsymbol{a}^{(j)} = \boldsymbol{P} \boldsymbol{a}^{(j)}\,,$$

where we define the vector $\boldsymbol{a}^{(j)} = (a_1^{(j)}, \ldots, a_M^{(j)})^T$. Note that $a_k^{(j)} = 0$ if $k$ is any other absorbing state, and $a_j^{(j)} = 1$.

**Example 6.8.9** (Solving the tennis problem)
We now have the necessary tools to tackle the motivating example from the beginning

of this section. Let $a_i$ be the probability of being absorbed in state 1 given that the state started at chain $i$. Starting from deuce, the probability that Player A wins the game is $a_3$. We have

$$\boldsymbol{a}^{(1)} = \boldsymbol{P}\boldsymbol{a}^{(1)} = \begin{pmatrix} 1 & 0 & 0 & 0 & 0 \\ p & 0 & 1-p & 0 & 0 \\ 0 & p & 0 & 1-p & 0 \\ 0 & 0 & p & 0 & 1-p \\ 0 & 0 & 0 & 0 & 1 \end{pmatrix} \begin{pmatrix} a_1^{(1)} \\ a_2^{(1)} \\ a_3^{(1)} \\ a_4^{(1)} \\ a_5^{(1)} \end{pmatrix}$$

$$\Leftrightarrow \left.\begin{aligned} a_1^{(1)} &= a_1^{(1)} \\ pa_1^{(1)} + (1-p)a_3^{(1)} &= a_2^{(1)} \\ pa_2^{(1)} + (1-p)a_4^{(1)} &= a_3^{(1)} \\ pa_3^{(1)} + (1-p)a_5^{(1)} &= a_4^{(1)} \\ a_5^{(1)} &= a_5^{(1)} \end{aligned}\right\}.$$

Using the boundary conditions $a_1^{(1)} = 1$ and $a_5^{(1)} = 0$, we obtain

$$\left.\begin{aligned} p + (1-p)a_3^{(1)} &= a_2^{(1)} \\ pa_2^{(1)} + (1-p)a_4^{(1)} &= a_3^{(1)} \\ pa_3^{(1)} &= a_4^{(1)} \end{aligned}\right\} \Rightarrow p\left[p + (1-p)a_3^{(1)}\right] + (1-p)pa_3^{(1)} = a_3^{(1)},$$

and, thus,

$$a_3^{(1)} = \frac{p^2}{1 - 2p(1-p)}.$$

When $p = 0.5$ we have $a_3^{(1)} = 0.5$, as expected. Notice, however, that Player A wins the game with probability 0.692 if $p = 0.6$, and probability 0.9 if $p = 0.75$.

By a similar argument, we can find the expected number of steps for the chain to be absorbed. If $A$ is the set of all absorbing states, we define $e_i^A$ to be the expected number of steps for the chain to reach $A$, starting from state $i$. Clearly, if $i \in A$, we have $e_i^A = 0$. If $i \notin A$ we have

$$e_i^A = \sum_{k=1}^{M} p_{i,k} e_k^A + 1 \Leftrightarrow \tilde{\boldsymbol{e}}^A = \tilde{\boldsymbol{P}}\boldsymbol{e}^A + \boldsymbol{1},$$

where $\tilde{\boldsymbol{P}}$ and $\tilde{\boldsymbol{e}}^A$ are, respectively, $\boldsymbol{P}$ and $\boldsymbol{e}^A$ with the rows corresponding to states in $A$ deleted.

**Example 6.8.10** (Mean absorption time for the tennis problem)
The set of absorbing states is $A = \{1,5\}$, so we have

$$\tilde{\boldsymbol{e}}^A = \tilde{\boldsymbol{P}}\boldsymbol{e}^A + \mathbf{1} \Leftrightarrow \begin{pmatrix} e_2^A \\ e_3^A \\ e_4^A \end{pmatrix} = \begin{pmatrix} p & 0 & 1-p & 0 & 0 \\ 0 & p & 0 & 1-p & 0 \\ 0 & 0 & p & 0 & 1-p \end{pmatrix} \begin{pmatrix} e_1^A \\ e_2^A \\ e_3^A \\ e_4^A \\ e_5^A \end{pmatrix} + \begin{pmatrix} 1 \\ 1 \\ 1 \end{pmatrix}$$

$$\Leftrightarrow \left.\begin{aligned} pe_1^A + (1-p)e_3^A + 1 = e_2^A \\ pe_2^A + (1-p)e_4^A + 1 = e_3^A \\ pe_3^A + (1-p)e_5^A + 1 = e_4^A \end{aligned}\right\}.$$

Using the boundary conditions $e_1^A = e_5^A = 0$, we have

$$\left.\begin{aligned} (1-p)e_3^A + 1 = e_2^A \\ pe_2^A + (1-p)e_4^A + 1 = e_3^A \\ pe_3^A + 1 = e_4^A \end{aligned}\right\} \Rightarrow p[(1-p)e_3^A + 1] + (1-p)[pe_3^A + 1] + 1 = e_3^A,$$

which gives

$$e_3^A = \frac{2}{1-2p(1-p)}.$$

The mean absorption time is maximised when the players are evenly matched, that is, $p = 0.5$.

### 6.8.3 Periodicity

Consider the two chains in Figure 6.12, where $0 < p < 1$.

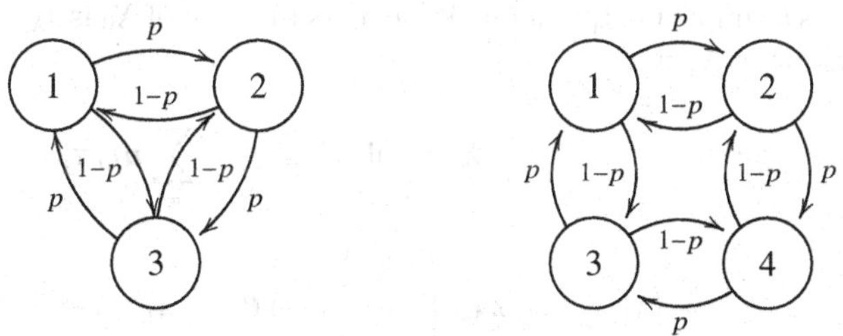


Figure 6.12 *An aperiodic (left) and a periodic (right) Markov chain.*

For the chain on the left, we have $p_{i,i}^{(n)} > 0$ for all $i \in S$ and $n \in \mathbb{Z}^+$, so the chain can return to its starting point in any positive number of steps. For the chain on the right, we have $p_{i,i}^{(n)} > 0$ for even $n$, and $p_{i,i}^{(n)} = 0$ for odd $n$. Every path that brings the chain back to its starting point has even length; this is an example of a **periodic** chain. Notice that this chain is irreducible (we can reach any state from any other state), but it is not regular, as the matrix $\boldsymbol{P}^n$ will contain zeroes for any value of $n$.

**Definition 6.8.11** (Period)
The **period** of a state $i$, denoted $d_i$, is defined as

$$d_i = \gcd\{n \in \mathbb{Z}^+ : p_{i,i}^{(n)} > 0\},$$

the greatest common divisor of the lengths of the paths that bring the chain back to $i$.

Periodicity is a class property; all states in a communicating class have the same period. A state with period equal to 1 is **aperiodic**. If all the states in a chain are aperiodic, the chain itself is said to be aperiodic. In the three-state chain above, all the states have period 1, and the chain is aperiodic. In the four-state chain, all the states have period 2, so the chain is periodic with period 2.

**Theorem 6.8.12** (Necessary and sufficient conditions for regularity)
*A finite-state Markov chain is regular if and only if it is both aperiodic and irreducible.*

### 6.8.4 *Limiting distribution*

We are often interested in the behaviour of a Markov chain as $n \longrightarrow \infty$, that is, the distribution of $X_n$ after the chain has been running for a long time. This is particularly important in the context of Markov chain Monte Carlo (MCMC), which we discuss in Chapter 12. For a wide class of chains, $X_n$ converges in distribution to some discrete probability distribution on $S$, known as the **limiting distribution**. In this section, we explain how we find this distribution for a given chain, and state the conditions for convergence. We begin with the concept of a **stationary distribution**.

**Definition 6.8.13** (Stationary distribution)
The vector of probabilities $\boldsymbol{\pi} = (\pi_1, \dots, \pi_M)$ is a stationary distribution of a Markov chain with transition matrix $\boldsymbol{P}$ if $\boldsymbol{\pi} = \boldsymbol{\pi}\boldsymbol{P}$.

Stationarity has a simple interpretation. If the mass function of $X_0$ is $f_{X_0}(i) = \pi_i$, then the mass function of $X_1$ is

$$\begin{aligned} f_{X_1}(i) &= \sum_{j=1}^{M} \mathrm{P}(X_1 = i | X_0 = j)\mathrm{P}(X_0 = j) = \sum_{j=1}^{M} p_{j,i}\pi_j \\ &= \begin{pmatrix} \pi_1 & \dots & \pi_M \end{pmatrix} \begin{pmatrix} p_{1,i} \\ \vdots \\ p_{M,i} \end{pmatrix} = [\boldsymbol{\pi}\boldsymbol{P}]_i = \pi_i \,. \end{aligned}$$

We can deduce that $X_2, X_3, \dots$ will also have the same mass function, so it appears that we have answered the initial question; as $n \longrightarrow \infty$, the limiting distribution of $X_n$ is $\boldsymbol{\pi}$.

**Example 6.8.14** (Finding limiting distributions)
Consider again the Markov chain from Example 6.8.6. We have

$$\boldsymbol{\pi}\boldsymbol{P} = \begin{pmatrix} \pi_1 & \pi_2 \end{pmatrix} \begin{pmatrix} 1 - p_1 & p_1 \\ p_2 & 1 - p_2 \end{pmatrix} = \begin{pmatrix} \pi_1(1 - p_1) + \pi_2 p_2 , & \pi_1 p_1 + \pi_2(1 - p_2) \end{pmatrix},$$

which we set equal to $\boldsymbol{\pi}$ to obtain

$$\left.\begin{aligned} \pi_1(1-p_1)+\pi_2 p_2 &= \pi_1 \\ \pi_1 p_1 + \pi_2(1-p_2) &= \pi_2 \end{aligned}\right\} \Rightarrow \left.\begin{aligned} -\pi_1 p_1 + \pi_2 p_2 &= 0 \\ \pi_1 p_1 - \pi_2 p_2 &= 0 \end{aligned}\right\}.$$

The two equations are the same, but we can use the condition $\pi_1 + \pi_2 = 1$ to find a unique solution,

$$\pi_1 = \frac{p_2}{p_1 + p_2} \quad \text{and} \quad \pi_2 = \frac{p_1}{p_1 + p_2}.$$

Notice that the rows of the matrix $\boldsymbol{P}^{\infty}$ are equal to the limiting distribution.

Unfortunately, stationary distributions are not necessarily unique; for example, for the chain in Figure 6.13, where $\boldsymbol{P}$ is the identity matrix, any $\boldsymbol{\pi} = (p, 1-p)$ with $\pi \in [0.1]$ is a stationary distribution.

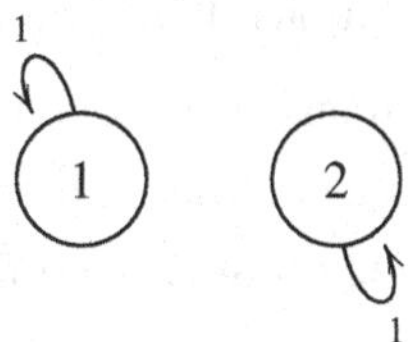


Figure 6.13 *Markov chain without a unique stationary distribution.*

However, we can establish necessary conditions for the existence of a unique stationary distribution.

**Theorem 6.8.15** (Fundamental theorem of Markov chains – finite-state case)
*Every finite-state, regular Markov chain has a unique stationary distribution, which is also its limiting distribution.*

We can establish a similar result for the general case (where the state space may be infinite), but we must first define the concepts of recurrence and transience.

### 6.8.5 *Recurrence and transience*

**Definition 6.8.16** (Recurrent and transient states)
A state $i$ of a Markov chain is **recurrent** if

$$f_{i,i} = \mathrm{P}(X_n = i \text{ for some } n \in \mathbb{Z}^+ | X_0 = i) = 1\,.$$

If this probability is less than 1, the state $i$ is **transient**.

In other words, state $i$ is recurrent if, starting from $i$, the chain is certain to return to $i$ at some point in the future. Recurrence and transience are closely linked to the concepts of **first-passage time** and **mean recurrence time**.

**Definition 6.8.17** (First-passage time and mean recurrence time)
If a chain starts from state $i$, the first-passage time to state $j$ is defined as

$$T_{i,j} = \min_n \{n \in \mathbb{Z}^+ : X_n = j | X_0 = i\},$$

that is, the first time the chain reaches state $j$. The mean recurrence time of state $i$ is then $\mu_i = \mathbb{E}(T_{i,i})$, the expected time it takes for the chain to first return to the starting state, $i$.

Transient states always have infinite mean recurrence time. Recurrent states with $\mu_i = \infty$ are called **null recurrent**, while those with $\mu_i < \infty$ are **positive recurrent**. We can work out $f_{i,i}$ and $\mu_i$ with the same approach we used for absorption probabilities and mean number of steps in section 6.8.2.

Recurrence is also a class property, that is, a communicating class can only contain one type of state, so the class itself can be said to be positive recurrent, null recurrent, or transient. The following proposition establishes some useful results.

**Lemma 6.8.18** (Recurrence and transience as class properties)

1. *All open communicating classes are transient.*
2. *All finite, closed communicating classes are positive recurrent.*

If all the states in a chain are recurrent/transient, we say that the chain itself is recurrent/transient. A chain is said to be **ergodic** if it is aperiodic, irreducible, and positive recurrent.

**Theorem 6.8.19** (Fundamental theorem of Markov chains – general case)
*Every ergodic Markov chain has a unique stationary distribution, which is also its limiting distribution.*

Finite-state, irreducible Markov chains are always positive recurrent, so we can deduce that regular chains (which are aperiodic and irreducible, from Theorem 6.8.12) are ergodic. In other words, Theorem 6.8.15 is the special case of Theorem 6.8.19 for finite-state chains.

### 6.8.6 *Continuous-time Markov chains*

A stochastic process $\{X(t) : t \geq 0\}$ with state space $S = \{1, 2, \ldots, M\}$ is a **continuous-time Markov chain** if, for all $n \in \mathbb{Z}^+$, $x_1, x_2, \ldots, x_n \in S$, and $0 \leq t_1 < t_2 < \ldots < t_n$, it holds that

$$\mathrm{P}(X(t_n) = x_n | X(t_{n-1}) = x_{n-1}, \ldots, X(t_1) = x_1) = \mathrm{P}(X(t_n) = x_n | X(t_{n-1}) = x_{n-1}).$$

If the chain is homogeneous (defined in the usual way) the transition probability, for $i, j \in S$ and $t, h \geq 0$, is given by

$$\mathrm{P}(X(t+h) = j | X(t) = i) = \begin{cases} 1 - g_{i,i}h + o(h) & \text{for } j = i, \\ g_{i,j}h + o(h) & \text{for } j \neq i, \end{cases}$$

where $g_{i,j} \geq 0$ is the $(i, j)^{\text{th}}$ entry of the **generator** matrix, $\boldsymbol{G}$, a matrix with rows that sum to 0. The similarity with Definition 6.7.7 is not coincidental; the Poisson process is a special case of a continuous-time Markov chain. Similarly to a Poisson process, a continuous-time Markov chain remains in each state for an exponentially distributed amount of time before jumping to the next state.

**Exercise 6.8**

1. Prove Propositions 6.8.3 and 6.8.7.
2. Find the 2- and 3-step transition matrices for the chain in the tennis example.
3. For the chain in Example 6.8.6, by carefully considering the different values $p_1$ and $p_2$ can take, find the possible limits for $\boldsymbol{P}$ as $n \longrightarrow \infty$.
4. A Markov chain has transition matrix

$$\boldsymbol{P} = \begin{pmatrix} 1 & 0 & 0 & 0 & 0 & 0 \\ 0 & 1/2 & 0 & 1/4 & 0 & 1/4 \\ 1/4 & 0 & 1/2 & 0 & 1/4 & 0 \\ 0 & 1/3 & 0 & 1/3 & 0 & 1/3 \\ 1/4 & 1/4 & 1/4 & 0 & 1/4 & 0 \\ 0 & 0 & 0 & 1 & 0 & 0 \end{pmatrix}$$

   (a) Find the communicating classes and determine whether each one is open or closed.
   (b) Find $a_i^{(1)}$, the probability of absorption at state $j$ given that the chain begins at state $i$, for $i = 1, 2, \ldots, 6$.
   (c) Verify, by direct calculation of the recurrence probabilities and mean recurrence times, that any open communicating classes are transient, while any closed ones are positive recurrent.

## 6.9 Further exercises

1. For the one-way ANOVA model in section 6.3.3, let $n_j$ be the number of observations in the sample that came from population $j$,

$$n_j = \sum_{i=1}^{n} x_{i,j} \,,$$

and define

$$\bar{Y}_j = \frac{1}{n_j} \sum_{i=1}^{n} x_{i,j} Y_i \,, \tag{6.34}$$

the mean of these observations; we refer to this as the $j^{\text{th}}$ **group mean**. Show that

$$\underbrace{\sum_{i=1}^{n} (Y_i - \bar{Y})^2}_{\substack{\text{Total sum} \\ \text{of squares}}} = \underbrace{\sum_{j=0}^{k-1} n_j (\bar{Y}_j - \bar{Y})^2}_{\substack{\text{Between-group} \\ \text{sum of squares}}} + \underbrace{\sum_{i=1}^{n} \left( Y_i - \sum_{j=0}^{k-1} x_{i,j} \bar{Y}_j \right)^2}_{\substack{\text{Within-group} \\ \text{sum of squares}}} . \tag{6.35}$$

[The total sum of squares is a measure of the variation in the response variable. The expression above shows that we can attribute some of this variation to the differences between the groups, and the rest to the differences within each group, which we typically consider to be random error.]

2. There is another way to formulate the logistic regression model, which uses the properties of the **logistic distribution**. If $Y \sim \text{Logistic}(\mu, \tau)$, then the CDF of $Y$ is

$$F_Y(y) = \left[1 + \exp\left(-\frac{y-\mu}{\tau}\right)\right]^{-1} \quad \text{for } y \in \mathbb{R}, \tag{6.36}$$

where $\mu$ is the location parameter and $\tau > 0$ is the scale parameter. Suppose that we have binary data $Y_1, \dots, Y_n$, where the value of $Y_i$ is dependent on the (unobserved) latent variable $Z_i \sim \text{Logistic}(\mathbf{x}_i\boldsymbol{\beta}, 1)$. As before, $\mathbf{x}_i$ denotes a vector of explanatory variables, and $\boldsymbol{\beta}$ are the model parameters. We observe $Y_i = 1$ if $Z_i > 0$, and $Y_i = 0$ otherwise. Show that this model is equivalent to the logistic regression model. What if the distribution of $Z_i$ were Normal with mean $\mathbf{x}_i\boldsymbol{\beta}$ and variance 1? Explain why we need to fix the variance parameter to 1.

3. Suppose we are interested in whether people at a particular city commute to work by public transport, private car, bike, or on foot. This is a categorical variable so we model it with the **categorical distribution**, which is the multinomial distribution with $n = 1$ (but slightly different notation). We write $Y_i = 1$ if the $i^{\text{th}}$ person takes public transport, $Y_i = 2$ if they use a private car, etc. The quantities of interest are $p_{ij} = P(Y_i = j)$, for $i = 1, \dots, n$ and $j = 1, \dots, k$, where $k$ is the number of categories. The link function is

$$\log\left(\frac{p_{ij}}{p_{i1}}\right) = \boldsymbol{x}_i\boldsymbol{\beta}_j \quad \text{for } j = 2, \dots, k, \tag{6.37}$$

where $\mathbf{x}_i$ is a vector of explanatory variables, and $\boldsymbol{\beta}_2, \dots, \boldsymbol{\beta}_k$ are vectors of parameters.

   (a) Invert the link function to obtain an expression for $p_{ij}$, where $j = 2, \dots, k$.
   (b) What is the corresponding expression for $p_{i1}$?
   (c) Extend your answer to part (a) to the case $j = 1$. You will need to define $\boldsymbol{\beta}_1$ appropriately.
   (d) Explain why we do not need to estimate parameters for $j = 1$.

4. The distribution of a random variable $Y$ has the memoryless property if

$$\text{P}(Y > x + y | Y > x) = S_Y(y) \text{ for all } x.$$

   (a) Show that if a distribution has constant hazard then it has the memoryless property. You will need to consider discrete and continuous cases separately.
   (b) Now suppose that the non-negative random variable $X$ has the memoryless property. If $X$ is discrete, show that is geometrically distributed; if it is continuous, show that it is exponentially distributed.

5. The random variable $Y$ has survival function

$$S_Y(y) = \begin{cases} \exp\left\{-\frac{\lambda}{\beta}(e^{\beta y} - 1)\right\} & \text{for } y > 0\,, \\ 1 & \text{otherwise,} \end{cases}$$

where $\lambda, \beta > 0$. This is the **Gompertz distribution**, commonly used in actuarial science as a model for adult lifespan.

(a) Find $F_Y(y)$, $f_Y(y)$, $H_Y(y)$, and $\lambda_Y(y)$.

(b) We observe survival times $(Y_i | \boldsymbol{X}_i = \boldsymbol{x}_i) \sim \text{Gompertz}(\lambda_i, \beta)$, for $i = 1, \ldots, n$, where $\lambda_i$ depends only on $\boldsymbol{x}_i$. Show that this is a proportional-hazards model and find the baseline hazard function.

6. A limitation of the Weibull distribution for time-to-event analysis is that the hazard function is strictly monotonic (or constant in the special case $\alpha = 1$). A common model for situations where we want the probability of failure to initially increase and later decrease is the **log-logistic distribution**. If $Y \sim \text{LogLogistic}(\mu, \tau)$ then the CDF of $X$ is

$$F_Y(y) = \begin{cases} \left[1 + \exp\left(-\frac{\log y - \mu}{\tau}\right)\right]^{-1} & \text{for } y > 0, \\ 0 & \text{otherwise,} \end{cases}$$

where $\mu \in \mathbb{R}$ and $\tau > 0$. The name of the distribution arises from the fact that $\log Y \sim \text{Logistic}(\mu, \tau)$ (see exercise 2 above).

(a) Find $S_Y(y)$ and $\lambda_Y(y)$.

(b) Show that we can choose $\tau$ so that the hazard function has the desired shape. Which value of $y$ maximises the hazard?

(c) We now want to introduce some covariates. Suppose that we observe survival times $Y_1, \ldots, Y_n$, where $(Y_i | \boldsymbol{X}_i = \boldsymbol{x}_i) \sim \text{LogLogistic}(\mu + \boldsymbol{x}_i \boldsymbol{\beta}, \tau)$ for $i = 1, \ldots, n$. Show that this is an accelerated-life model.

7. Let $\{\varepsilon_t\}$ be a sequence of $\text{IID}(0, \sigma^2)$ random variables. State which of the following processes are covariance stationary and find their ACF. Which of them are strictly stationary?

(a) $Y_t = \varepsilon_t + \varepsilon_{t-1}$,

(b) $Y_t = \varepsilon_0 \cos \lambda t + \varepsilon_1 \sin \lambda t$,

(c) $Y_t = Y_{t-1} + \varepsilon_t$,

(d) $Y_t = \varepsilon_0 + t\varepsilon_t$ ,

(e) $Y_t$ independent $\text{N}(1, 1)$ for $t$ odd, and $Y_t$ independent $\text{Exp}(1)$ for $t$ even.

8. A moving-average model of order $q$, denoted $\text{MA}(q)$, is

$$Y_t = \varepsilon_t + \theta_1 \varepsilon_{t-1} + \ldots + \theta_q \varepsilon_{t-q} \qquad \{\varepsilon_t\} \sim \text{IID}(0, \sigma_\varepsilon^2)\,.$$

Find a formula for the ACF and show that $\rho_Y(h) = 0$ for $|h| > q$.

9. You are trying to get from your hotel to the airport. Taxicabs that pass the hotel follow a Poisson process with rate $\lambda = 0.2$ per minute. There is also an airport shuttle which departs every 10 minutes, but you do not know the timetable. What is the probability a taxicab passes before the shuttle arrives? What if you have already been waiting for 2 minutes and neither vehicle has passed?

10. Consider a Poisson process $Y(t)$ with rate $\lambda$. If we know that $Y(s) = m$, where $m \geq 1$ and $s > 0$, find the conditional density of $T_n$ (the time of the $n^{\text{th}}$ arrival) for $n = 1, 2, \ldots, m$.

11. Cars that pass an intersection follow a Poisson process $Y(t)$ with rate $\lambda = 10$ per minute. The number of passengers in each car (excluding the driver) follows a Bin(4, 0.5) distribution. Work out the mean and standard deviation of the total number of people in cars who pass the intersection in one minute.

12. You draw a card at random from a standard deck of 52 cards and note its suit (♡, ♢, ♣, or ♠). This is your card for the first round of the game. In each subsequent round, you do the following:

    Step 1. Return your current card to the deck, shuffle it, and draw a new card.

    Step 2. If the card drawn is of a different suit to your current card, this is your new card. Otherwise, repeat Step 1.

    Suppose that the first card you drew was the two of hearts (2♡). What is the probability that the $n^{\text{th}}$ card is also a heart? What is the probability that it is a diamond?

13. You have 2 gold coins in your pocket and you want to play a series of games. In each game, if you bet $k$ coins, you win $k$ coins with probability $p$, and lose $k$ coins with probability $1 - p$. Your goal is to get to 5 coins. The game ends when you reach your goal or lose all your coins. You are considering the following strategies:

    Strategy A: Bet 1 coin every time.

    Strategy B: Bet all your coins or just enough to reach your goal – whichever is smaller. For example, if you had 3 coins, you would only bet 2.

    [This is an example of a **Gambler's ruin** problem.]

    (a) For each of the two strategies, work out the probability that you reach your goal.
    (b) Suppose that $p = 0.5$. Which strategy would you choose? What is the expected number of games for each strategy?
    (c) Which strategy is preferable if $p = 0.75$?

## 6.10 Chapter summary

- describe the difference between a model (represented by the sample) and data (the observed sample),
- set up a simple or multiple linear regression model,
- explain how a generalised linear model allows us to apply the ideas of linear modelling more widely,

- formulate a logistic or Poisson regression model,
- define the hazard function in the discrete and continuous case, and understand its significance,
- determine if a time series model is stationary and, if so, calculate its autocovariance,
- define a Poisson process and use it to compute probabilities relating to arrivals and arrival times,
- extend the basic Poisson process framework to the compound and non-homogeneous cases,
- state the Markov property,
- derive the Chapman-Kolmogorov transition equations, and
- compute absorption probabilities and mean time to absorption.

Chapter 7

# Sample moments and quantiles

As we discussed in Chapter 3, the key characteristics of a probability distribution are summarised by its moments, which we compute using the moment-generating function. In direct analogy, the characteristics of a sample – central tendency, spread, symmetry, tail fatness, and so on – are summarised by the sample moments.

In this chapter, we define sample moments, such as the sample mean, sample variance, sample skewness; and sample quantiles, such as the sample maximum and minimum. We work out the distribution of some of these quantities and discuss a remarkable result, the central limit theorem, which highlights the importance of the normal distribution in statistical inference.

Sample moments provide a natural mechanism for estimating their population counterparts. For example, we have often used the observed sample mean to estimate the unknown population mean. We can now answer the question of whether this is a reasonable thing to do. In later chapters, we will see that sample moments can also be used to estimate any distribution parameters, or to test distributional assumptions, such as multivariate normality.

## 7.1 Sample mean

We begin by considering the distributional properties of the first sample moment, the **sample mean**. Its definition is well known; we include it here for completeness.

**Definition 7.1.1** (Sample mean)
For a sample $Y_1, \dots, Y_n$ the sample mean, $\bar{Y}$, is given by

$$\bar{Y} = \frac{1}{n} \sum_{i=1}^{n} Y_i \,.$$

### 7.1.1 Mean and variance of the sample mean

The following proposition establishes the mean and variance of the sample mean. Note that neither of these properties requires us to make any assumption about the parametric form of the population distribution.

**Proposition 7.1.2** (Properties of the sample mean)
*Suppose that $Y_1, \ldots, Y_n$ is a random sample from a population distribution with mean $\mu$, then*

i. *the mean of the sample mean is the population mean,* $\mathbb{E}(\bar{Y}) = \mu$.

*If, in addition, the population distribution has finite variance,* $\sigma^2 < \infty$, *then*

ii. *the variance of the sample mean is given by* $\text{Var}(\bar{Y}) = \frac{\sigma^2}{n}$.

*Proof.*

i. The proof comes directly from the definition of the sample mean.

$$\begin{aligned} \mathbb{E}(\bar{Y}) = \mathbb{E}\left(\frac{1}{n}\sum_{i=1}^{n} Y_i\right) &= \frac{1}{n}\sum_{i=1}^{n}\mathbb{E}(Y_i) && \text{properties of expectation} \\ &= \frac{1}{n}\sum_{i=1}^{n}\mu && \text{identically distributed} \\ &= \mu\,. \end{aligned}$$

Note that we are only using the fact that the elements of a random sample are identically distributed with the population distribution. There is no need to assume independence to get this result.

ii. The result for the variance of the sample mean can also be proved directly.

$$\begin{aligned} \text{Var}(\bar{Y}) = \text{Var}\left(\frac{1}{n}\sum_{i=1}^{n} Y_i\right) &= \frac{1}{n^2}\sum_{i=1}^{n}\text{Var}(Y_i) && \text{by independence} \\ &= \frac{1}{n^2}\sum_{i=1}^{n}\sigma^2 && \text{identically distributed} \\ &= \frac{1}{n}\sigma^2. \end{aligned}$$

□

Notice that the variance of the sample mean decreases as the sample size increases, that is, taking averages reduces variability. This is a fundamental concept in statistics; for the sample mean to take an extreme value, the majority of the sample needs to take extreme values.

### 7.1.2 Central limit theorem

The **central limit theorem** is a remarkable result. One of its consequences is that the sample mean of a large random sample is approximately normally distributed, (almost) regardless of the distribution that we sample from. The mysterious appearance of the normal distribution in this context is widely used in inference. The proof we give relies on the existence of the moment-generating function of the population distribution. This is not a necessary condition for the central limit theorem to hold, but the proof is much more straightforward in this case.

**Theorem 7.1.3** (Central limit theorem)
*Suppose that* $\mathbf{Y} = (Y_1, \dots, Y_n)^T$ *is a random sample with* $\mathbb{E}(Y) = \mu$ *and* $\text{Var}(Y) = \sigma^2$. *Provided* $0 < \sigma^2 < \infty$, *then*

$$\frac{\bar{Y}_n - \mu}{\sqrt{\sigma^2/n}} \xrightarrow{d} \text{N}(0,1) \quad \textit{as} \quad n \to \infty\,.$$

*where* $\bar{Y}_n = \frac{1}{n}\sum_{j=1}^{n} Y_j$ *is the sample mean.*

*Proof of special case.*

As mentioned above, we only prove the case in which the moment-generating function of the $Y_j$ exists. If we define

$$Z_n = \frac{\bar{Y}_n - \mu}{\sqrt{\sigma^2/n}},$$

then we can readily show that

$$Z_n = \frac{S_n - n\mu}{\sqrt{n\sigma^2}},$$

where $S_n = \sum_{j=1}^{n} Y_j$ is the partial sum. We can get an expression for the moment-generating function of $Z$ in terms of the moment-generating function of $Y$ by using the fact that $m_{S_n}(t) = [m_Y(t)]^n$, so

$$\begin{aligned}
m_{Z_n}(t) = \mathbb{E}(e^{tZ_n}) &= \mathbb{E}\left\{\exp\left[t\left(\frac{S_n - n\mu}{\sqrt{n\sigma^2}}\right)\right]\right\} \\
&= \exp\left(-\frac{\sqrt{n}\mu t}{\sigma}\right)\mathbb{E}\left[\exp\left(\frac{t}{\sqrt{n\sigma^2}}S_n\right)\right] \\
&= \exp\left(-\frac{\sqrt{n}\mu t}{\sigma}\right) m_{S_n}\left(\frac{t}{\sqrt{n\sigma^2}}\right) \\
&= \exp\left(-\frac{\sqrt{n}\mu t}{\sigma}\right)\left[m_Y\left(\frac{t}{\sqrt{n\sigma^2}}\right)\right]^n.
\end{aligned}$$

Taking logs yields the cumulant-generating function,

$$K_{Z_n}(t) = -\frac{\sqrt{n}\mu t}{\sigma} + nK_Y\left(\frac{t}{\sqrt{n\sigma^2}}\right).$$

Now recall that, by the properties of cumulants,

$$K_Y(t) = \mu t + \sigma^2 t^2/2 + (\text{terms in } t^3 \text{ and higher order}).$$

Thus, focusing on the power of $\frac{1}{n}$ in the resulting terms, we have

$$\begin{aligned} K_{Z_n}(t) &= -\frac{\sqrt{n}\mu t}{\sigma} + n\left[\mu\frac{t}{\sqrt{n\sigma^2}} + \sigma^2\frac{t^2}{2n\sigma^2} + \left(\text{terms in } \left(\frac{1}{n}\right)^{3/2} \text{ and higher order}\right)\right] \\ &= \frac{t^2}{2} + \left(\text{terms in } \left(\frac{1}{n}\right)^{1/2} \text{ and higher order}\right). \end{aligned}$$

As $n \to \infty$ the terms involving $(1/n)^{1/2}$ and higher order will go to zero. Thus,

$$K_{Z_n}(t) \to t^2/2 \quad \text{as} \quad n \to \infty .$$

We conclude that $K_{Z_n}(t)$ converges to the cumulant-generating function of a standard normal. As the cumulant-generating function uniquely characterises the distribution of a random variable, this implies that $Z_n \xrightarrow{d} \text{N}(0,1)$ as $n \to \infty$. □

The central limit theorem is very appealing. It allows us to draw inference about the sample mean using the normal distribution regardless of the underlying population distribution. This is frequently exploited in interval estimation and hypothesis testing. However, the limitations of the theorem should be recognised. The central limit theorem does not tell us anything about the rate of convergence; we do not know how large a sample has to be for the normal to provide a reasonable approximation to the distribution of the sample mean. In fact, the answer to this question depends on the population distribution. If we are sampling from a normal population, the sample mean is normally distributed regardless of the sample size. At the other extreme, if we are sampling from a highly skewed discrete distribution, we would need a relatively large sample in order for the normal to provide a reasonable approximation to the sample mean.

**Exercise 7.1**

1. Suppose $Y_1, \ldots, Y_n$ is a random sample and $\bar{Y}_n = \frac{1}{n}\sum_{j=1}^{n} Y_j$ is the sample mean. In addition, suppose that $\mathbb{E}(Y) = \mu$, and the moment-generating function $m_Y$ is well defined. Using cumulant-generating functions, prove that

$$\bar{Y}_n \xrightarrow{d} \mu \quad \text{as} \quad n \to \infty .$$

Why can we deduce that $\bar{Y}_n \xrightarrow{p} \mu$? [This is known as the **weak law of large numbers**.]

2. Let $\boldsymbol{Y} = (Y_1, \ldots, Y_n)^T$ be a random sample from a positive-valued distribution. Explain why, for large $n$, the distribution of the sample product $\prod_{i=1}^{n} Y_i$ can be approximated by the lognormal (see Exercise 3.6). What conditions do the moments of $Y$ need to satisfy?

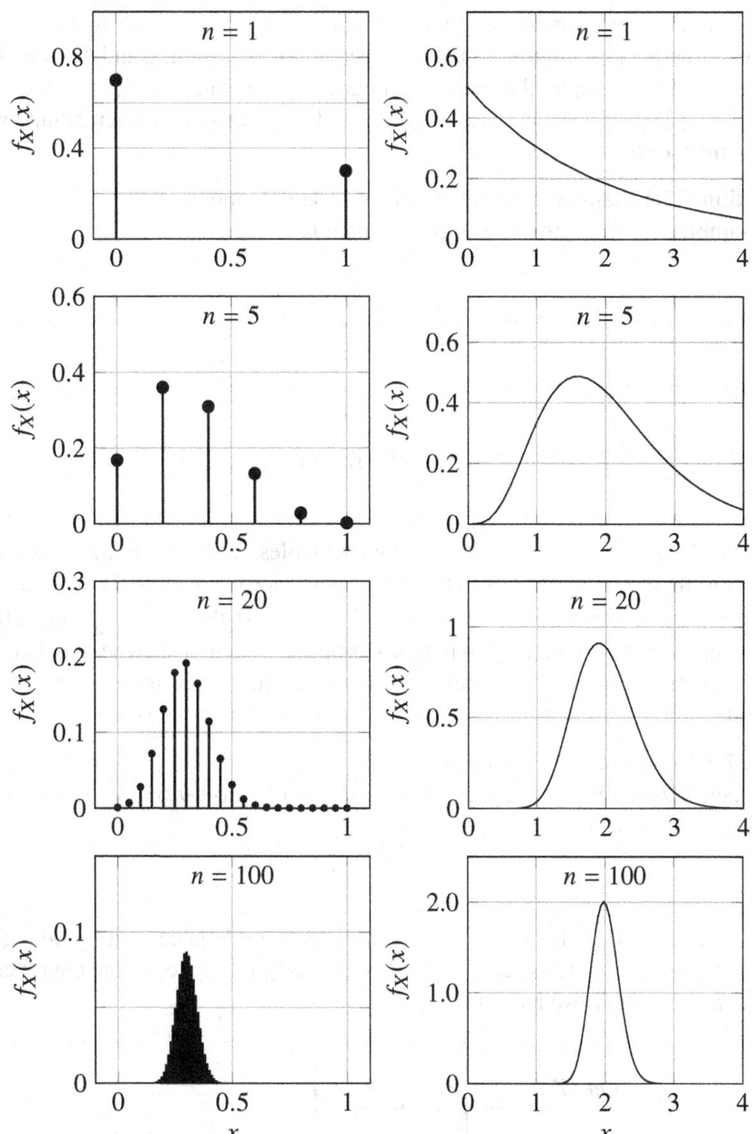


Figure 7.1 *The central limit theorem in practice. The plots show the distribution of the mean of a random sample from a* Bernoulli(0.3) *distribution (left) and an* Exp(0.5) *distribution (right) as the sample size (n) increases.*

## 7.2 Higher-order sample moments

In Chapter 3 we defined the moments of a probability distribution as expectations; the $r^{\text{th}}$ moment of the random variable $X$ is the expectation of $X^r$, and so on. For a sample, the operation that is analogous to taking expectations is summing and dividing by the sample size. For example, the sample analogue of the mean, $\mu = \mathbb{E}(Y)$, is the sample mean, $\bar{Y} = \frac{1}{n}\sum_{i=1}^{n} Y_i$. We can use this idea to define **sample moments** and **central sample moments.**

**Definition 7.2.1** (Sample moments and central sample moments)
For a sample $Y_1, \dots, Y_n$, the $r^{\text{th}}$ sample moment is

$$m'_r = \frac{1}{n}\sum_{i=1}^{n} Y_i^r,$$

and the $r^{\text{th}}$ central sample moment is

$$m_r = \frac{1}{n}\sum_{i=1}^{n}(Y_i - m'_1)^r.$$

Note that the sample moments are random variables. The sample mean is the first sample moment, $\bar{Y} = m'_1$, however, the sample variance is not quite the second central sample moment; if we define $S^2 = \frac{1}{n-1}\sum_{i=1}^{n}(Y_i - \bar{Y})^2$ then $S^2 = \frac{n}{n-1}m_2$. Higher-order sample moments are used to measure properties such as skewness and kurtosis. The following claim is very useful in establishing the properties of central sample moments.

**Claim 7.2.2** (Assuming zero mean)
*When determining the properties of central sample moments, we may assume a population mean of zero.*

*Proof.*

Let $\mathbf{Z} = (Z_1, \dots, Z_n)^T$ be a transformation of the sample generated by taking $Z_i = Y_i - \bar{Y}$ for $i = 1, \dots, n$. Thus, $\mathbb{E}(Z_i) = 0$ for all $i$. Using $m_r(\mathbf{Z})$ to denote the $r^{\text{th}}$ central sample moment of $\mathbf{Z}$, we have, for $r \geq 1$,

$$\begin{aligned}
nm_r(\mathbf{Z}) &= \sum_{i=1}^{n}\left[Z_i - \frac{1}{n}\sum_{j=1}^{n} Z_j\right]^r \\
&= \sum_{i=1}^{n}\left[Y_i - \frac{1}{n}\sum_{j=1}^{n} Y_j - \frac{1}{n}\sum_{j=1}^{n}(Y_j - \bar{Y})\right]^r \\
&= \sum_{i=1}^{n}\left[Y_i - \frac{1}{n}\sum_{j=1}^{n} Y_j\right]^r = nm_r(\mathbf{Y}),
\end{aligned}$$

where $m_r(\boldsymbol{Y})$ is the $r^{\text{th}}$ sample central moment of $\boldsymbol{Y}$. Thus, any statements made about the sample central moments of $\boldsymbol{Z}$ (with population mean zero) are equally true for $\boldsymbol{Y}$. □

### 7.2.1 Sample variance

For any sample of two or more observations, we can generate the **sample variance**.

**Definition 7.2.3** (Sample variance)
For a sample $Y_1, \ldots, Y_n$ with $n > 1$, the sample variance, $S^2$, is given by

$$S^2 = \frac{1}{n-1} \sum_{i=1}^{n} (Y_i - \bar{Y})^2.$$

Two things to note about the sample variance:

i. As is usual practice, we write the sample variance as a function of the sample mean. However, it is a function of sample value alone, since

$$S^2 = \frac{1}{n-1} \sum_{i=1}^{n} \left( Y_i - \frac{1}{n} \sum_{j=1}^{n} Y_j \right)^2,$$

which can be evaluated from a sample.

ii. The sample variance is generated by summing the squares of the differences between the individual sample members and the sample mean, then dividing by $n-1$. The reason for dividing by $n-1$ is made clear in Proposition 7.2.5.

The following lemma establishes useful facts about the sample variance that are simple consequences of the properties of summation.

**Lemma 7.2.4** (Four representations of sample variance)
*Let $Y_1, Y_2, \ldots$ be a sequence of random variables. Defining $\bar{Y}_n = \frac{1}{n}\sum_{i=1}^{n} Y_i$ and $S_n^2 = \frac{1}{n-1}\sum_{i=1}^{n}(Y_i - \bar{Y}_n)^2$, then for $n = 2, 3, \ldots$,*

*i.* $(n-1)S_n^2 = \sum_{i=1}^{n} Y_i(Y_i - \bar{Y}_n)$,

*ii.* $(n-1)S_n^2 = \sum_{i=1}^{n} Y_i^2 - n\bar{Y}_n^2$,

*iii.* $(n-1)S_n^2 = \left[\sum_{i=2}^{n}(Y_i - \bar{Y}_n)\right]^2 + \sum_{i=2}^{n}(Y_i - \bar{Y}_n)^2$,

*and for $n = 3, 4, \ldots$,*

*iv.* $(n-1)S_n^2 = (n-2)S_{n-1}^2 + \frac{n-1}{n}(Y_n - \bar{Y}_{n-1})^2$.

*Proof.*

The proof of i. and ii. is part of Exercise 7.2. For iii., first note that

$$\sum_{i=1}^{n}(Y_i - \bar{Y}_n) = 0 \Rightarrow (Y_1 - \bar{Y}_n) = -\sum_{i=2}^{n}(Y_i - \bar{Y}_n),$$

so

$$(n-1)S_n^2 = (Y_1 - \bar{Y}_n)^2 + \textstyle\sum_{i=2}^n (Y_i - \bar{Y}_n)^2 = \left[\sum_{i=2}^n (Y_i - \bar{Y}_n)\right]^2 + \sum_{i=2}^n (Y_i - \bar{Y}_n)^2,$$

as required.

For iv. we prove directly,

$$\begin{aligned}(n-1)S_n^2 &= \sum_{i=1}^n [Y_i - \bar{Y}_n]^2 \\ &= \sum_{i=1}^n \left[(Y_i - \bar{Y}_{n-1}) - \frac{1}{n}(Y_n - \bar{Y}_{n-1})\right]^2 \\ &= \sum_{i=1}^n \left[(Y_i - \bar{Y}_{n-1})^2 - \frac{2}{n}(Y_i - \bar{Y}_{n-1})(Y_n - \bar{Y}_{n-1}) + \frac{1}{n^2}(Y_n - \bar{Y}_{n-1})^2\right].\end{aligned}$$

Now using the fact that $\sum_{i=1}^{n-1}(Y_i - \bar{Y}_{n-1}) = \sum_{i=1}^{n-1} Y_i - (n-1)\bar{Y}_{n-1} = 0$ yields

$$\begin{aligned}(n-1)S_n^2 &= \sum_{i=1}^{n-1}(Y_i - \bar{Y}_{n-1})^2 + (Y_n - \bar{Y}_{n-1})^2 - \frac{2}{n}(Y_n - \bar{Y}_{n-1})^2 + \frac{1}{n}(Y_n - \bar{Y}_{n-1})^2 \\ &= (n-2)S_{n-1}^2 + \frac{n-1}{n}(Y_n - \bar{Y}_{n-1})^2.\end{aligned}$$

□

**Proposition 7.2.5** (Properties of sample variance)
*Suppose that $Y_1, \dots, Y_n$ is a random sample from a population with mean $\mu$ and finite variance $\sigma^2 < \infty$. If $S^2 = \frac{1}{n-1}\sum_{i=1}^n (Y_i - \bar{Y})^2$ is the sample variance then*

*i. the mean of the sample variance is the population variance, $\mathbb{E}(S^2) = \sigma^2$.*

*If, in addition, the population distribution has finite fourth central moment, $\mu_4 < \infty$, then*

*ii. the variance of the sample variance is given by*

$$\text{Var}(S^2) = \frac{1}{n}\left(\mu_4 - \frac{n-3}{n-1}\mu_2^2\right) = \frac{1}{n}\left(\mu_4 - \frac{n-3}{n-1}\sigma^4\right).$$

*Proof.*

Throughout this proof we will use Claim 7.2.2 and assume, without loss of generality, that $\mathbb{E}(Y) = 0$ and thus $\mathbb{E}(Y^2) = \text{Var}(Y) = \sigma^2$. We will also use a neat trick and split $Y_i - \bar{Y}$ into independent terms,

$$Y_i - \bar{Y} = Y_i - \frac{1}{n}\sum_{k=1}^n Y_k = \left(Y_i - \frac{1}{n}Y_i\right) - \frac{1}{n}\sum_{k \neq i} Y_k = V_i - W_i, \tag{7.1}$$

where $V_i = Y_i - \frac{1}{n}Y_i = \frac{n-1}{n}Y_i$ and $W_i = \frac{1}{n}\sum_{k\neq i} Y_k$. We can now use part i. of Lemma 7.2.4 to write

$$(n-1)S^2 = \sum_{i=1}^{n} Y_i(V_i - W_i)\,.$$

Note that, using the fact that $\mathbb{E}(Y) = 0$ and thus $\mathbb{E}(Y_i) = 0$ for all $i$, we have $\mathbb{E}(V_i) = 0$ and $\mathbb{E}(W_i) = 0$. The $V_i$s and the $W_i$s are identically distributed as a consequence of the $Y_i$s being identically distributed. Using the fact that $Y_i \perp\!\!\!\perp Y_j$ for $i \neq j$, we have $Y_i \perp\!\!\!\perp W_i$, $V_i \perp\!\!\!\perp W_i$ and, for $i \neq j$, $Y_i \perp\!\!\!\perp V_j$.

i. We prove the first result directly,

$$\begin{aligned}
\mathbb{E}[(n-1)S^2] &= \mathbb{E}\left[\sum_{i=1}^{n} Y_i(V_i - W_i)\right] \\
&= \sum_{i=1}^{n} [\mathbb{E}(Y_iV_i) - \mathbb{E}(Y_iW_i)] \\
&= \sum_{i=1}^{n} \mathbb{E}(Y_iV_i) && Y_i \perp\!\!\!\perp W_i \text{ and } \mathbb{E}(Y_i) = \mathbb{E}(W_i) = 0 \\
&= \sum_{i=1}^{n} \frac{n-1}{n}\mathbb{E}(Y_i^2) \\
&= \frac{n-1}{n} n\sigma^2 = (n-1)\sigma^2 \\
\Rightarrow \mathbb{E}(S^2) &= \sigma^2.
\end{aligned}$$

ii. Using standard results for variance and the unbiasedness established above, we have $\text{Var}(S^2) = \mathbb{E}(S^4) - [\mathbb{E}(S^2)]^2 = \mathbb{E}(S^4) - \sigma^4$. Thus, our problem is one of calculating $\mathbb{E}(S^4)$. We have

$$\begin{aligned}
&\mathbb{E}\{(n-1)^2S^4\} \\
&= \mathbb{E}\left\{\left[\sum_{i=1}^{n} Y_i(Y_i - \bar{Y})\right]^2\right\} \\
&= \sum_{i=1}^{n}\sum_{j=1}^{n} \mathbb{E}[Y_i(V_i - W_i)Y_j(V_j - W_j)] \\
&= \sum_{i=1}^{n}\sum_{j=1}^{n} [\mathbb{E}(Y_iV_iY_jV_j) - \mathbb{E}(Y_iV_iY_jW_j) - \mathbb{E}(Y_iW_iY_jV_j) + \mathbb{E}(Y_iW_iY_jW_j)].
\end{aligned}$$

When $i = j$, the second and third terms of the summand are zero, since $W_i \perp\!\!\!\perp Y_i$, $W_i \perp\!\!\!\perp V_i$, and $\mathbb{E}(W_i) = 0$. When $i \neq j$, the second term is zero, since $Y_j \perp\!\!\!\perp Y_i$, $Y_j \perp\!\!\!\perp V_i$, $Y_j \perp\!\!\!\perp W_j$, and $\mathbb{E}(Y_j) = 0$. An identical argument (swapping $i$ and $j$) shows that the third term is also zero when $i \neq j$. Using the fact that we have identically distributed sequences,

$$\begin{aligned}
&\mathbb{E}\{(n-1)^2 S^4\} \\
&= \sum_{i=1}^{n} \{\mathbb{E}(Y_i^2 V_i^2) + \mathbb{E}(Y_i^2)\,\mathbb{E}(W_i^2)\} + \sum_{i \neq j} \sum_{j=1}^{n} \{[\mathbb{E}(Y_i V_i)]^2 + [\mathbb{E}(Y_i W_j)]^2\} \\
&= n\left\{\frac{(n-1)^2}{n^2}\mu_4 + \frac{n-1}{n^2}\mu_2^2\right\} + n(n-1)\left\{\frac{(n-1)^2}{n^2}\mu_2^2 + \frac{1}{n^2}\mu_2^2\right\} \\
\Rightarrow \mathbb{E}(S^4) &= \frac{1}{n}\mu_4 + \frac{n^2-2n+3}{n(n-1)}\mu_2^2 \\
\Rightarrow \mathrm{Var}(S^2) &= \frac{1}{n}\mu_4 + \frac{n^2-2n+3}{n(n-1)}\mu_2^2 - \mu_2^2 = \frac{1}{n}\left(\mu_4 - \frac{n-3}{n-1}\sigma^4\right).
\end{aligned}$$

□

We have derived the mean and variance of the sample mean and sample variance. In practice, it is also useful to know something about the association between the sample mean and the sample variance.

**Proposition 7.2.6** (Association of sample mean and sample variance)
*Suppose that $Y_1, \dots, Y_n$ is a random sample from a population with finite variance, $\sigma^2 < \infty$, then*

i. *$\bar{Y}$ is uncorrelated with $(Y_j - \bar{Y})$ for $j = 1, \dots, n$.*

*If, in addition, the population distribution has finite third central moment, $\mu_3 < \infty$, then*

ii. *The covariance between the sample mean and sample variance is a function of the third central moment,* $\mathrm{Cov}(\bar{Y}, S^2) = \frac{1}{n}\mu_3$.

*Proof.*

We use the approach and terminology put forward in the proof of 7.2.5 and assume, without loss of generality, that the population mean is zero, that is, $\mathbb{E}(Y) = 0$.

i. We know that $\mathbb{E}(\bar{Y}) = 0$ and $\mathbb{E}(Y_i - \bar{Y}) = 0$, so to show that $\bar{Y}$ and $(Y_i - \bar{Y})$ are uncorrelated we just need to show that $\mathbb{E}[\bar{Y}(Y_i - \bar{Y})] = 0$. We have

$$\begin{aligned}
\mathbb{E}\left[\bar{Y}(Y_i - \bar{Y})\right] &= \mathbb{E}\left[\frac{1}{n}\sum_{j=1}^{n} Y_j(V_i - W_i)\right] = \frac{1}{n}\sum_{j=1}^{n}[\mathbb{E}(Y_j V_i) - \mathbb{E}(Y_j W_i)] \\
&= \frac{1}{n}\left[\mathbb{E}(Y_i V_i) - \sum_{j \neq i}\mathbb{E}(Y_j W_i)\right] = \frac{1}{n}\left[\frac{n-1}{n}\sigma^2 - (n-1)\frac{1}{n}\sigma^2\right] = 0
\end{aligned}$$

ii. We can now prove the covariance result directly:

$$\begin{aligned}\text{Cov}(\bar{Y}, S^2) &= \mathbb{E}(\bar{Y}S^2) - \mathbb{E}(\bar{Y})\,\mathbb{E}(S^2) = \mathbb{E}(\bar{Y}S^2)\\ &= \mathbb{E}\left[\left(\frac{1}{n}\sum_{i=1}^{n} Y_i\right)\left(\frac{1}{n-1}\sum_{j=1}^{n} Y_j(V_j - W_j)\right)\right]\\ &= \frac{1}{n(n-1)}\sum_{i=1}^{n}\sum_{j=1}^{n}[\mathbb{E}(Y_iY_jV_j) - \mathbb{E}(Y_iY_jW_j)]\,.\end{aligned}$$

When $i \neq j$, both terms in the summand are zero; the first because $Y_j \perp\!\!\!\perp Y_i$ and $Y_j \perp\!\!\!\perp V_j$, the second because $Y_j \perp\!\!\!\perp Y_i$ and $Y_j \perp\!\!\!\perp W_j$. When $i = j$, the second term is zero since $W_i \perp\!\!\!\perp Y_i$. In conclusion,

$$\text{Cov}(\bar{Y}, S^2) = \frac{1}{n(n-1)}\sum_{i=1}^{n}\mathbb{E}(Y_i^2V_i) = \frac{1}{n(n-1)}n\frac{n-1}{n}\mu_3 = \frac{1}{n}\mu_3\,.$$

□

We deduce that, when sampling from a symmetric distribution (the normal, for example), the sample mean and sample variance are uncorrelated.

### 7.2.2 *Joint sample moments*

Joint sample moments provide information about the dependence structure between two variables. We begin with some definitions.

**Definition 7.2.7** (Joint sample moments and joint central sample moments)
Suppose we have a random sample of pairs $\{(X_1, Y_1), \ldots, (X_n, Y_n)\}$. The $(r,s)^{\text{th}}$ **joint sample moment** is

$$m'_{r,s} = \frac{1}{n}\sum_{i=1}^{n} X_i^r Y_i^s\,,$$

and the $(r,s)^{\text{th}}$ **joint central sample moment** is

$$m_{r,s} = \frac{1}{n}\sum_{i=1}^{n}(X_i - \bar{X})^r(Y_i - \bar{Y})^s\,.$$

**Definition 7.2.8** (Sample covariance)
Given a random sample of pairs, $\{(X_1, Y_1), \ldots, (X_n, Y_n)\}$, the **sample covariance** is defined as

$$c_{X,Y} = \frac{1}{n-1}\sum_{i=1}^{n}(X_i - \bar{X})(Y_i - \bar{Y}) = \frac{n}{n-1}m_{1,1}\,,$$

or, alternatively,

$$c_{X,Y} = \frac{1}{n-1}\left(\sum_{i=1}^{n} X_iY_i - n\bar{X}\bar{Y}\right).$$

Proving the equivalence of these two forms is part of Exercise 7.2.

**Claim 7.2.9** (Properties of sample covariance)
*If $\{(X_1,Y_1),\ldots,(X_n,Y_n)\}$ is a random sample of pairs, the sample covariance has the following properties:*

*i.* $c_{X,Y} = c_{Y,X}$ .

*ii. If $U_i = a + bX_i$ and $V_i = c + dY_i$ for $i = 1,\ldots,n$, where a, b, c, and d are real constants, then $c_{U,V} = bd\, c_{X,Y}$ .*

*iii. If $W_i = X_i + Y_i$, then $S^2_W = S^2_X + S^2_Y + 2c_{X,Y}$, where $S^2_X$, $S^2_Y$, and $S^2_W$ denote the sample variance of the X, Y, and W values, respectively.*

Showing that these properties hold is part of Exercise 7.2.

**Definition 7.2.10** (Sample correlation)
Given a random sample of pairs $\{(X_1,Y_1),\ldots,(X_n,Y_n)\}$, the **sample correlation** is defined as

$$r_{X,Y} = \frac{c_{X,Y}}{\sqrt{S^2_X S^2_Y}},$$

where $S^2_X$ and $S^2_Y$ denote the sample variance of the $X$ and $Y$ values, respectively.

Sample correlation measures the degree of linear association present in the sample (Figures 7.2 and 7.3). We can prove that, in direct analogy to the correlation between two random variables, sample correlation is scaled so that $|r_{X,Y}| \leq 1$.

**Theorem 7.2.11** (Values taken by sample correlation)
*Given a random sample of pairs, $\{(X_1,Y_1),\ldots,(X_n,Y_n)\}$, it holds that*

$$-1 \leq r_{X,Y} \leq 1.$$

*If and only if we can find constants a and $b \neq 0$ such that $Y_i = a+bX_i$ for $i = 1,2,\ldots,n$, then*

$$r_{X,Y} = \begin{cases} 1 & \text{if } b > 0, \\ -1 & \text{if } b < 0. \end{cases}$$

The proof of this theorem is part of Exercise 7.2. It follows similar steps as the proof of Theorem 4.3.8.

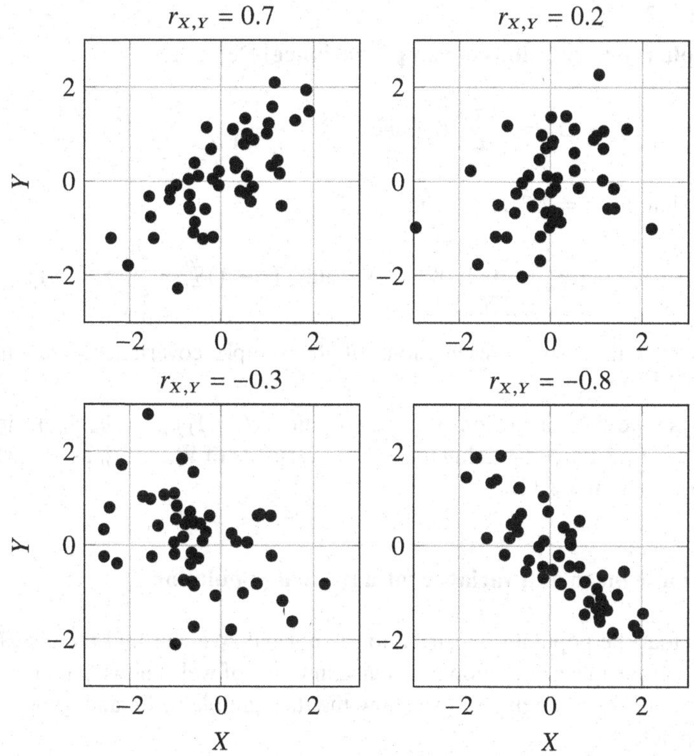


Figure 7.2 *Each plot shows a random sample of* 40 *pairs* $\{X_i, Y_i\}$ *with the given sample correlation. As* $r_{X,Y}$ *approaches* $+1/-1$, *the points get closer to a straight line with positive/negative slope.*

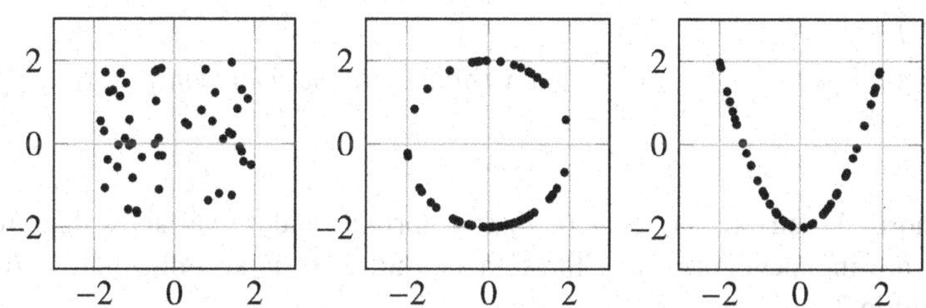


Figure 7.3 *Correlation only measures the degree of* linear *association. Each of the three plots above shows a random sample of* 40 *pairs with sample correlation zero; in two of them there is an obvious (nonlinear) pattern.*

**Exercise 7.2**

1. (**Simple representations of sample variance**) Defining

$$\bar{Y}_n = \frac{1}{n}\sum_{i=1}^{n} Y_i \quad \text{and} \quad S_n^2 = \frac{1}{n-1}\sum_{i=1}^{n}(Y_i - \bar{Y}_n)^2,$$

   show that, for $n = 2, 3, \dots$, we have

$$(n-1)S_n^2 = \sum_{i=1}^{n} Y_i(Y_i - \bar{Y}_n) \quad \text{and} \quad (n-1)S_n^2 = \sum_{i=1}^{n} Y_i^2 - n\bar{Y}_n^2.$$

2. Show that the two representations of the sample covariance given in Definition 7.2.8 are equivalent.
3. Suppose we have a random sample of pairs, $\{(X_1, Y_1), \dots, (X_n, Y_n)\}$, and define $V_i = Y_i - bX_i$. By considering the sample variance of $V_1, \dots, V_n$ as a function of $b$, prove Theorem 7.2.11.

## 7.3 Sample mean and variance for a normal population

Suppose that the population distribution is normal. We can find the distribution of both the sample mean and sample variance in terms of well-known parametric forms. Perhaps more surprisingly, we can show that the sample mean and sample variance are independent.

**Lemma 7.3.1** (Independence of sample mean and variance for normal population)
*Suppose that $Y_1, \dots, Y_n$ is a random sample from a normal population, that is $Y \sim \mathrm{N}(\mu, \sigma^2)$. Then,*

*i. the sample mean, $\bar{Y}$, and terms of the form $(Y_j - \bar{Y})$ are independent, $\bar{Y} \perp\!\!\!\perp (Y_j - \bar{Y})$ for $j = 1, \dots, n$,*

*ii. the sample mean and the sample variance are independent, $\bar{Y} \perp\!\!\!\perp S^2$.*

*Proof.*

For a normal distribution, $\mu_3 = 0$. Jointly normal random variables that are uncorrelated are independent. Using these facts, the result is an immediate corollary of Proposition 7.2.6. □

In constructing the sample variance, we consider sums of the squares of a sequence of random variables. If the variables are standard normal, the sum of the squares has a **chi-squared distribution** (chi is pronounced like "eye" preceded by a "k").

**Definition 7.3.2** (Chi-squared distribution)
Suppose that $Z_1, \dots, Z_n$ is a random sample from a standard normal distribution. If we define $V = \sum_{j=1}^{n} Z_j^2$, then $V$ has a chi-squared distribution on $n$ degrees of freedom. This is denoted $V \sim \chi_n^2$.

A consequence of the definition is that the chi-squared distribution has the additive property. If $U \sim \chi^2_m$, $V \sim \chi^2_n$, and $U \perp\!\!\!\perp V$, then $U + V \sim \chi^2_{m+n}$.

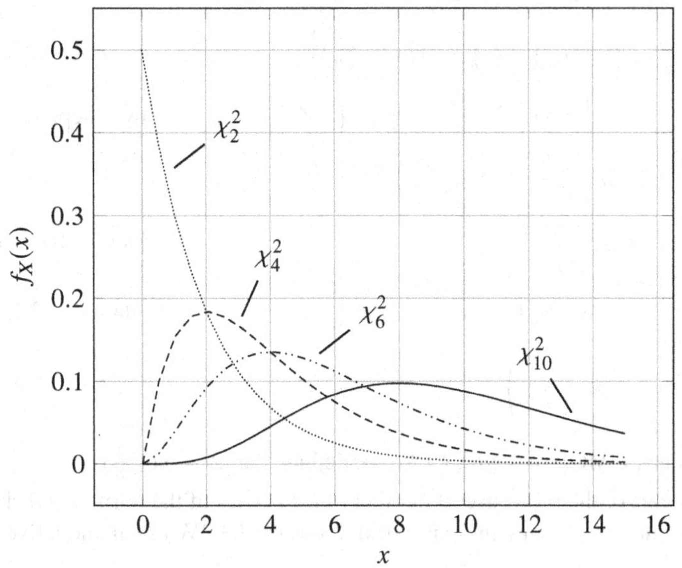


Figure 7.4 *The $\chi^2_n$ density function for various values of n.*

We are now in a position to derive the distributions of the sample mean and sample variance for a normal population. We start by considering the standard normal.

**Theorem 7.3.3** (Distribution of sample mean and variance for standard normal) *Suppose that $Z_1, \ldots, Z_n$ is a random sample from a population with a standard normal distribution, that is, $Z \sim \mathrm{N}(0,1)$. If we define $\bar{Z} = \frac{1}{n}\sum_{i=1}^{n} Z_i$ and $S^2 = \frac{1}{n-1}\sum_{i=1}^{n}(Z_i - \bar{Z})^2$, then*

*i. the sample mean is normally distributed, $\bar{Z} \sim \mathrm{N}(0, \frac{1}{n})$,*

*ii. $(n-1)$ times the sample variance has a chi-squared distribution, $(n-1)S^2 \sim \chi^2_{n-1}$.*

*Proof.*

i. We consider the moment-generating function of $\bar{Z}$. If we can show that this is the MGF of an $\mathrm{N}(0, \frac{1}{n})$ random variable, then (by the uniqueness of MGFs) we have completed this part of the proof.

$$
\begin{aligned}
m_{\bar{Z}}(t) &= \mathbb{E}\left[\exp(t\bar{Z})\right] = \mathbb{E}\left[\exp\left\{\frac{t}{n}(Z_1 + \ldots + Z_n)\right\}\right] \\
&= \mathbb{E}\left[\exp\left(\frac{t}{n}Z_1\right)\ldots\exp\left(\frac{t}{n}Z_n\right)\right] \\
&= \mathbb{E}\left[\exp\left(\frac{t}{n}Z_1\right)\right]\ldots\mathbb{E}\left[\exp\left(\frac{t}{n}Z_n\right)\right] && \text{by independence} \\
&= m_{Z_1}\left(\frac{t}{n}\right)\ldots m_{Z_n}\left(\frac{t}{n}\right) \\
&= \left[m_Z\left(\frac{t}{n}\right)\right]^n && \text{identically distributed} \\
&= \left[\exp\left(\frac{1}{2}\left(\frac{t}{n}\right)^2\right)\right]^n && \text{since } Z \sim \text{N}(0,1) \\
&= \exp\left(\frac{1}{n}\frac{t^2}{2}\right).
\end{aligned}
$$

This is the MGF of a $\text{N}(0, \frac{1}{n})$, so we deduce that $\bar{Z} \sim \text{N}(0, \frac{1}{n})$.

ii. This part of the proof uses a number of properties of the normal and chi-squared distributions. Proving these is part of Exercise 7.3. We use an inductive argument on

$$(k-1)S_k^2 = \sum_{i=1}^{k}(Z_i - \bar{Z}_k)^2,$$

where $\bar{Z}_k = \frac{1}{k}\sum_{i=1}^{k} Z_i$. Assume that

$$(k-2)S_{k-1}^2 \sim \chi_{k-2}^2.$$

From part iv. of Lemma 7.2.4,

$$(k-1)S_k^2 = (k-2)S_{k-1}^2 + \frac{k-1}{k}(Z_k - \bar{Z}_{k-1})^2.$$

We then have

$$Z_k \sim \text{N}(0,1) \text{ and } \bar{Z}_{k-1} \sim \text{N}\left(0, \frac{1}{k-1}\right) \Rightarrow (Z_k - \bar{Z}_{k-1}) \sim \text{N}\left(0, \frac{k}{k-1}\right),$$

and thus, by definition,

$$\frac{k-1}{k}(Z_k - \bar{Z}_{k-1})^2 \sim \chi_1^2.$$

From Lemma 7.3.1 and the properties of a random sample, $(Z_k - \bar{Z}_{k-1})^2$ and $S_{k-1}^2$ are functions of sets of random variables that are independent of each other. Functions of independent variables are also independent, so we have $(Z_k - \bar{Z}_{k-1})^2 \perp\!\!\!\perp S_{k-1}^2$. By the additive property of $\chi^2$ distributions, if our hypothesis is true for $S_{k-1}^2$, it is also true for $S_k^2$.

In order to complete the proof, we need to establish that the result holds for $S_2^2$. From part iv. of Lemma 7.2.4 or direct calculation,

$$S_2^2 = \frac{1}{2}(Z_2 - \bar{Z}_1)^2 = \frac{1}{2}(Z_2 - Z_1)^2.$$

Since $(Z_2 - Z_1) \sim \text{N}(0, 2)$, we can deduce that $S_2^2 \sim \chi_1^2$, and thus our hypothesis is true for $S_2^2$. We established above that

$$(k-2)S_{k-1}^2 \sim \chi_{k-2}^2 \Rightarrow (k-1)S_k^2 \sim \chi_{k-1}^2,$$

so, by induction, our hypothesis is true for $S_k^2$, where $k = 2, \ldots, n$.

□

Extending the results of Theorem 7.3.3 to the case of a general normal distribution is straightforward.

**Corollary 7.3.4** (Distribution of sample mean and variance for general normal) *Suppose that $Y_1, \ldots, Y_n$ is a random sample from a normal population with mean $\mu$ and variance $\sigma^2$, that is, $Y \sim \text{N}(\mu, \sigma^2)$. Then*

*i.* $\bar{Y} \sim \text{N}(\mu, \frac{\sigma^2}{n})$,

*ii.* $(n-1)S^2/\sigma^2 \sim \chi_{n-1}^2$.

*Proof.*

The basis of this proof is that, if $Z_i = (Y_i - \mu)/\sigma$, then $Z_i \sim \text{N}(0, 1)$. The result then follows directly from the fact that

$$\bar{Z} = (\bar{Y} - \mu)/\sigma,$$

and

$$S_Z^2 = \frac{1}{n-1}\sum_{i=1}^{n}(Z_i - \bar{Z})^2 = \frac{1}{\sigma^2}\frac{1}{n-1}\sum_{i=1}^{n}(Y_i - \bar{Y})^2 = \frac{1}{\sigma^2}S^2.$$

□

**Exercise 7.3**

1. Suppose that $Y_1 \sim \text{N}(\mu_1, \sigma_1^2)$, $Y_2 \sim \text{N}(\mu_2, \sigma_2^2)$, and $Y_1 \perp\!\!\!\perp Y_2$. Show that

$$Y_1 + Y_2 \sim \text{N}(\mu_1 + \mu_2, \sigma_1^2 + \sigma_2^2).$$

2. (**Connection between chi-squared and Gamma**) Suppose that $V \sim \chi_k^2$. Show that the moment-generating function of $V$ is

$$m_V(t) = (1 - 2t)^{-k/2},$$

and hence explain why the chi-squared distribution is a special case of the Gamma distribution. For which value of $k$ is $\chi_k^2$ a special case of the exponential distribution?

3. Suppose that $Y_1 \sim \chi^2_{d_1}$, $Y_2 \sim \chi^2_{d_2}$, and $Y_1 \perp\!\!\!\perp Y_2$. Show that

$$Y_1 + Y_2 \sim \chi^2_{d_1+d_2}.$$

## 7.4 Sample quantiles and order statistics

Mean and variance are not the only measures of central tendency and spread. Commonly quoted alternative measures include the median and interquartile range, both of which are calculated from quantiles. Quantile-based statistics are generally more **robust** than moment-based statistics; in other words, quantile-based statistics are less sensitive to extreme observations.

For a random variable $Y$ with cumulative distribution function $F_Y$, the $\alpha$-**quantile** is defined as the smallest value, $q_\alpha$, such that $F_Y(q_\alpha) = \alpha$. The **median** of a distribution is the point that has half the mass above and half the mass below it, $\text{med}(Y) = q_{0.5}$. The **interquartile range** is the difference between the upper and lower quartiles, $\text{IQR}(Y) = q_{0.75} - q_{0.25}$. In the context of inference, the median, interquartile range, and other functions of quantiles may be viewed as population parameters to be estimated. An obvious starting point for estimating population quantiles is to define sample quantiles. In turn, sample quantiles are most readily constructed from **order statistics**.

**Definition 7.4.1** (Order statistics)
For a sample, $Y_1, \ldots, Y_n$, the order statistics, $Y_{(1)}, \ldots, Y_{(n)}$, are the sample values placed in ascending order. Thus, $Y_{(i)}$ is the $i^{\text{th}}$ smallest value in our sample.

A simple consequence of our definition is that order statistics are random variables, $Y_{(1)}, \ldots, Y_{(n)}$, such that

$$Y_{(1)} \leq Y_{(2)} \leq \ldots \leq Y_{(n)}.$$

Definition 7.4.1 does odd things with random variables. In particular, we say that

$$Y_{(1)}, \ldots, Y_{(n)}$$

arise from putting

$$Y_1, \ldots, Y_n$$

in ascending order. We can make this more precise using the definition of random variables as functions from the sample space, $\Omega$, to the real line, $\mathbb{R}$. The order statistics, $Y_{(1)}, \ldots, Y_{(n)}$, are the functions such that, for every $\omega \in \Omega$,

$$Y_{(1)}(\omega), \ldots, Y_{(n)}(\omega)$$

is an ordering of $Y_1(\omega), \ldots, Y_n(\omega)$ satisfying

$$Y_{(1)}(\omega) \leq Y_{(2)}(\omega) \leq \ldots \leq Y_{(n)}(\omega).$$

It is perhaps more intuitively appealing to think about order statistics in terms of repeated samples. Suppose that we have $m$ observed samples,

$$y_1^{(j)}, \dots, y_n^{(j)} \text{ for } j = 1, \dots, m.$$

Each of these is just a collection of numbers that can be sorted into ascending order; the $j^{\text{th}}$ observed sample sorted into ascending order is denoted

$$y_{(1)}^{(j)}, \dots, y_{(n)}^{(j)}.$$

The minimum values from our $m$ samples,

$$y_{(1)}^{(1)}, \dots, y_{(1)}^{(m)},$$

can be thought of as instances of $Y_{(1)}$. The second-smallest values,

$$y_{(2)}^{(1)}, \dots, y_{(2)}^{(m)}$$

can be thought of as instances of $Y_{(2)}$, and so on. Note that $Y_{(i)}$ does not have the population distribution. For example, $Y_{(1)}$ is the smallest element in the sample, so $\mathbb{E}(Y_{(1)}) < \mathbb{E}(Y)$ for any non-degenerate distribution.

The $i^{\text{th}}$ observed order statistic, $y_{(i)}$, is the $i^{\text{th}}$ smallest value in the observed sample. This means that $y_{(i)}$ might be taken to be a reasonable estimate of the point that has $i/n$ of the mass of the population distribution below it. This idea is the basis of our definition of sample quantiles. We restrict this definition to $\alpha \in (\frac{1}{2n}, 1 - \frac{1}{2n})$. This allows us to make a definition with appealing symmetry properties (see Exercise 7.4) and reflects the fact that the extent to which we can estimate the extremes of a distribution is restricted by the sample size.

**Definition 7.4.2** (Sample quantiles)
Suppose $Y_1, \dots, Y_n$ is a sample and let $Q_\alpha$ denote the **sample $\alpha$-quantile**. We define $Q_\alpha$ in two pieces; first for $\alpha = 1/2$ and then for $\alpha \neq 1/2$.

$$Q_{0.5} = \begin{cases} Y_{((n+1)/2)} & \text{if } n \text{ odd,} \\ \frac{1}{2}(Y_{(n/2)} + Y_{((n/2)+1)}) & \text{if } n \text{ even.} \end{cases}$$

$$Q_\alpha = \begin{cases} Y_{(\{n\alpha\})} & \text{if } \frac{1}{2n} < \alpha < \frac{1}{2}, \\ Y_{(n+1-\{n(1-\alpha)\})} & \text{if } \frac{1}{2} < \alpha < 1 - \frac{1}{2n}. \end{cases}$$

Here $\{k\}$ denotes $k$ rounded to the nearest integer in the usual way.

Note that $Q_{0.5}$ is the sample median. If the location of the median falls between two order statistics (that is, when $n$ is even) we interpolate. For other sample quantiles we simply find the nearest order statistic in the sense defined in 7.4.2.

### 7.4.1 Sample minimum and sample maximum

By Definition 7.4.1, $Y_{(1)}$ is the smallest value in the sample and $Y_{(n)}$ is the largest,

$$\begin{aligned} Y_{(1)} &= \min_{1 \leq i \leq n} (Y_1, \dots, Y_n), \\ Y_{(n)} &= \max_{1 \leq i \leq n} (Y_1, \dots, Y_n). \end{aligned}$$

For a random sample, the cumulative distribution functions of both of these statistics are easy to derive.

**Proposition 7.4.3** (Distribution function of sample minimum and maximum)
*Suppose that $Y_1, \dots, Y_n$ is a random sample from a population with cumulative distribution function $F_Y$, that is, $Y \sim F_Y$. Then the cumulative distribution functions of the sample minimum and sample maximum are given by*

$$\begin{aligned} F_{Y_{(1)}}(y) &= P(Y_{(1)} \leq y) = 1 - [1 - F_Y(y)]^n, \\ F_{Y_{(n)}}(y) &= P(Y_{(n)} \leq y) = [F_Y(y)]^n. \end{aligned}$$

*Proof.*

We give a proof for the maximum; the minimum case is part of Exercise 7.4. Consider the event $Y_{(n)} \leq y$. This occurs when the largest value in the sample is less than or equal to $y$. By definition of $Y_{(n)}$ as the largest value, if $Y_{(n)} \leq y$, then $Y_i \leq y$ for all $i$ (and vice versa). Thus,

$$\begin{aligned} F_{Y_{(n)}}(y) &= P(Y_{(n)} \leq y) = P(Y_1 \leq y, \dots, Y_n \leq y) \\ &= P(Y_1 \leq y) \dots P(Y_n \leq y) && \text{by independence} \\ &= [P(Y \leq y)]^n && \text{identically distributed} \\ &= [F_Y(y)]^n. \end{aligned}$$

□

Proposition 7.4.3 holds for both discrete and continuous distributions. However, the mass functions of order statistics for discrete distributions and the density functions of order statistics for continuous distributions arise in rather different ways. Suppose, first of all, that the population distribution is discrete, that is, $Y$ is a discrete random variable. The support of the mass function of $Y$ will be a countable set of real numbers, $\{a_1, a_2, \dots\}$. We can assume, without loss of generality, that the indexing is chosen so that $a_1 \leq a_2 \leq \dots$ The mass function is found by taking differences; for example, the sample maximum has mass function

$$f_{Y_{(n)}}(y) = \begin{cases} [F_Y(a_k)]^n - [F_Y(a_{k-1})]^n & \text{for } y = a_k,\ k = 1, 2, \dots, \\ 0 & \text{otherwise.} \end{cases}$$

In the continuous case we derive the density by differentiation, for example,

$$f_{Y_{(n)}}(y) = \frac{d}{dy} F_{Y_{(n)}}(y) = n f_Y(y) [F_Y(y)]^{n-1}, \tag{7.2}$$

where $f_Y(y) = \frac{d}{dy}F_Y(y)$ is the population density. The support of the density of the maximum will be the same as the support of the population density.

**Example 7.4.4** (Maximum of Unif$[0,\theta]$ distribution)
Suppose that $\boldsymbol{Y} = (Y_1, \dots, Y_n)^T$ is a random sample from a Unif$[0,\theta]$ distribution. The density of an individual sample member is

$$f_Y(y) = \begin{cases} \frac{1}{\theta} & \text{for } 0 \le y \le \theta, \\ 0 & \text{otherwise.} \end{cases}$$

The associated cumulative distribution function is

$$F_Y(y) = \frac{y}{\theta} \quad \text{for } 0 \le y \le \theta\,.$$

An expression for the cumulative distribution function of the sample maximum follows directly from Proposition 7.4.3,

$$F_{Y_{(n)}}(y) = \left(\frac{y}{\theta}\right)^n \quad \text{for } 0 \le y \le \theta\,.$$

The density of the maximum can be calculated directly by differentiation or by substitution in equation (7.2),

$$f_{Y_{(n)}}(y) = \frac{n}{\theta}\left(\frac{y}{\theta}\right)^{n-1} \quad \text{for } 0 \le y \le \theta\,.$$

You should confirm that this is a valid density function.

### 7.4.2 *Distribution of $i^{th}$ order statistic*

We can extend the argument used in subsection 7.4.1 to derive a general expression for the cumulative distribution function of the $i^{\text{th}}$ order statistic.

**Proposition 7.4.5**
*Suppose that $Y_1, \dots, Y_n$ is a random sample from a population with cumulative distribution function $F_Y$, that is, $Y \sim F_Y$. Then the cumulative distribution function of the $i^{th}$ order statistic is given by*

$$F_{Y_{(i)}}(y) = \sum_{j=i}^{n} \binom{n}{j} F_Y(y)^j (1 - F_Y(y))^{n-j}\,.$$

*Proof.*

Consider the event that exactly $j$ values in the sample are less than or equal to $y$. Since they are identically distributed, for each individual sample value, the probability of being less than or equal to $y$ is $F_Y(y) = \text{P}(Y \le y)$. As they are independent, the probability that exactly $j$ are less than or equal to $y$ is given by the binomial mass function,

$$\binom{n}{j} F_Y(y)^j [1 - F_Y(y)]^{n-j}\,.$$

Now consider $Y_{(i)} \leq y$, the event that the $i^{\text{th}}$ smallest value in the sample is less than or equal to $y$. This occurs when at least $i$ individual sample values are less than or equal to $y$. Thus,

$$F_{Y_{(i)}}(y) = \mathrm{P}(Y_{(i)} \leq y) = \sum_{j=i}^{n} \binom{n}{j} F_Y(y)^j (1 - F_Y(y))^{n-j}.$$

□

For the discrete case, the mass function is derived using Proposition 7.4.5 and differencing. In the continuous case, it is possible to derive a neater form for the density.

**Proposition 7.4.6** (Density of $i^{\text{th}}$ order statistic)
*Suppose that $Y_1, \ldots, Y_n$ is a random sample from a population with continuous cumulative distribution function $F_Y$ and density function $f_Y$. The density of the $i^{\text{th}}$ order statistic is given by*

$$f_{Y_{(i)}}(y) = \frac{n!}{(i-1)!(n-i)!} f_Y(y)[F_Y(y)]^{i-1}[1 - F_Y(y)]^{n-i}.$$

*The support of $f_{Y_{(i)}}$ is the same as the support of $f_Y$.*

*Proof.*
We start by considering the definition of the density as a derivative,

$$f_{Y_{(i)}}(y) = \frac{d}{dy} F_{Y_{(i)}}(y) = \lim_{h \to 0} \frac{F_{Y_{(i)}}(y+h) - F_{Y_{(i)}}(y)}{h} = \lim_{h \to 0} \frac{\mathrm{P}(y < Y_{(i)} \leq y+h)}{h}.$$

Consider $y < Y_{(i)} \leq y + h$, the event that the $i^{\text{th}}$ smallest value falls in the interval $(y, y+h]$. We divide the real line into three intervals,

$$(-\infty, y],\ (y, y+h],\ \text{and}\ (y+h, \infty).$$

The probabilities that an individual sample value falls into these intervals are, respectively,

$$F_Y(y),\ F_Y(y+h) - F_Y(y),\ \text{and}\ 1 - F_Y(y+h).$$

In order for $y < Y_{(i)} \leq y + h$ to occur, we need at least one observation in $(y, y+h]$, at most $i - 1$ observations in $(-\infty, y]$, and at most $n - i$ in $(y + h, \infty)$. More specifically, if there are $j \geq 1$ observations in $(y, y+h]$ and $k$ observations in $(-\infty, y]$, the event $y < Y_{(i)} \leq y + h$ occurs when $i - j \leq k \leq i - 1$. Taking into account the fact that $k$ cannot be less than 0 nor greater than $n - j$, the probability of $y < Y_{(i)} \leq y + h$ is given by a sum of values from the **trinomial** mass function (the multinomial for $k = 3$ – see Exercise 4.9 question 12),

$$\begin{aligned}&\mathrm{P}(y < Y_{(i)} \leq y+h)\\ &= \sum_{j=1}^{n} \sum_{k=\max(i-j,0)}^{\min(i-1,n-j)} \frac{n!}{k!j!(n-k-j)!} F_Y(y)^k [F_Y(y+h) - F_Y(y)]^j [1 - F_Y(y+h)]^{n-k-j}.\end{aligned} \tag{7.3}$$

Now consider $[F_Y(y+h) - F_Y(y)]^j$, the central term in this expression . By definition of $f_Y$ as the derivative of $F_Y$, we have

$$F_Y(y+h) - F_Y(y) = hf_Y(y) + o(h),$$

and thus,

$$[F_Y(y+h) - F_Y(y)]^j = o(h) \text{ for } j > 1 .$$

We conclude that when we divide by $h$ and take the limit as $h \to 0$ in (7.3), the only term that does not go to zero is that corresponding to $j = 1$. In other words, as $h$ is small, we can ignore the possibility that there are two or more observations in the interval $(y, y+h]$. See Figure 7.5 for an illustration.

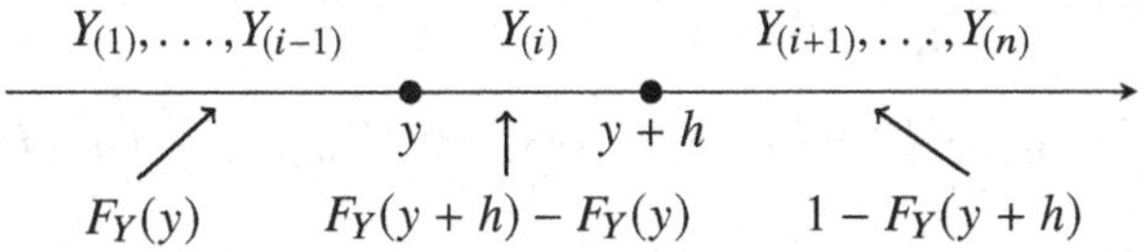

Figure 7.5 *Dividing the real line into three intervals to work out the distribution of $Y_{(i)}$. There must be $i-1$ observations in the leftmost interval, 1 in the middle one, and $n-i$ in the rightmost one. The probability that an observation falls in a given interval is shown below the line. We ignore the possibility that there are two or more observations in the middle interval.*

If $j = 1$ we must have $k = i - 1$, and the limit simplifies to

$$\begin{aligned}
&\lim_{h\to 0} \frac{P(y < Y_{(i)} \leq y+h)}{h} \\
&\quad = \lim_{h\to 0} \frac{n!}{(i-1)!1!(n-i)!} F_Y(y)^{i-1} \frac{[F_Y(y+h) - F_Y(y)]}{h} [1 - F_Y(y+h)]^{n-i}.
\end{aligned}$$

The result follows from the properties of limits of functions, since

$$\lim_{h\to 0} \frac{F_Y(y+h) - F_Y(y)}{h} = f_Y(y),$$

by definition, and

$$\lim_{h\to 0} F_Y(y+h) = F_Y(y),$$

by right continuity. □

The order statistics from a random sample are not independent. We can derive the joint density of $Y_{(i)}$ and $Y_{(j)}$ using an argument similar to that used in the proof of 7.4.6.

**Proposition 7.4.7** (Joint density of order statistics)
*Suppose that $Y_1, \ldots, Y_n$ is a random sample from a population with continuous cumulative distribution function $F_Y$ and density function $f_Y$. The joint density of the $i^{th}$*

*and $j^{th}$ order statistics is given by*

$$f_{Y_{(i)},Y_{(j)}}(v,w) = \frac{n!}{(i-1)!(j-i-1)!(n-j)!} \cdot f_Y(v) f_Y(w) [F_Y(v)]^{i-1} [F_Y(w) - F_Y(v)]^{j-i-1} [1 - F_Y(w)]^{n-j},$$

*The support is the region where $v < w$, $f_Y(v) > 0$, and $f_Y(w) > 0$.*

*Proof.*

We start from the definition of $f_{Y_{(i)},Y_{(j)}}$ as the derivative of $F_{Y_{(i)},Y_{(j)}}$,

$$\begin{aligned} f_{Y_{(i)},Y_{(j)}}(v,w) &= \frac{\partial}{\partial w}\frac{\partial}{\partial v} F_{Y_{(i)},Y_{(j)}}(v,w) \\ &= \lim_{h\to 0} \frac{1}{h^2}[F_{Y_{(i)},Y_{(j)}}(v+h,w+h) - F_{Y_{(i)},Y_{(j)}}(v+h,w) - F_{Y_{(i)},Y_{(j)}}(v,w+h) + F_{Y_{(i)},Y_{(j)}}(v,w)] \\ &= \lim_{h\to 0} \frac{1}{h^2}[\mathrm{P}(v < Y_{(i)} \le v+h, w < Y_{(j)} \le w+h)]. \end{aligned}$$

Consider the term $\mathrm{P}(v < Y_{(i)} \le v+h, w < Y_{(j)} \le w+h)$, the probability that the $i^{\text{th}}$ order statistic is in $(v, v+h]$ and the $j^{\text{th}}$ order statistic is in $(w, w+h]$. If we divide the real line up into five intervals,

$$(-\infty, v],\ (v, v+h],\ (v+h, w],\ (w, w+h],\ \text{and}\ (w+h, \infty],$$

the probabilities of an individual sample value falling into these intervals are, respectively,

$$F_Y(v),\ F_Y(v+h) - F_Y(v),\ F_Y(w) - F_Y(v+h),\ F_Y(w+h) - F_Y(w),\ \text{and}\ 1 - F_Y(w+h).$$

The probability $\mathrm{P}(v < Y_{(i)} \le v+h, w < Y_{(j)} \le w+h)$ is then a sum of terms from a multinomial mass function (with $n$ trials and five possible outcomes for each trial). Using an argument similar to that given in the proof of Proposition 7.4.6, when we divide $\mathrm{P}(v < Y_{(i)} \le v+h, w < Y_{(j)} \le w+h)$ by $h^2$ and take the limit as $h \to 0$, the only term that will be non-zero is that corresponding to having exactly one observation in $(v, v+h]$, and exactly one observation in $(w, w+h]$. Thus,

$$\begin{aligned} &\lim_{h\to 0} \frac{1}{h^2}[\mathrm{P}(v < Y_{(i)} \le v+h, w < Y_{(j)} \le w+h)] \\ &= \lim_{h\to 0} \left\{ \frac{n!}{(i-1)!1!(j-i-1)!1!(n-j)!} [F_Y(v)]^{i-1} \frac{[F_Y(v+h) - F_Y(v)]}{h} \right. \\ &\quad \left. \cdot [F_Y(w) - F_Y(v+h)]^{j-i-1} \frac{[F_Y(w+h) - F_Y(w)]}{h} [1 - F_Y(w+h)]^{n-j} \right\}. \end{aligned}$$

The result follows by the properties of limits of functions. □

**Example 7.4.8** (Density of $i^{\text{th}}$ order statistic for Unif$[0, \theta]$)
Suppose that $\boldsymbol{Y} = (Y_1, \dots, Y_n)^T$ is a random sample from a Unif$[0, \theta]$ distribution. We can derive the density of the $i^{\text{th}}$ order statistic by direct application of 7.4.6,

$$f_{Y_{(i)}}(y) = \frac{n!}{(i-1)!(n-i)!} \frac{1}{\theta} \left(\frac{y}{\theta}\right)^{i-1} \left(1 - \frac{y}{\theta}\right)^{n-i} \quad \text{for } 0 \leq y \leq \theta .$$

This is an example of the **Beta distribution**. In general, if $X \sim \text{Beta}(\alpha, \beta)$, where $\alpha, \beta > 0$ are shape parameters, the density of $X$ is

$$f_X(x) = \begin{cases} \frac{1}{B(\alpha,\beta)} x^{\alpha-1}(1-x)^{\beta-1} & \text{for } 0 \leq x \leq 1, \\ 0 & \text{otherwise,} \end{cases}$$

where $B(\alpha, \beta) = \Gamma(\alpha)\Gamma(\beta)/\Gamma(\alpha + \beta)$ is known as the **Beta function**. We deduce that, in this case, $(Y_{(i)}/\theta) \sim \text{Beta}(i, n + 1 - i)$.

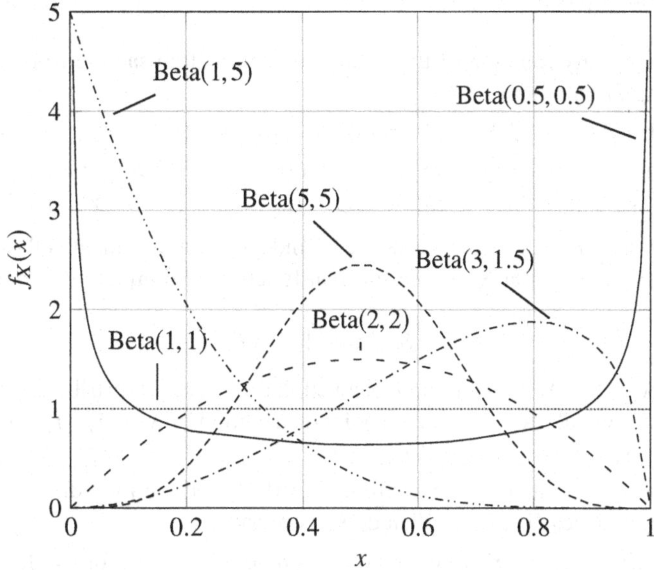


Figure 7.6 *The* Beta$(\alpha, \beta)$ *density function for various values of* $\alpha$ *and* $\beta$.

**Exercise 7.4**

1. (**Symmetry of sample quantiles**) Show that, using Definition 7.4.2, sample quantiles have the property that if $Q_\alpha = Y_{(i)}$ then $Q_{1-\alpha} = Y_{(n+1-i)}$. In other words, if $Q_\alpha$ is given by the $i^{\text{th}}$ smallest value, then $Q_{1-\alpha}$ is given by the $i^{\text{th}}$ largest value.
2. (**Deriving distribution and density of sample minimum**) Suppose that $Y_1, \dots, Y_n$ is a random sample from a continuous population distribution. Prove that

$$F_{Y_{(1)}}(y) = 1 - [1 - F_Y(y)]^n,$$

and hence derive an expression for the density of the sample minimum.

3. Let $X \sim \text{Beta}(\alpha, \beta)$, with density function

$$f_X(x) = \begin{cases} \frac{1}{B(\alpha,\beta)} x^{\alpha-1}(1-x)^{\beta-1} & \text{for } 0 \leq x \leq 1, \\ 0 & \text{otherwise.} \end{cases}$$

   (a) Show that $\mathbb{E}(X) = \alpha/(\alpha+\beta)$.
   (b) If $\alpha, \beta > 1$, show that the **mode** of the distribution (that is, the value of $x$ that maximises $f_X(x)$) is $(\alpha-1)/(\alpha+\beta-2)$.

4. Show that the minimum of $n$ independent and identically distributed exponential random variables is also exponentially distributed. Prove that this is also the case when the variables are independent but not identically distributed, i.e. when $Y_i \sim \text{Exp}(\lambda_i)$, for $i = 1, \ldots, n$.

## 7.5 Further exercises

1. Find the density function of the minimum of a random sample from each of the following distributions.
   (a) Pareto: $f_Y(y) = a/(1+y)^{a+1}$ for $0 < y < \infty$, $a > 0$.
   (b) Weibull: $f_Y(y) = c\tau y^{\tau-1}\exp(-cy^{\tau})$ for $0 < y < \infty$, $\tau > 0$, $c > 0$.
   (c) Extreme value: $f_Y(y) = \lambda e^y \exp(-\lambda e^y)$ for $-\infty < y < \infty$, $\lambda > 0$.

2. Suppose that $Y_1, Y_2, \ldots$ is a sequence of independent and identically distributed random variables. Let $S_k$ be the survival function of $\min_{1 \leq i \leq k} Y_i$. Show that, for $m, n \in \mathbb{Z}^+$,

$$S_{mn}(x) = [S_m(x)]^n.$$

3. (a) Suppose that $Y_1, Y_2, Y_3$ are independent and identically distributed exponential random variables. Work out the joint distribution of $X_1 = Y_1/(Y_1+Y_2+Y_3)$ and $X_2 = (Y_1+Y_2)/(Y_1+Y_2+Y_3)$.
   (b) If $U_1, U_2$ is a random sample from Unif$[0,1]$, find the joint distribution of the order statistics $U_{(1)}, U_{(2)}$. What do you observe?

4. Let $\{(X_1, Y_1), \ldots, (X_n, Y_n)\}$ be a random sample of pairs. If, for $i = 1, \ldots, n$, $U_i = a + bX_i$ and $V_i = c + dY_i$, where $a$, $b$, $c$, and $d$ are real constants, show that $r_{U,V} = \text{sign}(bd) r_{X,Y}$. [We define $\text{sign}(u) = 1$ if $u > 0$ and $\text{sign}(u) = -1$ if $u < 0$.]

5. Suppose that $X_1, \ldots, X_n$ is a random sample from an Exp(1) distribution. Give an expression for the CDF of $Y = X_{(n)} - \log n$ as $n \to \infty$. [This is an example of the **Gumbel distribution**.]

6. Let $Y_1, \ldots, Y_n$ be a random sample from a Bernoulli($p$) distribution. Explain why $Y_{(i)} \sim \text{Bernoulli}(q_i)$ and give an expression for the probability $q_i$.

7. Let $Y_1, \ldots, Y_n$ be a random sample from a Exp($\lambda$) distribution. Derive an expression for the joint density of the sample minimum, $Y_{(1)}$, and the sample range, $R = Y_{(n)} - Y_{(1)}$. What do you observe? Show that $R$ has the same distribution as the

maximum of an exponential random sample of size $n - 1$. Hence, by recursion, prove that

$$\mathbb{E}(Y_{(n)}) = \left(\frac{1}{n} + \frac{1}{n-1} + \ldots + \frac{1}{1}\right)\lambda^{-1}.$$

8. (**Distributional results for one-way ANOVA**) For the one-way ANOVA model described in section 6.3.3 and Exercise 6.3, prove that the between-group sum of squares is independent of the within-group sum of squares. If the true group means are equal (that is, if $\beta_1 = \ldots = \beta_{k-1} = 0$) what is the distribution of each sum of squares?

## 7.6 Chapter summary

On completion of this chapter you should be able to:

- find the mean and variance of the sample mean,
- prove the central limit theorem in the special case where the moment-generating function is well defined,
- describe practical uses and limitations of the central limit theorem,
- find the mean of the sample variance,
- use the trick of splitting $(Y_i - \bar{Y})$ into independent terms,
- derive the distributions of the sample mean and sample variance for a normal population,
- use order statistics to construct sample quantiles,
- derive the distribution of the sample minimum and sample maximum for a random sample from a specified distribution, and
- use the joint density of order statistics in simple cases.

Chapter 8

# Estimation, testing, and prediction

Statistics is fundamentally about using data to answer inferential questions. These questions are usually hard to answer just by looking at the raw data, so our instinct is to summarise the information available in the observed sample. We might do this by computing summary statistics, such as the sample mean, sample variance, sample minimum and maximum, and so on. Each of these statistics reduces the observed sample to a single number; this process, referred to as data reduction, is the first key topic of this chapter.

We introduce three basic procedures for inference that are the topics of subsequent chapters: point estimation, interval estimation, and hypothesis testing. In all of these procedures, statistics play a dominant role. The inferential methods in this chapter are firmly rooted in classical inference; an alternative framework, Bayesian inference, is discussed in Chapter 11.

## 8.1 Functions of a sample

### *8.1.1 Statistics*

The formal definition of a statistic is surprisingly loose: any function of the sample is a statistic. The function may be scalar-valued or vector-valued. As the sample is a random vector, a statistic is also a random vector.

**Definition 8.1.1** (Statistic)
For a sample, $\mathbf{Y} = (Y_1, \dots, Y_n)^T$, a **statistic**, $\mathbf{U} = \mathbf{h}(\mathbf{Y})$, is a random vector that is a function of the sample and known constants alone. Given an observed sample, $\mathbf{y} = (y_1, \dots, y_n)^T$, we can compute the observed value of a statistic, $\mathbf{u} = \mathbf{h}(\mathbf{y})$.

Loose definitions crop up repeatedly in statistical inference. Our approach is to define concepts very broadly, then specify desirable characteristics that will only be found in members of a much smaller subset. Statistics are useful as devices for data reduction when their dimension is smaller than that of the sample. In particular, we will often consider scalar statistics, that is, statistics that reduce the sample to a single random variable.

The distribution of the statistic $\boldsymbol{U}$ is often referred to as the **sampling distribution** of $\boldsymbol{U}$. Although a statistic is a function of the sample and known constants alone, the distribution of a statistic may depend on unknown parameters. The observed value of a statistic, $\boldsymbol{u}$, is just a vector of real numbers. Just as the observed sample, $\boldsymbol{y}$, is thought of as one instance of the sample, $\boldsymbol{Y}$, the value $\boldsymbol{u} = \boldsymbol{h}(\boldsymbol{y})$ is taken to be one instance of the statistic $\boldsymbol{U} = \boldsymbol{h}(\boldsymbol{Y})$.

In some cases, we can use the distributional results from previous chapters to work out the sampling distributions of statistics of interest. In others, the mathematics involved are not tractable and we need to resolve to simulation-based approaches; these are discussed in Chapter 12.

### 8.1.2 *Pivotal functions*

Consider the following situation: if $\boldsymbol{Y}$ is a random sample from an $\mathrm{N}(\mu, 1)$ distribution and $\bar{Y}$ is the sample mean, we know that $\sqrt{n}(\bar{Y} - \mu) \sim \mathrm{N}(0, 1)$ is a function of $\boldsymbol{Y}$ and $\mu$ whose distribution does not depend on $\mu$. This is an example of a **pivotal function**. Pivotal functions play a fundamental (or, even, pivotal) role in the construction of confidence intervals. We start with a more formal definition.

**Definition 8.1.2** (Pivotal function)
Consider a sample $\boldsymbol{Y}$ and a scalar parameter $\theta$. Let $g(\boldsymbol{Y}, \theta)$ be a function of $\boldsymbol{Y}$ and $\theta$ that does not involve any unknown parameter other than $\theta$. We say that $g(\boldsymbol{Y}, \theta)$ is a pivotal function if its distribution does not depend on $\theta$.

Note that a pivotal function defines a random variable, say $W = g(\boldsymbol{Y}, \theta)$. By definition, the distribution of $W$ does not depend on $\theta$.

We illustrate the use of pivotal functions with examples. Before we start on the examples, we introduce a distribution that plays an important role in both interval estimation and hypothesis testing.

**Definition 8.1.3** ($t$-distribution)
Suppose that $Z$ has a standard normal distribution and $V$ has a chi-squared distribution on $k$ degrees of freedom, that is, $Z \sim \mathrm{N}(0, 1)$ and $V \sim \chi^2_k$. Suppose also that $Z$ and $V$ are independent. If we define

$$T = \frac{Z}{\sqrt{V/k}},$$

then $T$ has a ***t*-distribution** on $k$ degrees of freedom, denoted $T \sim t_k$. The density function of the $t$-distribution is

$$f_T(t) = \frac{\Gamma\left(\frac{k+1}{2}\right)}{\sqrt{k\pi}\,\Gamma\left(\frac{k}{2}\right)} \left(1 + \frac{t^2}{k}\right)^{-(k+1)/2} \quad \text{for } -\infty < t < \infty\,.$$

Deriving this density is part of Exercise 8.1.

The density function of the $t$-distribution is a bell-shaped curve centred on the origin. It is characterised by having fatter tails (higher kurtosis) than the density of the standard normal distribution. This excess kurtosis tends to zero as the degrees of freedom tend to infinity. For values of $k$ larger than 30, the $t_k$ is practically indistinguishable from the standard normal (Figure 8.1).

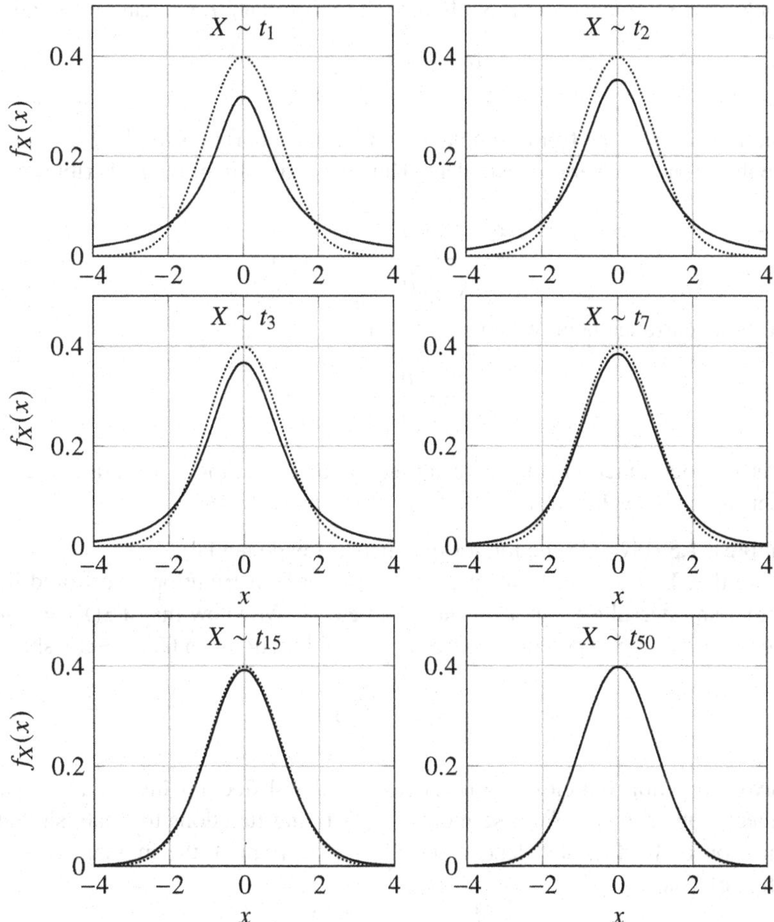


Figure 8.1 *The $t_k$ density function for various values of k, compared to the* N(0, 1) *density function (dotted).*

The $t$-distribution is often referred to as Student's $t$. Note the capitalisation and the location of the apostrophe; in this context, "Student" is a proper noun. In fact, Student was the pseudonym of William Sealy Gosset, an employee of the Guinness brewery whose contributions to statistics were motivated by a desire to improve the quality of beer (if only we all had such noble aspirations).

**Example 8.1.4** (Pivotal function for mean of normal – variance unknown)
Consider a sample $\boldsymbol{Y}$ from an $\mathrm{N}(\mu, \sigma^2)$ distribution. We know from Corollary 7.3.4 that

$$\frac{\bar{Y} - \mu}{\sqrt{\sigma^2/n}} \sim \mathrm{N}(0, 1)\,.$$

However, this does not constitute a pivotal function for $\mu$, as it involves another unknown parameter, namely $\sigma^2$. Suppose that we replace the variance, $\sigma^2$, by its estimator, the sample variance, $S^2$. Referring to Corollary 7.3.4 again we see that

$$\frac{(n-1)S^2}{\sigma^2} \sim \chi^2_{n-1}\,.$$

Furthermore, we know from Lemma 7.3.1 that, for a normal sample, $\bar{Y}$ and $S^2$ (and hence their functions) are independent. Thus, by definition of the $t$-distribution,

$$\frac{(\bar{Y} - \mu) \Big/ \sqrt{\sigma^2/n}}{\sqrt{S^2/\sigma^2}} \sim t_{n-1}\,.$$

With some rearrangement, we conclude that

$$\frac{\bar{Y} - \mu}{\sqrt{S^2/n}} \sim t_{n-1}\,.$$

This is a pivotal function since $S^2$ is a function of $\boldsymbol{Y}$, and the distribution, $t_{n-1}$, does not depend on the value of $\mu$.

**Example 8.1.5** (Pivotal function for parameter of exponential)
Suppose that $\boldsymbol{Y}$ is a random sample from an $\mathrm{Exp}(\lambda)$ distribution. We would like to construct an interval estimator for the parameter $\lambda$. We know that $\lambda\,\mathbb{E}(Y) = 1$ so we might consider $\lambda\bar{Y}$ as a potential source of pivotal functions. In fact, we can show that

$$W = \lambda \sum_{i=1}^{n} Y_i$$

is a pivotal function. It is clear from the definition that $W$ does not involve any unknown parameters other than $\lambda$. We use moment-generating functions to establish that the distribution of $W$ does not depend on $\lambda$. As $Y \sim \mathrm{Exp}(\lambda)$, the moment-generating function of $Y$ is

$$M_Y(t) = (1 - t/\lambda)^{-1}.$$

The moment-generating function of $W$ is

$$\begin{aligned}
M_W(t) = \mathbb{E}[\exp(tW)] &= \mathbb{E}\left[\exp\left(t\lambda \sum_{i=1}^{n} Y_i\right)\right] \\
&= \{\mathbb{E}\,[\exp(t\lambda Y)]\}^n && \text{by independence} \\
&= \{M_Y(t\lambda)\}^n \\
&= (1-t)^{-n}. && \text{by definition of } M_Y
\end{aligned}$$

The distribution of a random variable is completely characterised by its moment-generating function. As the moment-generating function of $W$ does not depend on $\lambda$, we conclude that the distribution of $W$ does not depend on $\lambda$, so $W$ is pivotal.

**Exercise 8.1**

1. (**Density of the $t$-distribution**) Find the density function of $X = \sqrt{V/k}$, where $V \sim \chi^2_k$. Now suppose that $Z \sim N(0,1)$ and $Z, X$ are independent. Using the formula for the density of a ratio of two random variables (see Exercise 4.6), show that $T = Z/X$ has the density function given in Definition 8.1.3.
2. (**Pivotal function for parameter of uniform distribution**) Suppose that $X_1, \cdots, X_n$ is a random sample from the Unif$[0, \theta]$ distribution. Find a pivotal quantity that is a function of $X_{(n)}$.

## 8.2 Point estimation

Consider a scalar parameter $\theta$; here, $\theta$ is just a single unknown number. Any method for estimating the value of $\theta$ based on a sample of size $n$ can be represented as a function $h : \mathbb{R}^n \to \mathbb{R}$. When this function is applied to the observed sample it yields a **point estimate**, $h(\boldsymbol{y})$. This estimate is just a number. In order to gain an insight into the properties of the estimation method, we consider applying the function to the sample. The resulting random variable, $h(\boldsymbol{Y})$, is referred to as a **point estimator**. Notice the rather subtle distinction here between an estimat<u>or</u>, which is a statistic (random variable), and an estimat<u>e</u>, which is an observed value of a statistic (just a number).

It is clear that any point estimator is a statistic. In fact, this association goes in both directions, as the following definition makes clear.

**Definition 8.2.1** (Point estimator)
Any scalar statistic may be taken to be a point estimator for a parameter, $\theta$. An observed value of this statistic is referred to as a point estimate.

Definition 8.2.1 seems rather loose; we do not mention anything about restricting our attention to point estimators that are likely to yield estimates close to the true value. In subsection 8.2.1 we introduce the concepts of **bias, mean squared error**, and **consistency**. These concepts allow us to formalise "likely to yield estimates close to the true value" as a desirable property of a point estimator. Some commonly used point estimators are given in the following example.

**Example 8.2.2** (Some well-known point estimators)
Consider an observed sample $\boldsymbol{y} = (y_1, \ldots, y_n)^T$ that we view as an instance of the sample $\boldsymbol{Y} = (Y_1, \ldots, Y_n)^T$. An obvious statistic to calculate is the sample mean. The observed value of the sample mean is

$$\bar{y} = \frac{1}{n} \sum_{j=1}^{n} y_j ,$$

which is clearly a function of the observed sample. Applying the same function to the sample, we get

$$\bar{Y} = \frac{1}{n} \sum_{j=1}^{n} Y_j .$$

The sample mean is the usual estimator for the population mean. We will consider its properties as an estimator using the results established in section 7.1.

In Table 8.1 we list some frequently used statistics, their observed sample values, and the population characteristics for which they are used as point estimators. The terminology gets a little bit cumbersome, but we will try to stick with "sample mean" and "sample variance" for statistics, and "observed sample mean" and "observed sample variance" for instances of these statistics. As always, we use $y_{(i)}$ to denote the $i^{\text{th}}$ smallest value when $y_1, \dots, y_n$ are placed in ascending order.

| | statistic ($U$) | observed value ($u$) | estimator for |
|---|---|---|---|
| sample mean | $\bar{Y} = \frac{1}{n}\sum_{j=1}^{n} Y_j$ | $\bar{y} = \frac{1}{n}\sum_{j=1}^{n} y_j$ | pop. mean |
| sample variance | $S^2 = \frac{1}{n-1}\sum_{j=1}^{n}(Y_j - \bar{Y})^2$ | $s^2 = \frac{1}{n-1}\sum_{j=1}^{n}(y_j - \bar{y})^2$ | pop. variance |
| 1st order statistic | $Y_{(1)} = \min_{1 \le j \le n} Y_j$ | $y_{(1)} = \min_{1 \le j \le n} y_j$ | pop. minimum |
| $n^{\text{th}}$ order statistic | $Y_{(n)} = \max_{1 \le j \le n} Y_j$ | $y_{(n)} = \max_{1 \le j \le n} y_j$ | pop. maximum |
| sample median | $Y_{((n+1)/2)}$ | $y_{((n+1)/2)}$ | pop. median |

Table 8.1 *Some frequently used point estimators, observed values, and population quantities that they are used to estimate.*

In general, we may have more than one parameter of interest. These are grouped together into a vector, $\boldsymbol{\theta} = (\theta_1, \dots, \theta_r)^T$. This vector can either be viewed as a single vector parameter or as a vector of scalar parameters. The statistic used as a point estimator for $\boldsymbol{\theta}$ is an $r \times 1$ random vector, and the corresponding point estimate will be an $r \times 1$ vector of real numbers (a point in $\mathbb{R}^r$). The orientation of $\boldsymbol{\theta}$ is often unimportant; row vectors and column vectors serve our purpose equally well. In these cases, we may drop the transpose and write $\boldsymbol{\theta} = (\theta_1, \dots, \theta_r)$. As an example of a vector of parameters, suppose that our population is normally distributed with unknown mean, $\mu$, and variance, $\sigma^2$, that is, $Y \sim \mathrm{N}(\mu, \sigma^2)$. We might take $\boldsymbol{\theta} = (\mu, \sigma^2)$. A point estimator for $\boldsymbol{\theta}$ would be $\hat{\boldsymbol{\theta}} = (\bar{Y}, S^2)$.

Before discussing the properties of point estimators, we reiterate the importance of the distinction between estimators and estimates. Suppose that a statistic $U = h(\boldsymbol{Y})$ is a point estimator of a parameter $\theta$. Consider the observed sample

$$\boldsymbol{y}^{(1)} = (y_1^{(1)}, \dots, y_n^{(1)})^T,$$

from which we evaluate the sample value of the statistic,

$$u^{(1)} = h(\boldsymbol{y}^{(1)}).$$

We could take another sample,

$$y^{(2)} = (y_1^{(2)}, \ldots, y_n^{(2)})^T,$$

and work out the corresponding value of the statistic,

$$u^{(2)} = h(y^{(2)}).$$

Since $u^{(2)}$ is based on a new sample, it will in general be a different value to $u^{(1)}$. We can repeat this process to build up a collection of values of the statistic, $u^{(1)}, \ldots, u^{(m)}$. Each of these values is an estimate of $\theta$, and each one can be viewed as an instance of the estimator $U$. Thus, the estimator is a random variable from whose distribution the estimates are sampled; the estimator captures our uncertainty about the values of our estimates. It is the properties of an estimator that determine how useful it is. We might expect a good point estimator to have most of its mass concentrated around the value that we are trying to estimate. In general, this will ensure that the estimator has a high probability of yielding a value close to the parameter value.

### *8.2.1 Bias, variance, and mean squared error*

For many distributions, mass is concentrated around the centre. In order to identify good estimators, we may insist that the centre of the distribution of the estimator is close to our parameter $\theta$. Our usual measure of central tendency of a distribution is the mean. If the mean of an estimator is equal to the parameter value, we say that the estimator is **unbiased**.

**Definition 8.2.3** (Bias)
Suppose that $U$ is a statistic. The **bias** of $U$ as a point estimator of $\theta$ is

$$\operatorname*{Bias}_{\theta}(U) = \operatorname*{\mathbb{E}}_{\theta}(U - \theta).$$

We say that $U$ is an unbiased estimator of $\theta$ if $\text{Bias}_\theta(U) = 0$, that is, if $\mathbb{E}_\theta(U) = \theta$.

The $\theta$ subscript on $\text{Bias}_\theta(U)$ indicates that we are considering the bias of $U$ as an estimator of $\theta$. We also use a $\theta$ subscript on the expectation and probability operators in this context. This is because the distribution of a statistic may depend on unknown parameters; we could denote this by $U \sim F_U(.;\theta)$, where $F_U(u;\theta) = P_\theta(U \leq u)$, and so

$$\operatorname*{\mathbb{E}}_{\theta}(U) = \begin{cases} \sum_u u f_U(u;\theta) & \text{for the discrete case,} \\ \int_{-\infty}^{\infty} u f_U(u;\theta) du & \text{for the continuous case,} \end{cases}$$

where $f_U$ is the mass or density function associated with $F_U$. We will drop the subscript on expectation, variance, and probability operators when the connection between the parameters and the distribution of the estimator is obvious.

An estimator centred on the parameter value does not guarantee that we have a high probability of getting an estimate close to the parameter value. The problem here is that bias measures expected distance from $\theta$; a statistic that attributes high probability both to values that are much larger than $\theta$ and to values that are much smaller than $\theta$ is clearly undesirable but could still be unbiased. One way of dealing with this problem is to consider the expected squared distance, that is, the **mean squared error**.

**Definition 8.2.4** (Mean squared error)
Suppose that $U$ is a statistic. The mean squared error (MSE) of $U$ as an estimator of $\theta$ is given by

$$\underset{\theta}{\text{MSE}}(U) = \underset{\theta}{\mathbb{E}}\left[(U-\theta)^2\right].$$

As we might expect, the mean squared error of an estimator is related to its bias.

**Proposition 8.2.5** (Relationship between MSE, bias, and variance)
*Suppose that $U$ is a statistic, then, for all $\theta \in \Theta$,*

$$\underset{\theta}{\text{MSE}}(U) = [\underset{\theta}{\text{Bias}}(U)]^2 + \underset{\theta}{\text{Var}}(U).$$

*For an unbiased estimator, the mean squared error is equal to the variance.*

Proving Proposition 8.2.5 is part of Exercise 8.2. The proposition has an intuitive interpretation: an estimator with low bias and low variance is appealing. Low variance means that mass is concentrated around the centre of the distribution but, given that we also have low bias, the centre of the distribution is close to the parameter value. In summary, an estimator with low mean squared error will have a distribution whose mass is concentrated near $\theta$. In practice, low bias and low variance are often competing demands; to reduce bias we may have to accept higher variance, whereas to have an estimator with low variance we may have to introduce some bias. This is referred to as the **bias-variance tradeoff**.

**Proposition 8.2.6** (MSE of the sample mean and variance)
*Suppose that $\mathbf{Y} = (Y_1, \ldots, Y_n)^T$ is a random sample from a population distribution with mean $\mu$, variance $\sigma^2$, and fourth central moment $\mu_4$. The mean squared error of the sample mean and variance as estimators of the population mean and variance, respectively, are*

$$\underset{\mu}{\text{MSE}}(\bar{Y}) = \frac{\sigma^2}{n},$$
$$\underset{\sigma^2}{\text{MSE}}(S^2) = \frac{1}{n}\left(\mu_4 - \frac{n-3}{n-1}\sigma^4\right).$$

These results follow directly from Propositions 7.1.2 and 7.2.5. Notice that, as both estimators are unbiased, their MSE is equal to their respective variances.

### 8.2.2 *Consistency*

If we could take a sample of infinite size, we might hope that our estimator would be perfect. In other words, we would want our estimator to yield the true parameter value

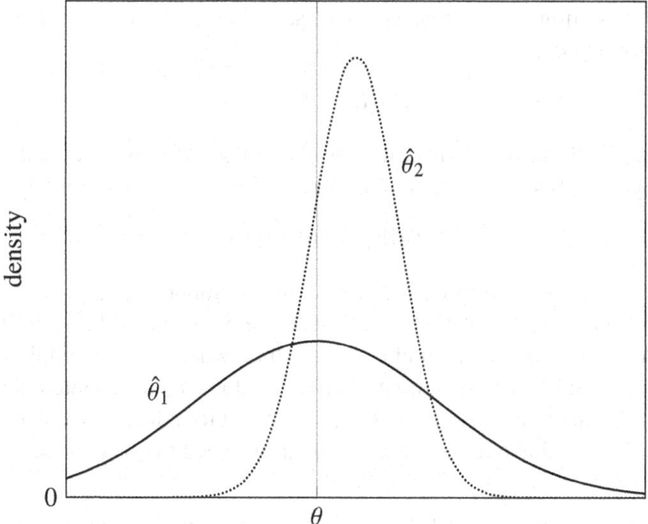


Figure 8.2 *Two estimators of the same parameter $\theta$. $\hat{\theta}_1$ has zero bias but high variance, so it may produce estimates very far from the true value of $\theta$. By contrast, $\hat{\theta}_2$ is biased but has much lower variance, so it would typically give more accurate estimates.*

with probability 1. This is the basic idea behind **consistency**. Consistency provides a formal expression of our hope that, as the sample size increases, the performance of our estimators improves. Some statisticians view consistency as so important that they will not even tentatively entertain an inconsistent estimator.

Consistency is an asymptotic property. In order to define consistency we first need to define a sequence of estimators. We have so far used $n$ to denote sample size, and in defining our estimators ignored the implicit dependence on $n$. Our estimators are defined in a way that is valid for any $n$, so it is easy to make the dependence on $n$ explicit and, thus, define a sequence of estimators. For example, if we define the sample mean for a sample of size $n$ to be $\bar{Y}_n = \frac{1}{n}\sum_{i=1}^{n} Y_i$, then $\bar{Y}_1 = Y_1$, $\bar{Y}_2 = \frac{1}{2}(Y_1 + Y_2)$, $\bar{Y}_3 = \frac{1}{3}(Y_1 + Y_2 + Y_3)$, and so on. For a general estimator, $U_n$, the subscript $n$ denotes the size of the sample on which the estimator is based. We then define $\{U_n : n = 1, 2, \dots\}$ to be a sequence of estimators; this is often denoted as just $\{U_n\}$ for brevity. We are now in a position to define consistency.

**Definition 8.2.7** (Consistency)
The sequence $\{U_n : n = 1, 2, \dots\}$ is a **consistent** sequence of estimators of the parameter $\theta$ if it converges in probability to $\theta$, that is, for every $\delta > 0$ and $\theta \in \Theta$,

$$\lim_{n\to\infty} P_\theta(|U_n - \theta| < \delta) = 1 .$$

In most situations the role of the sample size is obvious and we would just say that $U$ is a consistent estimator of $\theta$. It is worth examining Definition 8.2.7 in the context of the formal definition of convergence . This says that, given any $\varepsilon > 0$, we can find $N$ such that, for any $n > N$,

$$1 - P_\theta(|U_n - \theta| < \delta) < \varepsilon \, .$$

Rearranging the final statement using the properties of probability, our statement of consistency becomes: given any $\varepsilon > 0$, we can find $N$ such that, for any $n > N$,

$$P_\theta(|U_n - \theta| \geq \delta) < \varepsilon \, . \tag{8.1}$$

One possible intuitive interpretation is as follows. Suppose $\{U_n\}$ is consistent for $\theta$. This means that, for any open interval around the parameter value, say $(\theta - \delta, \theta + \delta)$, that we choose, we can find a number $N$ such that, whenever the sample size is larger than $N$, the probability of our estimator taking a value outside the interval is arbitrarily small. By arbitrarily small we mean that, if we want to reduce the probability that the estimator takes a value outside the interval, all we need to do is increase the value of $N$.

One of the consequences of (8.1) is an equivalent definition of a sequence of consistent estimators. The sequence $\{U_n\}$ is a consistent sequence of estimators of the parameter $\theta$ if, for every $\delta > 0$ and $\theta \in \Theta$,

$$\lim_{n \to \infty} P_\theta(|U_n - \theta| \geq \delta) = 0 \, . \tag{8.2}$$

Checking for consistency directly can be tedious. Fortunately, the following proposition provides another route.

**Proposition 8.2.8** (Relationship between MSE and consistency)
*Suppose that $\{U_n\}$ is a sequence of statistics and $\theta$ an unknown parameter. If, for all $\theta \in \Theta$,*

$$\lim_{n \to \infty} \operatorname*{MSE}_\theta(U_n) = 0 \, ,$$

*then $\{U_n\}$ is a consistent sequence of estimators for $\theta$.*

*Proof.*
If $\lim_{n \to \infty} \text{MSE}_\theta(U_n) = 0$ for all $\theta \in \Theta$ then, by definition,

$$\lim_{n \to \infty} \mathbb{E}_\theta[(U_n - \theta)^2] = 0 \text{ for all } \theta \in \Theta \, . \tag{8.3}$$

We now apply the Chebyshev inequality (Proposition 3.4.13) to the alternative definition of consistency (8.2) to obtain

$$P_\theta(|U_n - \theta| \geq \delta) \leq \frac{\mathbb{E}_\theta[(U_n - \theta)^2]}{\delta^2} \, . \tag{8.4}$$

From (8.3), we know that the right-hand side of (8.4) converges to zero for all $\theta \in \Theta$. The left-hand side of (8.4) is a non-negative number bounded above by something that converges to zero. Thus, $\lim_{n \to \infty} P_\theta(|U_n - \theta| \geq \delta) = 0$, the condition in (8.2) is satisfied, and $\{U_n\}$ is a consistent sequence of estimators for $\theta$. □

The condition for consistency given in Proposition 8.2.8 is sufficient but not necessary for consistency; there are consistent estimators whose MSE does not converge to zero. Taking into account the expression for MSE in terms of bias and variance given by Proposition 8.2.5 yields the following corollary.

**Corollary 8.2.9** (Relationship between consistency, bias, and variance)
*Suppose that $\{U_n\}$ is a sequence of statistics and $\theta$ an unknown parameter. If, for all $\theta \in \Theta$,*

$$\lim_{n\to\infty} \operatorname{Bias}_{\theta}(U_n) = 0 \text{ and } \lim_{n\to\infty} \operatorname{Var}_{\theta}(U_n) = 0,$$

*then $\{U_n\}$ is a consistent sequence of estimators for $\theta$.*

Consider the sample mean for $n$ observations, $\bar{Y}_n = \frac{1}{n}\sum_{i=1}^{n} Y_i$. This is our usual estimator of the population mean. As such, we might hope that the sample mean is a consistent estimator of the population mean. For a random sample, this is easy to establish. The result is sometimes referred to as the **weak law of large numbers**.

**Proposition 8.2.10** (Weak law of large numbers)
*If $\mathbf{Y} = (Y_1, \dots, Y_n)^T$ is a random sample with $\mathbb{E}(Y) = \mu < \infty$ and $\operatorname{Var}(Y) = \sigma^2 < \infty$, then the sample mean is a consistent estimator of the population mean, that is,*

$$\bar{Y}_n \xrightarrow{p} \mu \text{ as } n \to \infty,$$

*where $\bar{Y}_n = \frac{1}{n}\sum_{j=1}^{n} Y_j$ is the sample mean.*

*Proof.*

From Proposition 8.2.6,

$$\lim_{n\to\infty} \operatorname{MSE}_{\mu}(\bar{Y}_n) = \lim_{n\to\infty} \frac{1}{n}\sigma^2 = 0,$$

since $\sigma^2 < \infty$. By Proposition 8.2.8, $\bar{Y}_n$ is consistent for $\mu$ and, by definition, $\bar{Y}_n \xrightarrow{p} \mu$ as $n \to \infty$. □

If we make some slightly stronger assumptions, we can also show that the sample mean converges almost surely to the population mean. This result is referred to as the **strong law of large numbers** (the terminology is slightly unfortunate). Furthermore, we can deduce from Proposition 8.2.6 that, as long as the fourth central moment is finite, the sample variance is a consistent estimator of the population variance.

### 8.2.3 *The method of moments*

In general, population moments are functions of parameters that we would like to estimate. Sample moments are statistics, that is, functions of the sample. By equating population moments with sample moments and solving the resulting set of simultaneous equations, we can generate estimators for the population parameters. These are referred to as **method-of-moments estimators**.

Consider a sample $\boldsymbol{Y} = (Y_1, \dots, Y_n)^T$ parameterised by $\boldsymbol{\theta} = (\theta_1, \dots, \theta_r)^T$. Method-of-moments estimators for $\theta_1, \dots, \theta_r$ would be generated by solving

$$\mu_i'(\hat{\theta}_1, \dots, \hat{\theta}_r) = m_i'(Y_1, \dots, Y_n) \quad \text{for} \quad i = 1, \dots, r$$

to give expressions for $\hat{\theta}_1, \dots, \hat{\theta}_r$ in terms of $Y_1, \dots, Y_n$. Here, $\mu_i'(\hat{\theta}_1, \dots, \hat{\theta}_r)$ is the $i^{\text{th}}$ population moment and $m_i'(Y_1, \dots, Y_n)$ is the $i^{\text{th}}$ sample moment. We use this notation to emphasise that $\mu_i'$ depends on the population parameters, while $m_i'$ is a function of the sample only.

The method of moments has the advantage of being easy to implement; for most of the probability distributions encountered in this book, the corresponding equations can be solved easily by hand. However, the resulting estimators often have undesirable properties. Beyond very simple cases, the main use of the method of moments is to provide starting values for other estimation procedures.

**Example 8.2.11** (Method of moments for normal)
Suppose that $\boldsymbol{Y} = (Y_1, \dots, Y_n)^T$ is a sample from an $\text{N}(\mu, \sigma^2)$ distribution, where both $\mu$ and $\sigma^2$ are unknown parameters. The method-of-moments estimators are found by solving

$$\begin{aligned}
\mu_1'(\hat{\mu}, \hat{\sigma}^2) &= m_1'(Y_1, \dots, Y_n), \\
\mu_2'(\hat{\mu}, \hat{\sigma}^2) &= m_2'(Y_1, \dots, Y_n).
\end{aligned}$$

This yields

$$\begin{aligned}
\hat{\mu} &= \bar{Y}, \\
\hat{\sigma}^2 + \hat{\mu}^2 &= \frac{1}{n}\sum_{i=1}^{n} Y_i^2 \Rightarrow \hat{\sigma}^2 = \frac{1}{n}\sum_{i=1}^{n}(Y_i - \bar{Y})^2.
\end{aligned}$$

Notice that the method-of-moments estimator of the variance is not the sample variance $S^2$. In fact, if $\hat{\sigma}^2$ is the method-of-moments estimator, then

$$\hat{\sigma}^2 = \frac{n-1}{n} S^2.$$

As such, $\hat{\sigma}^2$ is a *biased* estimator of the population variance.

**Example 8.2.12** (Method of moments for binomial)
Consider a sample $Y_1, \cdots, Y_n$ from a $\text{Bin}(r, p)$ distribution, where both $r$ and $p$ are unknown. We know that

$$\begin{aligned}
\mathbb{E}(Y) &= rp, \\
\mathbb{E}(Y^2) &= \text{Var}(Y) + \mathbb{E}(Y)^2 = rp(1-p) + r^2p^2.
\end{aligned}$$

The method-of-moments estimators are found by solving

$$\hat{r}\hat{p} = \bar{Y},$$
$$\hat{r}\hat{p}(1-\hat{p}) + \hat{r}^2\hat{p}^2 = \frac{1}{n}\sum_{i=1}^{n} Y_i^2.$$

Solving for $\hat{r}$ by substituting $\hat{p} = \bar{Y}/\hat{r}$ yields

$$\hat{r} = \frac{\bar{Y}^2}{\bar{Y} - \frac{1}{n}\sum(Y_i - \bar{Y})^2},$$
$$\hat{p} = \frac{\bar{Y}}{\hat{r}}.$$

Filling in the details is part of Exercise 8.2.

### 8.2.4 *Ordinary least squares*

We now consider an estimation technique that is most commonly used in the linear regression setup of section 6.3, and which you may well have already encountered. Recall that

$$\boldsymbol{Y} = \boldsymbol{X}\boldsymbol{\beta} + \sigma\boldsymbol{\varepsilon},$$

where $\boldsymbol{Y}$ is an $n \times 1$ response vector, $\boldsymbol{X}$ is an $n \times (p+1)$ matrix of explanatory variables (including a constant term), $\boldsymbol{\beta}$ is a $(p+1) \times 1$ parameter vector, and $\boldsymbol{\varepsilon}$ is an $n \times 1$ error vector. Given an estimate $\boldsymbol{b}$ for $\boldsymbol{\beta}$, the estimated value of the $i^{\text{th}}$ observation of the response variable is

$$\hat{Y}_i = \boldsymbol{x}_i^T \boldsymbol{b},$$

where $\boldsymbol{x}_i^T$ is the $i^{\text{th}}$ row of $\boldsymbol{X}$. In **ordinary least-squares** (OLS) estimation, we seek to minimise the **error sum of squares,**

$$S(\boldsymbol{b}) = \sum_{i=1}^{n}(Y_i - \hat{Y}_i)^2 = (\boldsymbol{Y} - \boldsymbol{X}\boldsymbol{b})^T(\boldsymbol{Y} - \boldsymbol{X}\boldsymbol{b}).$$

The least-squares estimator for $\boldsymbol{\beta}$ is then

$$\hat{\boldsymbol{\beta}} = \arg\min_{\boldsymbol{b}} S(\boldsymbol{b}).$$

This has a straightforward geometric interpretation: in the simple regression case, we seek to minimise the squared vertical distance between each point $(X_i, Y_i)$ and the regression line (Figure 8.3). In higher dimensions, this is replaced by a distance between a point in $\mathbb{R}^{p+1}$ and a $p$-dimensional hyperplane (Figure 8.4).

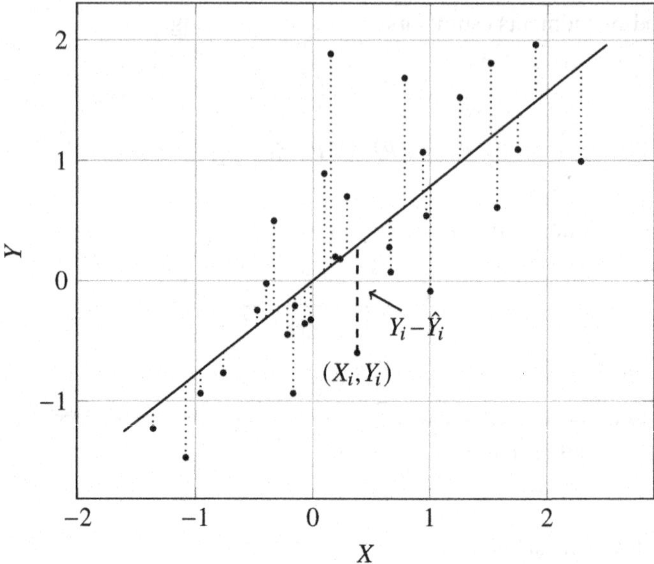


Figure 8.3 *Ordinary least squares in the simple regression case. We want to minimise the squared vertical distance between each point* $(X_i, Y_i)$ *and the regression line.*

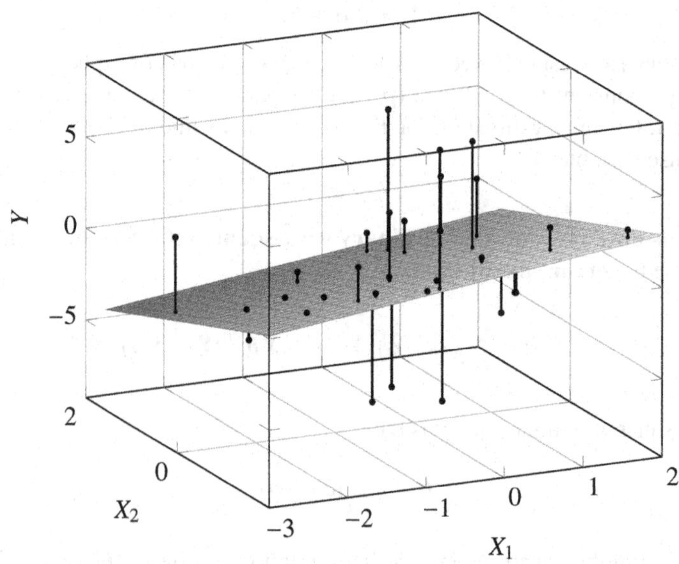


Figure 8.4 *Ordinary least squares with two explanatory variables. We now want to minimise the squared vertical distance between each point* $(X_{1,i}, X_{2,i}, Y_i)$ *and the regression plane.*

We compute $\hat{\boldsymbol{\beta}}$ by setting the derivative of $S(\boldsymbol{b})$ with respect to $\boldsymbol{b}$ equal to zero. We need a few standard results from matrix calculus; if $\boldsymbol{A}$ is an $r \times c$ constant matrix and $\boldsymbol{w}$ is a $c \times 1$ vector, we have

$$\frac{\partial \boldsymbol{A}\boldsymbol{w}}{\partial \boldsymbol{w}} = \boldsymbol{A}, \quad \frac{\partial \boldsymbol{w}^T \boldsymbol{A}}{\partial \boldsymbol{w}} = \boldsymbol{A}^T, \quad \text{and} \quad \frac{\partial \boldsymbol{w}^T \boldsymbol{A}\boldsymbol{w}}{\partial \boldsymbol{w}} = \boldsymbol{w}^T(\boldsymbol{A} + \boldsymbol{A}^T).$$

The derivative of $S(\boldsymbol{b})$ with respect to $\boldsymbol{b}$ is then

$$\begin{aligned}
\frac{\partial S(\boldsymbol{b})}{\partial \boldsymbol{b}} &= \frac{\partial}{\partial \boldsymbol{b}}(\boldsymbol{Y} - \boldsymbol{X}\boldsymbol{b})^T(\boldsymbol{Y} - \boldsymbol{X}\boldsymbol{b}) \\
&= \frac{\partial}{\partial \boldsymbol{b}}\left(\boldsymbol{Y}^T\boldsymbol{Y} - \boldsymbol{Y}^T\boldsymbol{X}\boldsymbol{b} - \boldsymbol{b}^T\boldsymbol{X}^T\boldsymbol{Y} + \boldsymbol{b}^T\boldsymbol{X}^T\boldsymbol{X}\boldsymbol{b}\right) \\
&= 0 - \boldsymbol{Y}^T\boldsymbol{X} - (\boldsymbol{X}^T\boldsymbol{Y})^T + \boldsymbol{b}^T[\boldsymbol{X}^T\boldsymbol{X} + (\boldsymbol{X}^T\boldsymbol{X})^T] \\
&= 2(-\boldsymbol{Y}^T\boldsymbol{X} + \boldsymbol{b}^T\boldsymbol{X}^T\boldsymbol{X}),
\end{aligned}$$

as $\boldsymbol{X}^T\boldsymbol{X}$ is symmetric. Setting the derivative equal to zero yields

$$\hat{\boldsymbol{\beta}}^T\boldsymbol{X}^T\boldsymbol{X} = \boldsymbol{Y}^T\boldsymbol{X} \Leftrightarrow \boldsymbol{X}^T\boldsymbol{X}\hat{\boldsymbol{\beta}} = \boldsymbol{X}^T\boldsymbol{Y} \Leftrightarrow \hat{\boldsymbol{\beta}} = (\boldsymbol{X}^T\boldsymbol{X})^{-1}\boldsymbol{X}^T\boldsymbol{Y},$$

assuming that the matrix $\boldsymbol{X}^T\boldsymbol{X}$ is of full rank and can be inverted. The second derivative is

$$\frac{\partial^2 S(\boldsymbol{b})}{\partial \boldsymbol{b}^T \partial \boldsymbol{b}} = \frac{\partial}{\partial \boldsymbol{b}^T} 2(-\boldsymbol{Y}^T\boldsymbol{X} + \boldsymbol{b}^T\boldsymbol{X}^T\boldsymbol{X}) = 2\boldsymbol{X}^T\boldsymbol{X},$$

which is a positive-definite matrix; for a vector-valued function, this is the equivalent of a positive second derivative, so we conclude that $\hat{\boldsymbol{\beta}}$ minimises $S(\boldsymbol{b})$.

Notice that $\hat{\boldsymbol{\beta}}$ is a linear function of $\boldsymbol{Y}$. Later in this section, we show that $\hat{\boldsymbol{\beta}}$ is unbiased and has the lowest variance of any linear unbiased estimator of $\boldsymbol{\beta}$.

We can write

$$\hat{\boldsymbol{Y}} = \boldsymbol{X}\hat{\boldsymbol{\beta}} = \boldsymbol{X}(\boldsymbol{X}^T\boldsymbol{X})^{-1}\boldsymbol{X}^T\boldsymbol{Y} = \boldsymbol{H}\boldsymbol{Y},$$

where $\boldsymbol{H} = \boldsymbol{X}(\boldsymbol{X}^T\boldsymbol{X})^{-1}\boldsymbol{X}^T$ is known as the **hat matrix**; it maps the observed values, $\boldsymbol{Y}$, to the estimated values, $\hat{\boldsymbol{Y}}$. The hat matrix is **idempotent**, that is, $\boldsymbol{H}^2 = \boldsymbol{H}$.

**Proposition 8.2.13**

*If we assume that $\boldsymbol{X}$ is non-random, and the errors have mean $\mathbb{E}(\boldsymbol{\varepsilon}) = 0$ and covariance matrix $\text{Var}(\boldsymbol{\varepsilon}) = \boldsymbol{I}_n$, then the least-squares estimator, $\hat{\boldsymbol{\beta}}$,*

*i. is unbiased, $\mathbb{E}(\hat{\boldsymbol{\beta}}) = \boldsymbol{\beta}$,*
*ii. has covariance matrix $\text{Var}(\hat{\boldsymbol{\beta}}) = (\boldsymbol{X}^T\boldsymbol{X})^{-1}\sigma^2$.*

*If, in addition, we assume that the errors are multivariate normal, then $\hat{\boldsymbol{\beta}}$ is also multivariate normal.*

Proving this proposition is part of Exercise 8.2.

**Theorem 8.2.14** (Gauss-Markov theorem)
*Assume again that $\boldsymbol{X}$ is non-random, and the errors have mean $\mathbb{E}(\boldsymbol{\varepsilon}) = 0$ and covariance matrix $\mathrm{Var}(\boldsymbol{\varepsilon}) = \boldsymbol{I}$. If $\hat{\boldsymbol{\beta}}$ is the least-squares estimator of $\boldsymbol{\beta}$, and $\hat{\boldsymbol{\beta}}^* \neq \hat{\boldsymbol{\beta}}$ is any other unbiased estimator of $\boldsymbol{\beta}$ that is a linear function of $\boldsymbol{Y}$, then*

$$\mathrm{Var}(\hat{\boldsymbol{\beta}}^*) - \mathrm{Var}(\hat{\boldsymbol{\beta}})$$

*is a positive definite matrix.*

*Proof.*

The estimator $\hat{\boldsymbol{\beta}}^*$ is linear, so we can write $\hat{\boldsymbol{\beta}}^* = \boldsymbol{CY}$ for some $(p+1) \times n$ matrix $\boldsymbol{C}$. It has expectation

$$\mathbb{E}(\hat{\boldsymbol{\beta}}^*) = \mathbb{E}(\boldsymbol{CY}) = \boldsymbol{CX\beta}$$

for all $\boldsymbol{\beta}$, and is unbiased, so $\boldsymbol{CX} = \boldsymbol{I}_{p+1}$. Its variance is

$$\mathrm{Var}(\hat{\boldsymbol{\beta}}^*) = \mathrm{Var}(\boldsymbol{CY}) = \sigma^2 \boldsymbol{CC}^T.$$

We now define the $(p+1) \times n$ matrix $\boldsymbol{D} = \boldsymbol{C} - (\boldsymbol{X}^T\boldsymbol{X})^{-1}\boldsymbol{X}^T$. Notice that

$$\boldsymbol{DX} = (\boldsymbol{C} - (\boldsymbol{X}^T\boldsymbol{X})^{-1}\boldsymbol{X}^T)\boldsymbol{X} = \boldsymbol{CX} - (\boldsymbol{X}^T\boldsymbol{X})^{-1}\boldsymbol{X}^T\boldsymbol{X} = \boldsymbol{I}_{p+1} - \boldsymbol{I}_{p+1} = \boldsymbol{0},$$

so we have

$$\begin{aligned}\boldsymbol{CC}^T &= [\boldsymbol{D} + (\boldsymbol{X}^T\boldsymbol{X})^{-1}\boldsymbol{X}^T][\boldsymbol{D} + (\boldsymbol{X}^T\boldsymbol{X})^{-1}\boldsymbol{X}^T]^T \\ &= \boldsymbol{DD}^T + (\boldsymbol{X}^T\boldsymbol{X})^{-1}\boldsymbol{X}^T\boldsymbol{X}(\boldsymbol{X}^T\boldsymbol{X})^{-1} = \boldsymbol{DD}^T + (\boldsymbol{X}^T\boldsymbol{X})^{-1}.\end{aligned}$$

Using the properties in Proposition 8.2.13, we can write

$$\mathrm{Var}(\hat{\boldsymbol{\beta}}^*) - \mathrm{Var}(\hat{\boldsymbol{\beta}}) = \sigma^2[\boldsymbol{DD}^T + (\boldsymbol{X}^T\boldsymbol{X})^{-1}] - \sigma^2(\boldsymbol{X}^T\boldsymbol{X})^{-1} = \sigma^2\boldsymbol{DD}^T.$$

The matrix $\boldsymbol{D}$ cannot be zero (as that would imply $\hat{\boldsymbol{\beta}}^* = \hat{\boldsymbol{\beta}}$), so $\boldsymbol{DD}^T$ must be positive definite. □

Note the significance of this theorem: if the difference between the covariance matrices is positive definite then, in the matrix sense, the variance of $\hat{\boldsymbol{\beta}}^*$ is higher than the variance of $\hat{\boldsymbol{\beta}}$. We say that the least-squares estimator is, thus, the **best linear unbiased estimator** (BLUE).

**Exercise 8.2**

1. (**Proving relationship between MSE, bias, and variance**) For a statistic $U$ show that $\mathrm{MSE}_\theta(U) = [\mathrm{Bias}_\theta(U)]^2 + \mathrm{Var}_\theta(U)$ for all $\theta \in \Theta$.
2. Suppose that $Y_1, \ldots, Y_n$ is a random sample from a normally distributed population with mean $\mu$ and variance 1, that is, $Y \sim \mathrm{N}(\mu, 1)$. Consider the following statistics:
   i. $U_1 = Y_1 + (Y_2 - Y_3)^2 - 2$ as an estimator of $\mu$,
   ii. $U_2 = \bar{Y}^2 - \frac{1}{n}$ as an estimator of $\mu^2$,

For each estimator determine whether they are

(a) unbiased;

(b) consistent.

[You may assume that, if $Y_n$ converges in probability to $y$, and $g$ is a continuous function, then $g(Y_n)$ converges in probability to $g(y)$.]

3. Suppose that $\boldsymbol{Y} = (Y_1, \dots, Y_n)^T$ is a random sample from a Bernoulli($p$) distribution, and define $S = \sum_{i=1}^n Y_i$. Show that

$$\hat{\theta} = \frac{S(S-1)}{n(n-1)}$$

is an unbiased estimator for $\theta = p^2$. Is this estimator consistent?

4. Suppose that $\boldsymbol{Y} = (Y_1, \dots, Y_n)^T$ is a sample from a Bin($r, p$) distribution where $r$ and $p$ are unknown. Carefully show that the method-of-moments estimators of $r$ and $p$ are

$$\hat{r} = \frac{\bar{Y}^2}{\bar{Y} - \frac{1}{n}\sum_i (Y_i - \bar{Y})^2} \quad \text{and} \hat{p} = \frac{\bar{Y}}{\hat{r}}.$$

5. Prove Proposition 8.2.13.

## 8.3 Interval estimation

In the previous section we described point estimators, which provide a single value for the parameter of interest. Another important class of inferential methods are **interval estimators**. As the name suggests, an interval estimator provides a range of possible values for our unknown parameter, rather than just a single point. Interval estimates are widely used but also widely misinterpreted. Interval estimators can be viewed as a specific case of a **confidence set**. We start by clarifying some terminology.

Recall that a point estimator is a statistic, that is, a function of the sample. By definition, a point estimator is a random variable. When we replace the sample by the observed sample the result is an estimate, that is, just a number. The situation for interval estimators is completely analogous. An interval estimator is a random interval; the end-points of the interval are statistics. When we replace the sample by the observed sample we end up with an interval estimate that is simply a section of the real line.

**Definition 8.3.1** (Interval estimator)
Suppose that we have a sample $\boldsymbol{Y}$ parameterised by $\theta$. Let $U_1 = h_1(\boldsymbol{Y})$ and $U_2 = h_2(\boldsymbol{Y})$ be statistics with $U_1 \le U_2$. The random interval $[U_1, U_2]$ is an interval estimator for $\theta$. If the observed sample is $\boldsymbol{y}$, then the observed values of the statistics are $u_1 = h_1(\boldsymbol{y})$ and $u_2 = h_2(\boldsymbol{y})$. The interval $[u_1, u_2]$ is an interval estimate for $\theta$.

As with the definition of a point estimator, the definition of an interval estimator is rather loose; any scalar statistics $U_1$ and $U_2$ will do for lower and upper end points, provided $U_1 \le U_2$. The obvious question is then what constitutes a good interval estimator. We have to balance two competing demands: we would like our interval estimate to be as narrow as possible, but we would also like the probability that it covers the true value to be high. The balance between **interval length** and **coverage probability** is discussed in section 8.3.1.

**Example 8.3.2** (Interval estimator for mean of normal – variance known)
Suppose that we have a random sample, $\boldsymbol{Y} = (Y_1, \ldots, Y_n)^T$, from an $\mathrm{N}(\mu, \sigma^2)$ distribution, so $Y_i \sim \mathrm{N}(\mu, \sigma^2)$ for $i = 1, \ldots, n$. We are interested in an interval estimator for $\mu$ assuming that $\sigma^2$ is known. We know that the sample mean is also normally distributed,

$$\bar{Y} \sim \mathrm{N}(\mu, \sigma^2/n),$$

and thus,

$$\frac{\bar{Y} - \mu}{\sqrt{\sigma^2/n}} \sim \mathrm{N}(0, 1).$$

The quantiles of a standard normal are well known; if $Z \sim \mathrm{N}(0, 1)$ then

$$\mathrm{P}(Z \le -1.96) = 0.025 \text{ and } \mathrm{P}(Z \ge 1.96) = 0.025,$$

and thus,

$$\mathrm{P}(-1.96 \le Z \le 1.96) = 0.95.$$

As $(\bar{Y} - \mu)/\sqrt{\sigma^2/n}$ has a standard normal distribution, we conclude that

$$\mathrm{P}\left(-1.96 \le \frac{\bar{Y} - \mu}{\sqrt{\sigma^2/n}} \le 1.96\right) = 0.95.$$

Rearranging this expression yields

$$\mathrm{P}\left(\bar{Y} - 1.96\frac{\sigma}{\sqrt{n}} \le \mu \le \bar{Y} + 1.96\frac{\sigma}{\sqrt{n}}\right) = 0.95.$$

Note that the random variable in this expression is $\bar{Y}$, not $\mu$. A sensible interval estimator for $\mu$ is then

$$\left[\bar{Y} - 1.96\frac{\sigma}{\sqrt{n}}, \bar{Y} + 1.96\frac{\sigma}{\sqrt{n}}\right].$$

The corresponding interval estimate is

$$\left[\bar{y} - 1.96\frac{\sigma}{\sqrt{n}}, \bar{y} + 1.96\frac{\sigma}{\sqrt{n}}\right], \tag{8.5}$$

where $\bar{y}$ is the observed sample mean.

### *8.3.1 Coverage probability and length*

A good interval estimator should have a high probability of containing the true value of the parameter. If this were our only criterion, we would always choose the interval $(-\infty, \infty)$, as this covers the true value with probability 1. Clearly, the interval $(-\infty, \infty)$ provides no useful information about plausible values of the parameter. In choosing an interval estimator, there is a tradeoff between the probability of covering the true value and the length of the interval; we would like coverage probability to be high and length to be small. As the probability of covering the true value may depend on the parameter, we make the distinction between **coverage probability** and **confidence coefficient**.

**Definition 8.3.3** (Coverage probability)
For an interval estimator $[U_1, U_2]$ of $\theta$, the coverage probability is the probability that the interval estimator covers $\theta$, that is, $\mathrm{P}(U_1 \leq \theta \leq U_2)$.

**Definition 8.3.4** (Confidence coefficient)
For an interval estimator $[U_1, U_2]$ of $\theta$, the confidence coefficient is the infimum over $\theta$ of the coverage probabilities, that is, $\inf_{\theta \in \Theta} \mathrm{P}(U_1 \leq \theta \leq U_2)$.

It is important to be clear that the random variables in $\mathrm{P}(U_1 \leq \theta \leq U_2)$ are $U_1$ and $U_2$, thus,

$$\mathrm{P}(U_1 \leq \theta \leq U_2) = \mathrm{P}[(U_1 \leq \theta) \cap (U_2 \geq \theta)] = 1 - \mathrm{P}(U_1 > \theta) - \mathrm{P}(U_2 < \theta),$$

since $U_1 \leq U_2$ implies that $U_1 > \theta$ and $U_2 < \theta$ are disjoint events.

In Example 8.3.2, we use the fact that $(\bar{Y} - \mu)/(\sqrt{\sigma^2/n})$ is a pivotal function of $\mu$, as its distribution is $\mathrm{N}(0, 1)$ regardless of the value of $\mu$. Thus, the coverage probability of the resulting interval estimator is always 0.95, and we can attribute a confidence coefficient of 0.95 to our interval estimator. In subsection 8.3.2 we formalise the procedure for obtaining confidence intervals from pivotal functions.

You will often see an interval estimator with confidence coefficient $(1-\alpha)$ referred to as a $100(1-\alpha)$% **confidence interval**. For example, if $\alpha = 0.05$ then the resulting interval estimator is often called the 95% confidence interval. In this context, the confidence coefficient (expressed as a percentage) is often referred to as the **confidence level**. We use the terms confidence interval and interval estimator interchangeably.

In general, the length of a confidence interval will be a random variable. One possible measure of the width of an interval is expected length.

**Definition 8.3.5** (Expected length)
Consider an interval estimator $[U_1, U_2]$. The **expected length** of the interval is defined as $\mathbb{E}(U_2 - U_1)$.

A desirable feature of an interval estimator is that the coverage probability is large for all values of $\theta$. The confidence coefficient represents the worst case scenario; by definition, for any value of $\theta$, the coverage probability will be at least as large as the confidence coefficient. We return to considering confidence intervals for the mean of a normal distribution to illustrate.

**Example 8.3.6** (Assessing the merits of several possible interval estimators)
Suppose that we have a random sample from an $N(\mu, 1)$ distribution, and we are interested in an interval estimator for $\mu$. Let $k_1$ and $k_2$ be non-negative finite constants, that is, $0 \leq k_1 < \infty$ and $0 \leq k_2 < \infty$. Any of the following is a valid interval estimator for $\mu$:

a. $[-k_1, k_2]$,
b. $[Y_1 - k_1, Y_1 + k_2]$,
c. $[\bar{Y} - k_1, \bar{Y} + k_2]$.

All of the intervals in this example are the same length, $k_1 + k_2$. In what follows, we work out the coverage probability and confidence coefficient associated with each one. A quick reminder of some notation: if $Z \sim N(0, 1)$, we use $\Phi$ to denote the cumulative distribution function of $Z$, so $\Phi(z) = P(Z \leq z)$.

a. This first interval does not depend on the sample. If $\mu$ is the true mean, there are two possible situations: either $\mu \in [-k_1, k_2]$ or $\mu \notin [-k_1, k_2]$. If $\mu \in [-k_1, k_2]$, then the coverage probability is 1, otherwise the coverage probability is 0. Thus, the confidence coefficient for this interval is 0.
b. We can use the fact that $Y_1 - \mu \sim N(0, 1)$ to work out the coverage probability directly. We have

$$\begin{aligned} P(Y_1 - k_1 \leq \mu \leq Y_1 + k_2) &= 1 - P(Y_1 - k_1 > \mu) - P(Y_1 + k_2 < \mu) \\ &= 1 - P(Y_1 - \mu > k_1) - P(Y_1 - \mu < -k_2) \\ &= \Phi(k_1) - \Phi(-k_2) \\ &= \Phi(k_1) + \Phi(k_2) - 1\,. \end{aligned}$$

This coverage probability does not depend on $\mu$, so the confidence coefficient, which is the infimum over $\mu$, is also $\Phi(k_1) + \Phi(k_2) - 1$.
c. Using the fact that $\sqrt{n}(\bar{Y} - \mu) \sim N(0, 1)$ and similar reasoning to that given in case b., we can show that the coverage probability is

$$P(Y_1 - k_1 \leq \mu \leq Y_1 + k_2) = \Phi(\sqrt{n}k_1) + \Phi(\sqrt{n}k_2) - 1\,.$$

As with case b., the coverage probability does not involve $\mu$, so the confidence coefficient is equal to the coverage probability.

It is clear that the first interval, with confidence coefficient 0, is of no practical value. Consider cases b. and c. Since $k_1$ is positive and $\Phi$ is a non-decreasing function, we have $\sqrt{n}k_1 \geq k_1$ and thus $\Phi(\sqrt{n}k_1) \geq \Phi(k_1)$ for all $n \geq 1$ (similarly for $k_2$). We conclude that

$$\Phi(\sqrt{n}k_1) + \Phi(\sqrt{n}k_2) - 1 \geq \Phi(k_1) + \Phi(k_2) - 1\,,$$

and thus, the confidence coefficient of the interval $[\bar{Y} - k_1, \bar{Y} + k_2]$ is larger than

| $\alpha_1$ | $\alpha_2$ | $z_{(1-\alpha_1)}$ | $z_{(1-\alpha_2)}$ | $\sqrt{n}$ length |
|---|---|---|---|---|
| 0.001 | 0.049 | 3.090 | 1.655 | 4.745 |
| 0.010 | 0.040 | 2.326 | 1.751 | 4.077 |
| 0.020 | 0.030 | 2.054 | 1.881 | 3.935 |
| 0.025 | 0.025 | 1.960 | 1.960 | 3.920 |

Table 8.2 *The length of 95% confidence intervals for the mean of a normal distribution for various lower and upper end points.*

that for $[Y_1 - k_1, Y_1 + k_2]$. If we had to choose between these intervals we would use $[\bar{Y} - k_1, \bar{Y} + k_2]$. As we will see in section 10.1, this is consistent with the principle of sufficiency ($\bar{Y}$ is sufficient for $\mu$).

In the previous example we considered three intervals of equal length and compared their confidence coefficients. In practice, the usual approach is the reverse: we fix the desired level of confidence and try to find the smallest corresponding interval. The following example illustrates.

**Example 8.3.7** (Shortest confidence interval for normal mean)
Suppose that we have a random sample from an $N(\mu, 1)$ distribution, and we want an interval estimator for $\mu$ with confidence coefficient $(1 - \alpha)$. A good place to start is with the pivotal function $\sqrt{n}(\bar{Y} - \mu)$, which is standard normal. Let $z_p$ denote the $p$-quantile of a standard normal, that is, $\Phi(z_p) = p$. If we choose $\alpha_1, \alpha_2 \geq 0$ such that $\alpha = \alpha_1 + \alpha_2$, we have

$$P(z_{\alpha_1} \leq \sqrt{n}(\bar{Y} - \mu) \leq -z_{\alpha_2}) = 1 - \alpha_1 - \alpha_2 = 1 - \alpha\,.$$

In other words, we split the total tail probability, $\alpha$, into a left tail probability, $\alpha_1$, and a right tail probability, $\alpha_2$. By rearrangement, and using the fact that $z_{(1-\alpha_1)} = -z_{\alpha_1}$, we can see that

$$\left[\bar{Y} - \frac{1}{\sqrt{n}} z_{(1-\alpha_2)}, \bar{Y} + \frac{1}{\sqrt{n}} z_{(1-\alpha_1)}\right] \tag{8.6}$$

is an interval estimator for $\mu$ with confidence coefficient $(1 - \alpha)$. The length of this interval is

$$\frac{1}{\sqrt{n}}(z_{(1-\alpha_1)} + z_{(1-\alpha_2)}).$$

If either $\alpha_1$ or $\alpha_2$ is 0, the length of the interval is infinite. Suppose that $\alpha = 0.05$, that is, we want a 95% confidence interval for $\mu$. Table 8.2 compares the length of the confidence interval defined by equation (8.6) for various possible values of $\alpha_1$ and $\alpha_2$. It is clear that the shortest interval is given by taking $\alpha_1 = \alpha_2$. This illustrates a general result that coincides with our intuition; for a given confidence coefficient, the shortest confidence intervals for the population mean of a normal distribution will be symmetric about the sample mean.

### 8.3.2 *Constructing interval estimators using pivotal functions*

Pivotal functions provide a simple mechanism for generating interval estimators with a given confidence coefficient. Suppose that we want an interval estimator for $\theta$ with confidence coefficient $1-\alpha$. We could use the following procedure:

1. Find a pivotal function $g(\boldsymbol{Y},\theta)$.
2. Use the distribution of the pivotal function to find values $w_1$ and $w_2$ such that
$$P(w_1 \leq g(\boldsymbol{Y},\theta) \leq w_2) = 1-\alpha\,. \tag{8.7}$$
[Note that $w_1$ and $w_2$ will depend only on $\alpha$.]
3. Manipulate the inequalities $g(\boldsymbol{Y},\theta) \geq w_1$ and $g(\boldsymbol{Y},\theta) \leq w_2$ to make $\theta$ the subject. This yields inequalities of the form
$$\theta \geq h_1(\boldsymbol{Y},w_1,w_2) \text{ and } \theta \leq h_2(\boldsymbol{Y},w_1,w_2),$$
for some functions $h_1$ and $h_2$.
4. We can now give
$$[h_1(\boldsymbol{Y},w_1,w_2), h_2(\boldsymbol{Y},w_1,w_2)]$$
as an interval estimator for $\theta$ with confidence coefficient $(1-\alpha)$. [In practice, the end-points of the interval are usually a function of one of $w_1$ or $w_2$ but not the other.]

If the pivotal happens to be a linear function of the unknown parameter, we can establish the optimal choice of $w_1$ and $w_2$ for a quite a wide class of distributions. We begin with a definition.

**Definition 8.3.8** (Unimodality)
Let $X$ be a random variable with mass/density function $f_X(x)$ and support $D$. We say that $X$ is (strictly) **unimodal** if there exists a value $x^*$ such that $f_X(x)$ is (strictly) increasing for $x < x^*$ and (strictly) decreasing for $x > x^*$, where $x \in D$. The value $x^*$ is the **mode** of the distribution.

We have encountered several of these distributions so far: $N(\mu,\sigma^2)$, $\text{Gamma}(\alpha,\lambda)$ for $\alpha > 1$, $t_k$, and $\chi^2_k$ for $k > 2$ are all strictly unimodal; $\text{Pois}(\lambda)$ is strictly unimodal if $\lambda \notin \mathbb{Z}^+$; and $\text{Bin}(n,p)$ is strictly unimodal if $0 < p < 1$ and $(n+1)p \notin \mathbb{Z}^+$.

**Example 8.3.9** (Shortest confidence interval from a linear pivotal)
Suppose that $W = g(\boldsymbol{Y},\theta)$ is a linear function of $\theta$, and $W$ is continuous and strictly unimodal, with mode $w^*$. The procedure for constructing a confidence interval from the pivotal function involves finding values $w_1$ and $w_2$ that satisfy (8.7). As $g(\boldsymbol{Y},\theta)$ is linear in $\theta$, we can write
$$g(\boldsymbol{Y},\theta) = a(\boldsymbol{Y}) + b(\boldsymbol{Y})\theta\,,$$
where $a(\boldsymbol{Y})$ and $b(\boldsymbol{Y})$ may depend on the sample, but not on $\theta$. Solving the inequalities

$W \geq w_1$ and $W \leq w_2$ for $\theta$, we obtain a confidence interval for $\theta$ of length $(w_2 - w_1)/|b(\boldsymbol{Y})|$. It is clearly in our interest to minimise the difference $(w_2 - w_1)$.

Now suppose we can find values $w_1$, $w_2$ that satisfy $w_1 < w^* < w_2$ and $f_W(w_1) = f_W(w_2)$. Notice that, by the symmetry of the normal distribution, the quantiles we use in Example 8.3.2 satisfy these conditions. We can readily show that we can do no better. Let $k = w_2 - w_1$ and define the function

$$p_k(w) = \mathrm{P}(w \leq W \leq w + k),$$

the probability than an interval of length $k$ with lower endpoint $w$ covers $W$. Notice that $p_k(w_1) = \mathrm{P}(w_1 \leq W \leq w_2) = 1 - \alpha$. The derivative of this function is

$$p_k'(w) = \frac{d}{dw} \int_w^{w+k} f_W(w)dw = f_W(w + k) - f_W(w),$$

from the fundamental theorem of calculus. There is a stationary point at $w = w_1$, as $p_k'(w_1) = f_W(w_2) - f_W(w_1) = 0$. The second derivative at that point is

$$p_k''(w_1) = f_W'(w_2) - f_W'(w_1).$$

The unimodality of $f_W$ implies that $f_W'(w_2) < 0$ and $f_W'(w_1) > 0$, so $p_k(w)$ has a maximum at $w = w_1$, which means that any interval of length $k$ other than $(w_1, w_2)$ will cover $W$ with probability less than $1-\alpha$. From this we can deduce that no interval of length less than $k$ can cover $W$ with the required probability.

If $f_W$ is also symmetric (as in the normal case), the resulting estimator is an equal-tail interval, in the sense that $\mathrm{P}(W < w_1) = \mathrm{P}(W > w_2) = \alpha/2$.

**Example 8.3.10** (Interval estimator for mean of normal – variance unknown)
Suppose that we have a random sample $\boldsymbol{Y} = (Y_1, \dots, Y_n)^T$ from an $\mathrm{N}(\mu, \sigma^2)$ distribution, but the variance $\sigma^2$ is unknown. We can use the pivotal function from Example 8.1.4, that is,

$$\frac{\bar{Y} - \mu}{\sqrt{S^2/n}} \sim t_{n-1}\,.$$

This pivotal function is linear in $\mu$, and the $t$-distribution is strictly unimodal and symmetric. It follows that we should choose

$$\mathrm{P}\left( t_{n-1,\alpha/2} \leq \frac{\bar{Y} - \mu}{\sqrt{S^2/n}} \leq t_{n-1,1-\alpha/2} \right) = 1 - \alpha\,, \tag{8.8}$$

where $t_{n-1,p}$ denotes the $p$-quantile of a $t_{n-1}$ distribution. By symmetry of the $t$-distribution we have

$$t_{n-1,\alpha/2} = -t_{n-1,1-\alpha/2}\,.$$

We now rearrange expression (8.8) to make $\mu$ the subject; this yields the interval estimator

$$\left[ \bar{Y} - t_{n-1,1-\alpha/2}\frac{S}{\sqrt{n}}, \bar{Y} + t_{n-1,1-\alpha/2}\frac{S}{\sqrt{n}} \right].$$

The corresponding interval estimate is

$$\left[\bar{y} - t_{n-1,1-\alpha/2}\frac{s}{\sqrt{n}}, \bar{y} + t_{n-1,1-\alpha/2}\frac{s}{\sqrt{n}}\right], \tag{8.9}$$

where $\bar{y}$ is the observed sample mean, and $s$ the observed sample standard deviation. As in Example 8.3.2, the confidence interval is symmetric about the sample mean.

The $t$-distribution has fatter tails than the standard normal so, for a 95% confidence interval, the value $t_{n-1,0.975}$ will be larger than 1.96, the corresponding quantile of $N(0, 1)$. In other words, all else being equal, the confidence interval (8.9) will be longer than the corresponding interval (8.5). This is a result of the additional uncertainty introduced by estimating $\sigma^2$. As the sample size increases, we know that $t_{n-1}$ converges to $N(0, 1)$, so $t_{n-1,0.975} \to 1.96$ and the two estimates are indistinguishable from each other. This is unsurprising; $S^2$ is a consistent estimator of $\sigma^2$, so it will converge to the true value as the sample size increases.

**Example 8.3.11**

Suppose that $\boldsymbol{Y}$ is a random sample from an $\text{Exp}(\lambda)$ distribution. In Example 8.1.5 we establish that $W = \lambda \sum_{i=1}^{n} Y_i$ is a pivotal function of $\lambda$, with moment-generating function

$$M_W(t) = (1 - t)^{-n}.$$

In order to use $W$ to construct interval estimators, we need the parametric form of the distribution of $W$. Recall from Exercise 7.3 that, if $V \sim \chi^2_k$, then

$$M_V(t) = (1 - 2t)^{-k/2}.$$

Comparing moment-generating functions, we can see that

$$2W \sim \chi^2_{2n}.$$

This pivotal function is linear in $\lambda$ and, if $n > 1$, its distribution is strictly unimodal. From Example 8.3.9, constructing the optimal confidence interval involves finding values $w_1$ and $w_2$ such that

$$P\left(w_1 \leq 2\lambda \sum_{i=1}^{n} Y_i \leq w_2\right) = 1 - \alpha \quad \text{and} \quad f_{2W}(w_1) = f_{2W}(w_2),$$

where $f_{2W}$ is the density of a chi-squared distribution on $2n$ degrees of freedom. There is no closed-form solution to these equations, so we usually resort to constructing an equal-tail interval instead (Figure 8.5).

Exploiting the parametric form of $2W$, we have

$$P\left(\chi^2_{2n,\alpha/2} \leq 2\lambda \sum_{i=1}^{n} Y_i \leq \chi^2_{2n,(1-\alpha/2)}\right) = 1 - \alpha,$$

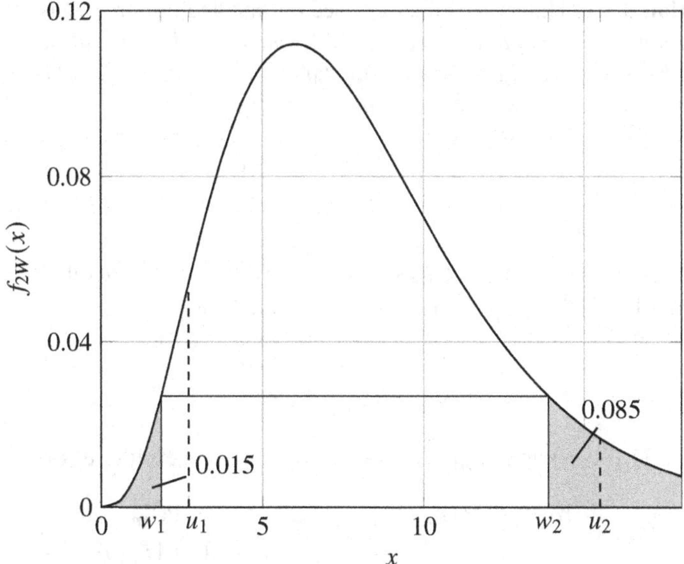


Figure 8.5 *Optimal and equal-tail confidence intervals for the rate of an Exponential, with* $n = 4$ *and* $\alpha = 0.1$. *The distribution of the pivotal function is* $2W \sim \chi^2_8$. *The values* $(u_1, u_2) = (2.73, 15.51)$ *give equal tail probabilities of 0.05, whereas* $(w_1, w_2) = (1.87, 13.89)$ *correspond to the optimal confidence interval with* $f_{2W}(w_1) = f_{2W}(w_2)$. *Notice that the two intervals are quite different when the sample size is low.*

where $\chi^2_{k,p}$ is the $p$-quantile of a chi-squared distribution on $k$ degrees of freedom. Rearranging the inequalities, we conclude that

$$\left[\frac{\chi^2_{2n,\alpha/2}}{2\sum_{i=1}^n Y_i}, \frac{\chi^2_{2n,1-\alpha/2}}{2\sum_{i=1}^n Y_i}\right],$$

is an interval estimator for $\lambda$ with confidence coefficient $(1 - \alpha)$.

### *8.3.3 Constructing interval estimators using order statistics*

In the examples we have considered so far, we have given confidence intervals for the mean, variance, and other specific parameters from well-known distributions. Another useful class of population parameters are the quantiles. In section 7.4 we show that order statistics provide a natural mechanism for inference about population quantiles. It is unsurprising then that interval estimators for quantiles are also based on order statistics.

We start by establishing the coverage probability associated with an interval based on the first and last order statistics as an interval estimator for a scalar parameter $\theta$.

**Proposition 8.3.12** (Interval estimator based on first and last order statistics)
*Consider a random sample $\boldsymbol{Y} = (Y_1, \dots, Y_n)^T$ from a distribution with scalar parameter $\theta$. Let $F_Y$ be the cumulative distribution function of $Y$ and $Y_{(i)}$ the $i^{th}$ order statistic. The probability that the random interval $[Y_{(1)}, Y_{(n)})$ covers $\theta$ is*

$$1 - [1 - F_Y(\theta)]^n - [F_Y(\theta)]^n.$$

*Proof.*

Using the properties of order statistics, we know that if $\{Y_{(1)} > \theta\}$ then $\{Y_i > \theta\}$ for all $i$. Similarly, if $\{Y_{(n)} \leq \theta\}$ then $\{Y_i \leq \theta\}$ for all $i$. Thus,

$$\begin{aligned} \mathrm{P}(Y_{(1)} > \theta) &= [1 - F_Y(\theta)]^n, \\ \mathrm{P}(Y_{(n)} \leq \theta) &= [F_Y(\theta)]^n. \end{aligned}$$

The events $\{Y_{(1)} > \theta\}$ and $\{Y_{(n)} \leq \theta\}$ are clearly mutually exclusive, so

$$\begin{aligned} \mathrm{P}((Y_{(1)} > \theta) \cup (Y_{(n)} \leq \theta)) &= \mathrm{P}(Y_{(1)} > \theta) + \mathrm{P}(Y_{(n)} \leq \theta) \\ &= [1 - F_Y(\theta)]^n + [F_Y(\theta)]^n. \end{aligned}$$

Using de Morgan's laws, the result follows,

$$\begin{aligned} \mathrm{P}((Y_{(1)} \leq \theta) \cap (Y_{(n)} > \theta)) &= 1 - \mathrm{P}((Y_{(1)} > \theta) \cup (Y_{(n)} \leq \theta)) \\ &= 1 - [1 - F_Y(\theta)]^n - [F_Y(\theta)]^n. \end{aligned}$$

□

Suppose that the parameter of interest is the $\beta$-quantile, $\theta = q_\beta$, where $q_\beta$ is the point satisfying $F_Y(q_\beta) = \beta$. The coverage probability given by Proposition 8.3.12 is then a confidence coefficient. The following corollary gives details.

**Corollary 8.3.13** ($[Y_{(1)}, Y_{(n)})$ as interval estimator for a quantile)
*Consider a random sample $\boldsymbol{Y} = (Y_1, \dots, Y_n)^T$. Let $q_\beta$ be the $\beta$-quantile of $Y$. The random interval $[Y_{(1)}, Y_{(n)})$ is an interval estimator for $q_\beta$ with confidence coefficient*

$$1 - (1 - \beta)^n - \beta^n.$$

In practice, we would like to be able to provide an interval estimator for $q_\beta$ for any given confidence coefficient. We gain greater flexibility by considering the interval estimator associated with general order statistics $Y_{(i)}$ and $Y_{(j)}$, where $i < j$. The following proposition indicates how coverage probability is calculated.

**Proposition 8.3.14** ($[Y_{(i)}, Y_{(j)})$ as an interval estimator)
*Consider a random sample $\boldsymbol{Y} = (Y_1, \dots, Y_n)^T$ from a distribution with scalar parameter $\theta$. Let $F_{Y_{(i)}}$ be the cumulative distribution function of the $i^{th}$ order statistic. If $i < j$, the probability that the random interval $[Y_{(i)}, Y_{(j)})$ covers $\theta$ is*

$$F_{Y_{(i)}}(\theta) - F_{Y_{(j)}}(\theta).$$

The proof follows the same reasoning as Proposition 8.3.12; this is part of Exercise 8.3. The values from the cumulative distribution function of order statistics can be evaluated using Proposition 7.4.5, for example,

$$F_{Y_{(i)}}(\theta) = \sum_{k=i}^{n} \binom{n}{k} [F_Y(\theta)]^k [1 - F_Y(\theta)]^{n-k}.$$

This calculation involves a sum of binomial probabilities, and can be more conveniently expressed in terms of a binomial distribution function. If we define $X \sim \text{Bin}(n, F_Y(\theta))$, then

$$F_{Y_{(i)}}(\theta) = \mathrm{P}(Y_{(i)} \leq \theta) = \mathrm{P}(X \geq i) = 1 - F_X(i-1)\,.$$

Substituting this expression into the result from Proposition 8.3.14, we conclude that the probability that the random interval $[Y_{(i)}, Y_{(j)})$ covers $\theta$ is

$$F_X(j-1) - F_X(i-1)\,,$$

where $X \sim \text{Bin}(n, F_Y(\theta))$. Once again, this result is most usefully applied to quantiles; if $\theta = q_\beta$, then $X \sim \text{Bin}(n, \beta)$.

We illustrate these ideas by constructing interval estimators for the median.

**Example 8.3.15** (Interval estimators for the median)
Consider a random sample $\boldsymbol{Y} = (Y_1, \ldots, Y_n)^T$ from any distribution. The median is the 0.5-quantile, $q_{0.5}$. By Corollary 8.3.13, the interval with end-points given by the sample minimum and maximum is an interval estimator for the median with confidence coefficient

$$1 - \left(\frac{1}{2}\right)^n - \left(\frac{1}{2}\right)^n = 1 - \left(\frac{1}{2}\right)^{n-1} = \frac{2^{n-1}-1}{2^{n-1}}\,.$$

For a sample of size 4, this confidence coefficient is $15/16 = 0.9375$. Clearly, in a reasonable size sample, the probability that $[Y_{(1)}, Y_{(n)}]$ covers the median will be very close to 1.

Suppose that we want an interval estimator for the median with confidence coefficient $1 - \alpha$. From Proposition 8.3.14 and subsequent discussion, we want to find $i$ and $j$ such that

$$F_X(j-1) - F_X(i-1) = 1 - \alpha\,,$$

where $X \sim \text{Bin}(n, 0.5)$. For example, if $n = 100$ then

$$\begin{aligned} \mathrm{P}(X \leq 39) &= 0.0176\,,\\ \mathrm{P}(X \leq 59) &= 0.9716\,, \end{aligned}$$

so $[Y_{(40)}, Y_{(60)})$ provides a confidence interval for the median with coverage coefficient 0.954.

### *8.3.4 Confidence sets*

In the bulk of our examples we will construct confidence intervals. However, it is important to understand that a confidence interval can be viewed as a special case of a **confidence set**. Confidence sets are useful in two contexts:

i. if we are unsure that the result of a procedure is an interval,
ii. if we have a vector of parameters, in which case we may refer to our confidence set as a **confidence region**.

A confidence set with confidence coefficient $1-\alpha$ for a vector parameter $\boldsymbol{\theta} \in \boldsymbol{\Theta}$ is defined as the random set $C(\boldsymbol{Y}) \subseteq \boldsymbol{\Theta}$ where

$$\mathrm{P}(\boldsymbol{\theta} \in C(\boldsymbol{Y})) = 1 - \alpha \,.$$

Note that in the above expression the order in which the variables appear can cause confusion; $C(\boldsymbol{Y})$ is the random variable here (it is a function of the sample). In this instance, for an observed sample $\boldsymbol{y}$ we would have the observed confidence set $C(\boldsymbol{y})$.

**Exercise 8.3**

1. The confidence level for a confidence interval is sometimes described as the probability that the true parameter value lies in the interval. Why is this slightly misleading?
2. Using a computer package or statistical tables, check the entries in Table 8.2.
3. Suppose $\boldsymbol{Y} = (Y_1, \dots, Y_n)^T$ is a random sample from an $\mathrm{N}(\mu, \sigma^2)$ distribution. Find a pivotal function for $\sigma^2$ in the case where $\mu$ is known, and the case where $\mu$ is unknown. In each case derive an interval estimator for $\sigma^2$ with confidence coefficient $1-\alpha$ and give an expression for the expected length of the interval.
4. Suppose that $\boldsymbol{Y} = (Y_1, \dots, Y_n)^T$ is a random sample from a Bernoulli($p$) distribution. Using the normal approximation to the binomial, find a function that is approximately pivotal for $p$. Hence, construct an interval estimator for $p$ with approximate confidence coefficient 0.95.
5. Prove Proposition 8.3.14.
6. Suppose that $\boldsymbol{Y} = (Y_1, \dots, Y_n)^T$ is a random sample from any distribution. Find interval estimators with confidence coefficients of approximately 0.95 for the upper quartile, $q_{0.75}$, when $n = 100$.

## 8.4 Hypothesis testing

A hypothesis test starts with a statement about the population; this statement is referred to as the **null hypothesis**. A hypothesis test determines whether the sample provides evidence to reject the null hypothesis. The basis for making this inference is a comparison of the observed value of a **test statistic** with the sampling distribution of the statistic when we assume the null hypothesis is true. A perfect hypothesis test would always reject a false null hypothesis and never reject a true one. In practice we have to strike a balance between the probability of rejecting a false hypothesis and the probability of rejecting a true one.

The formal setup for hypothesis testing has been extensively criticised and is somewhat out of fashion in modern statistics. However, the ideas presented here are still widely used even if the precise formality is not made explicit. News sites are full of reports on food scares, exam performance, causes of disease, and so on. These reports contain statements of the form, "there is no evidence of a risk to health", or "the change in performance is statistically significant", or "factor $x$ is associated with disease $y$". These statements are rooted in ideas of hypothesis testing and draw loosely on its terminology. Even if we have doubts about the value of formal testing methodologies, it is of fundamental importance that these methodologies and their limitations are properly understood. One important role for professional statisticians is to warn against naive interpretation (and identify gross misinterpretation) of hypothesis tests.

### *8.4.1 Statistical hypotheses*

The hypothesis tests we consider in this section are **parametric**, in the sense that the hypotheses are statements about the value of a distribution parameter; the form of the distribution itself is assumed to be known. By contrast, **nonparametric** tests do not require any assumptions about the form of the distribution. We have encountered some nonparametric techniques earlier in this book, such as the quantile-based confidence intervals in section 8.3.3.

In general, a parametric hypothesis test takes the form

$$H_* \;:\; \boldsymbol{\theta} \in \boldsymbol{\Theta}_* \text{ where } \boldsymbol{\Theta}_* \subset \boldsymbol{\Theta}. \tag{8.10}$$

$H_*$ is just a convenient label for our hypothesis. In saying $\boldsymbol{\theta} \in \boldsymbol{\Theta}_*$, where $\boldsymbol{\Theta}_* \subset \boldsymbol{\Theta}$, we are proposing that the true parameter value, $\boldsymbol{\theta}$, lies in a specific subset, $\boldsymbol{\Theta}_*$, of the parameter space, $\boldsymbol{\Theta}$. For scalar parameter $\theta$ and known constant $k$, forms of hypothesis that we might consider include $\theta = k$, $\theta < k$, $\theta \leq k$, $\theta > k$, and $\theta \geq k$. Note that, consistent with (8.10), $\theta = k$ could be written $\theta \in \{k\}$, $\theta < k$ could be written $\theta \in (-\infty, k)$, $\theta \leq k$ could be written $\theta \in (-\infty, k]$, and so on. In this brief illustration we have actually described two distinct forms of hypothesis. The following definition clarifies.

**Definition 8.4.1** (Simple and composite hypotheses)
A **simple hypothesis** gives an exact specification of the parameter value $\boldsymbol{\theta}$. Thus, a simple hypothesis will take the form $\boldsymbol{\theta} = \boldsymbol{k}$ for some known constant vector $\boldsymbol{k}$ of the same dimension as $\boldsymbol{\theta}$. Any hypothesis that is not simple is referred to as a **composite hypothesis**.

In the scalar parameter case it is clear that $\theta = k$ is a simple hypothesis. The other four examples do not give a specific value for $\theta$ so $\theta < k$, $\theta \leq k$, $\theta > k$, and $\theta \geq k$

are all composite hypotheses. In general, a simple hypothesis will propose that $\boldsymbol{\theta}$ is a member of a set with a single element, $\boldsymbol{\theta} \in \{\boldsymbol{k}\}$. A composite hypothesis is anything that proposes that $\boldsymbol{\theta}$ is a member of a set with more than one element, that is, $\boldsymbol{\theta} \in \boldsymbol{\Theta}_*$ where $|\boldsymbol{\Theta}_*| > 1$. The cardinality of $\boldsymbol{\Theta}_*$ may be finite, as in $\boldsymbol{\theta} \in \{\boldsymbol{k}_1, \boldsymbol{k}_2\}$ or $\boldsymbol{\theta} \in \{\boldsymbol{k}_1, \ldots, \boldsymbol{k}_n\}$; it may be countably infinite, as in $\boldsymbol{\theta} \in \{\boldsymbol{k}_1, \boldsymbol{k}_2, \ldots\}$; or it may be uncountably infinite, as in $\theta \in (k_1, k_2)$, $\theta \in (-\infty, k]$, or $\boldsymbol{\theta} \in \{\boldsymbol{x} : |\boldsymbol{x}| \leq k\}$.

In classical hypothesis testing, two hypotheses are put forward. The **null hypothesis** is the hypothesis that we would like to test; the **alternative hypothesis** informs the manner in which we construct the test of the null.

1. The null hypothesis is usually conservative, in the sense that it reflects established thinking. The null often takes the form "no change", "no difference", or "no effect". We will adopt the convention that the null hypothesis is labelled $H_0$.
2. The alternative reflects our suspicion about values that the parameter might take. The alternative is usually composite and often takes a form that can be interpreted as "there is a change", "there is a difference", or "there is an effect". We will use $H_1$ to label the alternative hypothesis.

We can now give a formal statement of a general hypothesis test. For a parameter $\theta$, a hypothesis test takes the form

$$\begin{aligned} H_0 &: \boldsymbol{\theta} \in \boldsymbol{\Theta}_0 , \\ H_1 &: \boldsymbol{\theta} \in \boldsymbol{\Theta}_1 . \end{aligned}$$

where $\boldsymbol{\Theta}_0 \subset \boldsymbol{\Theta}$, $\boldsymbol{\Theta}_1 \subset \boldsymbol{\Theta}$, and $\boldsymbol{\Theta}_0 \cap \boldsymbol{\Theta}_1 = \varnothing$. Note that we insist that $\boldsymbol{\Theta}_0$ and $\boldsymbol{\Theta}_1$ are mutually exclusive, so there is no way that $H_0$ and $H_1$ can both be true. However, there is no requirement for these sets to be exhaustive, that is, we do not require that $\boldsymbol{\Theta}_0 \cup \boldsymbol{\Theta}_1 = \boldsymbol{\Theta}$.

### *8.4.2 Decision rules*

Classical hypothesis testing involves a binary choice. The options available to us are:

a. reject the null hypothesis,
b. do not reject the null hypothesis.

Teachers of statistics are fond of pointing out that the statement *"do not reject the null hypothesis"* should not be treated as equivalent to *"accept the null"* (students are equally fond of ignoring this advice). This is a somewhat pedantic point, however it is important to realise that all a statistical procedure can do is check whether the available evidence is consistent with the null hypothesis. On the same theme, *"reject the null"* should not be interpreted as *"accept the alternative"*.

The evidence we use to make the decision whether or not to reject the null is the observed sample $\boldsymbol{y} = (y_1, \ldots, y_n)^T$. The sample is used to construct a **test statistic**, and the decision about $H_0$ is made on the basis of this statistic. For now, we will summarise the decision process in a single function, $\phi$, where

$$\phi(\mathbf{y}) = \begin{cases} 1 & \text{when } H_0 \text{ rejected,} \\ 0 & \text{when } H_0 \text{ not rejected.} \end{cases}$$

The function $\phi$ is referred to as a **decision rule**. The effect of $\phi$ is to partition the space of possible observed sample values into two sets: $R = \{\mathbf{y} : \phi(\mathbf{y}) = 1\}$ and $R^c = \{\mathbf{y} : \phi(\mathbf{y}) = 0\}$. The set $R$, which corresponds to observed sample values for which $H_0$ will be rejected, is often referred to as the **rejection region** or **critical region**. For a given test, the decision function and critical region are equivalent, in the sense that if either one is known then the other is completely determined. If the critical region $R$ is specified, we can write down the decision function

$$\phi(\mathbf{y}) = \begin{cases} 1 & \text{if } \mathbf{y} \in R, \\ 0 & \text{if } \mathbf{y} \notin R. \end{cases}$$

Similarly, for a specific decision function $\phi$, the critical region is given by

$$R = \{\mathbf{y} \in \mathbb{R}^n : \phi(\mathbf{y}) = 1\}.$$

### 8.4.3 *Types of error and the power function*

In any non-trivial decision problem, there is a chance that we make the wrong decision. In hypothesis testing two wrong decisions are possible. These are usually stated as:

a. Type I error : rejecting $H_0$ when $H_0$ is true,

b. Type II error : failing to reject $H_0$ when $H_0$ is false.

One possible mechanism for assessing the performance of a test is to consider the probability that the test makes an error. The probability of a type I error is associated with the **significance level** and the **size** of the test. These terms are often used interchangeably, although the meanings are distinct, as the following definition makes clear.

**Definition 8.4.2** (Significance level and size)
Consider testing the null hypothesis $H_0 : \boldsymbol{\theta} \in \boldsymbol{\Theta}_0$. The test has significance level $\alpha$ if

$$\sup_{\boldsymbol{\theta} \in \boldsymbol{\Theta}_0} P_{\boldsymbol{\theta}}(H_0 \text{ rejected}) \leq \alpha.$$

The test has size $\alpha$ if

$$\sup_{\boldsymbol{\theta} \in \boldsymbol{\Theta}_0} P_{\boldsymbol{\theta}}(H_0 \text{ rejected}) = \alpha.$$

The distinction is rather subtle and will not be important for the initial examples that we consider, as size can readily be calculated. However, in some more complex situations, size is not readily computed and we have to settle for significance level. Some notes on size and significance level follow.

1. A test of size $\alpha$ has significance level $\alpha$. The converse is not true.
2. A test with significance level $\alpha$ will always be at least as conservative as a test of size $\alpha$.
3. If the null hypothesis is simple (as is often the case) then the size of a test is the probability of a type I error.

Point 3 above relates to the fact that we can only specify a precise probability for type I errors if the null hypothesis is simple. Similarly, we can only specify a precise probability for type II errors if the alternative is simple. The fact that the alternative hypothesis is usually composite motivates the definition of the **power function.**

**Definition 8.4.3** (Power function)
For $\boldsymbol{\theta} \in \boldsymbol{\Theta}$ the power function is defined as

$$\beta(\boldsymbol{\theta}) = P_{\boldsymbol{\theta}}(H_0 \text{ rejected}),$$

that is, the probability that the null hypothesis is rejected if the true parameter value is $\boldsymbol{\theta}$.

The **power** of the test under a specific alternative, $\boldsymbol{\theta}_1 \in \boldsymbol{\Theta}_1$, is defined as $\beta(\boldsymbol{\theta}_1)$, that is, the value of the power function at the specific value, $\boldsymbol{\theta}_1$. The relationship with type II error is then clear, as

$$P_{\boldsymbol{\theta}_1}(\text{type II error}) = P_{\boldsymbol{\theta}_1}(H_0 \text{ not rejected}) = 1 - \beta(\boldsymbol{\theta}_1).$$

For specific parameter values under the null, the power function gives the probability of a type I error. If our test is of size $\alpha$, then it is clear from the definition of size that $\beta(\boldsymbol{\theta}) \leq \alpha$ for $\boldsymbol{\theta} \in \boldsymbol{\Theta}_0$. In fact, significance level and size can be defined in terms of the power function; if a test has significance level $\alpha$, then

$$\sup_{\boldsymbol{\theta} \in \boldsymbol{\Theta}_0} \beta(\boldsymbol{\theta}) \leq \alpha,$$

and, if the test has size $\alpha$, then

$$\sup_{\boldsymbol{\theta} \in \boldsymbol{\Theta}_0} \beta(\boldsymbol{\theta}) = \alpha.$$

### 8.4.4 *Basic ideas in constructing tests*

An intuitive description of the perfect hypothesis test is easy to think up; this test would never reject a true null hypothesis but would always reject a false one. Put another way, we would want to reject a true $H_0$ with probability 0 and reject a false $H_0$ with probability 1. The power function of this test would be

$$\beta(\boldsymbol{\theta}) = \begin{cases} 0 & \text{if } \boldsymbol{\theta} \in \boldsymbol{\Theta}_0, \\ 1 & \text{if } \boldsymbol{\theta} \notin \boldsymbol{\Theta}_0, \end{cases}$$

This perfect test has size 0 and power 1. It is clear that this ideal is impossible in any situation where there is uncertainty about the values of the parameters. In practice, there is a tradeoff between size and power.

The conventional approach to hypothesis testing is to control the size of the test. This is equivalent to controlling the probability of rejecting a true null hypothesis. If the consequences of rejecting a true null hypothesis are very severe, then we might fix the size of the test to be very small. If we are more relaxed about rejecting a true null then we might choose a larger value for the size. Often, size is fixed at 0.05, a situation referred to as "testing at a 5% level". Tests of size 0.1 and 0.01 are also frequently used. For a test of a given size we would like to maximise the power, that is, to maximise the probability of rejecting the null hypothesis when it is false.

### *8.4.5 Conclusions and p-values from tests*

In principle, there are two possible conclusions of a hypothesis test: we either reject the null hypothesis or we do not reject it. In practice, some information about the weight of evidence against the null hypothesis is desirable. This may take the form of the observed sample value of the test statistic. Another useful way in which the information can be presented is as a ***p*-value**.

**Definition 8.4.4** ($p$-value)
Consider a hypothesis test for a scalar parameter $\theta$ with test statistic $U = h(\boldsymbol{Y})$. Suppose, without loss of generality, that the null is rejected when $U$ takes large values. For an observed sample $\boldsymbol{y}$, the corresponding $p$-value is

$$p(\boldsymbol{y}) = \sup_{\theta \in \Theta_0} \mathrm{P}(U \geq h(\boldsymbol{y})) .$$

Clearly, $0 \leq p(\boldsymbol{y}) \leq 1$.

For a simple null hypothesis, we can remove the $\sup_{\theta \in \Theta_0}$ from the definition and say that the $p$-value is the probability that we got the result that we did from the sample, or a more extreme result. If we have a fixed significance level $\alpha$, then we can describe the critical region for a test as

$$R = \{\boldsymbol{y} : p(\boldsymbol{y}) \leq \alpha\} .$$

Put loosely, if there is only a small probability of observing a result at least as extreme as that observed in the sample, then we reject the null hypothesis.

**Example 8.4.5** (Testing the mean of a normal when variance known)
Suppose that we have a random sample $\boldsymbol{Y} = (Y_1, \ldots, Y_n)^T$ from an $\mathrm{N}(\mu, \sigma^2)$ distribution where $\sigma^2$ is known. We would like to test

$$H_0 : \mu = \mu_0,$$
$$H_1 : \mu < \mu_0.$$

Following the basic framework we have established, we can construct a test of size $\alpha$ with critical region

$$R = \left\{ \boldsymbol{y} : \bar{y} < \mu_0 - z_{(1-\alpha)} \frac{\sigma}{\sqrt{n}} \right\}.$$

Notice that, since $z_\alpha = -z_{(1-\alpha)}$, we can write this critical region as

$$R = \left\{ \boldsymbol{y} : \bar{y} < \mu_0 + z_\alpha \frac{\sigma}{\sqrt{n}} \right\}.$$

The power function for the test is given by

$$\begin{aligned} \beta(\mu) &= P_\mu(\boldsymbol{Y} \in R) \\ &= \mathrm{P}\left( \frac{\bar{Y} - \mu_0}{\sqrt{\sigma^2/n}} < z_\alpha \right) \\ &= \mathrm{P}\left( \frac{\bar{Y} - \mu}{\sqrt{\sigma^2/n}} < z_\alpha - \frac{\mu - \mu_0}{\sqrt{\sigma^2/n}} \right) \\ &= \Phi\left( z_\alpha - \frac{\mu - \mu_0}{\sqrt{\sigma^2/n}} \right), \end{aligned}$$

where $\Phi$ is the cumulative distribution function of a standard normal. As we might expect, the power is a strictly decreasing function of $\mu$ that takes the value $\alpha$ when $\mu = \mu_0$ (Figure 8.6).

The $p$-value associated with the test is

$$p(\boldsymbol{y}) = \Phi\left( \frac{\bar{y} - \mu_0}{\sqrt{\sigma^2/n}} \right).$$

Again, as we might expect, small $p$-values arise when $\bar{y}$ is much smaller than $\mu_0$ (Figure 8.7).

**Example 8.4.6** (Testing the mean of a normal when variance unknown)
Suppose that we have a random sample $\boldsymbol{Y} = (Y_1, \ldots, Y_n)^T$ from an $\mathrm{N}(\mu, \sigma^2)$ distribution where both $\mu$ and $\sigma^2$ are unknown. We would like to test

$$\begin{aligned} H_0 &: \mu = \mu_0, \\ H_1 &: \mu > \mu_0. \end{aligned}$$

In order to construct a test we will exploit the pivotal function derived in Example 8.1.4; under $H_0$,

$$\frac{\bar{Y} - \mu_0}{\sqrt{S^2/n}} \sim t_{n-1}.$$

The critical region for a test of size $\alpha$ is given by

$$R = \left\{ \boldsymbol{y} : \bar{y} > \mu_0 + t_{n-1,1-\alpha} \frac{s}{\sqrt{n}} \right\}.$$

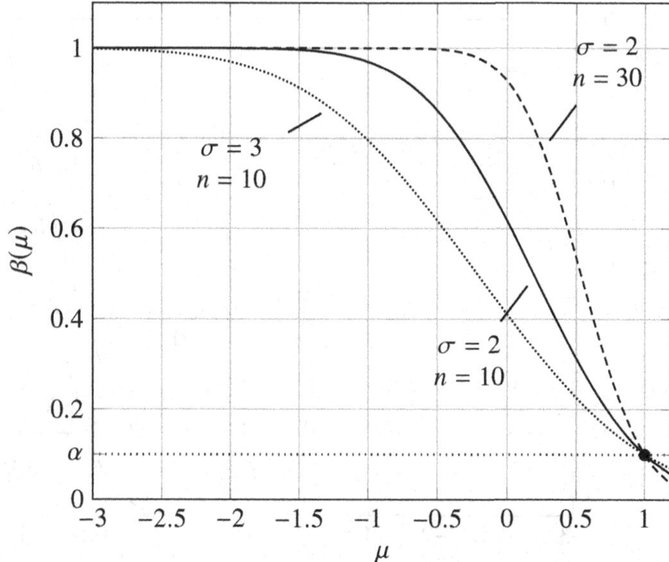


Figure 8.6 *Example of this power function with* $\mu_0 = 1$ *and* $\alpha = 0.1$, *for various values of n and* $\sigma$. *Notice that the curve is flatter (lower power) for higher* $\sigma$ *and steeper (higher power) for larger n.*

**Exercise 8.4**

1. Someone tells you that the $p$-value is the probability that the null hypothesis is true. How would you explain why this view is mistaken?
2. Consider the pivotal functions of section 8.1.2. How can these be used in hypothesis testing?
3. In the hypothesis test from Example 8.4.5, for a given value of $\mu$ (and assuming $\mu_0$ is fixed and unknown), how could you increase $\beta(\mu)$, the power of the test?

## 8.5 Prediction

So far, we have focused on inference for unknown parameters. In many problems in statistics our real interest lies in generating sensible estimates for potential sample members that we have not observed; these may include missing values and future values. We will start by supposing that we have a random sample, $\boldsymbol{X} = (X_1, \dots, X_n)^T$, and we are interested in generating an estimator of a random variable $Y$ based on this sample. In other words, we want to find a statistic $h(\boldsymbol{X})$ that provides us with a reasonable estimator for $Y$. Clearly, there must be some association between $\boldsymbol{X}$ and $Y$ for this problem to be interesting. An obvious question arises: how do we judge what constitutes a good estimator? We can borrow an idea from point estimation and use mean squared error,

$$\underset{Y}{\text{MSE}}(h(\boldsymbol{X})) = \mathbb{E}\left[(Y - h(\boldsymbol{X}))^2\right],$$

where the expectation is taken with respect to the joint distribution of $\boldsymbol{X}$ and $Y$.

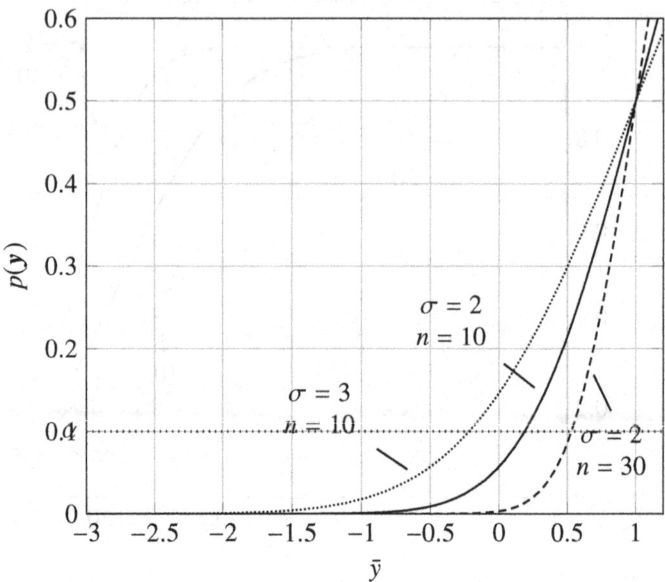


Figure 8.7 *The p-value of this test with* $\mu_0 = 1$ *and* $\alpha = 0.1$. *The critical region is where* $p(\boldsymbol{y}) < \alpha$, *so we require lower (more extreme) values of* $\bar{y}$ *to reject* $H_0$ *when* $\sigma$ *is higher or* $n$ *is smaller.*

An intuitively reasonable choice of estimator is the conditional expectation $\mathbb{E}(Y|\boldsymbol{X})$. In fact, this turns out to be optimal in the mean squared error context. Before we prove optimality, we establish an important property of conditional expectation that has a geometric interpretation.

**Lemma 8.5.1** (Conditional expectation as projection)
*For any reasonably well-behaved function* $g : \mathbb{R}^n \to \mathbb{R}$, *the random variable* $g(\boldsymbol{X})$ *is uncorrelated with* $Y - \mathbb{E}(Y|\boldsymbol{X})$.

*Proof.*

By iterated expectations, we know that $\mathbb{E}[Y - \mathbb{E}(Y|\boldsymbol{X})] = 0$. Thus, the covariance between $g(\boldsymbol{X})$ and $Y - \mathbb{E}(Y|\boldsymbol{X})$ can be written as the expectation of their product,

$$\begin{aligned}
\text{Cov}[g(\boldsymbol{X}), Y - \mathbb{E}(Y|\boldsymbol{X})] &= \mathbb{E}[g(\boldsymbol{X})(Y - \mathbb{E}(Y|\boldsymbol{X}))] \\
&= \mathbb{E}\{\mathbb{E}[g(\boldsymbol{X})(Y - \mathbb{E}(Y|\boldsymbol{X}))|\boldsymbol{X}]\} \\
&= \mathbb{E}\{g(\boldsymbol{X})\,\mathbb{E}[(Y - \mathbb{E}(Y|\boldsymbol{X}))|\boldsymbol{X}]\} \\
&= \mathbb{E}\{g(\boldsymbol{X})[\mathbb{E}(Y|\boldsymbol{X}) - \mathbb{E}(Y|\boldsymbol{X})]\} \\
&= 0.
\end{aligned}$$

□

We are now in a position to prove that the conditional expectation provides the minimum-mean-square estimator (MMSE).

**Proposition 8.5.2** (Minimum-mean-square estimator)
*The conditional expectation, $\mathbb{E}(Y|\boldsymbol{X})$, is the function of $\boldsymbol{X}$ that has minimum mean squared error as an estimator of $Y$.*

*Proof.*

Let $h(\boldsymbol{X})$ be any reasonably well-behaved function of $\boldsymbol{X}$. Our method of proof is to show that the mean squared error of $h(\boldsymbol{X})$ is smallest when $h(\boldsymbol{X}) = \mathbb{E}(Y|\boldsymbol{X})$. We have

$$\begin{aligned}\underset{Y}{\text{MSE}}(h(\boldsymbol{X})) &= \mathbb{E}\left[(Y - h(\boldsymbol{X}))^2\right] \\ &= \mathbb{E}\left[\{(Y - \mathbb{E}(Y|\boldsymbol{X})) + (\mathbb{E}(Y|\boldsymbol{X}) - h(\boldsymbol{X}))\}^2\right] \\ &= \mathbb{E}\left[(Y - \mathbb{E}(Y|\boldsymbol{X}))^2\right] + \mathbb{E}\left[(Y - \mathbb{E}(Y|\boldsymbol{X}))(\mathbb{E}(Y|\boldsymbol{X}) - h(\boldsymbol{X}))\right] \\ &\quad + \mathbb{E}\left[(\mathbb{E}(Y|\boldsymbol{X}) - h(\boldsymbol{X}))^2\right]. \end{aligned} \tag{8.11}$$

The term $\mathbb{E}(Y|\boldsymbol{X}) - h(\boldsymbol{X})$ is a function of $\boldsymbol{X}$ alone, so (by Lemma 8.5.1) it is uncorrelated with $Y - \mathbb{E}(Y|\boldsymbol{X})$. We conclude that the middle term in equation (8.11) is zero, and thus

$$\underset{Y}{\text{MSE}}(h(\boldsymbol{X})) = \mathbb{E}\left[(Y - \mathbb{E}(Y|\boldsymbol{X}))^2\right] + \mathbb{E}\left[(\mathbb{E}(Y|\boldsymbol{X}) - h(\boldsymbol{X}))^2\right].$$

As the expectation of a non-negative random variable,

$$\mathbb{E}\left[(\mathbb{E}(Y|\boldsymbol{X}) - h(\boldsymbol{X}))^2\right] \geq 0.$$

We conclude that $\text{MSE}_Y(h(\boldsymbol{X}))$ is minimised by taking $h(\boldsymbol{X}) = \mathbb{E}(Y|\boldsymbol{X})$. □

Conditional expectation has a number of desirable properties. In particular, it is unbiased and has mean squared error that is less than or equal to the variance of $Y$. These properties are summarised by the following proposition; the proof is part of Exercise 8.5.

**Proposition 8.5.3** (Properties of MMSE)
*The MMSE of $Y$, $\mathbb{E}(Y|\boldsymbol{X})$,*

*i. is unbiased,* $\mathbb{E}[Y - \mathbb{E}(Y|\boldsymbol{X})] = 0$.

*ii. has mean squared error* $\text{MSE}_Y[\mathbb{E}(Y|\boldsymbol{X})] = \text{Var}(Y) - \text{Var}(\mathbb{E}(Y|\boldsymbol{X}))$.

In many instances, the conditional expectation $\mathbb{E}(Y|\boldsymbol{X})$ is hard to construct. However, if we restrict our attention to estimators that are linear functions of the sample, the estimator that minimises mean squared error can be expressed in terms of the joint moments of $\boldsymbol{X}$ and $Y$. This estimator is referred to as the **minimum-mean-square linear estimator** (MMSLE). We will use the following notation:

$$\begin{aligned} \boldsymbol{\mu}_X &= \mathbb{E}(\boldsymbol{X}), \quad \mu_Y = \mathbb{E}(Y), \\ \boldsymbol{\Sigma}_X &= \text{Var}(\boldsymbol{X}), \quad \Sigma_Y = \text{Var}(Y), \quad \boldsymbol{\Sigma}_{YX} = \text{Cov}(Y, \boldsymbol{X}). \end{aligned}$$

Note that $\text{Cov}(\boldsymbol{X}, Y) = \boldsymbol{\Sigma}_{XY} = \boldsymbol{\Sigma}_{YX}^T$. We start by considering the zero-mean case.

**Lemma 8.5.4** (MMSLE zero-mean case)
*Suppose that* $\boldsymbol{\mu}_X = 0$ *and* $\mu_Y = 0$. *If*

$$\tilde{Y} = \boldsymbol{\Sigma}_{YX}\boldsymbol{\Sigma}_X^{-1}\boldsymbol{X},$$

*then* $\tilde{Y}$ *is the minimum-mean-square linear estimator of* $Y$ *based on* $\boldsymbol{X}$. *The mean squared error of this estimator is*

$$\operatorname*{MSE}_{Y}(\tilde{Y}) = \Sigma_Y - \boldsymbol{\Sigma}_{YX}\boldsymbol{\Sigma}_X^{-1}\boldsymbol{\Sigma}_{XY}\,.$$

*Proof.*

Consider the estimator $\boldsymbol{a}^T\boldsymbol{X}$, which is the linear function of $\boldsymbol{X}$ generated by the vector of coefficients $\boldsymbol{a} = (a_1, \ldots, a_n)^T$. The mean squared error of this estimator is

$$\begin{aligned}\operatorname*{MSE}_{Y}(\boldsymbol{a}^T\boldsymbol{X}) &= \mathbb{E}\left[(Y - \boldsymbol{a}^T\boldsymbol{X})(Y - \boldsymbol{a}^T\boldsymbol{X})^T\right] \\ &= \mathbb{E}\left[YY^T - \boldsymbol{a}^T\boldsymbol{X}Y^T - Y\boldsymbol{X}^T\boldsymbol{a} + \boldsymbol{a}^T\boldsymbol{X}\boldsymbol{X}^T\boldsymbol{a}\right] \\ &= \Sigma_Y - \boldsymbol{a}^T\boldsymbol{\Sigma}_{XY} - \boldsymbol{\Sigma}_{YX}\boldsymbol{a} + \boldsymbol{a}^T\boldsymbol{\Sigma}_X\boldsymbol{a}^T. \qquad (8.12)\end{aligned}$$

In order to find the MMSLE, we need the value of $\boldsymbol{a}$ that minimises $\mathrm{MSE}_Y(\boldsymbol{a}^T\boldsymbol{X})$. To calculate this value, we differentiate (8.12) with respect to $\boldsymbol{a}$, set the derivative equal to zero, and solve for $\boldsymbol{a}$. Using the matrix calculus results from subsection 8.2.4, we have

$$\begin{aligned}&\frac{\partial}{\partial \boldsymbol{a}}\operatorname*{MSE}_{Y}(\boldsymbol{a}^T\boldsymbol{X}) = -\boldsymbol{\Sigma}_{XY} - \boldsymbol{\Sigma}_{XY} + \boldsymbol{a}^T(\boldsymbol{\Sigma}_X + \boldsymbol{\Sigma}_X^T) \\ &-2\boldsymbol{\Sigma}_{YX} + 2\boldsymbol{a}^T\boldsymbol{\Sigma}_X\,.\end{aligned}$$

Setting this derivative equal to zero and solving yields

$$\boldsymbol{a} = \boldsymbol{\Sigma}_X^{-1}\boldsymbol{\Sigma}_{XY}, \qquad (8.13)$$

and, thus,

$$\tilde{Y} = \boldsymbol{a}^T\boldsymbol{X} = \boldsymbol{\Sigma}_{YX}\boldsymbol{\Sigma}_X^{-1}\boldsymbol{X}\,.$$

The expression for the mean squared error is an immediate consequence of substituting from equation (8.13) in equation (8.12). □

The general case is now a straightforward consequence of Lemma 8.5.4. The proof is part of Exercise 8.5.

**Proposition 8.5.5** (MMSLE general case)
*The minimum-mean-square linear estimator of* $Y$ *based on a sample* $\boldsymbol{X}$ *is given by*

$$\tilde{Y} = \mu_Y + \boldsymbol{\Sigma}_{YX}\boldsymbol{\Sigma}_X^{-1}(\boldsymbol{X} - \boldsymbol{\mu}_X)\,.$$

*The mean squared error of this estimator is*

$$\operatorname*{MSE}_{Y}(\tilde{Y}) = \Sigma_Y - \boldsymbol{\Sigma}_{YX}\boldsymbol{\Sigma}_X^{-1}\boldsymbol{\Sigma}_{XY}\,.$$

Furthermore, if $\boldsymbol{X}$ and $Y$ are jointly normally distributed, the MMSLE and its MSE are identical to the mean and variance of $Y$ conditional on $\boldsymbol{X} = \boldsymbol{x}$ (Proposition 5.7.2). In other words, in the normal case, the MMSLE is also the MMSE.

Proposition 8.5.5 has a strong connection with linear regression models. For the model $Y = \boldsymbol{X}\boldsymbol{\beta} + \sigma\varepsilon$, the least-squares estimator of $Y$ is

$$\hat{Y} = \boldsymbol{X}\hat{\boldsymbol{\beta}} = \boldsymbol{X}\underbrace{(\boldsymbol{X}^T\boldsymbol{X})^{-1}}_{\boldsymbol{S}_X}\underbrace{\boldsymbol{X}^T Y}_{\boldsymbol{S}_{XY}},$$

where $\boldsymbol{S}_X$ and $\boldsymbol{S}_{XY}$ are the sample analogues of $\boldsymbol{\Sigma}_X$ and $\boldsymbol{\Sigma}_{XY}$, respectively.

**Exercise 8.5**

1. Prove Proposition 8.5.3 and hence show that
$$\operatorname*{MSE}_{Y}(\mathbb{E}(Y|\boldsymbol{X})) = \mathbb{E}(\text{Var}(Y|\boldsymbol{X})).$$
2. Prove Proposition 8.5.5.

## 8.6 Further exercises

1. Suppose that $Y_1, \dots, Y_n$ is a random sample from $\text{N}(\mu, 1)$. Define $X_n$ to be a discrete random variable with support $\{0, n\}$ and mass function
$$f_X(x) = \begin{cases} 1 - \frac{1}{n} & \text{for } x = 0, \\ \frac{1}{n} & \text{for } x = n. \end{cases}$$
Show that the intuitively unappealing estimator
$$U_n = \bar{Y} + X_n$$
is consistent for $\mu$. What is the limit of the mean squared error of this estimator as $n \to \infty$?
2. Suppose that $Y_1, \dots, Y_n$ is a sample from a population with mean $\mu$ and variance $\sigma^2$. The sample is not random, $\text{Cov}(Y_i, Y_j) = \rho\sigma^2$ for $i \neq j$. Let $U = \sum_{i=1}^n a_i Y_i$. Give the condition on the constants $a_1, \dots, a_n$ for $U$ to be an unbiased estimator of $\mu$. Under this condition, calculate $\text{MSE}_\mu(U)$.
3. Find the bias and mean squared error for the method-of-moments estimator of the variance,
$$\hat{\sigma}^2 = \frac{n-1}{n} S^2,$$
based on a random sample from a normal distribution. On the basis of MSE, would you prefer to use $\hat{\sigma}^2$ or $S^2$ to estimate $\sigma^2$? Now consider all estimators of the form
$$\hat{\sigma}_k^2 = \frac{1}{k}\sum_{i=1}^n (Y_i - \bar{Y})^2.$$
Which value of $k$ minimises the MSE of $\hat{\sigma}_k^2$ as an estimator of $\sigma^2$?

4. You are watching a marathon race; from your vantage point, you observe $k$ competitors run past, with numbers $X_1, X_2, \dots, X_k$. Suppose that the runners are randomly assigned numbers $1, 2, \dots, n$, where $n$ is the total number of participants. Find the method-of-moments estimator of $n$. Is this a sensible estimator?

5. Consider a random sample $\boldsymbol{Y} = (Y_1, \dots, Y_n)^T$ from an $N(\mu, \sigma^2)$ distribution where $\mu$ is known and $\sigma^2$ is unknown. Two possible interval estimators for $\sigma^2$ are

$$\left[\frac{1}{a_1}\sum_{i=1}^{n}(Y_i - \bar{Y})^2, \frac{1}{a_2}\sum_{i=1}^{n}(Y_i - \bar{Y})^2\right], \text{ and}$$

$$\left[\frac{1}{b_1}\sum_{i=1}^{n}(Y_i - \mu)^2, \frac{1}{b_2}\sum_{i=1}^{n}(Y_i - \mu)^2\right],$$

where $a_1$, $a_2$, $b_1$, and $b_2$ are constants. For $n = 10$, find the values of $a_1$, $a_2$, $b_1$, and $b_2$ that give intervals with confidence coefficient 0.9. Compare the expected lengths of the intervals and comment.

6. If $V \sim \chi^2_n$, $W \sim \chi^2_m$, and $V$ and $W$ are independent, then the ratio

$$\frac{V/n}{W/m}$$

has an **$F$-distribution** on $(n, m)$ degrees of freedom (denoted $F_{n,m}$). Suppose that $X_1, \dots, X_n$ and $Y_1, \dots, Y_m$ are random samples from $N(\mu_X, \sigma_X^2)$ and $N(\mu_Y, \sigma_Y^2)$, respectively, where $\mu_X$ and $\mu_Y$ are known. Find a pivotal function for the ratio $\sigma_X^2/\sigma_Y^2$ and hence give an expression for an interval estimator for $\sigma_X^2/\sigma_Y^2$ with confidence coefficient $1 - \alpha$.

7. (**$F$-statistic for one-way ANOVA**) For the one-way ANOVA model described in section 6.3.3 and Exercise 6.3, we typically want to test the null hypothesis $H_0 : \beta_1 = \dots = \beta_{k-1} = 0$ (all group means are equal) against the alternative $H_1$ : "at least one $\beta_j \neq 0$". Use the distributional results from Exercise 7.5 to show that, under this null hypothesis, the ratio of the between-group sum of squares over the within-group sum of squares (scaled by an appropriate constant) follows an $F$-distribution. What are its degrees of freedom? How would you perform the test?

8. Consider a random sample from an $N(\mu, \sigma^2)$ distribution with $\sigma^2$ known. Show that the random set

$$\left(-\infty, \bar{Y} - 0.126\frac{\sigma}{\sqrt{n}}\right] \cup \left[\bar{Y} + 0.126\frac{\sigma}{\sqrt{n}}, \infty\right)$$

is a confidence set for the mean with confidence coefficient 0.95. Would you use this in practice?

9. Let $X_1, \cdots, X_n$ be a random sample from the Unif$[0, \theta]$ distribution. How would you construct a confidence interval for $\theta$?

## 8.7 Chapter summary

On completion of this chapter you should be able to:

- describe how statistics play a fundamental role in data reduction,
- write down pivotal functions for standard cases,
- distinguish between estimators and estimates,
- explain the association between bias, variance, and mean squared error,
- prove that, under certain conditions, the sample mean and sample variance are consistent estimators of the population mean and population variance,
- explain the distinction between an interval estimate and an interval estimator,
- interpret interval estimates,
- define the terms coverage probability and coverage coefficient,
- calculate the expected length of an interval estimator,
- use pivotal functions to derive confidence intervals,
- compute the confidence coefficient for an interval estimator based on order statistics.
- explain the distinction between simple and composite hypotheses, and the distinction between size and significance level of a test,
- define type I and type II errors, and describe the association between these errors and the notions of size and power,
- discuss the use of $p$-values in hypothesis testing, and
- construct hypothesis tests using pivotal functions.

Chapter 9

# Likelihood-based inference

In parametric statistics, we sometimes denote the joint density of the sample by $f_{\boldsymbol{Y}}(\boldsymbol{y};\theta)$, in order to make explicit its dependence on the unknown parameter $\theta$. We can treat this density as a function of $\theta$, known as the likelihood. Despite the fact that density and likelihood have exactly the same functional form, it is often useful to work with likelihood, particularly when we are interested in what happens across a range of parameter values.

In this chapter, we take a more technical approach to parameter estimation and hypothesis testing, based on the likelihood and its functions, such as the score and information. We introduce maximum-likelihood estimation, a key inferential technique, and explore the properties of the resulting estimators. In addition, we discuss various techniques for likelihood maximisation, such as the Newton-Raphson method and the EM algorithm.

We also look at some types of likelihood-based hypothesis tests – the likelihood-ratio, score, and Wald tests – which are widely applicable, as they do not require us to work out the exact distribution of the test statistics.

## 9.1 Likelihood function and log-likelihood function

We start by considering a sample $\boldsymbol{Y} = (Y_1, \dots, Y_n)^T$ from a population with a known parametric form (distribution) with a single, unknown, scalar parameter $\theta$.

**Definition 9.1.1** (Likelihood function)
Consider a sample $\boldsymbol{Y} = (Y_1, \dots, Y_n)^T$ with density $f_{\boldsymbol{Y}}(\boldsymbol{y};\theta)$. The **likelihood (function)** has the same functional form as the joint probability mass/density, but is viewed as a function of $\theta$ rather than $\boldsymbol{y}$. We use the notation $L_{\boldsymbol{Y}}$ to denote likelihood,

$$L_{\boldsymbol{Y}}(\theta;\boldsymbol{y}) = f_{\boldsymbol{Y}}(\boldsymbol{y};\theta).$$

We may sometimes refer to the likelihood as $L_{\boldsymbol{Y}}(\theta)$ when the dependence on $\boldsymbol{y}$ is not important, or as $L(\theta;\boldsymbol{y})$ when this can be done without ambiguity. It is important to understand that the likelihood is not a mass/density as it does not sum/integrate to 1 over $\theta$. However, the likelihood value $L_{\boldsymbol{Y}}(\theta; \boldsymbol{y})$ may be used as a measure of the plausibility of the parameter value $\theta$ for a given set of observations $\boldsymbol{y}$.

We will use likelihood in a number of ways. In addition to the obvious applications (maximum-likelihood estimation in section 9.3 and likelihood-ratio tests in section 9.4), likelihood arises during the discussion of unbiased estimators in section 10.2. In the remainder of this section we introduce a number of definitions, and prove results that will be used in several different contexts. The first of these definitions is the **log-likelihood function**. As the name suggests, this is just the (natural) logarithm of the likelihood function.

**Definition 9.1.2** (Log-likelihood function)
If $L_{\boldsymbol{Y}}$ is a likelihood function, we define the log-likelihood function $\ell_{\boldsymbol{Y}}$ as

$$\ell_{\boldsymbol{Y}}(\theta; \boldsymbol{y}) \equiv \log L_{\boldsymbol{Y}}(\theta; \boldsymbol{y}),$$

wherever $L_{\boldsymbol{Y}}(\theta; \boldsymbol{y}) \neq 0$.

Taking logs is a monotone transformation; the location of the maxima and minima of a function are not changed by a log transformation. However, the functional form of the log-likelihood is often more convenient to work with than that of the likelihood.

*Simplification for random samples*

If $\boldsymbol{Y} = (Y_1, \ldots, Y_n)^T$ is a random sample, we can express the likelihood and log-likelihood in terms of the shared marginal mass/density, $f_Y(y; \theta)$. We denote the likelihood and log-likelihood associated with a single observation as $L_Y(\theta; y)$ and $\ell_Y(\theta; y)$. Note the importance of the distinction between the vector $\boldsymbol{Y}$, denoting the whole sample, and the scalar $Y$, representing a single observation. For a random sample, we can exploit independence to write the joint density as a product of the marginal masses/densities. Since mass/density and likelihood have the same functional form, we have

$$L_{\boldsymbol{Y}}(\theta; \boldsymbol{y}) = \prod_{i=1}^{n} L_Y(\theta; y_i).$$

Taking logs yields

$$\ell_{\boldsymbol{Y}}(\theta; \boldsymbol{y}) = \sum_{i=1}^{n} \ell_Y(\theta; y_i). \tag{9.1}$$

In most of the cases we consider, the sample will be random and the likelihood can be written as a product of individual likelihood functions.

*Vector parameter case*

Now consider the case where there is more than one unknown parameter. The likelihood can be viewed as a function of a single vector parameter or as a function of several scalar parameters; in practice, these two viewpoints are equivalent. Suppose that we have a sample $\boldsymbol{Y} = (Y_1, \ldots, Y_n)^T$ and parameter(s) $\boldsymbol{\theta} = (\theta_1, \ldots, \theta_r)$. The likelihood and log-likelihood can be written as

$$L_{\boldsymbol{Y}}(\boldsymbol{\theta}; \boldsymbol{y}) = L_{\boldsymbol{Y}}(\theta_1, \ldots, \theta_r; \boldsymbol{y}),$$
$$\ell_{\boldsymbol{Y}}(\boldsymbol{\theta}; \boldsymbol{y}) = \ell_{\boldsymbol{Y}}(\theta_1, \ldots, \theta_r; \boldsymbol{y}).$$

If $\boldsymbol{Y} = (Y_1, \ldots, Y_n)^T$ is a random sample then

$$L_{\boldsymbol{Y}}(\boldsymbol{\theta}; \boldsymbol{y}) = \prod_{i=1}^{n} L_Y(\theta_1, \ldots, \theta_r; y_i),$$
$$\ell_{\boldsymbol{Y}}(\boldsymbol{\theta}; \boldsymbol{y}) = \sum_{i=1}^{n} \ell_Y(\theta_1, \ldots, \theta_r; y_i).$$

Some examples follow.

**Example 9.1.3** (Exponential likelihood)
Suppose $\boldsymbol{Y} = (Y_1, \ldots, Y_n)^T$ is a random sample from an Exp($\lambda$) distribution. The density function of each individual sample member is

$$f_Y(y_i; \lambda) = \lambda \exp(-\lambda y_i) \text{ for } y_i \geq 0.$$

By independence, the joint density of the sample is

$$\begin{aligned} f_{\boldsymbol{Y}}(\boldsymbol{y}; \lambda) &= \prod_{i=1}^{n} \lambda \exp(-\lambda y_i) \\ &= \lambda^n \exp\left(-\lambda \sum_{i=1}^{n} y_i\right) \\ &= \lambda^n \exp(-\lambda n \bar{y}) \text{ for } y_1 \geq 0, \ldots, y_n \geq 0. \end{aligned}$$

The likelihood has exactly the same functional form as the density, so

$$L_{\boldsymbol{Y}}(\lambda; \boldsymbol{y}) = \lambda^n \exp(-\lambda n \bar{y}) \text{ for } \lambda > 0.$$

Taking logs yields the log-likelihood,

$$\ell_{\boldsymbol{Y}}(\lambda; \boldsymbol{y}) = n \log(\lambda) - n\lambda\bar{y} \text{ for } \lambda > 0.$$

**Example 9.1.4** (Normal likelihood)
Suppose that $\boldsymbol{Y} = (Y_1, \ldots, Y_n)^T$ is a random sample from an N($\mu, \sigma^2$) distribution. We define the parameter vector $\boldsymbol{\theta} = (\mu, \sigma^2)$. The likelihood function is then

$$L_{\boldsymbol{Y}}(\boldsymbol{\theta}; \boldsymbol{y}) = L_{\boldsymbol{Y}}(\mu, \sigma^2; \boldsymbol{y}) = \left(\frac{1}{\sqrt{2\pi\sigma^2}}\right)^n \exp\left[-\frac{1}{2\sigma^2} \sum_{i=1}^{n} (y_i - \mu)^2\right].$$

Taking logs yields the log-likelihood,

$$\ell_{\boldsymbol{Y}}(\boldsymbol{\theta}; \boldsymbol{y}) = \ell_{\boldsymbol{Y}}(\mu, \sigma^2; \boldsymbol{y}) = -\frac{n}{2} \log(2\pi) - \frac{n}{2} \log(\sigma^2) - \frac{1}{2\sigma^2} \sum_{i=1}^{n} (y_i - \mu)^2.$$

**Exercise 9.1**

1. Find the likelihood and log-likelihood function of the parameters for a random sample of size $n$ from each of the following distributions:
   (a) Bernoulli: $f_Y(y) = p^y(1-p)^y$ for $y = 0, 1$ and $0 < p < 1$ .
   (b) Pareto: $f_Y(y) = a/(1+y)^{a+1}$ for $0 < y < \infty$ and $a > 0$ .
   (c) Weibull: $f_Y(y) = c\tau y^{\tau-1}\exp(-cy^\tau)$ for $0 < y < \infty$, $\tau > 0$ and $c > 0$ .
2. Let $X_1, \ldots, X_n$ be a random sample from a Unif$[0, \theta]$ distribution, with density function

$$f_X(x) = \frac{1}{\theta}\mathbf{1}_{[0,\theta]}(x) .$$

   Give an expression for the joint density that is a function of $\theta$ and $y_{(n)}$ (the sample maximum) only. Sketch the likelihood function.

## 9.2 Score and information

In section 9.3 we will be maximising the likelihood function, so it is natural that we are interested in derivatives of the likelihood. In fact, we define the **score** to be the first derivative of the log-likelihood. Working with derivatives of the log-likelihood has a number of advantages that will become apparent later on.

**Definition 9.2.1** (Score function)
The score function associated with the log-likelihood $\ell_Y(\theta; \boldsymbol{y})$ is defined as

$$s_Y(\theta; \boldsymbol{y}) = \frac{\partial}{\partial\theta}\ell_Y(\theta; \boldsymbol{y}) .$$

By the properties of logs and the chain rule, we have

$$s_Y(\theta; \boldsymbol{y}) = \frac{\frac{\partial}{\partial\theta}L_Y(\theta; \boldsymbol{y})}{L_Y(\theta; \boldsymbol{y})} . \tag{9.2}$$

The score is defined to make dependence on $\boldsymbol{y}$ explicit. Of course we could apply this function to the random variable $\boldsymbol{Y}$. It turns out that the expected value of $s_Y(\theta; \boldsymbol{Y})$ is zero. In proving this result, and in subsequent proofs, we assume $\boldsymbol{Y}$ is continuous. Similar proofs hold for the discrete case.

**Lemma 9.2.2** (A property of the score function)
*Assuming sufficient regularity to allow us to differentiate under integral signs,* $\mathbb{E}[s_Y(\theta; \boldsymbol{Y})] = 0$.

*Proof.*

To avoid cluttering the notation, we will drop the $\boldsymbol{Y}$ subscript for this proof, so we write $L(\theta; \mathbf{y}) = L_{\mathbf{Y}}(\theta; \mathbf{y})$, $f(\mathbf{y}; \theta) = f_{\mathbf{Y}}(\mathbf{y}; \theta)$ and $s(\theta; \mathbf{y}) = s_{\mathbf{Y}}(\theta; \mathbf{y})$. We have

$$\begin{aligned}
\mathbb{E}[s(\theta; \mathbf{Y})] &= \int_{\mathbb{R}^n} s(\theta; \mathbf{y}) f(\mathbf{y}; \theta) d\mathbf{y} \\
&= \int_{\mathbb{R}^n} \frac{\frac{\partial}{\partial\theta} L(\theta; \mathbf{y})}{L(\theta; \mathbf{y})} f(\mathbf{y}; \theta) d\mathbf{y} && \text{from (9.2)} \\
&= \int_{\mathbb{R}^n} \frac{\partial}{\partial\theta} f(\mathbf{y}; \theta) d\mathbf{y} && \text{since } f(\mathbf{y}; \theta) = L(\theta; \mathbf{y}) \\
&= \frac{d}{d\theta} \int_{\mathbb{R}^n} f(\mathbf{y}; \theta) d\mathbf{y} \,.
\end{aligned}$$

We know that $\int_{\mathbb{R}^n} f(\mathbf{y}; \theta) d\mathbf{y} = 1$, so the derivative of $\int_{\mathbb{R}^n} f(\mathbf{y}; \theta) d\mathbf{y}$ with respect to $\theta$ is zero. Thus, $\mathbb{E}[s(\theta; \mathbf{Y})] = 0$. □

For a given sample $\mathbf{y}$, we can plot the likelihood $L_{\mathbf{Y}}(\theta; \mathbf{y})$ (or equivalently the log-likelihood $\ell_{\mathbf{Y}}(\theta; \mathbf{y})$) against $\theta$. Consider two extreme cases:

i. The likelihood has a sharp peak: this suggests that a small subset of values of $\theta$ are far more plausible than other values.
ii. The likelihood is flat: a large set of values of $\theta$ are equally plausible.

In the first case the likelihood provides us with a lot of information; it allows us to narrow down plausible values of $\theta$ to a small subset. In the second case, looking at the likelihood does not help us make inferences about $\theta$. The first case is characterised by a sharp peak in the likelihood. This arises when the likelihood changes rapidly with $\theta$, that is, the absolute values of the first derivatives are large. These ideas can be formalised by defining the concept of **Fisher information**. In fact, Fisher information is found by looking at the square (rather than absolute value) of the derivative of the log-likelihood, and taking expectations across sample values.

**Definition 9.2.3** (Fisher information)
Suppose that $\mathbf{Y} = (Y_1, \ldots, Y_n)^T$ is a sample from a population whose distribution is parameterised by $\theta$, and that $\ell_{\mathbf{Y}}(\theta; \mathbf{y})$ is the log-likelihood function. The Fisher information about parameter $\theta$ in the sample $\mathbf{Y}$ is defined as

$$\mathcal{I}_{\mathbf{Y}}(\theta) = \mathbb{E}\left[\left(\frac{\partial}{\partial\theta} \ell_{\mathbf{Y}}(\theta; \mathbf{Y})\right)^2\right].$$

Many texts drop the "Fisher" and refer to this quantity simply as the information.

The quantity defined by 9.2.3 is sometimes referred to as **total Fisher information** (or just total information). This is to indicate that we are looking at information in the entire sample. We will encounter Fisher information associated with individual sample members shortly.

Fisher information is a quantity that you might see expressed in a number of different ways. From the definition of the score, we have

$$\mathcal{I}_Y(\theta) = \mathbb{E}\left[s_Y(\theta; Y)^2\right].$$

Perhaps more usefully, we can exploit the fact that $\mathbb{E}[s_Y(\theta; Y)] = 0$, and thus

$$\text{Var}[s_Y(\theta; Y)] = \mathbb{E}[s_Y(\theta; Y)^2],$$

to write

$$\mathcal{I}_Y(\theta) = \text{Var}\left[s_Y(\theta; Y)\right].$$

Another form of Fisher information, that is often computationally convenient, is given by the following lemma.

**Lemma 9.2.4** (Alternative form for Fisher information)
*Consider $Y = (Y_1, \ldots, Y_n)^T$, a sample from a population whose distribution is parameterised by $\theta$, and suppose that $\ell_Y(\theta; y)$ is the log-likelihood function. An alternative representation of the Fisher information about parameter $\theta$ in the sample $Y$ is*

$$\mathcal{I}_Y(\theta) = -\mathbb{E}\left[\frac{\partial^2}{\partial\theta^2}\ell_Y(\theta; Y)\right].$$

*Proof.*

This result is obtained by differentiating $\mathbb{E}[s(\theta; Y)]$ with respect to $\theta$. Once again, in the interests of uncluttered notation we will drop $Y$ subscripts on the density and the score. We have

$$\begin{aligned}
&\frac{d}{d\theta}\mathbb{E}[s(\theta; Y)] \\
&= \frac{d}{d\theta}\int_{\mathbb{R}^n} s(\theta; y) f(y; \theta) dy \\
&= \int_{\mathbb{R}^n} \frac{d}{d\theta}\left(s(\theta; y) f(y; \theta)\right) dy \\
&= \int_{\mathbb{R}^n} \left[\left(\frac{\partial}{\partial\theta}s(\theta; y)\right) f(y; \theta) + s(\theta; y)\left(\frac{\partial}{\partial\theta}f(y; \theta)\right)\right] dy && \text{product rule} \\
&= \int_{\mathbb{R}^n} \left[\frac{\partial}{\partial\theta}s(\theta; y) + s(\theta; y)^2\right] f(y; \theta) dy && \text{by 9.2} \\
&= \mathbb{E}\left[\frac{\partial}{\partial\theta}s(\theta; Y) + s(\theta; Y)^2\right].
\end{aligned}$$

By Lemma 9.2.2, $\mathbb{E}[s(\theta; Y)] = 0$, so $\frac{d}{d\theta}\mathbb{E}[s(\theta; Y)] = 0$. We thus have

$$\mathbb{E}\left[s(\theta; Y)^2\right] = -\mathbb{E}\left[\frac{\partial}{\partial\theta}s(\theta; Y)\right].$$

Recalling that $s(\theta; y)$ is defined as $\frac{\partial}{\partial\theta}\ell_Y(\theta; y)$ gives the final result. □

We now have two key expressions for Fisher information, and several possible variations using the score; it is important to understand that these all refer to the same quantity. From the original definition (9.2.3) we have

$$\mathcal{I}_{\boldsymbol{Y}}(\theta) = \mathbb{E}\left[\left(\frac{\partial}{\partial\theta}\ell_{\boldsymbol{Y}}(\theta;\boldsymbol{Y})\right)^2\right] = \mathbb{E}\left[s_{\boldsymbol{Y}}(\theta;\boldsymbol{Y})^2\right] = \text{Var}\left[s_{\boldsymbol{Y}}(\theta;\boldsymbol{Y})\right],$$

and from the alternative form given by Lemma 9.2.4,

$$\mathcal{I}_{\boldsymbol{Y}}(\theta) = -\,\mathbb{E}\left[\frac{\partial^2}{\partial\theta^2}\ell_{\boldsymbol{Y}}(\theta;\boldsymbol{Y})\right] = -\,\mathbb{E}\left[\frac{\partial}{\partial\theta}s_{\boldsymbol{Y}}(\theta;\boldsymbol{Y})\right].$$

The quantity

$$\hat{\mathcal{I}}_{\boldsymbol{Y}}(\theta;\boldsymbol{Y}) = -\frac{\partial^2}{\partial\theta^2}\ell_{\boldsymbol{Y}}(\theta;\boldsymbol{Y})$$

is known as the **observed information**. It is the sample analogue of the Fisher information, in the sense that it is a function of the random sample $\boldsymbol{Y}$. Notice that

$$\mathbb{E}\left[\hat{\mathcal{I}}_{\boldsymbol{Y}}(\theta;\boldsymbol{Y})\right] = \mathcal{I}_{\boldsymbol{Y}}(\theta).$$

*Simplification for random samples*

In the previous subsection we saw that if $\boldsymbol{Y} = (Y_1,\dots,Y_n)^T$ is a random sample, we can express the likelihood as a product of individual likelihoods, and the log-likelihood as a sum of individual log-likelihoods. Similar results hold for the score and information. We denote the score and information associated with a single observation as $s_Y(\theta;y)$ and $\mathcal{I}_Y(\theta)$ respectively. Differentiating (9.1) yields

$$s_{\boldsymbol{Y}}(\theta;\boldsymbol{y}) = \sum_{i=1}^{n} s_Y(\theta;y_i).$$

We can also readily show that

$$\mathcal{I}_{\boldsymbol{Y}}(\theta) = n\mathcal{I}_Y(\theta). \tag{9.3}$$

Proving this relationship is part of Exercise 9.2. Equation (9.3) indicates that, for a random sample, total Fisher information is just the Fisher information for a single observation multiplied by the sample size; in other words, information is additive, and every observation contains the same amount of information about the unknown parameter.

*Vector parameter case*

Suppose that we have a vector of parameters $\boldsymbol{\theta} = (\theta_1,\dots,\theta_r)$. In this case the score will be an $r\times 1$ vector, and the Fisher information will be an $r\times r$ matrix. We introduce the following definitions:

**Definition 9.2.5** (Score vector and information matrix)
Consider the sample $\boldsymbol{Y} = (Y_1, \dots, Y_n)^T$ with log-likelihood $\ell_{\boldsymbol{Y}}(\boldsymbol{\theta}; \boldsymbol{Y})$. The **score vector** is defined as

$$s_{\boldsymbol{Y}}(\boldsymbol{\theta}; \boldsymbol{y}) = \nabla_{\boldsymbol{\theta}} \ell_{\boldsymbol{Y}}(\boldsymbol{\theta}; \boldsymbol{y}) = \left( \frac{\partial}{\partial \theta_1} \ell_{\boldsymbol{Y}}(\boldsymbol{\theta}; \boldsymbol{y}), \dots, \frac{\partial}{\partial \theta_r} \ell_{\boldsymbol{Y}}(\boldsymbol{\theta}; \boldsymbol{y}) \right)^T,$$

and the information matrix is given by

$$\boldsymbol{\mathcal{I}}_{\boldsymbol{Y}}(\boldsymbol{\theta}) = \mathbb{E}[s_{\boldsymbol{Y}}(\boldsymbol{\theta}; \boldsymbol{Y}) s_{\boldsymbol{Y}}(\boldsymbol{\theta}; \boldsymbol{Y})^T].$$

The symbol $\nabla$ denotes the **del** operator, which is a generalisation of the derivative to higher dimensions.

Consider the function $g : \mathbb{R}^k \to \mathbb{R}$, so if $\mathbf{x} = (x_1, \dots, x_k)^T$ is a $k$-dimensional vector, $g(\mathbf{x})$ is a scalar. We use the del operator to represent the partial derivatives of $g(\mathbf{x})$ with respect to each element of $\mathbf{x}$, that is,

$$\nabla_{\mathbf{x}} g(\mathbf{x}) = \left( \frac{\partial}{\partial x_1} g(\mathbf{x}), \dots, \frac{\partial}{\partial x_k} g(\mathbf{x}) \right)^T.$$

Notice that $\nabla_{\mathbf{x}} g(\mathbf{x})$ has the same dimension as $\mathbf{x}$.

The $(i, j)^{\text{th}}$ element of the information matrix is

$$[\boldsymbol{\mathcal{I}}_{\boldsymbol{Y}}(\boldsymbol{\theta})]_{i,j} = \mathbb{E}\left[ \frac{\partial}{\partial \theta_i} \ell_{\boldsymbol{Y}}(\boldsymbol{\theta}; \boldsymbol{Y}) \frac{\partial}{\partial \theta_j} \ell_{\boldsymbol{Y}}(\boldsymbol{\theta}; \boldsymbol{Y}) \right].$$

Proving that this relationship holds is part of Exercise 9.3.

We can derive properties for the score vector and information matrix that are very similar to those given in the scalar case; the following proposition summarises them.

**Proposition 9.2.6**
*Consider the sample* $\boldsymbol{Y} = (Y_1, \dots, Y_n)^T$ *parameterised by* $\boldsymbol{\theta}$, *with log-likelihood* $\ell_{\boldsymbol{Y}}(\boldsymbol{\theta}; \boldsymbol{Y})$. *The following relationships hold:*

*i.* $\mathbb{E}[s_{\boldsymbol{Y}}(\boldsymbol{\theta}; \boldsymbol{Y})] = \boldsymbol{0}$,
*ii.* $\boldsymbol{\mathcal{I}}_{\boldsymbol{Y}}(\boldsymbol{\theta}) = \text{Var}[s_{\boldsymbol{Y}}(\boldsymbol{\theta}; \boldsymbol{Y})]$,
*iii.* $\boldsymbol{\mathcal{I}}_{\boldsymbol{Y}}(\boldsymbol{\theta}) = -\mathbb{E}\left[\nabla_{\boldsymbol{\theta}} \nabla_{\boldsymbol{\theta}}^T \ell_{\boldsymbol{Y}}(\boldsymbol{\theta}; \boldsymbol{Y})\right] = -\mathbb{E}[\nabla_{\boldsymbol{\theta}} s_{\boldsymbol{Y}}(\boldsymbol{\theta}; \boldsymbol{Y})]$.

The third of these statements says that the $(i, j)^{\text{th}}$ element of the information matrix can be written as

$$[\boldsymbol{\mathcal{I}}_{\boldsymbol{Y}}(\boldsymbol{\theta})]_{i,j} = -\mathbb{E}\left[ \frac{\partial^2}{\partial \theta_i \partial \theta_j} \log f_{\boldsymbol{Y}}(\boldsymbol{Y}; \boldsymbol{\theta}) \right].$$

Proving Proposition 9.2.6 is part of Exercise 9.2.

**Example 9.2.7** (Score and information for exponential)
Suppose $Y = (Y_1, \dots, Y_n)^T$ is a random sample from an Exp($\lambda$) distribution. By differentiating the expression for the log-likelihood found in Example 9.1.3, we obtain the score,

$$s_Y(\lambda; y) = \frac{n}{\lambda} - \sum_{i=1}^{n} y_i \,.$$

This is a case in which the information is much easier to derive using the alternative form given by Lemma 9.2.4,

$$\frac{\partial}{\partial \lambda} s_Y(\lambda; y) = -\frac{n}{\lambda^2},$$

so

$$\mathcal{I}_Y(\lambda) = \mathbb{E}\left[-\frac{\partial}{\partial \lambda} s_Y(\lambda; Y)\right] = \frac{n}{\lambda^2}\,.$$

**Example 9.2.8** (Score and information for mean of normal with known variance)
Consider $Y = (Y_1, \dots, Y_n)^T$, a random sample from an N($\mu, \sigma^2$) distribution where $\sigma^2$ is known. The score is then just the scalar function given by differentiating the log-likelihood from Example 9.1.4 with respect to $\mu$,

$$s_Y(\mu; y) = \frac{\partial}{\partial \mu} \ell(\mu; y) = \frac{1}{\sigma^2} \sum_{i=1}^{n} (y_i - \mu)\,.$$

Using Lemma 9.2.4, the information is

$$\mathcal{I}_Y(\mu) = \mathbb{E}\left[-\frac{\partial}{\partial \mu} s_Y(\mu; y)\right] = \mathbb{E}\left[\frac{1}{\sigma^2} \sum_{i=1}^{n} 1\right] = \frac{n}{\sigma^2}\,.$$

This expression is intuitively appealing as it says that the information about $\mu$ available from the sample increases with the sample size and decreases as the variance grows.

**Exercise 9.2**

1. Given a random sample $Y = (Y_1, \dots, Y_n)^T$, give the likelihood, log-likelihood, score, and information for the parameter in each of the following cases:
   (a) $Y \sim$ Pois($\lambda$),
   (b) $Y \sim$ N($0, \sigma^2$),
   (c) $Y \sim$ Geometric($p$),
   (d) $Y \sim$ N($\mu, \sigma^2$), with both $\mu$ and $\sigma^2$ unknown.
2. Suppose that $Y = (Y_1, \dots, Y_n)^T$ is a random sample parameterised by $\theta$. Show that

$$\mathcal{I}_Y(\theta) = n\mathcal{I}_Y(\theta)$$

where $\mathcal{I}_Y(\theta)$ is the total information and $\mathcal{I}_Y(\theta)$ is the information associated with a single observation.

3. Given the setup in Example 9.2.7, derive the information using Definition 9.2.3. Given the setup in Example 9.2.8, derive the information using the expression given in 9.2.4. In each case, show that you arrive at the same result starting from the information associated with a single observation.
4. Prove Proposition 9.2.6.

## 9.3 Maximum-likelihood estimation

Recall from section 9.1 that the likelihood is used as a measure of the plausibility of parameter values for a given sample. It is therefore reasonable to take as a point estimator the value that maximises the likelihood function. Such estimators are referred to (unsurprisingly) as **maximum-likelihood estimators**. We start by considering the scalar parameter case.

**Definition 9.3.1** (Maximum-likelihood estimate)
Consider a sample $\boldsymbol{Y}$ parameterised by $\theta$, and let $L_{\boldsymbol{Y}}$ be the likelihood function. Given an observed sample $\boldsymbol{y}$, the **maximum-likelihood estimate** (MLE) of $\theta$ is the value, $\hat{\theta}$, that maximises $L_{\boldsymbol{Y}}(\theta; \boldsymbol{y})$ as a function of $\theta$.

Some notes on maximum-likelihood estimates and maximum-likelihood estimators follow.

1. The notation is rather unfortunate; $\hat{\theta}$ defined above is the maximum-likelihood estimate, which is just a number. However, by convention, $\hat{\theta}$ is also used to denote the corresponding statistic. We attempt to clarify below.
   - If $\hat{\theta} = h(\boldsymbol{y})$, then $\hat{\theta}$ is the maximum-likelihood estimat<u>e</u> of $\theta$. This is the point estimate (a number) that we calculate in practice given an observed sample $\boldsymbol{y}$.
   - If $\hat{\theta} = h(\boldsymbol{Y})$, then $\hat{\theta}$ is the maximum-likelihood estimat<u>or</u> of $\theta$. This is a function of the sample. It is the theoretical quantity that we consider when determining the properties of maximum likelihood.

   It should be clear from the context when $\hat{\theta}$ is an estimate and when it is an estimator.
2. Let $\Theta$ be the parameter space. The maximum-likelihood estimate of $\theta$ given an observed sample $\boldsymbol{y}$ is the value $\hat{\theta}$ such that
$$L_{\boldsymbol{Y}}(\hat{\theta}; \boldsymbol{y}) = \sup_{\theta \in \Theta} L_{\boldsymbol{Y}}(\theta; \boldsymbol{y}).$$
   Notice that, if $\Theta$ is an open set, this definition admits the possibility that $\hat{\theta}$ is not a member of $\Theta$. We will deal with regular cases in which this issue does not arise.
3. The definition of the maximum-likelihood estimate ensures that no other possible parameter value has a larger likelihood. If $\hat{\theta}$ is the maximum-likelihood estimate based on an observed sample $\boldsymbol{y}$, then
$$L_{\boldsymbol{Y}}(\theta; \boldsymbol{y}) \leq L_{\boldsymbol{Y}}(\hat{\theta}; \boldsymbol{y}),$$
   for all $\theta \in \Theta$.

4. We can also generate the maximum-likelihood estimator by maximising the log-likelihood. Taking logs is a monotone transformation, so the value of $\theta$ that maximises the log-likelihood, $\ell_Y(\theta; Y)$, is the same as the value that maximises the likelihood, $L_Y(\theta; Y)$. Thus, if $\hat{\theta}$ is the maximum-likelihood estimate of $\theta$ based on an observed sample $y$, then
$$\ell_Y(\hat{\theta}; y) = \sup_{\theta \in \Theta} \ell_Y(\theta; y).$$
5. The maximum-likelihood estimate is the value at which the likelihood function attains its maximum. It is often (but not always) the case that we can find the maximum-likelihood estimate for $\theta$ given $y$ by solving
$$\left.\frac{\partial}{\partial \theta} L_Y(\theta; y)\right|_{\theta=\hat{\theta}} = 0.$$
for $\hat{\theta}$. By point 4. above we could also maximise the log-likelihood. We exploit the definition of the score as the derivative of the log-likelihood. In regular cases, the maximum-likelihood estimate $\hat{\theta}$ satisfies
$$s_Y(\hat{\theta}; y) = 0.$$

Maximum likelihood is a widely used inferential technique. In subsection 9.3.1 we describe a number of desirable properties to which maximum likelihood owes its popularity. Before doing so we consider the generalisation to a vector of parameters.

*Vector parameter case*

The definitions given above translate naturally to the case where we have more than one parameter. Consider a sample $\boldsymbol{Y} = (Y_1, \ldots, Y_n)^T$ parameterised by $\boldsymbol{\theta} = (\theta_1, \ldots, \theta_r)$, where $\boldsymbol{\theta}$ takes a value in the parameter space $\Theta$. The maximum-likelihood estimate $\hat{\boldsymbol{\theta}}$ is the value satisfying
$$L_Y(\hat{\boldsymbol{\theta}}; y) = \sup_{\boldsymbol{\theta} \in \boldsymbol{\Theta}} L_Y(\boldsymbol{\theta}, \boldsymbol{Y}),$$
or equivalently
$$\ell_Y(\hat{\boldsymbol{\theta}}; y) = \sup_{\boldsymbol{\theta} \in \boldsymbol{\Theta}} \ell_Y(\boldsymbol{\theta}; \boldsymbol{Y}).$$
In regular cases, the maximum-likelihood estimates satisfy
$$s_Y(\hat{\boldsymbol{\theta}}; y) = \boldsymbol{0}. \tag{9.4}$$

Equation (9.4) actually defines a system of $r$ equations. In most cases of practical interest, this system has no closed-form solution and we require numerical methods to find maximum-likelihood estimates; this is the subject of subsection 9.3.2.

**Example 9.3.2** (Maximum-likelihood estimation for $\mathrm{N}(\mu, \sigma^2)$)
Suppose $\boldsymbol{Y} = (Y_1, \ldots, Y_n)^T$ is a random sample from an $\mathrm{N}(\mu, \sigma^2)$ distribution where both $\mu$ and $\sigma^2$ are unknown, and we have observed $\boldsymbol{y} = (y_1, \ldots, y_n)^T$. We can use the

expression for the density function given at the end of Example 10.1.9 to write the log-likelihood as

$$\ell(\mu, \sigma^2; y) = -\frac{n}{2}\log(2\pi) - \frac{n}{2}\log(\sigma^2) - \frac{n}{2\sigma^2}(\bar{y} - \mu)^2 - \frac{(n-1)}{2\sigma^2}s^2 . \tag{9.5}$$

In the normal case, we can derive closed-form expressions for the maximum-likelihood estimates by setting derivatives of the log-likelihood equal to zero. We start with the mean,

$$\frac{\partial}{\partial \mu}\ell(\mu, \sigma^2; y) = \frac{n}{\sigma^2}(\bar{y} - \mu),$$

so

$$\left.\frac{\partial}{\partial \mu}\ell(\mu, \sigma^2; y)\right|_{\mu=\hat{\mu}} = 0 \quad \Rightarrow \quad \hat{\mu} = \bar{y} .$$

We can be sure that this is a maximum as the likelihood is a quadratic in $\mu$, and the coefficient of $\mu^2$ is negative (think of the shape of the curve defined by $y = -x^2$). Substituting back into equation (9.5) yields the **profile** log-likelihood,

$$\ell(\hat{\mu}, \sigma^2; y) = -\frac{n}{2}\log(2\pi) - \frac{n}{2}\log(\sigma^2) - \frac{(n-1)}{2\sigma^2}s^2 .$$

Differentiating with respect to $\sigma^2$ yields

$$\frac{\partial}{\partial \sigma^2}\ell(\hat{\mu}, \sigma^2; y) = -\frac{n}{2\sigma^2} + \frac{n-1}{2(\sigma^2)^2}s^2 ,$$

so

$$\left.\frac{\partial}{\partial \sigma^2}\ell(\hat{\mu}, \sigma^2; y)\right|_{\sigma^2=\hat{\sigma}^2} = 0 \quad \Rightarrow \quad \hat{\sigma}^2 = \frac{n-1}{n}s^2 .$$

We verify that this is a maximum by checking that the second derivative (with respect to $\sigma^2$) of the profile log-likelihood is negative.

We conclude that the maximum-likelihood estimators of the mean and variance for a normally distributed population are

$$\hat{\mu} = \bar{Y} \quad \text{and} \quad \hat{\sigma}^2 = \frac{n-1}{n}S^2 .$$

These are identical to the method-of-moments estimators. Note that the maximum-likelihood estimator of the variance is biased.

### 9.3.1 *Properties of maximum-likelihood estimates*

We will consider the properties of the maximum-likelihood estimator of a scalar parameter $\theta$. Some of these results extend naturally to the vector parameter case. We start with a result of fundamental importance.

**Proposition 9.3.3** (Consistency of maximum-likelihood estimator)
*Under weak regularity conditions, the maximum-likelihood estimator $\hat{\theta}$ is a consistent estimator of $\theta$.*

Neither the regularity conditions required nor the proof of this proposition are particularly enlightening, so we do not include them here. We will be satisfied to assume that all the cases we deal with are sufficiently regular for consistency of the maximum-likelihood estimator to hold.

Consistency does not give us any clues about the sampling distributions of maximum-likelihood estimators. However, we can use the central limit theorem and the law of large numbers to establish that maximum-likelihood estimators are asymptotically normal. A rigorous proof is beyond the scope of this text, but we provide a sketch that exploits some of the techniques frequently used in rigorous proof of asymptotic results.

**Proposition 9.3.4** (Asymptotic normality of maximum-likelihood estimator)
*Under weak regularity conditions, the maximum-likelihood estimator $\hat{\theta}$ satisfies*

$$\sqrt{n}(\hat{\theta} - \theta) \xrightarrow{d} \mathrm{N}(0, \mathcal{I}_Y^{-1}(\theta)) .$$

*Sketch proof.*

Rather than attempt a rigorous proof, we will indicate why this result is reasonable for a random sample. We start by considering the score function evaluated at the maximum-likelihood estimator. Taylor series expansion of $s_{\mathbf{Y}}(\hat{\theta}; \mathbf{Y})$ around the true value $\theta$ yields

$$s_{\mathbf{Y}}(\hat{\theta}; \mathbf{Y}) = s_{\mathbf{Y}}(\theta; \mathbf{Y}) + (\hat{\theta} - \theta)\frac{\partial}{\partial\theta}s(\theta; \mathbf{Y}) + O((\hat{\theta} - \theta)^2),$$

where $O((\hat{\theta} - \theta)^2)$ denotes terms that are at least quadratic in $(\hat{\theta} - \theta)$. As $\hat{\theta}$ is a consistent estimator of $\theta$ we might hope that these terms would contribute little to the distribution of probability associated with $s_{\mathbf{Y}}(\hat{\theta}; \mathbf{Y})$, so it is reasonable to assume that

$$s_{\mathbf{Y}}(\hat{\theta}; \mathbf{Y}) \approx s_{\mathbf{Y}}(\theta; \mathbf{Y}) + (\hat{\theta} - \theta)\frac{\partial}{\partial\theta}s_{\mathbf{Y}}(\theta; \mathbf{Y}) .$$

In regular cases the score evaluated at $\hat{\theta}$ is zero, that is,

$$s_{\mathbf{Y}}(\hat{\theta}; \mathbf{Y}) = 0,$$

and thus,

$$s_{\mathbf{Y}}(\theta; \mathbf{Y}) + (\hat{\theta} - \theta)\frac{\partial}{\partial\theta}s_{\mathbf{Y}}(\theta; \mathbf{Y}) \approx 0 .$$

Rearranging and multiplying both sides by $\sqrt{n}$ yields

$$\sqrt{n}(\hat{\theta} - \theta) \approx \frac{s_{\mathbf{Y}}(\theta; \mathbf{Y}) \,/\, \sqrt{n}}{-\frac{\partial}{\partial\theta}s_{\mathbf{Y}}(\theta; \mathbf{Y}) \Big/ n} \qquad (9.6)$$

In this sketch proof we assume that we have a random sample, so the score function can be written as a sum of the marginal scores,

$$s_{\mathbf{Y}}(\theta; \mathbf{Y}) = \sum_{i=1}^{n} s_Y(\theta; Y_i).$$

Since the $Y_i$s are independent and identically distributed, the random variables $s_Y(\theta; Y_i)$ must also be independent and identically distributed. We also have, from section 9.2,

$$\mathbb{E}(s_Y(\theta; Y_i)) = 0 \quad \text{and} \quad \text{Var}(s_Y(\theta; Y_i)) = \mathcal{I}_Y(\theta).$$

With some additional weak assumptions, the conditions for the central limit theorem are met, so

$$\frac{s_{\mathbf{Y}}(\theta; \mathbf{Y})}{\sqrt{n}} = \sqrt{n}\frac{\sum_{i=1}^{n} s_Y(\theta; Y_i)}{n} \xrightarrow{d} \text{N}(0, \mathcal{I}_Y(\theta)).$$

We now turn our attention to the denominator of (9.6). This is the sample mean of a sequence of independent and identically distributed random variables of the form

$$-\frac{\partial}{\partial \theta} s_Y(\theta; Y).$$

From section 9.2 we know that

$$\mathbb{E}\left(-\frac{\partial}{\partial \theta} s_Y(\theta; Y)\right) = \mathcal{I}_Y(\theta),$$

so by the law of large numbers, the denominator of (9.6) converges to $\mathcal{I}_Y(\theta)$. The result follows immediately. □

An immediate corollary of this result is that in large samples we can use the normal distribution to approximate the sampling distribution of the maximum-likelihood estimator.

**Corollary 9.3.5** (Large-sample distribution of maximum-likelihood estimator)
*If the sample size n is large, the maximum-likelihood estimator $\hat{\theta}$ is approximately* $\text{N}(\theta, \mathcal{I}_{\mathbf{Y}}^{-1}(\theta))$.

We are often interested in some function of the parameter, say $g(\theta)$. Given an estimator $\hat{\theta}$ of $\theta$, it is tempting to use $g(\hat{\theta})$ as an estimator of $g(\theta)$. This is sometimes referred to as a **plug-in estimator**. In general, plug-in estimators do not preserve the properties of the original estimator; for example, the sample variance $S^2$ is unbiased for $\sigma^2$, but $S$ is not unbiased for $\sigma$. A useful property of maximum likelihood is that plug-in estimators are also maximum-likelihood estimators. In other words, if $\hat{\theta}$ is the maximum-likelihood estimator of $\theta$, then $g(\hat{\theta})$ is the maximum-likelihood estimator of $g(\theta)$.

If $g$ is one-to-one this is easy to establish. Suppose that $\lambda = g(\theta)$ for some one-to-one function $g$. We can reparameterise using the fact that $\theta = g^{-1}(\lambda)$. The likelihood function can then be expressed in terms of $\lambda$,

$$L_Y(\theta; Y) = L_Y(g^{-1}(\lambda); Y).$$

This attains its maximum at $\hat{\theta}$ so $g^{-1}(\hat{\lambda}) = \hat{\theta}$, which implies $\hat{\lambda} = g(\hat{\theta})$.

In general we are interested in functions that are not one-to-one. In this case we have to be clear about exactly what is meant by the maximum-likelihood estimator of $\lambda$ where $\lambda = g(\theta)$. The following definition clarifies.

**Definition 9.3.6** (Induced likelihood)
Suppose that $Y = (Y_1, \ldots, Y_n)^T$ is a sample with likelihood function $L_Y$ parameterised by $\theta$, and let $\lambda = g(\theta)$ where $g : \Theta \to \Lambda$. The **induced likelihood** for $\lambda$ given observed sample $y$ is

$$L'_Y(\lambda; y) = \sup_{\{\theta : g(\theta) = \lambda\}} L_Y(\theta; y).$$

for all $\lambda \in \Lambda$.

Induced likelihood is a simple concept; if $g : \Theta \to \Lambda$ is not a one-to-one function, then for a given value $\lambda \in \Lambda$ there may exist different parameter values $\theta_1, \theta_2, \ldots, \theta_m \in \Theta$ with $g(\theta_1) = g(\theta_2) = \ldots = g(\theta_m) = \lambda$. The induced likelihood $L'_Y(\lambda; y)$ is just the likelihood $L_Y(\theta; y)$ maximised over the set $\{\theta_1, \theta_2, \ldots, \theta_m\}$.

Note that here $\Theta$ is the parameter space and $\Lambda$ is the domain of the function $g$. It is then natural to define the maximum-likelihood estimator $\hat{\lambda}$ as the value that maximises $L'_Y(\lambda; y)$ as a function of $\lambda$. Using this setup we can readily prove the general invariance properties of maximum-likelihood estimators.

**Proposition 9.3.7** (Invariance of MLE)
*If $\hat{\theta}$ is the maximum-likelihood estimator of $\theta$ then, for any function $g(\theta)$, the maximum-likelihood estimator of $g(\theta)$ is $g(\hat{\theta})$.*

*Proof.*

Given the definition of induced likelihood given above, we just need to show

$$L'_Y(g(\hat{\theta}); Y) = L'_Y(\hat{\lambda}; Y).$$

It is clear from the definition that the induced likelihood has the same maxima as the likelihood,

$$\begin{aligned}
L'_Y(\hat{\lambda}; Y) &= \sup_{\lambda \in \Lambda} L'_Y(\lambda; Y) && \text{by definition of } \hat{\lambda} \\
&= \sup_{\lambda \in \Lambda} \sup_{\{\theta : g(\theta) = \lambda\}} L_Y(\theta; Y) && \text{by definition of induced likelihood} \\
&= \sup_{\theta \in \Theta} L_Y(\theta; Y) \\
&= L_Y(\hat{\theta}; Y).
\end{aligned}$$

If we restrict our attention to the set of parameters $\{\theta : g(\theta) = g(\hat{\theta})\}$ then $\hat{\theta}$ is in this set so

$$\begin{aligned} L_{\boldsymbol{Y}}(\hat{\theta};\boldsymbol{Y}) &= \sup_{\theta\in\Theta} L_{\boldsymbol{Y}}(\theta;\boldsymbol{Y}) \\ &= \sup_{\{\theta:g(\theta)=g(\hat{\theta})\}} L_{\boldsymbol{Y}}(\theta;\boldsymbol{Y}) \\ &= L'_{\boldsymbol{Y}}(g(\hat{\theta});\boldsymbol{Y}). && \text{by definition of induced likelihood.} \end{aligned}$$

We conclude that $L'_{\boldsymbol{Y}}(g(\hat{\theta});\boldsymbol{Y}) = L'_{\boldsymbol{Y}}(\hat{\lambda};\boldsymbol{Y})$, and the result is proved. □

**Example 9.3.8** (Asymptotic pivots from maximum likelihood)
Suppose that $\hat{\theta}$ is the maximum-likelihood estimator of a scalar parameter $\theta$. We know that, under mild regularity conditions,

$$\hat{\theta} \xrightarrow{d} \mathrm{N}(\theta, \mathcal{I}_{\boldsymbol{Y}}^{-1}(\theta)),$$

and that, for large $n$,

$$\sqrt{\mathcal{I}_{\boldsymbol{Y}}(\theta)}(\hat{\theta}-\theta) \overset{\text{appr.}}{\sim} \mathrm{N}(0,1).$$

As $\sqrt{\mathcal{I}_{\boldsymbol{Y}}(\theta)}(\hat{\theta}-\theta)$ converges to a distribution that does not depend on $\theta$, we say that $\sqrt{\mathcal{I}_{\boldsymbol{Y}}(\theta)}(\hat{\theta}-\theta)$ is asymptotically pivotal for $\theta$. Constructing a confidence interval for $\theta$ in a specific case is part of Exercise 9.3.

### 9.3.2 *Numerical maximisation of likelihood*

In practice, numerical optimisation techniques are often required to find maximum-likelihood estimates. We will consider a vector of parameters $\boldsymbol{\theta}$ and the regular case where the maximum-likelihood estimate of $\boldsymbol{\theta}$ given observed sample $\boldsymbol{y}$ satisfies

$$\boldsymbol{s}_{\boldsymbol{Y}}(\hat{\boldsymbol{\theta}};\boldsymbol{y}) = \boldsymbol{0}.$$

The score function features heavily in this section. In order to make the notation less cumbersome, we will drop the $\boldsymbol{Y}$ subscript and the $\boldsymbol{y}$ argument from the score, so

$$\boldsymbol{s}(\boldsymbol{\theta}) = \boldsymbol{s}_{\boldsymbol{Y}}(\boldsymbol{\theta},\boldsymbol{y}) = \nabla_{\boldsymbol{\theta}}\ell_{\boldsymbol{Y}}(\boldsymbol{\theta};\boldsymbol{y}),$$

where $\nabla_{\boldsymbol{\theta}}$ refers to differentiation with respect to a vector in the sense described in section 9.2. We will use $\dot{\boldsymbol{s}}$ to refer to the matrix given by the first derivative of the transpose of the score, so

$$\dot{\boldsymbol{s}}(\boldsymbol{\theta}) = \nabla_{\boldsymbol{\theta}}\boldsymbol{s}(\boldsymbol{\theta})^T = \nabla_{\boldsymbol{\theta}}\nabla_{\boldsymbol{\theta}}^T\ell_{\boldsymbol{Y}}(\boldsymbol{\theta};\boldsymbol{y}).$$

Suppose that our current estimate of $\boldsymbol{\theta}$ is $\boldsymbol{\theta}^*$. We would like to find the maximum-likelihood estimator $\hat{\boldsymbol{\theta}}$. We know that, in regular cases, $\hat{\boldsymbol{\theta}}$ is the solution of $\boldsymbol{s}(\hat{\boldsymbol{\theta}}) = 0$.

We use a Taylor series expansion of $\boldsymbol{s}(\hat{\boldsymbol{\theta}})$ around $\boldsymbol{\theta}^*$ to give us an idea of how we might improve our estimate. Ignoring terms of second order or higher,

$$\boldsymbol{s}(\hat{\boldsymbol{\theta}}) \approx \boldsymbol{s}(\boldsymbol{\theta}^*) + \dot{\boldsymbol{s}}(\boldsymbol{\theta})(\hat{\boldsymbol{\theta}} - \boldsymbol{\theta}^*).$$

Since $\boldsymbol{s}(\hat{\boldsymbol{\theta}}) = \mathbf{0}$, we can rearrange to obtain

$$\hat{\boldsymbol{\theta}} \approx \boldsymbol{\theta}^* - \dot{\boldsymbol{s}}(\boldsymbol{\theta}^*)^{-1}\boldsymbol{s}(\boldsymbol{\theta}^*). \tag{9.7}$$

Clearly, if $\boldsymbol{\theta}^*$ is a long way from $\hat{\boldsymbol{\theta}}$, the higher-order terms (that is, terms in $(\hat{\boldsymbol{\theta}} - \boldsymbol{\theta}^*)^2$) will not be small, and the approximation given by (9.7) will be poor. In practice, (9.7) is used as part of an iterative process, known as the **Newton-Raphson scheme**.

Suppose that $\boldsymbol{\theta}^*_k$ is our current estimate; we define the next estimate as

$$\boldsymbol{\theta}^*_{k+1} = \boldsymbol{\theta}^*_k - \dot{\boldsymbol{s}}(\boldsymbol{\theta}^*_k)^{-1}\boldsymbol{s}(\boldsymbol{\theta}^*_k).$$

The iterations are initialised with some sensible estimate $\boldsymbol{\theta}^*_0$. We may, for example, use a method-of-moments estimate as the initial value for the procedure. The iterative steps are then repeated for $k = 1, 2, \ldots$ in the hope that the sequence $\boldsymbol{\theta}^*_1, \boldsymbol{\theta}^*_2, \ldots,$ will converge to the maximum-likelihood estimate.

The **method of scoring** is an alternative numerical procedure. It is a modification of the Newton-Raphson scheme, which exploits the result

$$\mathbb{E}(-\dot{\boldsymbol{s}}(\boldsymbol{\theta})) = \mathcal{I}_{\boldsymbol{Y}}(\boldsymbol{\theta}),$$

established in section 9.2. Replacing the term $-\dot{\boldsymbol{s}}(\boldsymbol{\theta}^*_k)$ in the Newton-Raphson scheme by its expected value, the iterations become

$$\boldsymbol{\theta}^*_{k+1} = \boldsymbol{\theta}^*_k + \mathcal{I}_{\boldsymbol{Y}}(\boldsymbol{\theta}^*_k)^{-1}\boldsymbol{s}(\boldsymbol{\theta}^*_k).$$

The information matrix does not depend on the observed sample and is often simpler to evaluate than the derivative of the score.

Numerical procedures are not guaranteed to find the maximum-likelihood estimate, that is, we cannot be certain of locating the global maximum of the likelihood function. A necessary condition for finding the maximum-likelihood estimate is that the iterative procedure converges. Convergence is judged using the difference between consecutive estimates. We set some arbitrary tolerance $\delta > 0$ and say that the procedure has converged if we can find a $j$ for which

$$|\boldsymbol{\theta}^*_j - \boldsymbol{\theta}^*_{j-1}| < \delta.$$

We take $\hat{\boldsymbol{\theta}} = \boldsymbol{\theta}^*_j$ to be the maximum-likelihood estimate. All numerical procedures suffer from a number of potential drawbacks related to the following two key issues:

i. The procedure may not converge. This is a particular problem when dealing with relatively flat likelihood surfaces.

ii. Even if the procedure converges, it may not converge to the correct value. For example, we may find a local maximum rather than the global maximum.

Clearly there is a balance here; if we set $\delta$ too small, the procedure may take a long time to converge or not converge at all. On the other hand, if we set $\delta$ too large, we increase our chances of converging on a value that is not close to the maximum. A value for $\delta$ is chosen using a number of considerations including the accuracy of the computer being used, the nature of the problem at hand, and – all importantly – what has worked in the past.

Comparison of the relative merits of the Newton-Raphson scheme and the method of scoring are far from straightforward. If both schemes converge, then Newton-Raphson usually converges faster. However, the method of scoring is often less sensitive to the choice of initial values. In practice slightly more sophisticated variants of the Newton-Raphson scheme (such as the Broyden-Fletcher-Goldfarb-Shanno algorithm and related methods) are used to find maximum-likelihood estimates.

### 9.3.3 EM algorithm

The **EM algorithm** is an iterative process for finding maximum-likelihood estimates. Each iteration of the algorithm has two stages. The letters "E" and "M" each refer to one stage: E for expectation and M for maximization. In describing the EM algorithm we will make extensive use of conditioning. In this context, density function notation, $f_{\boldsymbol{Y}}(\boldsymbol{y};\boldsymbol{\theta})$, is clearer than the likelihood function notation $L_{\boldsymbol{Y}}(\boldsymbol{\theta};\boldsymbol{y})$. We will use the former with the understanding that the functional form of the likelihood and the density are identical.

In many problems, a natural model specification involves **latent** or **hidden** variables. As the names suggest, these are variables that we do not observe directly. For convenience, we will group all non-observed variables as a single variable $\boldsymbol{Z}$. We refer to $(\boldsymbol{Y},\boldsymbol{Z})$ as the complete data. It is often the case that it is easier to find maximum-likelihood estimators for the complete data, $(\boldsymbol{Y},\boldsymbol{Z})$, than it is for the sample, $\boldsymbol{Y}$. However, these estimators are functions of variables that we do not observe. The EM algorithm gets around this problem by taking expectations with respect to the hidden variables.

We begin by giving a brief description of the algorithm, then go on to show that this algorithm will give a sequence of parameter estimates that will never reduce the value of the likelihood function. We use $\boldsymbol{\theta}$ to denote a generic parameter value, and $\boldsymbol{\theta}_j^*$ to denote our estimate at the $j^{\text{th}}$ step of the iterative process. The notation is by necessity rather cumbersome, but is somewhat reduced by first defining the **log-likelihood ratio** for the complete data case,

$$K(\boldsymbol{\theta};\boldsymbol{\theta}_j^*,\boldsymbol{y},\boldsymbol{z}) = \log \frac{f_{\boldsymbol{Y},\boldsymbol{Z}}(\boldsymbol{y},\boldsymbol{z};\boldsymbol{\theta})}{f_{\boldsymbol{Y},\boldsymbol{Z}}(\boldsymbol{y},\boldsymbol{z};\boldsymbol{\theta}_j^*)},$$

and the expectation of the complete data log-likelihood ratio conditioned on the observed data,

$$\begin{aligned} J(\boldsymbol{\theta}; \boldsymbol{\theta}_j^*, \boldsymbol{y}) &= \mathop{\mathbb{E}}_{\boldsymbol{\theta}_j^*}(K(\boldsymbol{\theta}; \boldsymbol{\theta}_j^*, \boldsymbol{Y}, \boldsymbol{Z}) | \boldsymbol{Y} = \boldsymbol{y}) \\ &= \int_{\mathbb{R}^n} K(\boldsymbol{\theta}; \boldsymbol{\theta}_j^*, \boldsymbol{y}, \boldsymbol{z}) f_{\boldsymbol{Z}|\boldsymbol{Y}}(\boldsymbol{z}|\boldsymbol{y}; \boldsymbol{\theta}_j^*) d\boldsymbol{y} \, . \end{aligned}$$

Note that we view $K$ and $J$ primarily as functions of the $\boldsymbol{\theta}$. The function $J$ provides our objective function (that is, the function we wish to minimise) at each step of the iterative process. Given some initial parameter value $\boldsymbol{\theta}_1^*$, the $j^{\text{th}}$ step in the EM-algorithm consists of two stages:

i. Expectation: calculate $J(\boldsymbol{\theta}; \boldsymbol{\theta}_j^*, \boldsymbol{y})$ .
ii. Maximisation: set $\boldsymbol{\theta}_{j+1}^* = \arg\max_{\boldsymbol{\theta}} J(\boldsymbol{y}; \boldsymbol{\theta}, \boldsymbol{\theta}_j^*)$ .

We can describe the algorithm somewhat more informally. Given a starting estimate for the parameter $\boldsymbol{\theta}$, we:

i. find the expected values of the missing data, treating the parameter estimate as the true value;
ii. estimate the parameter, treating the expected values of the missing data as the true values.

The two steps are repeated until convergence. We will now show that the EM algorithm results in a sequence of estimates with non-decreasing likelihood values. We begin by proving a useful result.

**Proposition 9.3.9** (Gibbs inequality)
*If $X$ is a random variable with density function $f_X$ and support $A \subset \mathbb{R}$, and $f_Y$ is any other density function with the same support, then*

$$D(f_X, f_Y) = \int_A \log\left(\frac{f_X(x)}{f_Y(x)}\right) f_X(x) dx \geq 0 \, .$$

*This quantity is known as the* ***Kullback-Leibler divergence*** *of $f_Y$ from $f_X$.*

*Proof.*
Define the random variable $Y = f_Y(X)/f_X(X)$. We have

$$\int_A \log\left(\frac{f_X(x)}{f_Y(x)}\right) f_X(x) dx = \int_A \left\{-\log\left(\frac{f_Y(x)}{f_X(x)}\right)\right\} f_X(x) dx = \mathbb{E}(-\log Y) \, .$$

The function $g(y) = -\log y$ is convex, so we can apply Jensen's inequality,

$$\begin{aligned} \mathbb{E}(-\log Y) \geq -\log \mathbb{E}(Y) &= -\log \int_A \frac{f_Y(x)}{f_X(x)} f_X(x) dx \\ &= -\log \int_A f_Y(x) dx = -\log 1 = 0 \, , \end{aligned}$$

and the result is proved. □

The Kullback-Leibler divergence is a measure of the difference of $f_Y$ from the "true" density $f_X$. As $g(y) = -\log y$ is strictly convex for all $y > 0$, the divergence is equal to zero only when the two distributions are the same.

**Proposition 9.3.10** (EM-algorithm estimates)
*Consider the sequence of EM-algorithm estimates* $\boldsymbol{\theta}_1^*, \boldsymbol{\theta}_2^*, \ldots$ *If* $L_{\boldsymbol{Y}}(\boldsymbol{\theta}; \boldsymbol{y})$ *is the likelihood function, then* $L_{\boldsymbol{Y}}(\boldsymbol{\theta}_{j+1}^*; \boldsymbol{y}) \geq L_{\boldsymbol{Y}}(\boldsymbol{\theta}_j^*; \boldsymbol{y})$ *for* $j = 1, 2, \ldots$.

*Proof.*

In this proof, we drop subscripts on densities and likelihoods to avoid cluttering the notation. Consider the log-likelihood ratio for consecutive parameter estimates in the complete data case,

$$\begin{aligned}
K(\boldsymbol{\theta}_{j+1}^*; \boldsymbol{\theta}_j^*, \boldsymbol{Y}, \boldsymbol{Z}) &= \log \frac{f(\boldsymbol{Y}, \boldsymbol{Z}; \boldsymbol{\theta}_{j+1}^*)}{f(\boldsymbol{Y}, \boldsymbol{Z}; \boldsymbol{\theta}_j^*)} \\
&= \log \frac{f(\boldsymbol{Z}|\boldsymbol{Y}; \boldsymbol{\theta}_{j+1}^*) f(\boldsymbol{Y}; \boldsymbol{\theta}_{j+1}^*)}{f(\boldsymbol{Z}|\boldsymbol{Y}; \boldsymbol{\theta}_j^*) f(\boldsymbol{Y}; \boldsymbol{\theta}_j^*)} \\
&= \log \frac{f(\boldsymbol{Z}|\boldsymbol{Y}; \boldsymbol{\theta}_{j+1}^*)}{f(\boldsymbol{Z}|\boldsymbol{Y}; \boldsymbol{\theta}_j^*)} + \log \frac{f(\boldsymbol{Y}; \boldsymbol{\theta}_{j+1}^*)}{f(\boldsymbol{Y}; \boldsymbol{\theta}_j^*)} \\
&= \log \frac{f(\boldsymbol{Z}|\boldsymbol{Y}; \boldsymbol{\theta}_{j+1}^*)}{f(\boldsymbol{Z}|\boldsymbol{Y}; \boldsymbol{\theta}_j^*)} + \log \frac{L(\boldsymbol{\theta}_{j+1}^*; \boldsymbol{Y})}{L(\boldsymbol{\theta}_j^*; \boldsymbol{Y})} \\
\Rightarrow \log \frac{L(\boldsymbol{\theta}_{j+1}^*; \boldsymbol{Y})}{L(\boldsymbol{\theta}_j^*; \boldsymbol{Y})} &= K(\boldsymbol{\theta}_{j+1}^*; \boldsymbol{\theta}_j^*, \boldsymbol{Y}, \boldsymbol{Z}) - \log \frac{f(\boldsymbol{Z}|\boldsymbol{Y}; \boldsymbol{\theta}_{j+1}^*)}{f(\boldsymbol{Z}|\boldsymbol{Y}; \boldsymbol{\theta}_j^*)} \\
&= K(\boldsymbol{\theta}_{j+1}^*; \boldsymbol{\theta}_j^*, \boldsymbol{Y}, \boldsymbol{Z}) + \log \frac{f(\boldsymbol{Z}|\boldsymbol{Y}; \boldsymbol{\theta}_j^*)}{f(\boldsymbol{Z}|\boldsymbol{Y}; \boldsymbol{\theta}_{j+1}^*)}.
\end{aligned}$$

Conditioning on $\boldsymbol{Y} = \boldsymbol{y}$ and taking expectation with respect to $Z$ (assuming parameter value $\boldsymbol{\theta}_j^*$) yields

$$\log \frac{L(\boldsymbol{\theta}_{j+1}^*; \boldsymbol{y})}{L(\boldsymbol{\theta}_j^*; \boldsymbol{y})} = J(\boldsymbol{\theta}_{j+1}^*; \boldsymbol{\theta}_j^*, \boldsymbol{y}) + \int_{-\infty}^{\infty} \log \frac{f(\boldsymbol{z}|\boldsymbol{y}; \boldsymbol{\theta}_j^*)}{f(\boldsymbol{z}|\boldsymbol{y}; \boldsymbol{\theta}_{j+1}^*)} f(\boldsymbol{z}|\boldsymbol{y}; \boldsymbol{\theta}_j^*) d\boldsymbol{z}. \tag{9.8}$$

We consider each term on the right-hand side of (9.8) in turn. For the first term, we know that $\boldsymbol{\theta}_{j+1}^*$ maximises $J(\boldsymbol{\theta}; \boldsymbol{\theta}_j^*, \boldsymbol{y})$ with respect to $\boldsymbol{\theta}$, so

$$J(\boldsymbol{\theta}_{j+1}^*; \boldsymbol{\theta}_j^*, \boldsymbol{y}) \geq J(\boldsymbol{\theta}_j^*; \boldsymbol{\theta}_j^*, \boldsymbol{y}) = 0.$$

The second term is the Kullback-Leibler divergence of $f(\boldsymbol{z}|\boldsymbol{y}; \boldsymbol{\theta}_{j+1}^*)$ from $f(\boldsymbol{z}|\boldsymbol{y}; \boldsymbol{\theta}_j^*)$, so it also greater than zero. We conclude that $\log(L(\boldsymbol{\theta}_{j+1}^*; \boldsymbol{y})/L(\boldsymbol{\theta}_j^*; \boldsymbol{y}))$ is non-negative. As log is a monotone function, this implies $L(\boldsymbol{\theta}_{j+1}^*; \boldsymbol{y}) \geq L(\boldsymbol{\theta}_j^*; \boldsymbol{y})$. □

The definition of the function $J$ in terms of the log-likelihood ratio makes the proof of

Proposition 9.3.10 more straightforward. In practice, the denominator plays no part in the maximisation and we may take

$$J(\boldsymbol{\theta}; \boldsymbol{\theta}_j^*, \boldsymbol{y}) = \mathbb{E}_{\boldsymbol{\theta}_j^*}(\log f(\boldsymbol{Y}, \boldsymbol{Z}; \boldsymbol{\theta}) | \boldsymbol{Y} = \boldsymbol{y}). \tag{9.9}$$

While Proposition 9.3.10 tells us that the EM algorithm will never give us a worse estimate that the starting value, it gives no guarantee of convergence. In fact, it can be established that the EM-algorithm estimates converge to the (incomplete-data) maximum-likelihood estimate. However, in some instances the rate of convergence may be slow. In practice, parameter estimation often exploits a combination of approaches. For example, we may use the method of moments to generate initial estimates, the EM algorithm to get close to the maximum-likelihood estimates, then some variant of the Newton-Raphson method for final refinement.

**Example 9.3.11** (Estimation for mixtures)
Consider a situation in which we know that our sample comes from two distinct populations, but we do not know whether any given sample member came from the first or second population. In addition, suppose that the populations have the same distributional form but with different parameters. For the sake of simplicity, we assume one half of our sample comes from a population with parameter $\phi_0$, and the other half comes from a population with parameter $\phi_1$. The likelihood for the observed data can be written as

$$f_{\boldsymbol{Y}}(\boldsymbol{y}; \phi_0, \phi_1) = \prod_{i=1}^{n} \left[ \frac{1}{2} g(y_i; \phi_0) + \frac{1}{2} g(y_i; \phi_1) \right],$$

where $g$ represents the common density function for the two populations. In general, this likelihood function is not easy to maximise. We use the EM approach and introduce a latent (unobserved) variable $\boldsymbol{Z}$. The obvious candidate for the latent variable in this case is an indicator for whether an observation comes from population 0 or population 1. For the $i^{\text{th}}$ sample member we denote this variable $Z_i$, so

$$Z_i = \begin{cases} 0 & \text{if } Y_i \text{ from population 0,} \\ 1 & \text{if } Y_i \text{ from population 1.} \end{cases}$$

We can set $Z_i \sim \text{Bernoulli}(0.5)$. The likelihood for the complete data, $(\boldsymbol{Y}, \boldsymbol{Z})$, is then

$$f_{\boldsymbol{Y},\boldsymbol{Z}}(\boldsymbol{y}, \boldsymbol{z}) = \prod_{i=1}^{n} g(y_i; \phi_0)^{1-z_i} g(y_i; \phi_1)^{z_i},$$

and thus the log-likelihood is

$$\log f_{\boldsymbol{Y},\boldsymbol{Z}}(\boldsymbol{y}, \boldsymbol{z}) = \sum_{i=1}^{n} (1 - z_i) \log g(y_i; \phi_0) + \sum_{i=1}^{n} z_i \log g(y_i; \phi_1).$$

If we define the parameter vector as $\boldsymbol{\theta} = (\phi_0, \phi_1)$, so $\boldsymbol{\theta}_j^* = (\phi_{0,j}^*, \phi_{1,j}^*)$ is our estimate at the $j^{\text{th}}$ step, equation (9.9) yields

$$\begin{aligned}
J(\boldsymbol{\theta};\boldsymbol{\theta}_j^*,\boldsymbol{y}) &= \mathop{\mathbb{E}}_{\boldsymbol{\theta}_j^*}\left[\sum_{i=1}^n (1-Z_i)\log g(Y_i;\phi_0) + \sum_{i=1}^n Z_i \log g(Y_i;\phi_1)\middle| \boldsymbol{Y}=\boldsymbol{y}\right] \\
&= \sum_{i=1}^n \left(1-\mathop{\mathbb{E}}_{\boldsymbol{\theta}_j^*}(Z_i|\boldsymbol{Y}=\boldsymbol{y})\right)\log g(y_i;\phi_0) + \sum_{i=1}^n \mathop{\mathbb{E}}_{\boldsymbol{\theta}_j^*}(Z_i|\boldsymbol{Y}=\boldsymbol{y})\log g(y_i;\phi_1)\,.
\end{aligned}$$

As each $Z_i$ is a Bernoulli random variable, we can write

$$\begin{aligned}
\mathop{\mathbb{E}}_{\boldsymbol{\theta}_j^*}(Z_i|\boldsymbol{Y}=\boldsymbol{y}) &= P_{\boldsymbol{\theta}_j^*}(Z_i=1|\boldsymbol{Y}=\boldsymbol{y}) \\
&= \frac{f_{\boldsymbol{Y},Z_i}(\boldsymbol{y},1;\boldsymbol{\theta}_j^*)}{\sum_z f_{\boldsymbol{Y},Z_i}(\boldsymbol{y},z;\boldsymbol{\theta}_j^*)} \\
&= \frac{f_{\boldsymbol{Y}}(\boldsymbol{y}|Z_i=1;\boldsymbol{\theta}_j^*)\mathrm{P}(Z_i=1)}{\sum_{j=0}^1 f_{\boldsymbol{Y}}(\boldsymbol{y}|Z_i=j;\boldsymbol{\theta}_j^*)\mathrm{P}(Z_i=j)} \\
&= \frac{g(y_i;\phi_{1,j}^*)\frac{1}{2}}{g(y_i;\phi_{0,j}^*)\frac{1}{2}+g(y_i;\phi_{1,j}^*)\frac{1}{2}} \\
&= \left(1+\frac{g(y_i;\phi_{0,j}^*)}{g(y_i;\phi_{1,j}^*)}\right)^{-1}.
\end{aligned}$$

We then proceed by substituting this expectation into the expression for $J(\boldsymbol{\theta};\boldsymbol{\theta}_j^*,\boldsymbol{y})$, and maximising with respect to $\boldsymbol{\theta}$.

**Example 9.3.12** (Exponential family)
Now suppose that the distribution of the complete data belongs to an exponential family (as in Example 3.3.24), so the likelihood is

$$f_{\boldsymbol{Y},\boldsymbol{Z}}(\boldsymbol{y},\boldsymbol{z}) = h(\boldsymbol{y},\boldsymbol{z})a(\boldsymbol{\theta})\exp\left[\sum_{i=1}^k \eta_i(\boldsymbol{\theta})t_i(\boldsymbol{y},\boldsymbol{z})\right],$$

and the log-likelihood is

$$\log f_{\boldsymbol{Y},\boldsymbol{Z}}(\boldsymbol{Y},\boldsymbol{Z}) = \log h(\boldsymbol{y},\boldsymbol{z}) + \log a(\boldsymbol{\theta}) + \sum_{i=1}^k \eta_i(\boldsymbol{\theta})t_i(\boldsymbol{y},\boldsymbol{z})\,.$$

The first term in this expression does not involve $\boldsymbol{\theta}$ and can be ignored. We then have

$$\begin{aligned}
J(\boldsymbol{\theta};\boldsymbol{\theta}_j^*,\boldsymbol{y}) &= \mathop{\mathbb{E}}_{\boldsymbol{\theta}_j^*}\left[\log a(\boldsymbol{\theta}) + \sum_{i=1}^k \eta_i(\boldsymbol{\theta})t_i(\boldsymbol{Y},\boldsymbol{Z}) + C\middle| \boldsymbol{Y}=\boldsymbol{y}\right] \\
&= \log a(\boldsymbol{\theta}) + \sum_{i=1}^k \eta_i(\boldsymbol{\theta})\mathop{\mathbb{E}}_{\boldsymbol{\theta}_j^*}\left[t_i(\boldsymbol{y},\boldsymbol{Z})|\boldsymbol{Y}=\boldsymbol{y}\right] + C\,.
\end{aligned}$$

We need to maximise this expression with respect to $\boldsymbol{\theta}$, so we set its derivative equal to zero to obtain

$$\frac{d}{d\boldsymbol{\theta}} \log a(\boldsymbol{\theta}) + \sum_{i=1}^{k} \left( \frac{d}{d\boldsymbol{\theta}} \eta_i(\boldsymbol{\theta}) \right) \mathbb{E}_{\boldsymbol{\theta}_j^*} [t_i(\boldsymbol{y}, \boldsymbol{Z}) | \boldsymbol{Y} = \boldsymbol{y}] = 0 .$$

If we had the complete data, we would find the MLE of $\boldsymbol{\theta}$ by setting the score function equal to zero, that is,

$$\frac{d}{d\boldsymbol{\theta}} \log a(\boldsymbol{\theta}) + \sum_{i=1}^{k} \left( \frac{d}{d\boldsymbol{\theta}} \eta_i(\boldsymbol{\theta}) \right) t_i(\boldsymbol{y}, \boldsymbol{z}) = 0 .$$

Thus, the EM algorithm involves simply replacing the unknown term $t_i(\boldsymbol{y}, \boldsymbol{z})$ by its expectation given the sample, $\boldsymbol{y}$.

**Exercise 9.3**

1. Find the maximum-likelihood estimator for $\theta$ given a random sample of size $n$ from a population with density function

$$f_Y(y; \theta) = \frac{2y}{\theta^2} \quad \text{for } 0 < y \leq \theta .$$

   Do not use calculus. Draw a picture of the log-likelihood.

2. Suppose that we have a random sample of size $n$ from the logistic distribution with density function

$$f_Y(y; \mu) = \frac{\exp(y - \mu)}{[1 + \exp(y - \mu)]^2} \quad \text{for} \quad -\infty < y < \infty.$$

   Find the log-likelihood function and derive the iterations for the method of scoring. [*Hint*: write quantities, as much as possible, in terms of the density function and the cumulative distribution function for the logistic.]

3. Suppose that $\boldsymbol{Y} = (Y_1, \ldots, Y_n)^T$ is a random sample from a Pois($\lambda$) distribution. Using the approximate normality of the maximum-likelihood estimator, construct an interval estimator for $\lambda$ with confidence coefficient 0.9.

4. Let $X_1, \ldots, X_n$ be a random sample from an Exp($\lambda$) distribution. Suppose the last $n - m$ observations ($m < n$) are censored in the sense that, for $j = m + 1, \ldots, n$, we only know $X_j \in [u_j, v_j]$ for some given $v_j > u_j > 0$, while $X_j$ itself is unknown. Describe an EM algorithm for estimating the parameter $\lambda$.

5. Let $X \sim \mathrm{N}(\mu, \sigma^2)$. Find the values of $\mu$ and $\sigma^2$ which minimise the Kullback-Leibler divergence of $f_X$ from $f_Y$, where

   (a) $Y \sim \text{Exp}(\lambda)$,

   (b) $Y \sim \text{Gamma}(\alpha, \lambda)$,

   (c) $Y \sim \text{Unif}[\alpha, \beta]$.

## 9.4 Likelihood-ratio test

Consider the hypothesis testing problem

$$H_0 : \boldsymbol{\theta} \in \boldsymbol{\Theta}_0 ,$$
$$H_1 : \boldsymbol{\theta} \notin \boldsymbol{\Theta}_0 .$$

We can view the null hypothesis as placing a constraint on the values that the parameters can take. In this setup, $\boldsymbol{\Theta}_1 = \boldsymbol{\Theta}_0^c$ and $\boldsymbol{\Theta}_0 \cup \boldsymbol{\Theta}_1 = \boldsymbol{\Theta}$, that is, the null and alternative cover the whole of the parameter space. The **likelihood ratio** provides us with an explicit test statistic for this situation; as the name suggests, this statistic is based on the likelihood function.

**Definition 9.4.1** (Likelihood-ratio test statistic)
Consider a sample $\boldsymbol{Y}$ with likelihood function $L_{\boldsymbol{Y}}$. The **likelihood-ratio test statistic** associated with testing $H_0 : \boldsymbol{\theta} \in \boldsymbol{\Theta}_0$ is

$$r(\boldsymbol{Y}) = \frac{\sup_{\boldsymbol{\theta} \in \boldsymbol{\Theta}} L_{\boldsymbol{Y}}(\boldsymbol{\theta}; \boldsymbol{Y})}{\sup_{\boldsymbol{\theta} \in \boldsymbol{\Theta}_0} L_{\boldsymbol{Y}}(\boldsymbol{\theta}; \boldsymbol{Y})},$$

where $\boldsymbol{\Theta}$ is the entire parameter space.

The likelihood-ratio test statistic is intimately linked with maximum likelihood. For an observed sample $\boldsymbol{y}$, the observed value of the likelihood-ratio test statistic is

$$r(\boldsymbol{y}) = \frac{\sup_{\boldsymbol{\theta} \in \boldsymbol{\Theta}} L_{\boldsymbol{Y}}(\boldsymbol{\theta}; \boldsymbol{y})}{\sup_{\boldsymbol{\theta} \in \boldsymbol{\Theta}_0} L_{\boldsymbol{Y}}(\boldsymbol{\theta}; \boldsymbol{y})}.$$

The numerator is the global maximum of the likelihood function; this is attained at the maximum-likelihood estimate $\hat{\boldsymbol{\theta}}$. The denominator is the maximum of the likelihood function subject to the constraint that $\boldsymbol{\theta} \in \boldsymbol{\Theta}_0$; we use $\hat{\boldsymbol{\theta}}_0$ to denote the value at which this constrained likelihood attains its maximum. Thus, we can write

$$r(\boldsymbol{y}) = \frac{L_{\boldsymbol{Y}}(\hat{\boldsymbol{\theta}}; \boldsymbol{y})}{L_{\boldsymbol{Y}}(\hat{\boldsymbol{\theta}}_0; \boldsymbol{y})}.$$

The likelihood-ratio test statistic is intuitively appealing. We can show that the observed value, $r(\boldsymbol{y})$, must be greater than or equal to 1 for all $\boldsymbol{y}$ (see Exercise 9.3). However, we will only reject the null if the observed value is *substantially* bigger than 1. Put loosely, we know that the constrained parameter estimate, $\hat{\boldsymbol{\theta}}_0$, will be worse than the unconstrained estimate, $\hat{\boldsymbol{\theta}}$, in the sense that the associated likelihood value will be smaller. However, we only reject our constraints if $\hat{\boldsymbol{\theta}}_0$ is substantially worse than $\hat{\boldsymbol{\theta}}$. An immediate question arises as to how to judge what constitutes "substantially worse". The obvious hypothesis-testing solution is to control the size of the test, that is, to choose a critical value $c$ such that

$$P_{\boldsymbol{\theta}_0}(r(\boldsymbol{Y}) > c) = \alpha .$$

The observed value of the likelihood-ratio test statistic is calculated using values of the likelihood function. These values often have to be found using a numerical optimisation procedure; it is no great surprise that the exact sampling distribution of the likelihood ratio is not usually known. However, under certain conditions, we can give an approximate distribution for $2 \log r(\boldsymbol{Y})$. Details of this are given in subsection 9.4.2. The next subsection deals with a common application of likelihood-ratio tests.

### 9.4.1 *Testing in the presence of nuisance parameters*

Likelihood-ratio tests are useful in a rather specific but very commonly occurring situation. We have the (by now familiar) setup of a sample $\boldsymbol{Y}$ parameterised by $\boldsymbol{\theta}$. Suppose that $\boldsymbol{\theta}$ can be split into two groups: the main parameters of interest, $\boldsymbol{\psi}$, and the remaining parameters, $\boldsymbol{\lambda}$, that are of little interest. The parameters $\boldsymbol{\lambda}$ are often referred to, rather unkindly, as **nuisance parameters**. Thus, we can write

$$\boldsymbol{\theta} = \begin{pmatrix} \boldsymbol{\psi} \\ \boldsymbol{\lambda} \end{pmatrix} = (\boldsymbol{\psi}^T, \boldsymbol{\lambda}^T)^T,$$

where the transposes in the final expression ensure that $\boldsymbol{\theta}$ is a column vector. In many instances, the orientation of the parameters is unimportant; we use $\boldsymbol{\theta}$, $\boldsymbol{\psi}$, and $\boldsymbol{\lambda}$ simply as lists of scalar parameters. For the remainder of this section, we will often drop the transposes and refer to

$$\boldsymbol{\theta} = (\boldsymbol{\psi}, \boldsymbol{\lambda}).$$

It will at times be convenient to view each of $\boldsymbol{\theta}$, $\boldsymbol{\psi}$, and $\boldsymbol{\lambda}$ as individual (vector) parameters.

Given this setup, in which $\boldsymbol{\psi}$ is the quantity of interest, we may want to test the hypothesis that $\boldsymbol{\psi}$ takes a particular value,

$$\begin{aligned} H_0 &: \boldsymbol{\psi} = \boldsymbol{\psi}_0, \\ H_1 &: \boldsymbol{\psi} \neq \boldsymbol{\psi}_0. \end{aligned} \tag{9.10}$$

It is clear that the alternative here is a composite hypothesis; it may not be immediately obvious that the null is also composite too. Recall that the model is parameterised by $\boldsymbol{\theta} = (\boldsymbol{\psi}, \boldsymbol{\lambda})$, so (9.10) could be stated as

$$\begin{aligned} H_0 &: \boldsymbol{\psi} = \boldsymbol{\psi}_0, \ \boldsymbol{\lambda} \in \boldsymbol{\Lambda}, \\ H_1 &: \boldsymbol{\psi} \neq \boldsymbol{\psi}_0, \ \boldsymbol{\lambda} \in \boldsymbol{\Lambda}, \end{aligned}$$

where $\boldsymbol{\Lambda}$ is the parameter space of $\boldsymbol{\lambda}$.

The null hypothesis can be viewed as placing a constraint on the model that we are fitting. In (9.10) this constraint is that the parameter $\boldsymbol{\psi}$ must take a particular value $\boldsymbol{\psi}_0$. The likelihood-ratio test statistic compares two models, with each model having certain advantages:

- The **unconstrained model** will fit better, in the sense that the associated likelihood will be larger. We write $\hat{\boldsymbol{\theta}} = (\hat{\boldsymbol{\psi}}, \hat{\boldsymbol{\lambda}})$ for the unconstrained maximum-likelihood estimate.
- The **constrained model** has fewer parameters to be estimated. If the dimension of $\boldsymbol{\psi}$ is $k$, then the values of the $k$ parameters contained in $\boldsymbol{\psi}$ are fixed; we only have to estimate $\boldsymbol{\lambda}$, where $\dim(\boldsymbol{\lambda}) < \dim(\boldsymbol{\theta})$. Under the constraint $\boldsymbol{\psi} = \boldsymbol{\psi}_0$, the maximum-likelihood estimate of $\boldsymbol{\lambda}$ is generated by maximising $L_{\boldsymbol{Y}}(\boldsymbol{\psi}_0, \boldsymbol{\lambda}; \boldsymbol{y})$ with respect to $\boldsymbol{\lambda}$. We denote the resulting estimate as $\hat{\boldsymbol{\lambda}}_0$. Thus, the constrained maximum-likelihood estimate of $\boldsymbol{\theta}$ is $\hat{\boldsymbol{\theta}}_0 = (\boldsymbol{\psi}_0, \hat{\boldsymbol{\lambda}}_0)$.

In many situations we test $H_0 : \boldsymbol{\psi} = 0$; setting $\boldsymbol{\psi} = 0$ in effect removes $k$ parameters from the model. There is a tradeoff here between number of parameters and goodness of fit. We can always improve goodness of fit by using a model with more parameters. However, models with too many parameters are likely to perform badly when they come to be used in prediction. One of the guiding principles of statistics is **parsimony**, that is, we should use the model that provides an adequate fit with the fewest parameters. The likelihood-ratio test is used in practical situations to help determine which features of a model are important.

**Example 9.4.2** (Testing mean of normal in presence of a nuisance parameter)
Suppose that we have a sample of independent random variables $\boldsymbol{Y} = (Y_1, \ldots, Y_n)^T$, where $Y_1 \sim \mathrm{N}(\lambda, 1)$, and $Y_j \sim \mathrm{N}(\lambda + \psi_j, 1)$ for $j = 2, \ldots, n$. We want to test for differences in the mean, that is,

$$\begin{aligned} H_0 &: \psi_j = 0, \quad \text{for all } j, \\ H_1 &: \psi_j \neq 0, \quad \text{for some } j. \end{aligned}$$

The null here is that all the observations have the same mean; the alternative is that at least one differs from the others. The likelihood function is

$$L_{\boldsymbol{Y}}(\psi_2, \ldots, \psi_n, \lambda; \boldsymbol{Y}) = \left(\frac{1}{\sqrt{2\pi}}\right)^n \exp\left\{-\frac{1}{2}(Y_1 - \lambda)^2 - \frac{1}{2}\sum_{i=2}^{n}(Y_i - \lambda - \psi_i)^2\right\}.$$

In the unconstrained case we are estimating $n$ means from $n$ data points, which will result in $\hat{\lambda} = Y_1$ and $\hat{\lambda} + \hat{\psi}_j = Y_j$ for $j = 2, \ldots, n$. Under the null, we have $\hat{\lambda}_0 = \bar{Y}$. The resulting likelihood-ratio test statistic is

$$r(\boldsymbol{Y}) = \frac{L_{\boldsymbol{Y}}(\hat{\psi}_2, \ldots, \hat{\psi}_n, \hat{\lambda}; \boldsymbol{Y})}{L_{\boldsymbol{Y}}(0, \ldots, 0, \hat{\lambda}_0; \boldsymbol{Y})} = \exp\left\{\frac{1}{2}\sum_{i=1}^{n}(Y_i - \bar{Y})^2\right\}. \tag{9.11}$$

Thus, we will reject the null hypothesis for large values of $\sum_{i=1}^n (Y_i - \bar{Y})^2$.

### 9.4.2 *Properties of the likelihood ratio*

In subsection 9.3.1 we established that maximum-likelihood estimators have a number of desirable properties. This section continues in a similar vein to establish the properties of likelihood-ratio tests. The likelihood-ratio test has a strong connection to the notion of sufficiency (more on this in the next chapter). In fact, we can show that the observed likelihood-ratio test statistic we would calculate using just the sufficient statistic is identical to that calculated from the full sample.

In the introduction to this section we mention the fact that, in general, the sampling distribution of the likelihood-ratio test statistic, $r(\boldsymbol{Y})$, is not known. We now give two results relating to the asymptotic distribution of $2\log r(\boldsymbol{Y})$. We will not prove either of these results, although some heuristic justification is provided by considering the normal case.

**Proposition 9.4.3** (Asymptotic distribution of likelihood ratio – simple case)
*Consider a random sample $\boldsymbol{Y}$ parameterised by a scalar parameter $\theta$. Let $r(\boldsymbol{Y})$ be the likelihood-ratio test statistic associated with testing $H_0 : \theta = \theta_0$. Under the null hypothesis,*

$$2\log r(\boldsymbol{Y}) \xrightarrow{d} \chi^2_1,$$

*provided certain regularity conditions hold.*

*Proof.*

Again, we give a sketch proof of this result. A rigorous proof requires Slutsky's theorem, which has to do with the convergence of sums or products of sequences of random variables.

Taylor series expansion of $\ell_{\boldsymbol{Y}}(\theta;\boldsymbol{Y})$ around $\hat{\theta}$ yields

$$\begin{aligned}\ell_{\boldsymbol{Y}}(\theta_0;\boldsymbol{Y}) &= \ell_{\boldsymbol{Y}}(\hat{\theta};\boldsymbol{Y}) + (\theta_0-\hat{\theta})\left.\frac{\partial \ell_{\boldsymbol{Y}}(\theta;\boldsymbol{Y})}{\partial\theta}\right|_{\theta=\hat{\theta}} + \frac{1}{2}(\theta_0-\hat{\theta})^2\left.\frac{\partial^2 \ell_{\boldsymbol{Y}}(\theta;\boldsymbol{Y})}{\partial\theta^2}\right|_{\theta=\hat{\theta}} + \dots \\ &= \ell_{\boldsymbol{Y}}(\hat{\theta};\boldsymbol{Y}) + (\theta_0-\hat{\theta})s_{\boldsymbol{Y}}(\hat{\theta};\boldsymbol{Y}) + \frac{1}{2}(\theta_0-\hat{\theta})^2\left[-\hat{I}_{\boldsymbol{Y}}(\hat{\theta};\boldsymbol{Y})\right] + \dots\end{aligned}$$

where the omitted terms are $O((\theta_0-\hat{\theta})^3)$. As $s_{\boldsymbol{Y}}(\hat{\theta};\boldsymbol{Y}) = 0$, we can rearrange this to obtain

$$2\log r(\boldsymbol{Y}) = 2[\ell_{\boldsymbol{Y}}(\hat{\theta};\boldsymbol{Y}) - \ell_{\boldsymbol{Y}}(\theta_0;\boldsymbol{Y})] \approx \hat{I}_{\boldsymbol{Y}}(\hat{\theta};\boldsymbol{Y})(\hat{\theta}-\theta_0)^2 .$$

Now recall, from Example 9.3.8, that

$$\sqrt{I_{\boldsymbol{Y}}(\theta)}(\hat{\theta}-\theta) \overset{\text{appr.}}{\sim} \mathrm{N}(0,1)$$

so

$$I_{\boldsymbol{Y}}(\theta)(\hat{\theta}-\theta)^2 \overset{\text{appr.}}{\sim} \chi^2_1 .$$

To complete the sketch proof, we use the fact that $\hat{I}_{\boldsymbol{Y}}(\hat{\theta};\boldsymbol{Y}) \xrightarrow{p} I_{\boldsymbol{Y}}(\theta)$ as $n \to \infty$. □

The number of degrees of freedom of the asymptotic $\chi^2$ distribution is related to the number of restrictions that we place on the model, as the following proposition makes clear.

**Proposition 9.4.4** (Asymptotic distribution of likelihood ratio – nuisance parameter case)
*Consider a random sample* $\boldsymbol{Y}$ *parameterised by* $\boldsymbol{\theta} = (\boldsymbol{\psi}, \boldsymbol{\lambda})$, *where* $\boldsymbol{\psi}$ *is of dimension* $k$. *Let* $r(\boldsymbol{Y})$ *be the likelihood-ratio test statistic associated with testing* $H_0 : \boldsymbol{\psi} = \boldsymbol{\psi}_0$. *Under the null hypothesis,*

$$2 \log r(\boldsymbol{Y}) \xrightarrow{d} \chi^2_k ,$$

*provided certain regularity conditions hold.*

**Example 9.4.5** (Distribution of likelihood-ratio test statistic in normal case)
Consider the setup in Example 9.4.2. From equation (9.11) it is clear that

$$2 \log r(\boldsymbol{Y}) = \sum_{i=1}^{n} (Y_i - \bar{Y})^2 .$$

The parameter of interest, $\boldsymbol{\psi}$, is of dimension $n - 1$. We know that in the normal case $\sum_{i=1}^{n} (Y_i - \bar{Y})^2 \sim \chi^2_{n-1}$. Thus, in this particular case the asymptotic distribution is also the exact sampling distribution of the test statistic for any sample size. The critical region for a test of size $\alpha$ would then be

$$R = \left\{ \boldsymbol{y} : \sum_{i=1}^{n} (y_i - \bar{y})^2 > \chi^2_{n-1,1-\alpha} \right\} .$$

If we were to reject the null hypothesis in this test, we would conclude that there is evidence to suggest that the mean of at least one observation differs from the others.

### 9.4.3 *Approximate tests*

Calculating the likelihood-ratio test statistic involves maximising the likelihood under both null and alternative models. In practice, this is not always feasible. The **score test** and **Wald test** provide appealing alternatives that can be viewed as approximations to the likelihood-ratio test. The basic setup is identical to that used in the likelihood-ratio test, that is, we test $H_0 : \boldsymbol{\psi} = \boldsymbol{\psi}_0$ vs. $H_1 : \boldsymbol{\psi} \neq \boldsymbol{\psi}_0$ in the presence of a nuisance parameter $\boldsymbol{\lambda}$. In addition, we assume that $\boldsymbol{\psi}$ is of length $k$, and that all the parameters can be grouped together as $\boldsymbol{\theta} = (\boldsymbol{\psi}, \boldsymbol{\lambda})$.

*Score test*

Suppose that the maximum-likelihood estimate under the null hypothesis, $\hat{\boldsymbol{\theta}}_0$, is available; we sometimes refer to this as the constrained MLE. The idea behind the score test is that the score function evaluated at $\hat{\boldsymbol{\theta}}$ (the unconstrained MLE) is equal to zero, so it would be sensible to reject $H_0$ if the score at $\hat{\boldsymbol{\theta}}_0$ is far from zero.

We can construct a test statistic based on $\hat{\boldsymbol{\theta}}_0$ whose distribution under the null is approximately that of $-2 \log r(\boldsymbol{Y})$. Let $\boldsymbol{s}_1(\boldsymbol{\theta}; \boldsymbol{Y})$ denote the first $k$ elements of the score vector, that is, the elements corresponding to the parameters in $\boldsymbol{\psi}$. The score test statistic is defined as

$$S = h(\boldsymbol{Y}; \hat{\boldsymbol{\theta}}_0) = \boldsymbol{s}_1(\hat{\boldsymbol{\theta}}_0; \boldsymbol{Y})^T [\mathcal{I}^{-1}(\hat{\boldsymbol{\theta}}_0)]_{11} \boldsymbol{s}_1(\hat{\boldsymbol{\theta}}_0; \boldsymbol{Y}), \tag{9.12}$$

where $[\mathcal{I}^{-1}(\boldsymbol{\theta})]_{11}$ is the $k \times k$ upper-left-hand block of the inverse of the information matrix, $\mathcal{I}^{-1}(\boldsymbol{\theta})$. This block is the variance of $\boldsymbol{s}_1(\boldsymbol{\psi}; \boldsymbol{Y})$, so the score test statistic is equal to the magnitude of the score function at $\hat{\boldsymbol{\theta}}_0$ standardised by its variance.

The distribution of $S$ is approximately $\chi^2_k$, and we reject $H_0$ for large values of $S$. Notice that we do not need to know the unconstrained MLE in order to perform this test.

*Wald test*

The Wald test is useful in the opposite situation, that is, when we have only the unconstrained MLE, $\hat{\boldsymbol{\theta}}$. This test is based on the idea that when $H_0$ is true we would expect the unconstrained MLE $\hat{\boldsymbol{\psi}}$ to not be too far from $\boldsymbol{\psi}_0$, the null value. The Wald test statistic is defined as

$$W = \left(\hat{\boldsymbol{\psi}} - \boldsymbol{\psi}_0\right)^T [\mathcal{I}^{-1}(\hat{\boldsymbol{\theta}})]_{11}^{-1} \left(\hat{\boldsymbol{\psi}} - \boldsymbol{\psi}_0\right). \tag{9.13}$$

From Proposition 9.3.4, the variance of $(\hat{\boldsymbol{\psi}} - \boldsymbol{\psi}_0)$ is approximately equal to $[\mathcal{I}^{-1}(\boldsymbol{\theta})]_{11}$. We can estimate this variance by evaluating the information at the MLE, $\hat{\boldsymbol{\theta}}$. The Wald test statistic is thus equal to the magnitude of the difference $(\hat{\boldsymbol{\psi}} - \boldsymbol{\psi}_0)$ standardised by its (estimated) variance. Like the score test statistic, The Wald test statistic also has a null distribution that is approximately $\chi^2_k$, and we reject $H_0$ for large values of $W$.

**Example 9.4.6** (Score and Wald tests in the normal case)
We return again to the setup of Examples 9.4.2 and 9.4.5. The random variables $\boldsymbol{Y} = (Y_1, \dots, Y_n)^T$ are independent, with $Y_1 \sim \mathrm{N}(\lambda, 1)$, and $Y_j \sim \mathrm{N}(\lambda + \psi_j, 1)$ for $j = 2, \dots, n$. We want to test

$$\begin{aligned} H_0 &: \psi_j = 0, \quad \text{for all } j, \\ H_1 &: \psi_j \neq 0, \quad \text{for some } j. \end{aligned}$$

The log-likelihood is

$$\ell_{\boldsymbol{Y}}(\boldsymbol{\psi}, \lambda; \boldsymbol{Y}) = -\frac{1}{2}(Y_1 - \lambda)^2 - \frac{1}{2}\sum_{i=2}^{n}(Y_i - \lambda - \psi_i)^2 + C,$$

so the elements of the score vector are

$$\begin{aligned} \frac{\partial}{\partial \psi_i} \ell_{\boldsymbol{Y}}(\boldsymbol{\psi}, \lambda; \boldsymbol{Y}) &= Y_i - \lambda - \psi_i, \text{ for } i = 2, \dots, n, \\ \frac{\partial}{\partial \lambda} \ell_{\boldsymbol{Y}}(\boldsymbol{\psi}, \lambda; \boldsymbol{Y}) &= (Y_1 - \lambda) + \sum_{i=2}^{n}(Y_i - \lambda - \psi_i). \end{aligned}$$

Differentiating again with respect to $\boldsymbol{\theta}$ and taking expectations yields

$$\mathbb{E}\left[-\frac{\partial^2}{\partial\psi_j\partial\psi_i}\ell_{\boldsymbol{Y}}(\boldsymbol{\psi},\lambda;\boldsymbol{Y})\right] = \begin{cases} 1 & \text{for } i = j\,, \\ 0 & \text{for } i \neq j\,, \end{cases}$$

$$\mathbb{E}\left[-\frac{\partial^2}{\partial\lambda\partial\psi_i}\ell_{\boldsymbol{Y}}(\boldsymbol{\psi},\lambda;\boldsymbol{Y})\right] = 1\,,$$

$$\mathbb{E}\left[-\frac{\partial^2}{\partial\lambda^2}\ell_{\boldsymbol{Y}}(\boldsymbol{\psi},\lambda;\boldsymbol{Y})\right] = n\,.$$

The information matrix is, thus,

$$\mathcal{I}(\boldsymbol{\psi},\lambda) = \begin{pmatrix} 1 & & & & 1 \\ & 1 & & & 1 \\ & & \ddots & & 1 \\ & & & 1 & 1 \\ 1 & 1 & 1 & 1 & n \end{pmatrix} = \begin{bmatrix} \boldsymbol{I}_{n-1} & \mathbf{1}_{n-1} \\ \mathbf{1}_{n-1}^T & n \end{bmatrix},$$

in block matrix form, where $\mathbf{1}_{n-1}$ is the $(n-1)\times 1$ vector of 1s. We can verify that its inverse is

$$\mathcal{I}^{-1}(\boldsymbol{\psi},\lambda) = \begin{bmatrix} \boldsymbol{I}_{n-1} + \mathbf{1}_{n-1}\mathbf{1}_{n-1}^T & -\mathbf{1}_{n-1} \\ -\mathbf{1}_{n-1}^T & 1 \end{bmatrix},$$

so $[\mathcal{I}^{-1}(\boldsymbol{\psi},\lambda)]_{11} = \boldsymbol{I}_{n-1} + \mathbf{1}_{n-1}\mathbf{1}_{n-1}^T$. The score test statistic is

$$\begin{aligned} S &= \boldsymbol{s}_1(\hat{\boldsymbol{\theta}}_0;\boldsymbol{Y})^T[\mathcal{I}^{-1}(\hat{\boldsymbol{\theta}}_0)]_{11}\boldsymbol{s}_1(\hat{\boldsymbol{\theta}}_0;\boldsymbol{Y}) \\ &= \boldsymbol{s}_1(\hat{\boldsymbol{\theta}}_0;\boldsymbol{Y})^T(\boldsymbol{I}_{n-1} + \mathbf{1}_{n-1}\mathbf{1}_{n-1}^T)\boldsymbol{s}_1(\hat{\boldsymbol{\theta}}_0;\boldsymbol{Y}) \\ &= \boldsymbol{s}_1(\hat{\boldsymbol{\theta}}_0;\boldsymbol{Y})^T\boldsymbol{s}_1(\hat{\boldsymbol{\theta}}_0;\boldsymbol{Y}) + [\mathbf{1}_{n-1}^T\boldsymbol{s}_1(\hat{\boldsymbol{\theta}}_0;\boldsymbol{Y})]^T[\mathbf{1}_{n-1}^T\boldsymbol{s}_1(\hat{\boldsymbol{\theta}}_0;\boldsymbol{Y})]\,. \end{aligned}$$

Under the null hypothesis, we have $\hat{\lambda}_0 = \bar{Y}$ and $\psi_j = 0$ for all $j$, so

$$\boldsymbol{s}_1(\hat{\boldsymbol{\theta}}_0;\boldsymbol{Y})^T\boldsymbol{s}_1(\hat{\boldsymbol{\theta}}_0;\boldsymbol{Y}) = \begin{pmatrix} Y_2-\bar{Y} \\ \vdots \\ Y_n-\bar{Y} \end{pmatrix}^T \begin{pmatrix} Y_2-\bar{Y} \\ \vdots \\ Y_n-\bar{Y} \end{pmatrix} = \sum_{i=2}^n (Y_i-\bar{Y})^2\,,$$

$$\mathbf{1}_{n-1}^T\boldsymbol{s}_1(\hat{\boldsymbol{\theta}}_0;\boldsymbol{Y}) = \begin{pmatrix} 1 \\ \vdots \\ 1 \end{pmatrix}^T \begin{pmatrix} Y_2-\bar{Y} \\ \vdots \\ Y_n-\bar{Y} \end{pmatrix} = \sum_{i=2}^n (Y_i-\bar{Y}) = -(Y_1-\bar{Y})\,,$$

and the score test statistic simplifies to

$$S = \sum_{i=2}^n (Y_i-\bar{Y})^2 + (Y_1-\bar{Y})^2 = \sum_{i=1}^n (Y_i-\bar{Y})^2\,,$$

which is the same as the likelihood-ratio test statistic.

Under the alternative hypothesis, we have $\psi_0 = 0$, $\hat{\lambda} = Y_1$, and $\hat{\psi}_j = Y_j - \hat{\lambda} = Y_j - Y_1$. The Wald test statistic is, thus,

$$W = \left(\hat{\psi} - \psi_0\right)^T [\mathcal{I}^{-1}(\hat{\theta})]_{11}^{-1} \left(\hat{\psi} - \psi_0\right) = \begin{pmatrix} Y_2 - Y_1 \\ \vdots \\ Y_n - Y_1 \end{pmatrix}^T (\boldsymbol{I}_{n-1} + \mathbf{1}_{n-1}\mathbf{1}_{n-1}^T)^{-1} \begin{pmatrix} Y_2 - Y_1 \\ \vdots \\ Y_n - Y_1 \end{pmatrix}.$$

which also simplifies to

$$W = \sum_{i=1}^{n} (Y_i - \bar{Y})^2 .$$

Filling in the details is part of Exercise 9.4.

**Exercise 9.4**

1. Briefly explain why $L_Y(\hat{\theta}; \boldsymbol{y})/L_Y(\hat{\theta}_0; \boldsymbol{y}) \geq 1$ for all $\boldsymbol{y}$.
2. Suppose that $X_1, \ldots, X_n$ and $Y_1, \ldots, Y_n$ are two independent random samples from two exponential distributions with mean $\theta_1$ and $\theta_2$, respectively. Construct the likelihood-ratio test for
$$H_0 : \theta_1 = \theta_2,$$
$$H_1 : \theta_1 \neq \theta_2.$$
Specify the asymptotic distribution of the test statistic under $H_0$.
3. A survey of the use of a particular product was conducted in four areas. 200 randomly selected potential users were interviewed in each area. The number of respondents from area $i$ who said that they use the product was $y_i$. Construct a likelihood-ratio test to determine whether the proportion of the population using the product is the same in each of the four areas. Carry out the test at the 5% level for the case $y_1 = 76$, $y_2 = 53$, $y_3 = 59$, and $y_4 = 48$.
4. Derive the Wald test statistic in Example 9.4.6. You may use the matrix inversion result
$$(\boldsymbol{I}_{n-1} + \mathbf{1}_{n-1}\mathbf{1}_{n-1}^T)^{-1} = \boldsymbol{I}_{n-1} - \frac{1}{n}\mathbf{1}_{n-1}\mathbf{1}_{n-1}^T .$$

## 9.5 Further exercises

1. Consider a distribution parameterised by $\theta$, where the log-likelihood for a single observation can be written as
$$\ell_Y(\theta; y) = a(\theta)t(y) + b(\theta) + c(y),$$
for some functions $a, t, b,$ and $c$ meeting regularity conditions. Using the properties of the score function find:
   (a) $\mathbb{E}(t(Y))$,
   (b) $\mathcal{I}_Y(\theta)$ (by differentiating twice),

(c) $\text{Var}(t(Y))$ (using part (b)),

in terms of $a$ and $b$ and their derivatives. Apply these results to the binomial distribution where the number of trials, $n$, is fixed and the parameter of interest is $p$, the probability of success.

2. If $\boldsymbol{\theta}$ is a vector of parameters, prove that the $(i,j)^{\text{th}}$ element of the information matrix is

$$[\boldsymbol{\mathcal{I}}_{\boldsymbol{Y}}(\boldsymbol{\theta})]_{i,j} = \mathbb{E}\left[\frac{\partial}{\partial\theta_i}\ell_{\boldsymbol{Y}}(\boldsymbol{\theta};\boldsymbol{Y})\frac{\partial}{\partial\theta_j}\ell_{\boldsymbol{Y}}(\boldsymbol{\theta};\boldsymbol{Y})\right].$$

3. Find the information matrix for $(p_1, p_2)$ associated with one observation, $(y_1, y_2)$, from the trinomial distribution with mass function

$$f_{Y_1,Y_2}(y_1,y_2) = \frac{n!}{y_1!y_2!(n-y_1-y_2)!}p_1^{y_1}p_2^{y_2}(1-p_1-p_2)^{n-y_1-y_2}.$$

4. (**Quasi-likelihood**) Let $X_1, \dots, X_n$ be a random sample from a continuous distribution with density $f_X$, defined on $(0,\infty)$. Suppose that the distribution of $X$ is unknown, and we wish to approximate it with the candidate model $Y \sim \text{Exp}(1/\theta)$. Consider the function

$$\ell^*(\theta) = \frac{1}{n}\sum_{i=1}^{n}\log f_Y(X_i|\theta)$$

(a) Show that $\ell^*(\theta)$ converges to $-D(f_x, f_y) + C$ as $n \to \infty$, where $D(f_x, f_y)$ is the Kullback-Leibler divergence of $f_Y$ from $f_X$, and $C$ is a quantity which does not depend on $\theta$.

(b) Find an estimator of $\theta$ by maximising $\ell^*(\theta)$. Explain why this is a sensible thing to do.

(c) What does the estimator converge to as $n \to \infty$?

[$\ell^*(\theta)$ is known as a **quasi-likelihood function**, and the resulting estimator is a **quasi-maximum-likelihood estimator** (QMLE). Under some regularity conditions, the QMLE converges to the value of $\theta$ which minimises the Kullback-Leibler divergence from the true model.]

5. (**Kullback-Leibler divergence for discrete distributions**) Suppose that $X$ and $Y$ are two random variables taking values in some countable subset $\{x_1, x_2, \dots\}$ of $\mathbb{R}$, and let $f_X$ and $f_Y$ be their respective probability mass functions. The Kullback-Leibler divergence of $f_Y$ from $f_X$ is defined as

$$D(f_x, f_y) = \sum_{i=1}^{\infty}\log\left(\frac{f_X(x_i)}{f_Y(x_i)}\right)f_X(x_i).$$

(a) Prove that this quantity is non-negative.

(b) If $Y$ follows a Geometric$(p)$ distribution and $f_X$ is a discrete distribution on $\{1, 2, \dots\}$ with mean $\mu$, find the value of $p$ that minimises the Kullback-Leibler divergence of $f_Y$ from $f_X$.

6. Suppose that we observe data $\boldsymbol{Y} = (Y_1, Y_2, Y_3)^T \sim \text{Multinom}(n, \boldsymbol{p})$, where the probabilities are

$$\boldsymbol{p} = \left(\frac{1}{2} + \frac{\theta}{4}, \frac{1-\theta}{2}, \frac{\theta}{4}\right)^T.$$

   (a) Write down the likelihood for this model. Can you find a closed form for the MLE of $\theta$?

   (b) Let the complete data be $\boldsymbol{X} = (X_0, X_1, X_2, X_3)^T \sim \text{Multinom}(n, \boldsymbol{p}')$, where

$$\boldsymbol{p}' = \left(\frac{1}{2}, \frac{\theta}{4}, \frac{1-\theta}{2}, \frac{\theta}{4}\right)^T.$$

   How would you define $\boldsymbol{Y}$ in terms of $\boldsymbol{X}$? Describe an EM algorithm for estimating $\theta$ using the complete-data likelihood.

7. Let $X_1, \dots, X_n$ be a random sample from the density function

$$f_X(x) = \begin{cases} \theta^2 x e^{-\theta x} & x > 0, \\ 0 & x \leq 0, \end{cases}$$

   where $\theta > 0$ is an unknown parameter.

   (a) Find the maximum-likelihood estimator for $\theta$.

   (b) Consider the situation where only the last $n - m$ observations $(m < n)$ of the sample are known explicitly, while for the first $m$ only their sum, $Z = X_1 + \cdots + X_m$, is known. Find the probability density function of $Z$ and determine the MLE of $\theta$ based on the observations $Z, X_{m+1}, \dots, X_n$.

   (c) Suppose now that the first $m$ observations are censored in the sense that, for $j = 1, \dots, m$, we only know $X_j \in [u_j, v_j]$ for some given $v_j > u_j > 0$, while $X_j$ itself is unknown. Describe an EM algorithm for estimating the parameter $\theta$.

8. Consider the simple linear regression model

$$Y_i = \beta_0 + \beta_1 X_i + \sigma\varepsilon_i, \quad \varepsilon_i \sim \text{IID}(0, 1), \quad \text{for } i = 1, \dots, n.$$

   Construct the Wald test statistic for $H_0 : \beta_1 = 2\beta_0$ vs. $H_1 : \beta_1 \neq 2\beta_0$.

9. Consider the multiple linear regression model

$$Y_i = \beta_0 + \beta_1 X_{i,1} + \ldots + \beta_p X_{i,p} + \sigma\varepsilon_i, \quad \varepsilon_i \sim \text{N}(0, 1), \quad \text{for } i = 1, \dots, n,$$

   which has $p$ explanatory variables. Suppose that we want to use the likelihood-ratio test to compare this model to a simpler model, which omits the first $q$ of these $p$ explanatory variables $(q < p)$. Write down the null and alternative hypotheses. What is the test statistic? When would we reject the null hypothesis?

## 9.6 Chapter summary

On completion of this chapter you should be able to:

- define likelihood, log-likelihood, score, and information,
- exploit the properties of random samples to write simpler forms of likelihood, log-likelihood, score, and information,
- calculate the score function and information for standard distributions,
- write down two distinct forms of information and understand the association between them,
- analytically derive maximum-likelihood estimators in simple situations,
- describe the large-sample properties of maximum-likelihood estimators,
- derive the updating step for the method of scoring to numerically approximate the maximum-likelihood estimates of the parameters associated with a specified distribution,
- explain the rationale behind the likelihood-ratio test,
- describe how the likelihood-ratio test is used in the presence of nuisance parameters, and
- construct likelihood-ratio tests for standard cases.

Chapter 10

# Inferential theory

The statistics discussed in Chapter 8 are a useful mechanism for data reduction, that is, summarising the key features of a sample, but the process clearly results in some loss of information. A good inferential procedure should provide us with an intelligible summary of the data while not throwing away too much relevant information.

In this chapter we discuss how we can construct statistics that summarise the data as efficiently as possible, and explore their properties. These statistics are closely linked to maximum-likelihood estimation, and are crucial in obtaining parameter estimators with desirable properties. Finally, we discuss a generic method for constructing optimal hypothesis tests.

## 10.1 Sufficiency

In section 8.1 we indicate that good inferential procedures should use statistics that summarise all the relevant information about a parameter of interest. In this section we make this notion of relevance more precise by introducing the concept of **sufficiency**.

### *10.1.1 Sufficient statistics and the sufficiency principle*

Consider a sample $\boldsymbol{Y}$ parameterised by $\boldsymbol{\theta}$. Sufficiency provides one possible definition of a statistic that summarises all the relevant information that $\boldsymbol{Y}$ contains about $\boldsymbol{\theta}$.

**Definition 10.1.1**
][Sufficiency] Suppose that $\boldsymbol{Y} = (Y_1, \dots, Y_n)^T$ is a sample. A statistic $\boldsymbol{U} = \boldsymbol{h}(\boldsymbol{Y})$ is a **sufficient statistic** for a parameter $\boldsymbol{\theta}$ if the conditional distribution of $\boldsymbol{Y}$ given $\boldsymbol{U}$ does not depend on $\boldsymbol{\theta}$.

The definition of sufficiency refers to the distribution of $\boldsymbol{Y}$. For this reason, a statistic is sometimes defined as being sufficient for a family of distributions, $\{F_{\boldsymbol{Y}}(.;\boldsymbol{\theta}) : \boldsymbol{\theta} \in \boldsymbol{\Theta}\}$. The statement that $\boldsymbol{U}$ is a sufficient statistic is then equivalent to saying that the conditional distribution function, $F_{Y|U}(y|u)$, is not a function of $\theta$. Equivalently, if $f_Y(.;\theta)$ is the mass/density function of $Y$, then $f_{Y|U}(y|u)$ is not a function of $\theta$.

The conditional mass/density function $f_{Y|U}(y|u)$ is slightly unusual; we are conditioning on a statistic, $U$, which is a function of the sample, $Y$. The properties of this type of conditioning are exploited to derive results for sufficient statistics. In general, these results are easier to derive in the discrete case, since conditional mass can be interpreted as a probability.

**Lemma 10.1.2** (Conditioning on a function of the sample – discrete case)
*Suppose that $Y$ is a sample from a discrete distribution and $U = h(Y)$ is a statistic. The conditional mass function of $Y$ given $U$ is then*

$$f_{Y|U}(y|u) = \begin{cases} f_Y(y)/f_U(u) & \text{when } h(y) = u \text{ and } f_U(u) \neq 0, \\ 0 & \text{otherwise.} \end{cases}$$

*Proof.*

We define $h(y) = u$, so

$$Y = y \Rightarrow h(Y) = h(y) \Rightarrow U = u .$$

In terms of events this means that

$$\{Y = y\} \subseteq \{U = u\},$$

and so, by elementary properties of probability, the joint mass function of $Y$ and $U$ is given by

$$f_{Y,U}(y,u) = \mathrm{P}(Y = y, U = u) = \mathrm{P}(Y = y) = f_Y(y) .$$

If $h(y) \neq u$ then $\{Y = y\} \cap \{U = u\} = \emptyset$, and so $f_{Y,U}(y,u) = 0$. The result follows from the definition of $f_{Y|U}(y|u)$ as a conditional probability. □

If we believe that a sufficient statistic for $\theta$ contains all the relevant information that the sample has to offer, it is reasonable to insist that any inferences we make about $\theta$ only use a sufficient statistic. This is referred to as the **sufficiency principle**. It is stated more formally below.

*Sufficiency principle*
If $U = h(Y)$ is a sufficient statistic for $\theta$, then inference about $\theta$ should depend on the sample, $Y$, only through $u = h(y)$.

Suppose that $U = h(Y)$ is a sufficient statistic for $\theta$. Consider two observed samples, $x$ and $y$. If the observed value of the sufficient statistic is the same for these two samples, that is, if $h(x) = h(y)$, the sufficiency principle tells us that inferences about $\theta$ based on observing $x$ should be the same as those based on observing $y$.

**Example 10.1.3** (Sufficient statistic for $p$ in repeated Bernoulli trials)
Suppose that $\boldsymbol{Y} = (Y_1, \dots, Y_n)^T$ is a random sample from a Bernoulli($p$) distribution. The marginal mass of each member of the sample is

$$f_Y(y) = \begin{cases} p & \text{for } y = 1, \\ 1 - p & \text{for } y = 0. \end{cases}$$

For any given observed sample, $\boldsymbol{y} = (y_1, \dots, y_n)^T$, the number of 1s is $u = \sum_{i=1}^{n} y_i$ and the number of 0s is $n - u$. By independence, the joint mass function of the sample is

$$f_{\boldsymbol{Y}}(\boldsymbol{y}) = p^u (1-p)^{n-u} \quad \text{where } u = \sum_{i=1}^{n} y_i\,.$$

Consider the statistic generated by summation of the sample, $U = \sum_{i=1}^{n} Y_i$. From the properties of repeated independent Bernoulli trials, we know that $U \sim \text{Bin}(n, p)$. Thus, the mass function of $U$ is

$$f_U(u) = \binom{n}{u} p^u (1-p)^{n-u} \quad \text{for } u = 0, \dots, n\,.$$

Using Lemma 10.1.2, we have

$$f_{\boldsymbol{Y}|U}(\boldsymbol{y}|u) = \frac{f_{\boldsymbol{Y}}(\boldsymbol{y})}{f_U(u)} = \frac{p^u(1-p)^{n-u}}{\binom{n}{u} p^u (1-p)^{n-u}} = \frac{1}{\binom{n}{u}}, \tag{10.1}$$

where $u = \sum_{i=1}^{n} y_i$ and $u \in \{0, \dots, n\}$. This makes sense intuitively; if there are $n$ trials and we know we have $u$ successes, there are $\binom{n}{u}$ ways of achieving these $n$ successes, and each one is equally probable. Equation (10.1) tells us that $f_{\boldsymbol{Y}|U}(\boldsymbol{y}|u)$ does not depend on $p$. We conclude that, by Definition 10.1.1, $U$ is sufficient for $p$.

We can try to understand this result in the context of the sufficiency principle. The principle tells us that, if we want to make inferences about $p$, we should only use the sample information via $U$. If observed samples give us the same value for $U$, we make the same inferences about $p$. Consider the obvious point estimator $U/n$, the proportion of successes in the sample. This estimator is clearly consistent with the principle of sufficiency. Suppose that $n = 10$ and we observe

$$\boldsymbol{x} = (0, 1, 1, 0, 1, 0, 1, 1, 1, 0)^T,$$
$$\boldsymbol{y} = (1, 1, 0, 0, 1, 1, 1, 0, 0, 1)^T.$$

In this case, $\boldsymbol{x} \neq \boldsymbol{y}$ but $\sum_i x_i = \sum_i y_i = 6$, so the observed sample value of the statistic $U$ is the same for $\boldsymbol{x}$ and $\boldsymbol{y}$. Our estimate for $p$ would be $\frac{6}{10} = \frac{3}{5}$ for both of these observed samples.

### *10.1.2 Factorisation theorem*

In applying the sufficiency principle there are two things that we often want to do:

i. find a sufficient statistic for a parameter,

ii. check whether a given statistic is sufficient.

The definition of sufficiency given by 10.1.1 is not well suited to either of these purposes. Fortunately, we can prove a result that provides a condition for a statistic to be sufficient. This condition can be checked by inspection of the joint density or mass function of the sample.

**Theorem 10.1.4** (Factorisation criterion)
*Let* $\boldsymbol{Y} = (Y_1, \ldots, Y_n)^T$ *be a sample with joint density or mass function* $f_{\boldsymbol{Y}}(\boldsymbol{y}; \boldsymbol{\theta})$. *The statistic* $\boldsymbol{U} = \boldsymbol{h}(\boldsymbol{Y})$ *is sufficient for the parameter* $\boldsymbol{\theta}$ *if and only if we can find functions* $b$ *and* $c$ *such that*

$$f_{\boldsymbol{Y}}(\boldsymbol{y}; \boldsymbol{\theta}) = b(\boldsymbol{h}(\boldsymbol{y}), \boldsymbol{\theta})c(\boldsymbol{y}).$$

for all $\boldsymbol{y} \in \mathbb{R}^n$ and $\boldsymbol{\theta} \in \boldsymbol{\Theta}$.

*Proof.*

We prove this theorem for the discrete case. The proof for the continuous case is more involved, since values of the density function are not probabilities.

Consider the "only if" part first. Suppose that $\boldsymbol{U} = \boldsymbol{h}(\boldsymbol{Y})$ is a sufficient statistic for $\boldsymbol{\theta}$. By definition of a sufficient statistic, the distribution of the sample conditioned on the value of the sufficient statistic does not depend on the parameter, thus,

$$f_{\boldsymbol{Y}|\boldsymbol{U}}(\boldsymbol{y}|\boldsymbol{u}) = c(\boldsymbol{y}),$$

for some function $c$. Using Lemma 10.1.2, we have

$$f_{\boldsymbol{Y}}(\boldsymbol{y}) = f_{\boldsymbol{Y}|\boldsymbol{U}}(\boldsymbol{y}|\boldsymbol{u}) f_{\boldsymbol{U}}(\boldsymbol{u}) = c(\boldsymbol{y}) f_{\boldsymbol{U}}(\boldsymbol{u}) \text{ when } \boldsymbol{u} = \boldsymbol{h}(\boldsymbol{y}). \tag{10.2}$$

Making the dependence on the parameter explicit, we can write

$$f_{\boldsymbol{U}}(\boldsymbol{u}; \boldsymbol{\theta}) = f_{\boldsymbol{U}}(\boldsymbol{h}(\boldsymbol{y}); \boldsymbol{\theta}) = b(\boldsymbol{h}(\boldsymbol{y}), \boldsymbol{\theta}) \text{ when } \boldsymbol{u} = \boldsymbol{h}(\boldsymbol{y}),$$

for some function $b$. Substituting into (10.2) yields

$$f_{\boldsymbol{Y}}(\boldsymbol{y}; \boldsymbol{\theta}) = b(\boldsymbol{h}(\boldsymbol{y}), \boldsymbol{\theta})c(\boldsymbol{y}),$$

as required.

We now consider the "if" part of the theorem. Suppose that we can find functions $b$ and $c$ such that $f_{\boldsymbol{Y}}(\boldsymbol{y}; \boldsymbol{\theta}) = b(\boldsymbol{h}(\boldsymbol{y}), \boldsymbol{\theta})c(\boldsymbol{y})$. We have

$$\begin{aligned}
f_{\boldsymbol{U}}(\boldsymbol{u}) &= \sum_{\boldsymbol{v}} f_{\boldsymbol{Y},\boldsymbol{U}}(\boldsymbol{v}, \boldsymbol{u}) && \text{properties of discrete distributions} \\
&= \sum_{\boldsymbol{v}:\boldsymbol{h}(\boldsymbol{v})=\boldsymbol{u}} f_{\boldsymbol{Y}}(\boldsymbol{v}) && \text{since } f_{\boldsymbol{Y},\boldsymbol{U}}(\boldsymbol{v}, \boldsymbol{u}) = f_{\boldsymbol{Y}}(\boldsymbol{v}) \text{ when } \boldsymbol{h}(\boldsymbol{v}) = \boldsymbol{u} \\
&= \sum_{\boldsymbol{v}:\boldsymbol{h}(\boldsymbol{v})=\boldsymbol{u}} b(\boldsymbol{h}(\boldsymbol{v}), \boldsymbol{\theta})c(\boldsymbol{v}) && \text{factorisation property} \\
&= b(\boldsymbol{u}, \boldsymbol{\theta}) \sum_{\boldsymbol{v}:\boldsymbol{h}(\boldsymbol{v})=\boldsymbol{u}} c(\boldsymbol{v}).
\end{aligned}$$

Exploiting Lemma 10.1.2, if $\boldsymbol{u} = \boldsymbol{h}(\boldsymbol{y})$ then

$$f_{\boldsymbol{Y}|\boldsymbol{U}}(\boldsymbol{y}|\boldsymbol{u}) = \frac{f_{\boldsymbol{Y}}(\boldsymbol{y})}{f_{\boldsymbol{U}}(\boldsymbol{u})} = \frac{b(\boldsymbol{h}(\boldsymbol{y}),\boldsymbol{\theta})c(\boldsymbol{y})}{b(\boldsymbol{u},\boldsymbol{\theta})\sum_{\{\boldsymbol{v}:\boldsymbol{h}(\boldsymbol{v})=\boldsymbol{u}\}} c(\boldsymbol{v})} = \frac{c(\boldsymbol{y})}{\sum_{\{\boldsymbol{v}:\boldsymbol{h}(\boldsymbol{v})=\boldsymbol{u}\}} c(\boldsymbol{v})}.$$

So, in full,

$$f_{\boldsymbol{Y}|\boldsymbol{U}}(\boldsymbol{y}|\boldsymbol{u}) = \begin{cases} c(\boldsymbol{y})/\sum_{\{\boldsymbol{v}:\boldsymbol{h}(\boldsymbol{v})=\boldsymbol{u}\}} c(\boldsymbol{v}) & \text{for } \boldsymbol{h}(\boldsymbol{y}) = \boldsymbol{u}, \\ 0 & \text{otherwise.} \end{cases}$$

Thus, the distribution of $\boldsymbol{Y}$ conditioned on the value of the statistic $\boldsymbol{U}$ does not depend on $\boldsymbol{\theta}$. From Definition 10.1.1, $\boldsymbol{U}$ is a sufficient statistic for $\boldsymbol{\theta}$. □

There is a fundamental association between likelihood-based inference and sufficiency. The following proposition describes one aspect of this association.

**Proposition 10.1.5** (Relationship between maximum likelihood and sufficiency)
*The maximum-likelihood estimate $\hat{\boldsymbol{\theta}}$ is a function of every sufficient statistic for $\boldsymbol{\theta}$.*

*Proof.*

This is an elementary consequence of the factorisation theorem. If $\boldsymbol{U} = \boldsymbol{h}(\boldsymbol{Y})$ is sufficient for $\boldsymbol{\theta}$, we know that the mass/density can be factorised as

$$f_{\boldsymbol{Y}}(\boldsymbol{y};\boldsymbol{\theta}) = b(\boldsymbol{h}(\boldsymbol{y}),\boldsymbol{\theta})c(\boldsymbol{y}).$$

The likelihood has the same functional form as $f_{\boldsymbol{Y}}$, so the log-likelihood can be written as

$$\ell(\boldsymbol{\theta}) = \log b(\boldsymbol{h}(\boldsymbol{y}),\boldsymbol{\theta}) + \log c(\boldsymbol{y}).$$

Maximising the log-likelihood with respect to $\boldsymbol{\theta}$ is thus equivalent to maximising the function $\log b(\boldsymbol{h}(\boldsymbol{y}),\boldsymbol{\theta})$, so the maximum-likelihood estimate depends on $\boldsymbol{y}$ only through $\boldsymbol{h}(\boldsymbol{y})$. We conclude that the maximum-likelihood estimate is a function of the sufficient statistic $\boldsymbol{u}$. □

**Proposition 10.1.6** (Likelihood ratio and sufficiency)
*Consider a sample $\boldsymbol{Y}$ parameterised by $\boldsymbol{\theta}$, and suppose that we are interested in testing $H_0 : \boldsymbol{\theta} \in \boldsymbol{\Theta}_0$. If $\boldsymbol{U}$ is sufficient for $\boldsymbol{\theta}$, then the likelihood-ratio statistic based on $\boldsymbol{U}$ is identical to that based on $\boldsymbol{Y}$.*

*Proof.*

First we need to be clear what we mean by the likelihood-ratio test based on $\boldsymbol{U}$. We define the associated test statistic as

$$r^*(\boldsymbol{u}) = \frac{\sup_{\boldsymbol{\theta}\in\boldsymbol{\Theta}} L_{\boldsymbol{U}}(\boldsymbol{\theta};\boldsymbol{u})}{\sup_{\boldsymbol{\theta}\in\boldsymbol{\Theta}_0} L_{\boldsymbol{U}}(\boldsymbol{\theta};\boldsymbol{u})},$$

where $L_U$ is the likelihood function of $\boldsymbol{U}$. If $\boldsymbol{U} = \boldsymbol{h}(\boldsymbol{Y})$, then in order to prove the proposition we need to show that

$$r^*(\boldsymbol{u}) = r^*(\boldsymbol{h}(\boldsymbol{y})) = r(\boldsymbol{y}) \text{ for all } \boldsymbol{y}.$$

We use the fact that, since $\boldsymbol{U}$ is sufficient for $\boldsymbol{\theta}$, the conditional mass/density $f_{\boldsymbol{Y}|\boldsymbol{U}}$ does not depend on $\boldsymbol{\theta}$. Thus, for any observed sample $\boldsymbol{y}$,

$$\begin{aligned}
r(\boldsymbol{y}) &= \frac{\sup_{\boldsymbol{\theta}\in\boldsymbol{\Theta}} L_{\boldsymbol{Y}}(\boldsymbol{\theta};\boldsymbol{y})}{\sup_{\boldsymbol{\theta}\in\boldsymbol{\Theta}_0} L_{\boldsymbol{Y}}(\boldsymbol{\theta};\boldsymbol{y})} \\
&= \frac{\sup_{\boldsymbol{\theta}\in\boldsymbol{\Theta}} f_{\boldsymbol{Y}}(\boldsymbol{y};\boldsymbol{\theta})}{\sup_{\boldsymbol{\theta}\in\boldsymbol{\Theta}_0} f_{\boldsymbol{Y}}(\boldsymbol{y};\boldsymbol{\theta})} && \text{definition of likelihood} \\
&= \frac{\sup_{\boldsymbol{\theta}\in\boldsymbol{\Theta}} f_{\boldsymbol{U}}(\boldsymbol{u};\boldsymbol{\theta}) f_{\boldsymbol{Y}|\boldsymbol{U}}(\boldsymbol{y}|\boldsymbol{u})}{\sup_{\boldsymbol{\theta}\in\boldsymbol{\Theta}_0} f_{\boldsymbol{U}}(\boldsymbol{u};\boldsymbol{\theta}) f_{\boldsymbol{Y}|\boldsymbol{U}}(\boldsymbol{y}|\boldsymbol{u})} && \text{by Lemma 10.1.2} \\
&= \frac{\sup_{\boldsymbol{\theta}\in\boldsymbol{\Theta}} f_{\boldsymbol{U}}(\boldsymbol{u};\boldsymbol{\theta})}{\sup_{\boldsymbol{\theta}\in\boldsymbol{\Theta}_0} f_{\boldsymbol{U}}(\boldsymbol{u};\boldsymbol{\theta})} && \text{since } f_{\boldsymbol{Y}|\boldsymbol{U}} \text{ does not depend on } \boldsymbol{\theta} \\
&= \frac{\sup_{\boldsymbol{\theta}\in\boldsymbol{\Theta}} L_{\boldsymbol{U}}(\boldsymbol{\theta};\boldsymbol{u})}{\sup_{\boldsymbol{\theta}\in\boldsymbol{\Theta}_0} L_{\boldsymbol{U}}(\boldsymbol{\theta};\boldsymbol{u})} && \text{definition of likelihood} \\
&= r^*(\boldsymbol{u}).
\end{aligned}$$

□

**Example 10.1.7** (Sufficient statistic for $p$ in repeated Bernoulli trials)
We know from Example 10.1.3 that, if $Y_1, \dots, Y_n$ is a random sample from a Bernoulli($p$) distribution, then $U = \sum_{i=1}^n Y_i$ is a sufficient statistic for $p$. Confirming this result using the factorisation theorem is rather trivial. The joint mass function is

$$f_{\boldsymbol{Y}}(\boldsymbol{y}) = p^u (1-p)^{n-u} \text{ where } u = \sum_{i=1}^n y_i,$$

which factorises as

$$f_{\boldsymbol{Y}}(\boldsymbol{y}) = b(u,p)\, c(\boldsymbol{y}),$$

where

$$b(u,p) = p^u (1-p)^{n-u} \text{ and } c(\boldsymbol{y}) = 1.$$

We conclude that $U = \sum_{i=1}^n Y_i$ is sufficient for $p$.

**Example 10.1.8** (Sufficient statistic for mean of normal)
Consider a random sample $\boldsymbol{Y} = (Y_1, \dots, Y_n)^T$ from an N($\mu$, 1) distribution, and define the sample mean $\bar{Y} = \frac{1}{n}\sum_{i=1} Y_i$. The joint density function for the sample is

$$f_{\boldsymbol{Y}}(\boldsymbol{y};\mu) = \left(\frac{1}{\sqrt{2\pi}}\right)^n \exp\left\{-\frac{1}{2}\sum_{j=1}^n (y_j - \mu)^2\right\}.$$

We can factorise this density using the fact that

$$\sum_{j=1}^{n}(y_j - \mu)^2 = n(\bar{y} - \mu)^2 + \sum_{j=1}^{n}(y_j - \bar{y})^2.$$

Proving this result is part of Exercise 10.1. The density can then be written as

$$f_{\boldsymbol{Y}}(\boldsymbol{y};\mu) = \left(\frac{1}{\sqrt{2\pi}}\right)^n \exp\left\{-\frac{n}{2}(\bar{y}-\mu)^2\right\} \exp\left\{-\frac{1}{2}\sum_{j=1}^{n}(y_j-\bar{y})^2\right\}.$$

In this factorisation, we have a term that depends on the parameter $\mu$ and the sample value of the statistic $\bar{y}$,

$$b(\bar{y},\mu) = \left(\frac{1}{\sqrt{2\pi}}\right)^n \exp\left\{-\frac{n}{2}(\bar{y}-\mu)^2\right\},$$

and another term that depends on $\boldsymbol{y}$ alone,

$$c(\boldsymbol{y}) = \exp\left\{-\frac{1}{2}\sum_{j=1}^{n}(y_j-\bar{y})^2\right\}.$$

Thus, by the factorisation criterion, $\bar{Y}$ is a sufficient statistic for $\mu$.

**Example 10.1.9** (Sufficient statistics for mean and variance of a normal)
Now suppose that $\boldsymbol{Y}$ is a random sample from an $\text{N}(\mu,\sigma^2)$ distribution where both $\mu$ and $\sigma^2$ are unknown. We define $\boldsymbol{\theta} = (\mu,\sigma^2)$ to be the vector of parameters of interest. Factorisation of the density is trivial since

$$\begin{aligned}
f_{\boldsymbol{Y}}(\boldsymbol{y};\mu,\sigma^2) &= \left(\frac{1}{\sqrt{2\pi\sigma^2}}\right)^n \exp\left[-\frac{1}{2\sigma^2}\sum_{i=1}^{n}(y_i-\mu)^2\right] \\
&= \left(\frac{1}{\sqrt{2\pi\sigma^2}}\right)^n \exp\left[-\frac{1}{2\sigma^2}\left(\sum_{i=1}^{n}y_i^2 - 2\mu\sum_{i=1}^{n}y_i + n\mu^2\right)\right] \\
&= b(\boldsymbol{h}(\boldsymbol{y}),\boldsymbol{\theta})c(\boldsymbol{y}).
\end{aligned}$$

We conclude that if

$$\boldsymbol{U} = \boldsymbol{h}(\boldsymbol{Y}) = \left(\sum_{i=1}^{n}Y_i, \sum_{i=1}^{n}Y_i^2\right),$$

then $\boldsymbol{U}$ is sufficient for $(\mu,\sigma^2)$. A one-to-one function of a sufficient statistic is sufficient (see Lemma 10.1.11). We know that

$$\bar{Y} = \frac{1}{n}\sum_{i=1}^{n}Y_i,$$

$$S^2 = \frac{1}{n-1}\left[\sum_{i=1}^{n}Y_i^2 - n\left(\sum_{i=1}^{n}Y_i\right)^2\right],$$

so $(\bar{Y}, S^2)$ is also sufficient for $(\mu, \sigma^2)$. This is also clear from the fact that we can factorise the density as

$$f_{\boldsymbol{Y}}(\boldsymbol{y}; \mu, \sigma^2) = \left(\frac{1}{\sqrt{2\pi\sigma^2}}\right)^n \exp\left[-\frac{1}{2\sigma^2}\left(n(\bar{y}-\mu)^2\right)\right] \exp\left[-\frac{1}{2\sigma^2}\left((n-1)s^2\right)\right].$$

Providing the details of this factorisation is part of Exercise 10.1.

**Example 10.1.10** (Sufficient statistics for uniform distribution)
Consider a random sample $\boldsymbol{Y}$ from a continuous Unif$[0, \theta]$ distribution. We have to proceed with caution since the support of this distribution depends on the parameter. An individual sample member has density function

$$f_Y(y_i) = \begin{cases} \frac{1}{\theta} & \text{for } y_i \in [0, \theta], \\ 0 & \text{otherwise.} \end{cases}$$

Using an indicator function will allow us to avoid this piecewise definition of the density,

$$f_Y(y_i) = \frac{1}{\theta}\mathbf{1}_{[0,\theta]}(y_i).$$

By independence,

$$f_{\boldsymbol{Y}}(\boldsymbol{y}) = \left(\frac{1}{\theta}\right)^n \prod_{i=1}^n \mathbf{1}_{[0,\theta]}(y_i).$$

This product of indicators only takes the value 1 if $y_i \in [0, \theta]$ for all $i$, otherwise it takes the value 0. The statement "$y_i \in [0, \theta]$ for all $i$" is equivalent to saying "$\min_i(y_i) \geq 0$ and $\max_i(y_i) \leq \theta$", thus,

$$\prod_{i=1}^n \mathbf{1}_{[0,\theta]}(y_i) = \mathbf{1}_{[0,\infty)}\left(\min_i(y_i)\right)\mathbf{1}_{(-\infty,\theta]}\left(\max_i(y_i)\right).$$

We can now factorise the density of the sample as

$$\begin{aligned} f_{\boldsymbol{Y}}(\boldsymbol{y}) &= \left[\left(\frac{1}{\theta}\right)^n \mathbf{1}_{(-\infty,\theta]}\left(\max_i(y_i)\right)\right]\mathbf{1}_{[0,\infty)}\left(\min_i(y_i)\right) \\ &= b\left(\max_i(y_i), \theta\right)c(\boldsymbol{y}). \end{aligned}$$

By the factorisation criterion, $Y_{(n)} = \max_i(Y_i)$ is sufficient for $\theta$. This is somewhat surprising; if we know the lower bound of a continuous uniform distribution, then the sample maximum is a sufficient statistic for the upper bound.

### 10.1.3 *Minimal sufficiency*

Finding a sufficient statistic is easy. The distribution of $\boldsymbol{Y}$ conditional on $\boldsymbol{Y} = \boldsymbol{y}$ is degenerate, with all of its mass at $\boldsymbol{y}$. As such, it does not depend on any parameters. We conclude that the whole sample $\boldsymbol{Y}$ is a sufficient statistic for any parameter. Clearly for a statistic to be of use as a point estimator, it must provide us with some data reduction, that is, its dimension must be smaller than the dimension of the sample. Ideally, we would like to define a sufficient statistic that allows for maximum data reduction without loss of information about the parameters. This is the basis for defining **minimal sufficiency**. Before we make the formal definition, it is important to clarify some ideas associated with functions of samples.

Data reduction involves taking functions of the sample, that is, calculating statistics. The output from a function cannot contain more information than the argument. For example, if

$$\bar{Y} = h(\boldsymbol{Y}) = \frac{1}{n}\sum_{i=1}^{n} Y_i,$$

then $\bar{Y}$ contains less information than $\boldsymbol{Y}$. Statistics can be functions of other statistics. For example, if $\boldsymbol{V}$ is the two-dimensional statistic

$$\boldsymbol{V} = (V_1, V_2) = \left(Y_1, \sum_{i=2}^{n} Y_i\right),$$

then, clearly, $\bar{Y}$ is a function of $\boldsymbol{V}$. We could write

$$\bar{Y} = g(\boldsymbol{V}) = \frac{1}{n}(V_1 + V_2).$$

As such, $\bar{Y}$ contains less information about the sample than $\boldsymbol{V}$. We can now prove the following results about functions of statistics and sufficiency.

**Lemma 10.1.11** (Functions of statistics and sufficiency)
*Suppose that* $\boldsymbol{V}$ *is a statistic. If* $\boldsymbol{U}$ *is a function of* $\boldsymbol{V}$ *alone, that is,* $\boldsymbol{U} = \boldsymbol{g}(\boldsymbol{V})$ *for some* $\boldsymbol{g}$*, then,*

i. $\boldsymbol{U}$ *is a statistic;*
ii. *if* $\boldsymbol{U}$ *is sufficient, then* $\boldsymbol{V}$ *is sufficient;*
iii. *if* $\boldsymbol{V}$ *is not sufficient, then* $\boldsymbol{U}$ *is not sufficient;*
iv. *if* $\boldsymbol{g}$ *is one-to-one and* $\boldsymbol{V}$ *is sufficient, then* $\boldsymbol{U}$ *is sufficient.*

Proving this lemma is part of Exercise 10.1.

We can apply the ideas set out in Lemma 10.1.11 to define minimal sufficiency. Suppose that $\boldsymbol{V}$ is sufficient; we can be certain that $\boldsymbol{U}$ does not contain more information than $\boldsymbol{V}$, since it is a function of $\boldsymbol{V}$ alone. In fact, some information may have been lost. However, if $\boldsymbol{U}$ is sufficient for the parameters of interest, we have not lost anything relevant in going from $\boldsymbol{V}$ to $\boldsymbol{U}$. This is the idea behind the (rather convoluted) definition of a **minimal sufficient statistic**.

**Definition 10.1.12** (Minimal sufficient statistic)
A sufficient statistic $\boldsymbol{U}$ is a minimal sufficient statistic if, for any other sufficient statistic $\boldsymbol{V}$, if holds that $\boldsymbol{U}$ is a function of $\boldsymbol{V}$.

The condition given for minimal sufficiency is hard to check directly, however the following observations are sometimes helpful.

1. If a sufficient statistic is scalar (that is, has dimension one), then it must be a minimal sufficient statistic.
2. Minimal sufficient statistics are not unique. However, if two statistics are minimal sufficient, they must have the same dimension.
3. A one-to-one function of a minimal sufficient statistic is also a minimal sufficient statistic.
4. Some caution is required as the dimension of the minimal sufficient statistic is not always the same as the dimension of the parameter of interest.

In thinking about minimal sufficiency, it is useful to consider the partition generated by a statistic. Suppose $\boldsymbol{U} = \boldsymbol{h}(\boldsymbol{Y})$ is a statistic and $\boldsymbol{h} : S \to T$ for some sets $S$ and $T$. Like any function, $\boldsymbol{h}$ defines a partition on its domain $S$; we say that $\boldsymbol{x}$ and $\boldsymbol{y}$ belong to the same element of the partition if $\boldsymbol{h}(\boldsymbol{x}) = \boldsymbol{h}(\boldsymbol{y})$. The elements of this partition can be indexed by the range $T$, so for $\boldsymbol{t} \in T$,

$$A_{\boldsymbol{t}} = \{\boldsymbol{y} \in S : \boldsymbol{h}(\boldsymbol{y}) = \boldsymbol{t}\}.$$

Proving that $\{A_{\boldsymbol{t}} : \boldsymbol{t} \in T\}$ forms a partition of $S$ is part of Exercise 10.1. A minimal sufficient statistic is a sufficient statistic for which the associated partition is as coarse as possible. To see this, consider some other sufficient statistic $\boldsymbol{V} = \boldsymbol{r}(\boldsymbol{Y})$. If the partition associated with $\boldsymbol{U}$ is as coarse as possible, any two points giving the same value of $\boldsymbol{V}$ must also give the same value of $\boldsymbol{U}$, that is,

$$\boldsymbol{r}(\boldsymbol{y}) = \boldsymbol{r}(\boldsymbol{x}) \Rightarrow \boldsymbol{h}(\boldsymbol{y}) = \boldsymbol{h}(\boldsymbol{x}).$$

Thus, we can find a function $\boldsymbol{g}$ such that $\boldsymbol{h}(\boldsymbol{y}) = \boldsymbol{g}(\boldsymbol{r}(\boldsymbol{y}))$ for all $\boldsymbol{y} \in S$ (see Exercise 10.1). We conclude that $\boldsymbol{U} = \boldsymbol{g}(\boldsymbol{V})$, and $\boldsymbol{U}$ is minimal sufficient. The idea of partitioning using sufficient statistics forms the basis of the proof of the following proposition.

**Proposition 10.1.13** (Characterisation of minimal sufficiency)
*Consider a sample $\boldsymbol{Y} = (Y_1, \ldots, Y_n)^T$ with joint mass/density function $f_{\boldsymbol{Y}}(\boldsymbol{y}; \theta)$. If we can find a function $\boldsymbol{h}$ such that*

$$\boldsymbol{h}(\boldsymbol{y}) = \boldsymbol{h}(\boldsymbol{x}) \Leftrightarrow f_{\boldsymbol{Y}}(\boldsymbol{y}; \boldsymbol{\theta}) = k(\boldsymbol{y}, \boldsymbol{x}) f_{\boldsymbol{Y}}(\boldsymbol{x}; \boldsymbol{\theta}),$$

*then $\boldsymbol{h}(\boldsymbol{Y})$ is a minimal sufficient statistic for $\boldsymbol{\theta}$. Here $k(\boldsymbol{y}, \boldsymbol{x})$ does not depend on $\boldsymbol{\theta}$.*

*Proof.*

We prove sufficiency first. Let $\boldsymbol{h} : S \to T$ and consider the partition generated by $\boldsymbol{h}$,

$$A_{\boldsymbol{t}} = \{\boldsymbol{y} \in S : \boldsymbol{h}(\boldsymbol{y}) = \boldsymbol{t}\}.$$

Let $\boldsymbol{x}(\boldsymbol{t}) \in S$ be a fixed point in $A_t$, so

$$\boldsymbol{x}(\boldsymbol{t}) = \boldsymbol{x}(\boldsymbol{h}(\boldsymbol{y})) \text{ for all } \boldsymbol{y} \in A_t\,.$$

We now define the functions $\boldsymbol{d} : S \to S$, $c : S \to \mathbb{R}$, and $b : T \to \mathbb{R}$, where

$$\begin{aligned} \boldsymbol{d}(\boldsymbol{y}) &= \boldsymbol{x}(\boldsymbol{h}(\boldsymbol{y})), \\ c(\boldsymbol{y}) &= k(\boldsymbol{y}, \boldsymbol{d}(\boldsymbol{y})), \text{ and} \\ b(\boldsymbol{t}; \boldsymbol{\theta}) &= f_{\boldsymbol{Y}}(\boldsymbol{x}(\boldsymbol{t}); \boldsymbol{\theta})\,. \end{aligned}$$

For any point $\boldsymbol{y} \in S$ we have $\boldsymbol{y} \in A_t$ where $\boldsymbol{h}(\boldsymbol{y}) = \boldsymbol{h}(\boldsymbol{x}(\boldsymbol{t}))$, and thus,

$$\begin{aligned} f_{\boldsymbol{Y}}(\boldsymbol{y}; \boldsymbol{\theta}) &= k(\boldsymbol{y}, \boldsymbol{x}(\boldsymbol{t})) f_{\boldsymbol{Y}}(\boldsymbol{x}(\boldsymbol{t}); \boldsymbol{\theta}) \\ &= k(\boldsymbol{y}, \boldsymbol{x}(\boldsymbol{h}(\boldsymbol{y}))) f_{\boldsymbol{Y}}(\boldsymbol{x}(\boldsymbol{h}(\boldsymbol{y})); \boldsymbol{\theta}) \\ &= k(\boldsymbol{y}, \boldsymbol{d}(\boldsymbol{y})) b(\boldsymbol{h}(\boldsymbol{y}); \boldsymbol{\theta}) \\ &= c(\boldsymbol{y}) b(\boldsymbol{h}(\boldsymbol{y}); \boldsymbol{\theta}), \end{aligned}$$

so $\boldsymbol{h}(\boldsymbol{Y})$ is sufficient for $\boldsymbol{\theta}$. In order to prove minimal sufficiency we consider some other sufficient statistic, $\boldsymbol{V} = \boldsymbol{g}(\boldsymbol{Y})$. By the factorisation theorem, we can find $c'$ and $b'$ such that

$$f_{\boldsymbol{Y}}(\boldsymbol{y}; \boldsymbol{\theta}) = c'(\boldsymbol{y}) b'(\boldsymbol{g}(\boldsymbol{y}), \boldsymbol{\theta})\,.$$

Then, if $\boldsymbol{g}(\boldsymbol{y}) = \boldsymbol{g}(\boldsymbol{x})$, we have

$$f_{\boldsymbol{Y}}(\boldsymbol{y}; \boldsymbol{\theta}) = c'(\boldsymbol{y}) b'(\boldsymbol{g}(\boldsymbol{x}), \boldsymbol{\theta}) = \frac{c'(\boldsymbol{y})}{c'(\boldsymbol{x})} f_{\boldsymbol{Y}}(\boldsymbol{x}; \boldsymbol{\theta})\,.$$

By the conditions in the statement of the proposition, this implies $\boldsymbol{h}(\boldsymbol{y}) = \boldsymbol{h}(\boldsymbol{x})$. Thus, we can find a function $\boldsymbol{r}$ such that $\boldsymbol{h}(\boldsymbol{y}) = \boldsymbol{r}(\boldsymbol{g}(\boldsymbol{y}))$ for all $\boldsymbol{y}$, and hence $\boldsymbol{U} = \boldsymbol{r}(\boldsymbol{V})$. We conclude that $\boldsymbol{U}$ is minimal sufficient. □

**Example 10.1.14** (Minimal sufficiency for parameters of normal)
Consider the case of a random sample $\boldsymbol{Y}$ from an $\mathrm{N}(\mu, \sigma^2)$ distribution. We want to know whether the sufficient statistic $(\sum_{i=1}^n Y_i, \sum_{i=1}^n Y_i^2)$ is minimally sufficient. Consider two observed samples $\boldsymbol{x}$ and $\boldsymbol{y}$; the ratio of the joint densities is

$$\frac{f_{\boldsymbol{Y}}(\boldsymbol{y}; \mu, \sigma^2)}{f_{\boldsymbol{Y}}(\boldsymbol{x}; \mu, \sigma^2)} = \exp\left\{-\frac{1}{2\sigma^2}\left[\left(\sum_{i=1}^n y_i^2 - 2\mu \sum_{i=1}^n y_i\right) - \left(\sum_{i=1}^n x_i^2 - 2\mu \sum_{i=1}^n x_i\right)\right]\right\}. \tag{10.3}$$

Filling in the details for this calculation is part of Exercise 10.1. We want the ratio defined by equation (10.3) to be constant as a function of $\mu$ and $\sigma^2$; this is true if and only if

$$\sum_{i=1}^n y_i^2 - 2\mu \sum_{i=1}^n y_i = \sum_{i=1}^n x_i^2 - 2\mu \sum_{i=1}^n x_i\,.$$

Viewing these expressions as polynomials in $\mu$ and comparing coefficients, we conclude that

$$f_{\boldsymbol{Y}}(\boldsymbol{y};\boldsymbol{\theta}) = k(\boldsymbol{y},\boldsymbol{x})f_{\boldsymbol{Y}}(\boldsymbol{x};\boldsymbol{\theta}) \Leftrightarrow \left(\sum_{i=1}^{n} y_i, \sum_{i=1}^{n} y_i^2\right) = \left(\sum_{i=1}^{n} x_i, \sum_{i=1}^{n} x_i^2\right).$$

By Proposition 10.1.13, $(\sum_{i=1}^{n} Y_i, \sum_{i=1}^{n} Y_i^2)$ is minimally sufficient for $(\mu, \sigma^2)$.

### *10.1.4 Application of sufficiency in point estimation*

In a previous section we discussed mean squared error as a measure of the quality of an estimator; our aim is to find estimators with as small an MSE as possible. The principle of sufficiency indicates that we should base inference on sufficient statistics. In this section we demonstrate that the sufficiency principle and the desire for low MSE can be combined in choosing estimators. The basic idea is as follows: suppose that $\boldsymbol{U}$ is an estimator for $\boldsymbol{\theta}$. It turns out that we can find a function of a sufficient statistic for $\boldsymbol{\theta}$ that is at least as good as $\boldsymbol{U}$ in terms of MSE. This is in some sense a free lunch; we can satisfy the principle of sufficiency without sacrificing mean squared error performance.

The result hinges on the following lemma. This is rather obvious and is often taken as read. We include it here to emphasise the role of sufficiency.

**Lemma 10.1.15** (Conditioning on a sufficient statistic)
*Suppose that $\boldsymbol{Y}$ is a sample, $\boldsymbol{U}$ is a statistic, and $\boldsymbol{S}$ is sufficient for $\boldsymbol{\theta}$. If we define $\boldsymbol{T} = \mathbb{E}(\boldsymbol{U}|\boldsymbol{S})$, then $\boldsymbol{T}$ is a statistic, that is, $\boldsymbol{T}$ does not depend on $\theta$.*

*Proof.*

Since $\boldsymbol{S}$ is sufficient we know, by definition, that the conditional density $f_{\boldsymbol{Y}|\boldsymbol{S}}(\boldsymbol{y}|\boldsymbol{s})$ does not depend on $\boldsymbol{\theta}$. The statistic $\boldsymbol{U}$ is some function of $\boldsymbol{Y}$, say $\boldsymbol{U} = \boldsymbol{h}(\boldsymbol{Y})$, so

$$\mathbb{E}(\boldsymbol{U}|\boldsymbol{S} = \boldsymbol{s}) = \mathbb{E}(\boldsymbol{h}(\boldsymbol{Y})|\boldsymbol{S} = \boldsymbol{s}) = \int_{\mathbb{R}^n} \boldsymbol{h}(\boldsymbol{y}) f_{\boldsymbol{Y}|\boldsymbol{S}}(\boldsymbol{y}|\boldsymbol{s}) d\boldsymbol{y}$$

does not depend on $\boldsymbol{\theta}$, and $\boldsymbol{T} = \mathbb{E}(\boldsymbol{U}|\boldsymbol{S})$ is a statistic. □

We can now prove the Rao-Blackwell theorem. This result shows that, for any point estimator $\boldsymbol{U}$, we can find another point estimator that is a function of a sufficient statistic, and is at least as good as $\boldsymbol{U}$ in terms of mean squared error.

**Theorem 10.1.16** (Rao-Blackwell)
*Suppose that $\boldsymbol{Y}$ is a sample, $\boldsymbol{U}$ is a statistic of the same dimension as $\boldsymbol{\theta}$, and $\boldsymbol{S}$ is sufficient for $\boldsymbol{\theta}$. If we define the statistic $\boldsymbol{T} = \mathbb{E}(\boldsymbol{U}|\boldsymbol{S})$, then* $\text{MSE}_{\boldsymbol{\theta}}(\boldsymbol{T}) \leq \text{MSE}_{\boldsymbol{\theta}}(\boldsymbol{U})$.

*Proof.*

The fact that $\boldsymbol{T}$ is a statistic comes from Lemma 10.1.15, which exploits the sufficiency of $\boldsymbol{S}$. In order to establish the MSE part of the result, we start by considering the variance of $\boldsymbol{U} - \boldsymbol{\theta}$ conditional on $\boldsymbol{S}$,

$$\text{Var}(\boldsymbol{U} - \boldsymbol{\theta}|\boldsymbol{S}) = \mathbb{E}[(\boldsymbol{U} - \boldsymbol{\theta})^2|\boldsymbol{S}] - \mathbb{E}(\boldsymbol{U} - \boldsymbol{\theta}|\boldsymbol{S})^2.$$

We know that $\text{Var}(\boldsymbol{U} - \boldsymbol{\theta}|\boldsymbol{S}) \geq 0$, so

$$\mathbb{E}[(\boldsymbol{U} - \boldsymbol{\theta})^2|\boldsymbol{S}] \geq \mathbb{E}(\boldsymbol{U} - \boldsymbol{\theta}|\boldsymbol{S})^2. \tag{10.4}$$

We can now prove the result directly,

$$\begin{aligned}
\underset{\boldsymbol{\theta}}{\text{MSE}}(\boldsymbol{U}) &= \mathbb{E}\left[(\boldsymbol{U} - \boldsymbol{\theta})^2\right] && \text{by definition} \\
&= \mathbb{E}\left\{\mathbb{E}\left[(\boldsymbol{U} - \boldsymbol{\theta})^2|\boldsymbol{S}\right]\right\} && \text{law of iterated expectations} \\
&\geq \mathbb{E}\left\{(\mathbb{E}\left[(\boldsymbol{U} - \boldsymbol{\theta})|\boldsymbol{S}\right])^2\right\} && \text{from 10.4} \\
&= \mathbb{E}\left\{(\mathbb{E}[\boldsymbol{U}|\boldsymbol{S}] - \boldsymbol{\theta})^2\right\} && \text{properties of expectation} \\
&= \mathbb{E}\left\{(\boldsymbol{T} - \boldsymbol{\theta})^2\right\} && \text{definition of } \boldsymbol{T} \\
&= \underset{\boldsymbol{\theta}}{\text{MSE}}(\boldsymbol{T}).
\end{aligned}$$

□

One implication of the Rao-Blackwell theorem is that we can improve (or at least not worsen) any point estimator by taking its expectation conditional on a sufficient statistic. Notice that the bias remains unchanged, as

$$\mathbb{E}(\boldsymbol{T}) = \mathbb{E}[\mathbb{E}(\boldsymbol{U}|\boldsymbol{S})] = \mathbb{E}(\boldsymbol{U}),$$

by iterated expectations.

We give an example of the application of Theorem 10.1.16 below. Although this example is rather trivial, it illustrates the principle that we can use sufficient statistics to improve estimators.

**Example 10.1.17** (Improving an estimator for the mean of $\text{N}(\mu, 1)$)
Suppose that $\boldsymbol{Y} = (Y_1, \ldots, Y_n)^T$ is a random sample from an $\text{N}(\mu, 1)$ distribution. Let $U = Y_1$ be our initial estimator for $\mu$. This is clearly unbiased, and

$$\underset{\mu}{\text{MSE}}(U) = \text{Var}(U) = \text{Var}(Y_1) = 1\,.$$

We know that $S = \sum_{i=1}^{n} Y_i$ is sufficient for $\mu$. We can show (see Exercise 10.1) that $(U, S)^T$ is jointly normally distributed, with

$$\begin{aligned}
\mathbb{E}\left[(U, S)^T\right] &= (\mu, n\mu)^T, \\
\text{Var}\left[(U, S)^T\right] &= \begin{pmatrix} 1 & 1 \\ 1 & n \end{pmatrix}.
\end{aligned}$$

From bivariate normal theory,

$$\begin{aligned}\mathbb{E}(U|S) &= \mathbb{E}(U) + \text{Cov}(U,S)\{\text{Var}(S)\}^{-1}(S - \mathbb{E}(S))\\ &= \mu + \frac{1}{n}(S - n\mu) = \frac{1}{n}S = \frac{1}{n}\sum_{i=1}^{n} Y_i = \bar{Y}.\end{aligned}$$

Notice that we could have also obtained this result by setting

$$\begin{aligned}\mathbb{E}(U|S) &= \mathbb{E}(Y_1|S)\\ &= \frac{1}{n}\left[\mathbb{E}(Y_1|S) + \mathbb{E}(Y_2|S) + \ldots + \mathbb{E}(Y_n|S)\right]\\ &= \frac{1}{n}\mathbb{E}\left(\sum_{i=1}^{n} Y_i \middle| S\right) = \frac{1}{n}\mathbb{E}(S|S) = \frac{1}{n}S.\end{aligned}$$

The Rao-Blackwell theorem tells us that $\bar{Y}$ is a better estimator of $\mu$ than $Y_1$ (or, at least, no worse). This is no great surprise, as we know that

$$\text{Var}(\bar{Y}) = \frac{\text{Var}(Y_1)}{n},$$

so, clearly,

$$\underset{\mu}{\text{MSE}}(\bar{Y}) \leq \underset{\mu}{\text{MSE}}(Y_1).$$

However, the result is still remarkable. Starting from any unbiased estimator and taking its expectation conditional on a sufficient statistic, we will end up with another unbiased estimator with variance at least as small as our initial statistic.

The Rao-Blackwell theorem has applications in sample surveys. In an adaptive sampling design, the procedure for selecting sample units depends on what values have already been observed during the survey. In this situation, it is often simpler to construct estimators that depend on the order in which the sample units were drawn. Applying the Rao-Blackwell theorem, we can obtain better estimators by taking the expectation of the original estimators over all possible permutations of the sample.

**Exercise 10.1**

1. Prove that, if $\bar{y} = \frac{1}{n}\sum_{i=1}^{n} y_i$, then
$$\sum_{j=1}^{n}(y_j - \mu)^2 = n(\bar{y} - \mu)^2 + \sum_{j=1}^{n}(y_j - \bar{y})^2,$$
and hence derive the factorisation of the normal density given at the end of Example 10.1.9.
2. Prove Lemma 10.1.11.
3. Given a random sample $\boldsymbol{Y} = (Y_1, \ldots, Y_n)^T$, use the factorisation criterion to find a simple sufficient statistic for the unknown parameter (or parameters) in each of the following cases:

(a) $Y \sim \text{Pois}(\lambda)$,
(b) $Y \sim \text{N}(0, \sigma^2)$,
(c) $Y \sim \text{Geometric}(p)$,
(d) $Y \sim \text{Unif}[a, b]$,
(e) $Y$ has density $f_Y(y; \theta) = \exp[-(y - \theta)]$ for $y > \theta$.

4. Consider the function $h : S \rightarrow T$ for arbitrary sets $S$ and $T$. For all $t \in T$, define

$$A_t = \{y \in S : h(y) = t\}.$$

Prove that $\{A_t : t \in T\}$ forms a partition of $S$.

5. Consider the functions $r : S \rightarrow Q$ and $h : S \rightarrow T$ for arbitrary sets $S$, $T$, and $Q$. Prove that, if

$$r(y) = r(x) \Rightarrow h(y) = h(x),$$

then we can find a function $g : Q \rightarrow T$ such that

$$h(y) = g(r(y)).$$

## 10.2 Variance of unbiased estimators

Unbiased estimators have the appealing property that their expected value is the true value of the parameter. If we restrict our attention to unbiased estimators, an obvious goal is to try to find the estimator with the smallest variance, that is, the **minimum-variance unbiased estimator** (MVUE). We start by proving a result that describes the best performance we can hope for from an unbiased estimator. Throughout what follows, we will assume that $\boldsymbol{Y} = (Y_1, \dots, Y_n)^T$ is a sample with joint density $f_{\boldsymbol{Y}}(\boldsymbol{y}; \theta)$, where $\theta$ is a scalar parameter; $\theta$ is a value somewhere in the parameter space $\Theta$.

**Theorem 10.2.1** (Cramér-Rao lower bound)
*Let $\boldsymbol{Y} = (Y_1, \dots, Y_n)^T$ be a sample parameterised by $\theta$ (a scalar), and $U = h(\boldsymbol{Y})$ an unbiased estimator of a function of the parameter, $g(\theta)$. Under regularity conditions (allowing us to differentiate under an integral sign), it holds that, for any $\theta \in \Theta$,*

$$\text{Var}(U) \geq \frac{\left[\frac{d}{d\theta} g(\theta)\right]^2}{\mathcal{I}_{\boldsymbol{Y}}(\theta)},$$

*where $\mathcal{I}_{\boldsymbol{Y}}(\theta)$ is the total Fisher information about $\theta$ in the sample $\boldsymbol{Y}$.*

*Proof.*

In the proof of Theorem 4.3.8 we established that, for random variables $X$ and $Y$,

$$\text{Cov}(X, Y)^2 \leq \text{Var}(X)\,\text{Var}(Y). \tag{10.5}$$

This is the variance/covariance form of the Cauchy-Schwarz inequality. We will deal with the continuous case, and drop $\boldsymbol{Y}$ subscripts on the density $f$ and score $s$. By unbiasedness, $\mathbb{E}_\theta[h(\boldsymbol{Y})] = g(\theta)$, so

$$g(\theta) = \int_{\mathbb{R}^n} h(\boldsymbol{y}) f(\boldsymbol{y};\theta) d\boldsymbol{y} .$$

Differentiating both sides with respect to $\theta$ yields

$$\begin{aligned} \frac{d}{d\theta} g(\theta) &= \int_{\mathbb{R}^n} h(\boldsymbol{y}) \frac{\partial}{\partial\theta} f(\boldsymbol{y};\theta) d\boldsymbol{y} \\ &= \int_{\mathbb{R}^n} h(\boldsymbol{y}) s(\theta;\boldsymbol{y}) f(\boldsymbol{y};\theta) d\boldsymbol{y} \qquad \text{as } s(\theta;\boldsymbol{y}) = \frac{1}{f(\boldsymbol{y};\theta)} \frac{\partial}{\partial\theta} f(\boldsymbol{y};\theta) \\ &= \mathbb{E}[h(\boldsymbol{Y}) s(\theta;\boldsymbol{Y})] . \end{aligned} \tag{10.6}$$

Recall from Lemma 9.2.2 that $\mathbb{E}[s(\theta;\boldsymbol{Y})] = 0$. As a consequence,

$$\text{Cov}[h(\boldsymbol{Y}), s(\theta;\boldsymbol{Y})] = \mathbb{E}[h(\boldsymbol{Y}) s(\theta;\boldsymbol{Y})] ,$$

and we can rewrite equation (10.6) as

$$\frac{d}{d\theta} g(\theta) = \text{Cov}[h(\boldsymbol{Y}), s(\theta;\boldsymbol{Y})] .$$

Applying inequality (10.5) then yields

$$\left[\frac{d}{d\theta} g(\theta)\right]^2 \leq \text{Var}[h(\boldsymbol{Y})] \, \text{Var}[s(\theta;\boldsymbol{Y})] .$$

From a simple rearrangement we get

$$\text{Var}[h(\boldsymbol{Y})] \geq \frac{\left[\frac{d}{d\theta} g(\theta)\right]^2}{\text{Var}[s(\theta;\boldsymbol{Y})]} .$$

Recalling that $U = h(\boldsymbol{Y})$ and $\mathcal{I}_{\boldsymbol{Y}}(\theta) = \text{Var}[s(\theta;\boldsymbol{Y})]$ gives the result. □

The general form of the Cramér-Rao theorem does not make any assumptions about the sample beyond some requirements for regularity of the joint distribution. In practice, we often assume that our sample is random. This leads to a substantial simplification in computation of the lower bound on variance of an unbiased estimator. Recall that in section 9.1 we found that, for a random sample $\boldsymbol{Y} = (Y_1, \ldots, Y_n)^T$,

$$\mathcal{I}_{\boldsymbol{Y}}(\theta) = n \mathcal{I}_Y(\theta) ,$$

where $\mathcal{I}_{\boldsymbol{Y}}(\theta)$ is the total information and $\mathcal{I}_Y(\theta)$ is information associated with a single observation (the distinction between vector $\boldsymbol{Y}$ and scalar $Y$ is important here). This leads directly to the following corollary, which establishes a lower bound on the variance of an unbiased estimator based on a random sample.

**Corollary 10.2.2**
*If $\boldsymbol{Y} = (Y_1, \ldots, Y_n)^T$ is a random sample and $U = h(\boldsymbol{Y})$ is an unbiased estimator of $g(\theta)$, then*

$$\text{Var}(U) \geq \frac{\left[\frac{d}{d\theta} g(\theta)\right]^2}{n \mathcal{I}_Y(\theta)},$$

*where $\mathcal{I}_Y(\theta)$ is the Fisher information about $\theta$ in a single sample member.*

Of course, in many instances we are interested in estimating $\theta$ itself rather than some function of $\theta$. This leads to further simplification of the lower bound. In particular, if $\hat{\theta}$ is an unbiased estimator of $\theta$ based on a random sample $(Y_1, \ldots, Y_n)^T$ then

$$\text{Var}(\hat{\theta}) \geq \frac{1}{n \mathcal{I}_Y(\theta)}. \tag{10.7}$$

Theorem 10.2.1 establishes a lower bound for the variance of an unbiased estimator. An obvious question is whether this bound is ever attained. If a statistic $U$ attains the lower bound, that is, if $U$ is an unbiased estimator of $g(\theta)$ and

$$\text{Var}(U) = \frac{\left[\frac{d}{d\theta} g(\theta)\right]^2}{\mathcal{I}_{\boldsymbol{Y}}(\theta)},$$

then $U$ must be a minimum-variance unbiased estimator of $g(\theta)$. However, the converse does not hold; there are instances in which the minimum-variance unbiased estimator does not attain the Cramér-Rao lower bound. An additional condition is required to ensure that the bound is attained.

**Proposition 10.2.3** (Attaining the Cramér-Rao bound)
*If $U = h(\boldsymbol{Y})$ is an unbiased estimator of $g(\theta)$, then $U$ attains the Cramér-Rao lower bound if and only if*

$$s(\theta; \boldsymbol{y}) = b(\theta)[h(\boldsymbol{y}) - g(\theta)],$$

*where $b(\theta)$ is a function involving the parameter $\theta$ but not $\boldsymbol{y}$.*

*Proof.*
Recall that the proof of the Cramér-Rao lower bound comes directly from the inequality $\text{Cov}(X,Y)^2 \leq \text{Var}(X)\,\text{Var}(Y)$. Recall also that we have equality if and only if $X$ and $Y$ are linearly related, that is, $Y = \alpha + \beta X$ for some constants $\alpha, \beta$ with $\beta \neq 0$. In the Cramér-Rao proof we use the fact that

$$\text{Cov}[h(\boldsymbol{Y}), s(\theta; \boldsymbol{Y})] \leq \text{Var}[h(\boldsymbol{Y})]\,\text{Var}[s(\theta; \boldsymbol{Y})].$$

Thus, in order to have equality we must have

$$s(\theta; \boldsymbol{Y}) = \alpha + \beta h(\boldsymbol{Y}),$$

In this instance, we know that $\mathbb{E}[s(\theta; \boldsymbol{Y})] = 0$ and $\mathbb{E}[h(\boldsymbol{Y})] = g(\theta)$. This implies that $\alpha + \beta g(\theta) = 0$ and thus $\frac{\alpha}{\beta} = -g(\theta)$. Using $\beta = b(\theta)$, to denote the possibility that the constant depends on $\theta$, gives us our result. □

So far we have considered a scalar parameter $\theta$. For a vector parameter, $\boldsymbol{\theta} = (\theta_1, \ldots, \theta_r)^T$, the variance of the estimator, $\text{Var}(\boldsymbol{U})$, and the information matrix, $\boldsymbol{I}_{\boldsymbol{Y}}(\boldsymbol{\theta})$, are both $r \times r$ matrices. The vector version of equation (10.7), that is, the Cramér-Rao bound for an estimator of $\boldsymbol{\theta}$ itself, is given by the following theorem.

**Theorem 10.2.4** (Cramér-Rao for a vector of parameters)
*If $\boldsymbol{Y} = (Y_1, \ldots, Y_n)^T$ is a random sample parameterised by $\boldsymbol{\theta} = (\theta_1, \ldots, \theta_r)^T$, and $\boldsymbol{U}$ is an unbiased estimator of $\boldsymbol{\theta}$, then*

$$\text{Var}(\boldsymbol{U}) \geq \{\boldsymbol{I}_{\boldsymbol{Y}}(\boldsymbol{\theta})\}^{-1},$$

*in the sense that* $\text{Var}(\boldsymbol{U}) - \{\boldsymbol{I}_{\boldsymbol{Y}}(\boldsymbol{\theta})\}^{-1}$ *is a non-negative-definite matrix.*

We omit the proof of this theorem.

**Example 10.2.5** (Bound on variance for estimator of sample mean of normal)
Suppose $\boldsymbol{Y} = (Y_1, \ldots, Y_n)^T$ is a random sample from an $\text{N}(\mu, \sigma^2)$ distribution where $\sigma^2$ is known. We show in Example 9.2.8 that the information about $\mu$ in the sample $\boldsymbol{Y}$ is

$$\mathcal{I}_{\boldsymbol{Y}}(\mu) = \frac{n}{\sigma^2}.$$

Here we are estimating the parameter (rather than a function of the parameter) using a random sample; by application of equation (10.7), if $U$ is unbiased for $\mu$ then

$$\text{Var}(U) \geq \frac{\sigma^2}{n}.$$

We know that this bound is attained since $\text{Var}(\bar{Y}) = \sigma^2/n$. We can readily check that $\bar{Y}$ meets the conditions set out in Proposition 10.2.3. From Example 9.2.8, the score is

$$s_{\boldsymbol{Y}}(\mu; \boldsymbol{y}) = \frac{1}{\sigma^2} \sum_{i=1}^{n} (y_i - \mu) = \frac{n}{\sigma^2}(\bar{y} - \mu),$$

which clearly meets the factorisation criterion.

**Exercise 10.2**

1. Find the Cramér-Rao bound for the variance of an unbiased estimator of $\lambda$ for a random sample $\boldsymbol{Y} = (Y_1, \ldots, Y_n)^T$ from the Gamma distribution with density function

$$f_Y(y) = \begin{cases} \frac{\lambda^r}{\Gamma(r)} y^{r-1} e^{-\lambda y} & \text{for } y > 0, \\ 0 & \text{otherwise,} \end{cases}$$

   where $\lambda > 0$ and $r$ is known.

2. Find the Cramér-Rao bound for the variance of an unbiased estimator of the square of the probability of success of a binomial distribution. Is it attainable?

3. Consider a random vector $\boldsymbol{Y}$, and let $U = h(\boldsymbol{Y})$ be an unbiased estimator for $g(\theta)$. The Cramér-Rao inequality can be written

$$\text{Var}(U) \geq \frac{[\frac{d}{d\theta} g(\theta)]^2}{\mathbb{E}\left([\frac{\partial}{\partial\theta} \ell_{\boldsymbol{Y}}(\theta; \boldsymbol{Y})]^2\right)},$$

Suppose that $\ell_Y(\theta; y)$ is the log-likelihood function.

(a) Show that the denominator on the right-hand side of the Cramér-Rao inequality can be replaced by

$$-\mathbb{E}\left[\left(\frac{\partial}{\partial\theta}\right)^2 \ell_Y(\theta; Y)\right].$$

(b) State the condition under which there exists an unbiased estimator for $g(\theta)$ whose variance attains the Cramér-Rao lower bound.

## 10.3 Most powerful tests

As we are going to be comparing tests, it is useful to be able to refer to the tests using a specific identifier. A test is characterised by the decision function or, equivalently, the critical region. For a test $S$, we use $\phi_S$ to denote the decision function, $R_S$ to denote the critical region, and $\beta_S$ to denote the power function. Note that $\beta_S(\boldsymbol{\theta})$ is the probability that $H_0$ is rejected by test $S$ if $\boldsymbol{\theta}$ is the true parameter value. The following lemma establishes the connection between the power function and decision function. It will be useful in proving the main result of this section.

**Lemma 10.3.1** (Relationship between decision rule and power function)
*If $\phi_S$ is the decision rule for a hypothesis test $S$, then the power function is given by*

$$\beta_S(\boldsymbol{\theta}) = \mathbb{E}_{\boldsymbol{\theta}}[\phi_S(\boldsymbol{Y})].$$

*Proof.*

The proof is direct.

$$\begin{aligned}
\beta_S(\boldsymbol{\theta}) &= P_{\boldsymbol{\theta}}(H_0 \text{ rejected by test } S) \\
&= P_{\boldsymbol{\theta}}(\boldsymbol{Y} \in R_S) \\
&= \int_{R_S} f_{\boldsymbol{Y}}(\boldsymbol{y}; \boldsymbol{\theta}) d\boldsymbol{y} \\
&= \int_{\mathbb{R}^n} \phi_S(\boldsymbol{y}) f_{\boldsymbol{Y}}(\boldsymbol{y}; \boldsymbol{\theta}) d\boldsymbol{y} && \text{since } \phi_S(\boldsymbol{y}) = 0 \text{ for } \boldsymbol{y} \notin R_S \\
&= \mathbb{E}_{\boldsymbol{\theta}}[\phi_S(\boldsymbol{Y})].
\end{aligned}$$

□

The starting point for developing the theory for choosing tests on the basis of their power is to consider two simple hypotheses. Suppose that we are testing

$$\begin{aligned}
H_0 &: \boldsymbol{\theta} = \boldsymbol{\theta}_0, \\
H_1 &: \boldsymbol{\theta} = \boldsymbol{\theta}_1.
\end{aligned}$$

In this situation, the power of a test $S$ is simply the power function evaluated at the value specified by the alternative, that is, $\beta_S(\boldsymbol{\theta}_1)$.

**Definition 10.3.2** (Most powerful test)
Suppose that we are testing $H_0 : \theta = \theta_0$ against $H_1 : \theta = \theta_1$. We say that $T$ is a **most powerful test** if $\beta_T(\theta_1) \geq \beta_S(\theta_1)$ for all tests $S$ such that $\beta_S(\theta_0) = \beta_T(\theta_0)$.

As we might expect, Definition 10.3.2 says that $T$ is a most powerful test if its power is greater than or equal to that of any other test of the same size. We will often refer to a most powerful test **of size** $\alpha$; the size being made explicit to emphasise its crucial role in the definition of a most powerful test.

We can now give a statement of one of the earliest results in hypothesis testing. The proof of the theorem is direct, if somewhat subtle in places. We give the proof followed by detailed notes to shed light on some of the more awkward steps. It is a good exercise to try to understand as much of the proof as possible before reading the notes. The notion of likelihood plays a fundamental role.

**Theorem 10.3.3** (Neyman-Pearson lemma)
*Suppose that $T$ is a test of size $\alpha$ for $H_0 : \theta = \theta_0$ against $H_1 : \theta = \theta_1$. If the critical region for $T$ is*

$$R_T = \{y \in \mathbb{R}^n : L_Y(\theta_1; y) - k_\alpha L_Y(\theta_0; y) > 0\},$$

*then $T$ is the most powerful test of size $\alpha$. Note that $L_Y$ is the likelihood function and $k_\alpha$ is a constant that depends on $\alpha$.*

*Proof.*

Let $S$ be a test of size $\alpha$ for $H_0 : \theta = \theta_0$ against $H_1 : \theta = \theta_1$. We start by considering the power of $S$ minus $k_\alpha$ times its size:

$$\begin{aligned}
\beta_S(\theta_1) - k_\alpha \beta_S(\theta_0) &= \int_{\mathbb{R}^n} \phi_S(y)[f_Y(y;\theta_1) - k_\alpha f_Y(y;\theta_0)]dy && \text{point 1 below} \\
&= \int_{\mathbb{R}^n} \phi_S(y)[L_Y(\theta_1;y) - k_\alpha L_Y(\theta_0;y)]dy \\
&\leq \int_{R_T} \phi_S(y)[L_Y(\theta_1;y) - k_\alpha L_Y(\theta_0;y)]dy && \text{point 2 below} \\
&\leq \int_{R_T} \phi_T(y)[L_Y(\theta_1;y) - k_\alpha L_Y(\theta_0;y)]dy && \text{point 3 below} \\
&= \int_{R_T} \phi_T(y)[f_Y(y;\theta_1) - k_\alpha f_Y(y;\theta_0)]dy \\
&= \int_{\mathbb{R}^n} \phi_T(y)[f_Y(y;\theta_1) - k_\alpha f_Y(y;\theta_0)]dy && \text{point 4 below} \\
&= \beta_T(\theta_1) - k_\alpha \beta_T(\theta_0)\,. && \text{point 1 again}
\end{aligned}$$

Since both $S$ and $T$ are of size $\alpha$, we know that $\beta_S(\theta_0) = \beta_T(\theta_0) = \alpha$, which implies that $\beta_S(\theta_1) \leq \beta_T(\theta_1)$. We conclude that $T$ is a most powerful test of size $\alpha$. □

**Notes on proof of Neyman-Pearson lemma**

1. From Lemma 10.3.1, for any test $S$,

$$\beta_S(\boldsymbol{\theta}) = \mathop{\mathbb{E}}_{\boldsymbol{\theta}}[\phi_S(\boldsymbol{Y})] = \int_{\mathbb{R}^n} \phi_S(\boldsymbol{y}) f_{\boldsymbol{Y}}(\boldsymbol{y};\boldsymbol{\theta}) d\boldsymbol{y}\,.$$

2. The statement of the theorem defines the region

$$R_T = \{\boldsymbol{y} \in \mathbb{R} : L_{\boldsymbol{Y}}(\boldsymbol{\theta}_1;\boldsymbol{y}) - k_\alpha L_{\boldsymbol{Y}}(\boldsymbol{\theta}_0;\boldsymbol{y}) > 0\}\,.$$

The decision function $\phi_S(\boldsymbol{y})$ only takes the values 0 and 1, so

$$\phi_S(\boldsymbol{y})[L_{\boldsymbol{Y}}(\boldsymbol{\theta}_1;\boldsymbol{y}) - k_\alpha L_{\boldsymbol{Y}}(\boldsymbol{\theta}_0;\boldsymbol{y})] \begin{cases} \geq 0 & \text{for any } \boldsymbol{y} \in R_T, \\ \leq 0 & \text{for any } \boldsymbol{y} \notin R_T. \end{cases}$$

An integral over just the region $R_T$ excludes all possible negative values of the function. Thus, the result will be at least as large as that given by integrating over the whole of $\mathbb{R}^n$.

3. By definition of the decision rule, $\phi_T(\boldsymbol{y}) = 1$ for all $\boldsymbol{y} \in R_T$. So, since any decision rule only takes the values 0 and 1, we have

$$\phi_T(\boldsymbol{y}) \geq \phi_S(\boldsymbol{y}) \text{ for all } \boldsymbol{y} \in R_T.$$

4. By definition of the decision rule, $\phi_T(\boldsymbol{y}) = 0$ for all $\boldsymbol{y} \notin R_T$. Thus integrating any product of $\phi_T(\boldsymbol{y})$ over $R_T$ will yield the same result as integrating over the whole of $\mathbb{R}^n$.

5. The $\alpha$ index on $k_\alpha$ is to make clear the dependence of the choice of $k_\alpha$ on the desired size of the test. First note that

$$P_{\boldsymbol{\theta}_0}(H_0 \text{ rejected by test } T) = P_{\boldsymbol{\theta}_0}(\boldsymbol{Y} \in R_T) = P_{\boldsymbol{\theta}_0}(L_{\boldsymbol{Y}}(\boldsymbol{\theta}_1;\boldsymbol{Y}) - k_\alpha L_{\boldsymbol{Y}}(\boldsymbol{\theta}_0;\boldsymbol{Y}) > 0)\,.$$

Then the value $k_\alpha$ is chosen to satisfy

$$P_{\boldsymbol{\theta}_0}\left(\frac{L_{\boldsymbol{Y}}(\boldsymbol{\theta}_1;\boldsymbol{Y})}{L_{\boldsymbol{Y}}(\boldsymbol{\theta}_0;\boldsymbol{Y})} > k_\alpha\right) = \alpha\,. \tag{10.8}$$

In equation (10.8), we are considering the ratio of the likelihood calculated at $\theta_1$ to the likelihood calculated at $\theta_0$. We encountered likelihood ratios in a more general context in section 9.4.

As it stands, the Neyman-Pearson lemma is of limited practical value; we rarely want to test against a simple alternative. The definition of a **uniformly most powerful test** (UMPT) extends the idea of most powerful tests to the composite case. Suppose that we want to test a simple null hypothesis against a composite alternative hypothesis,

$$\begin{aligned} H_0 &: \boldsymbol{\theta} = \boldsymbol{\theta}_0\,, \\ H_1 &: \boldsymbol{\theta} \in \boldsymbol{\Theta}_1\,. \end{aligned}$$

The Neyman-Pearson lemma no longer applies directly. However, in many instances we find that the most powerful test of size $\alpha$ for $H_0 : \boldsymbol{\theta} = \boldsymbol{\theta}_0$ against $H_1 : \boldsymbol{\theta} = \boldsymbol{\theta}_1$ is the same for every $\boldsymbol{\theta}_1 \in \boldsymbol{\Theta}_1$. In other words, there is a test $T$ which is the most powerful test of size $\alpha$ for every possible alternative value in $\boldsymbol{\Theta}_1$. Under these conditions, we say that $T$ is the uniformly most powerful test of size $\alpha$ for $H_0 : \boldsymbol{\theta} = \boldsymbol{\theta}_0$ against $H_1 : \boldsymbol{\theta} \in \boldsymbol{\Theta}_1$.

For a composite null hypothesis we need to take some care to ensure that the test is the appropriate size. The following example illustrates.

**Example 10.3.4** (Uniformly most powerful test for a scalar parameter)
Suppose that we have a test $S$ that is the most powerful test of size $\alpha$ for

$$H_0 : \theta = \theta_0 ,$$
$$H_1 : \theta = \theta_1 ,$$

where $\theta_1 > \theta_0$. Now consider the composite hypotheses

$$H_0 : \theta \leq \theta_0 ,$$
$$H_1 : \theta > \theta_1 .$$

The test $S$ is the uniformly most powerful test of size $\alpha$ for these hypotheses if:

i. the critical region for $S$ does not depend on the value of $\theta_1$.
ii. the probability of rejecting the null hypothesis is $\leq \alpha$ for any $\theta \leq \theta_0$, that is,

$$\beta_S(\theta) \leq \beta_S(\theta_0) \text{ for } \theta \leq \theta_0 .$$

Similar statements can be made for testing $H_0 : \theta \geq \theta_0$ against $H_1 : \theta < \theta_0$, and also in the vector parameter case.

**Example 10.3.5** (Uniformly most powerful test for normal mean)
Suppose that we have a random sample $\boldsymbol{Y} = (Y_1, \ldots, Y_n)^T$ from an $\mathrm{N}(\mu, \sigma^2)$ distribution where $\sigma^2$ is known. We start by considering testing the simple hypotheses

$$H_0 : \mu = \mu_0 ,$$
$$H_1 : \mu = \mu_1 .$$

The likelihood ratio from the Neyman-Pearson lemma is

$$\begin{aligned}
\frac{L_{\boldsymbol{Y}}(\mu_1; \boldsymbol{Y})}{L_{\boldsymbol{Y}}(\mu_0; \boldsymbol{Y})} &= \frac{\left(\frac{1}{\sqrt{2\pi\sigma^2}}\right)^n \exp\left[-\frac{1}{2\sigma^2}\sum_{i=1}^n (Y_i - \mu_1)^2\right]}{\left(\frac{1}{\sqrt{2\pi\sigma^2}}\right)^n \exp\left[-\frac{1}{2\sigma^2}\sum_{i=1}^n (Y_i - \mu_0)^2\right]} \\
&= \exp\left\{\frac{1}{2\sigma^2}\left[2n\bar{Y}\mu_1 - n\mu_1^2 - 2n\bar{Y}\mu_0 + n\mu_0^2\right]\right\} \\
&= \exp\left\{\frac{n}{2\sigma^2}\left[(\mu_0^2 - \mu_1^2) - 2\bar{Y}(\mu_0 - \mu_1)\right]\right\} \\
&= \exp\left\{\frac{n}{2\sigma^2}\left[(\mu_0 - \mu_1)(\mu_0 + \mu_1 - 2\bar{Y})\right]\right\} .
\end{aligned}$$

There are two distinct cases:

- $\mu_0 < \mu_1$, the likelihood ratio increases as $\bar{Y}$ increases,
- $\mu_0 > \mu_1$, the likelihood ratio increases as $\bar{Y}$ decreases.

In either case, we can construct a most powerful test based on $\bar{Y}$. For example, in the case $\mu_0 < \mu_1$, a most powerful test has critical region,

$$R = \left\{ \mathbf{y} : \bar{y} > \mu_0 + z_{1-\alpha} \frac{\sigma}{\sqrt{n}} \right\}. \tag{10.9}$$

This critical region does not depend on $\mu_1$, so the test is uniformly most powerful in testing

$$H_0 : \mu = \mu_0\,,$$
$$H_1 : \mu > \mu_0\,.$$

Notice that this does not work if the alternative is $H_1 : \mu \neq \mu_0$ (no UMPT exists in this case).

Now consider the case of a composite null,

$$H_0 : \mu \leq \mu_0\,,$$
$$H_1 : \mu > \mu_0\,.$$

In order to establish that the critical region given by equation (10.9) provides a uniformly most powerful test in this case, we need to show that the probability of rejecting the null is at most $\alpha$ for any $\mu \leq \mu_0$. We can see that this is true because

$$P_\mu(\mathbf{Y} \in R) = \Phi\left( z_\alpha + \frac{\mu - \mu_0}{\sqrt{\sigma^2/n}} \right),$$

and, if $\mu < \mu_0$, then the argument of $\Phi$ is less than $z_\alpha$. As $\Phi$ is a non-decreasing function, we conclude that the probability of rejecting the null is less than or equal to $\alpha$.

**Example 10.3.6** (Uniformly most powerful test for binomial proportion)
Suppose that $Y \sim \text{Bin}(n, p)$ and we want to test

$$H_0 : p = p_0\,,$$
$$H_1 : p > p_0\,.$$

We start by considering a simple alternative, $H_1 : p = p_1$, for $p_1 > p_0$. The likelihood ratio from the Neyman-Pearson lemma is

$$\frac{L_Y(p_1; y)}{L_Y(p_0; y)} = \frac{\binom{n}{y} p_1^y (1-p_1)^{n-y}}{\binom{n}{y} p_0^y (1-p_0)^{n-y}} = \left( \frac{p_1(1-p_0)}{p_0(1-p_1)} \right)^y \left( \frac{1-p_1}{1-p_0} \right)^n.$$

Since $p_1 > p_0$, we have $[p_1(1-p_0)]/[p_0(1-p_1)] > 1$. We conclude that the value of the likelihood increases as $y$ increases, and so we can construct a most powerful test that rejects $H_0$ when $y > k_\alpha$, for some $k_\alpha$. The value of $k_\alpha$ (and hence the form of the critical region) will depend only on the null distribution, $\text{Bin}(n, p_0)$. Hence, the critical region will not depend on $p_1$, and the test is uniformly most powerful.

**Exercise 10.3**

1. Suppose $\boldsymbol{Y} = (Y_1, \ldots, Y_n)^T$ is a random sample from an $\text{N}(0, \theta)$ distribution. Construct a most powerful test of
$$H_0 : \theta = 1\,,$$
$$H_1 : \theta = 2\,.$$
Find the power of the test of size 0.05 when $n = 10$.
2. Let $\boldsymbol{Y} = (Y_1, \ldots, Y_n)^T$ be a random sample from $\text{N}(0, \sigma^2)$. Find the uniformly most powerful test of
$$H_0 : \sigma^2 \leq \sigma_0^2\,,$$
$$H_1 : \sigma^2 > \sigma_0^2\,,$$
where $\sigma_0 > 0$ is a given constant.
3. Let $Y_1, \ldots, Y_n$ be a random sample from a uniform distribution on the interval $[0, \theta]$, where $\theta > 0$ is unknown. Find the most powerful test for the null hypothesis $H_0 : \theta = 1$ against the alternative hypothesis $H_1 : \theta = 4$ which never rejects a true null hypothesis. Find the power of this most powerful test when $n = 4$.

## 10.4 Further exercises

1. Suppose that $\boldsymbol{Y} = (Y_1, \ldots, Y_n)^T$ is a random sample from an $\text{N}(\mu, \sigma^2)$ distribution. Consider two observed samples $\boldsymbol{x}$ and $\boldsymbol{y}$. Carefully show that
$$\frac{f_{\boldsymbol{Y}}(\boldsymbol{y}; \mu, \sigma^2)}{f_{\boldsymbol{Y}}(\boldsymbol{x}; \mu, \sigma^2)} = \exp\left\{-\frac{1}{2\sigma^2}\left[\left(\sum_{i=1}^n y_i^2 - 2\mu \sum_{i=1}^n y_i\right) - \left(\sum_{i=1}^n x_i^2 - 2\mu \sum_{i=1}^n x_i\right)\right]\right\}.$$
2. Suppose that $\boldsymbol{Y} = (Y_1, \ldots, Y_n)^T$ is a random sample from an $\text{N}(\mu, 1)$ distribution. Let $U = Y_1$ and $S = \sum_{i=1}^n Y_i$. Show that $U$ and $S$ are jointly normally distributed.
3. Suppose that $\boldsymbol{Y} = (Y_1, \ldots, Y_n)^T$ is a random sample from a Bernoulli$(p)$ distribution. Starting with the estimator $\hat{p} = Y_1$, use the Rao-Blackwell theorem to find a better estimator.
4. Suppose that $Y_1, \ldots, Y_n$ is a random sample from a population with density
$$f_Y(y) = \frac{2y}{\theta^2} \text{ for } 0 < y \leq \theta\,,$$
where $\theta > 0$.

(a) Is the last order statistic $Y_{(n)}$ a sufficient statistic for $\theta$?

(b) Derive the density of $Y_{(n)}$.

(c) Find an unbiased estimator of $\theta$ that is a function of $Y_{(n)}$.

(d) By considering the condition that a function of $Y_{(n)}$ is unbiased for $\theta$, determine whether there is a better unbiased estimator for $\theta$.

5.(a) Explain what is meant by minimum-variance unbiased estimator.

(b) Let $Y_1, \ldots, Y_n$ be a random sample from the uniform distribution on the interval $[0, \theta]$, where $\theta > 0$ is an unknown parameter. Define $U = \max\{Y_1, \ldots, Y_n\}$.

i. Show that $U$ is a sufficient statistic for $\theta$.

ii. Show that $\hat{\theta} \equiv (n+1)U/n$ is an unbiased estimator of $\theta$. Find its variance.

iii. Show that $\hat{\theta}$ is a minimum-variance unbiased estimator for $\theta$.

6.(a) Let $Y_1, \ldots, Y_n$ be a random sample from Gamma$(k, \theta)$, with density function

$$f_Y(y) = \begin{cases} \frac{\theta^k}{(k-1)!} y^{k-1} \exp(-\theta y) & \text{for } x > 0, \\ 0 & \text{otherwise,} \end{cases}$$

where $k \geq 1$ is a known integer, and $\theta > 0$ is an unknown parameter.

i. Show that the moment-generating function of the sum $Y_1 + \ldots + Y_n$ is $M_Y(t) = (1 - t/\theta)^{-nk}$.

ii. Find the constant $c$ such that $c/\bar{Y}$ is an unbiased estimator for $\theta$, where $\bar{Y}$ is the sample mean. Calculate the variance of this estimator and compare it with the Cramér-Rao lower bound.

iii. Find the minimum-variance unbiased estimator for $\alpha = 1/\theta$.

7. Let $Y_1, \ldots, Y_n$ be a random sample from an exponential distribution with unknown mean $\theta > 0$.

(a) Find the maximum-likelihood estimator $\hat{\theta}$ for $\theta$.

(b) Is $\hat{\theta}$ a sufficient statistic for $\theta$?

(c) Show that $\hat{\theta}$ is an unbiased and consistent estimator.

State clearly any theorems you use to draw your conclusions.

8. Let $\hat{\theta}$ be the minimum-variance unbiased estimator for parameter $\theta$, and let $a$ and $b$ be known constants with $a \neq 0$. Show that $\hat{\mu} = a\hat{\theta} + b$ is the minimum-variance unbiased estimator for the parameter $\mu = a\theta + b$.

9.(a) Let $Y_1, \ldots, Y_n$ be a random sample, where $n > 1$.

i. Show that $\tilde{\theta} \equiv \mathbf{1}_{(-\infty,0)}(X_1)$ is an unbiased estimator for $\theta \equiv \mathrm{P}(Y_1 < 0)$. Compute Var$(\tilde{\theta})$.

ii. Explain why $\tilde{\theta}$ is a poor estimator. Find an estimator with smaller mean squared error.

(b) Let $Y_1, \ldots, Y_n$ be a random sample from a distribution with density function

$$f_Y(y; \alpha, \theta) = \begin{cases} \alpha e^{-\alpha(x-\theta)} & x > \theta, \\ 0 & x \leq \theta, \end{cases}$$

where $\alpha > 0$, $\theta > 0$ are parameters. Find the minimal sufficient statistic and the maximum-likelihood estimator for:

i. $\alpha$ when $\theta$ is known,
ii. $\theta$ when $\alpha$ is known,
iii. $(\alpha, \theta)$ when both $\alpha$ and $\theta$ are unknown.
iv. $\mu = 1/\alpha$ when $\theta$ is known. Is the maximum-likelihood estimator for $\mu$ also the minimum-variance unbiased estimator? Give reasons.

10. (a) State and prove the Neyman-Pearson lemma. Briefly discuss how a uniformly most powerful test for testing a simple null hypothesis against a one-sided composite alternative may be constructed based on this lemma.

(b) Let $Y_1, \ldots, Y_n$ be a random sample from a Poisson distribution with mean $\theta > 0$.

i. Find the maximum-likelihood estimator for $\theta$ and write down its asymptotic distribution.
ii. Construct the uniformly most powerful test for testing hypotheses

$$H_0 : \theta = \theta_0 ,$$
$$H_1 : \theta > \theta_0 .$$

Suppose that $n = 20$, the sample mean is 1.2, and $\theta_0 = 0.7$. Will you reject $H_0$ at a significance level of 1%?

11. (**Cramér-Rao lower bound for a biased estimator**) Let $Y_1, \ldots, Y_n$ be a random sample parameterised by $\theta$, a scalar parameter. Suppose that $\hat{\theta} = h(\boldsymbol{Y})$ is a biased estimator of $\theta$, with $\mathbb{E}(\hat{\theta}) = \theta + b(\theta)$.

(a) Show that $\text{Var}(\hat{\theta}) \geq (1 + \frac{d}{d\theta} b(\theta))^2 / \mathcal{I}_{\boldsymbol{Y}}(\theta)$.
(b) Obtain a lower bound for the mean squared error (MSE) of $\hat{\theta}$.
(c) Suppose that there exists an unbiased estimator, $\tilde{\theta}$, which attains the Cramér-Rao lower bound. Is it possible for $\hat{\theta}$ to have lower MSE than $\tilde{\theta}$? Explain your reasoning.

## 10.5 Chapter summary

On completion of this chapter you should be able to:

- define a sufficient statistic and a minimal sufficient statistic,
- use the factorisation theorem to check whether a statistic is sufficient,
- use conditioning on a sufficient statistic to improve an estimator,
- prove the Rao-Blackwell theorem,
- prove the Cramér-Rao inequality,
- use the Cramér-Rao inequality to determine the lower bound on the variance of unbiased estimators of the parameters of a specified distribution,
- determine whether an estimator is the minimum-variance unbiased estimator for the parameters of a specified distribution,

- describe the basic principles of maximum-likelihood estimation,
- state and prove the Neyman-Pearson lemma,
- construct most powerful tests for simple hypotheses, and
- determine whether a test is uniformly most powerful.

Chapter 11

# Bayesian inference

Techniques in classical statistics tend to involve averages over the population distribution for fixed values of the parameters. In reality we do not know the parameter values. The Bayesian approach attempts to model our uncertainty about parameters by treating them as random variables.

Bayesian statistics is a very broad and well-developed field. We do little more here than attempt to set out some of the practical aspects of a Bayesian approach. As indicated earlier, the authors, in common with many pragmatic statisticians, have little interest in philosophical debates about the relative merits of classical and Bayesian inference. Both approaches have substantial flaws; both contribute useful methods and interpretations to the statistician's toolbox.

In order to keep the terminology from becoming overcomplicated, we always refer to a single parameter. The case of several scalar parameters is covered by stacking into a single vector parameter. We will focus on the continuous case in our discussion, although much of it carries over to the discrete case, replacing integrals with summations. Bayesian inference methods often involve the computation of intractable integrals, but these can often be estimated efficiently and accurately through approximate numerical techniques, which we discuss in Chapter 12.

## 11.1 Prior and posterior distributions

A Bayesian approach requires us to specify a marginal distribution for the parameter. This distribution is referred to as the **prior**. Much of the controversy that has surrounded Bayesian statistics focuses on the prior. In its early incarnations, Bayesianism was strongly associated with a subjective interpretation of probability. To a subjectivist, probability is a reflection of our degree of belief, so different people may rationally ascribe different probabilities to the same event. The name prior reflects this heritage; in the subjective Bayesian world, the prior represents our beliefs about the distribution of the parameter prior to observing the data. Once we have the observations, the prior distribution is *updated* using Bayes' theorem. The resulting **posterior** distribution forms the basis for Bayesian inference.

Some notation helps to clarify this process. We drop our earlier convention of identifying with subscripts the random variable with which a distribution is associated. In the remainder of this chapter, we confine our attention to two vectors of random variables,

$$\text{the sample, } \boldsymbol{Y} = (Y_1, \ldots, Y_n)^T,$$
$$\text{the parameter, } \boldsymbol{\Theta} = (\Theta_1, \ldots, \Theta_r)^T.$$

We use $f$ to denote density functions associated with the sample. As with the classical frequentist approach, we assume that we know the sampling density $f(\boldsymbol{y}|\boldsymbol{\theta})$, that is, the density of $\boldsymbol{Y}$ given that $\boldsymbol{\Theta}$ takes a particular value $\boldsymbol{\theta}$. This is just the density function of the random variable $\boldsymbol{Y}|\boldsymbol{\Theta} = \boldsymbol{\theta}$. When we view the sampling density as a function of the parameter, $\boldsymbol{\theta}$, we refer to $f(\boldsymbol{y}|\boldsymbol{\theta})$ as the likelihood.

Density functions associated with the parameter are denoted using $\pi$. Specifically, we have

$$\text{the prior density, } \pi(\boldsymbol{\theta}),$$
$$\text{the posterior density, } \pi(\boldsymbol{\theta}|\boldsymbol{y}).$$

Notice that the prior density is the marginal density of $\boldsymbol{\Theta}$, and the posterior density is the density of the random variable $\boldsymbol{\Theta}|\boldsymbol{Y} = \boldsymbol{y}$. We use $\mathbb{E}(\cdot|\boldsymbol{y})$, instead of the usual $\mathbb{E}(\cdot|\boldsymbol{Y} = \boldsymbol{y})$, to denote expectations involving the posterior density.

For the joint density of $\boldsymbol{Y}$ and $\boldsymbol{\Theta}$, rather than introduce a new piece of notation, we use the likelihood (the sampling density) multiplied by the prior, $f(\boldsymbol{y}|\boldsymbol{\theta})\pi(\boldsymbol{\theta})$. Thus, the marginal density of $\boldsymbol{Y}$ is

$$f(\boldsymbol{y}) = \int_{-\infty}^{\infty} f(\boldsymbol{y}|\boldsymbol{\theta})\pi(\boldsymbol{\theta})d\boldsymbol{\theta}.$$

Bayes' theorem provides the connection between prior and posterior distributions,

$$\pi(\boldsymbol{\theta}|\boldsymbol{y}) = \frac{f(\boldsymbol{y}|\boldsymbol{\theta})\pi(\boldsymbol{\theta})}{f(\boldsymbol{y})}. \tag{11.1}$$

We view $\pi(\boldsymbol{\theta}|\boldsymbol{y})$ as a function of $\boldsymbol{\theta}$, and $f(\boldsymbol{y})$ as a function of $\boldsymbol{y}$ alone, so equation (11.1) is often written as

$$\pi(\boldsymbol{\theta}|\boldsymbol{y}) \propto f(\boldsymbol{y}|\boldsymbol{\theta})\pi(\boldsymbol{\theta}). \tag{11.2}$$

The proportionality relationship (11.2) is often remembered as *"posterior is proportional to likelihood times prior"*.

**Exercise 11.1**

1. Suppose we wish to estimate $\boldsymbol{\theta}$ by maximising the posterior $\pi(\boldsymbol{\theta}|\boldsymbol{y})$. Under what circumstances is the resulting estimate the same as the maximum-likelihood estimate of $\boldsymbol{\theta}$?

2. Let $\boldsymbol{y}_1, \boldsymbol{y}_2$ be two independent random samples from $f(\boldsymbol{y}|\boldsymbol{\theta})$, where $\boldsymbol{\theta}$ has prior $\pi(\boldsymbol{\theta})$. Consider two possible situations: (i) we observe $\boldsymbol{y}_1$ first, update the prior to $\pi(\boldsymbol{\theta}|\boldsymbol{y}_1)$, then observe $\boldsymbol{y}_2$ and update the distribution again; and (ii) we observe $\boldsymbol{y}_1, \boldsymbol{y}_2$ simultaneously and update the prior in one step. Show that we obtain the same posterior distribution of $\boldsymbol{\theta}$ in both cases.

## 11.2 Choosing a prior

We can see from equation (11.1) that the form of the posterior distribution is dependent on the choice of prior. The fact that something that we choose has an impact on our inferences is often cited as a weakness of the Bayesian approach. This criticism is somewhat spurious; frequentist inference also involves arbitrary choices, such as the choice of significance level in hypothesis testing. However, the practical problem of choosing a prior must be resolved if we are going to perform Bayesian inference. If we believe that the prior should reflect our degree of belief in the various possible values of the parameter, then it should be chosen by carefully eliciting and combining the opinions of rational experts. This approach is practical (and sensible) for problems where there is extensive domain-specific knowledge. In our discussion of choice of prior we assume that we do not have any well-formed opinions about the parameters, perhaps just a vague range of plausible values. We start by establishing some terminology.

Some of the priors that are used in practice are what is referred to as **improper**. The term "improper prior" brings to mind a monk who does not take his vows too seriously; the Bayesian use of the term is somewhat similar. It is possible to define priors that do not have proper densities (in the sense that they do not integrate to 1s), but still lead to proper posterior densities. The posterior forms the basis for inference so, provided the posterior has a proper density, we need not be too concerned if the prior is improper. Several examples of improper priors are given below.

One way that we that we might try to make a Bayesian analysis as objective as possible is to use a **non-informative** prior. The term non-informative is used in a variety of contexts; we use it to mean any prior that attempts to keep inclusion of subjective information about the parameters to a minimum. Another term which is often used for priors that contain little information and are dominated by the likelihood is **reference** priors. In other instances, we choose a particular type of prior for mathematical convenience; these are referred to as **conjugate** priors. We discuss the construction of reference and conjugate priors below.

### *11.2.1 Constructing reference priors*

We identify three ways in which a non-informative or reference prior may be constructed. Note that not all authors would agree with our classification and there are substantial overlaps between categories.

1. *Flat priors*: One possible mechanism for conveying prior ignorance is to use a flat prior. A flat prior assigns equal probability to all parameter values in its support. We could choose end points for the interval of plausible parameter values and then use a prior that is uniform on this interval. In some instances, there is an obvious range of values that the parameter might take. For example, if $Y|\Theta = \theta \sim \text{Bernoulli}(\theta)$, we know that $\theta \in [0, 1]$. In general, adopting this approach would require an arbitrary choice of interval end points. An alternative is to specify

$$\pi(\theta) = c \text{ for all } \theta .$$

Clearly, unless the support of $\theta$ is a finite interval, this is an improper prior; $\pi$ is not a valid density as $\int \pi(\theta) d\theta$ is not finite. However, we can still evaluate the posterior density from the product of the likelihood and the prior. In this case, as the prior is a fixed constant, the posterior density is proportional to the likelihood,

$$\pi(\boldsymbol{\theta}|\boldsymbol{y}) \propto f(\boldsymbol{y}|\boldsymbol{\theta}) .$$

2. *Priors from data-translated forms*: Suppose that we can express the likelihood in **data-translated form**,

$$f(\boldsymbol{y}|\theta) \propto r\left(g(\theta) - h(\boldsymbol{y})\right) ,$$

for some functions $r$, $g$, and $h$. Note that the proportionality here is in terms of $\theta$, thus the constants of proportionality may involve $\boldsymbol{y}$. Consider the reparameterisation given by $\psi = g(\theta)$. The impact of the data is to translate the likelihood relative to $\psi$. One possible choice is to use a flat prior for $\psi$. By the change-of-variables formula, the implied prior for $\Theta$ has the proportionality relationship,

$$\pi(\theta) \propto \left|\frac{\partial \psi}{\partial \theta}\right| .$$

If $g$ is the identity function, that is, if $\psi = \theta$, then using this criterion is equivalent to specifying a flat prior for $\theta$. If the relationship between $\theta$ and $\boldsymbol{y}$ in the likelihood takes the form of a ratio, we can still express the likelihood in data-translated form by setting

$$r\left(\frac{g(\theta)}{h(\boldsymbol{y})}\right) = r\left(\exp\left\{\log\left(\frac{g(\theta)}{h(\boldsymbol{y})}\right)\right\}\right) = r\left(\exp\left\{\log g(\theta) - \log h(\boldsymbol{y})\right\}\right) .$$

In this case, we would take $\psi = \log g(\theta)$, and our prior for $\theta$ would be

$$\pi(\theta) \propto \left|\frac{g'(\theta)}{g(\theta)}\right| .$$

3. *Transformation invariance and Jeffreys' rule*: Consider a situation where we are interested in a well-behaved monotone transformation of the original parameter. We denote the transformed parameters as $\Psi = g(\Theta)$. The implied prior for $\Psi$ is

$$\pi_{\Psi}(\psi) = \pi_{\Theta}(g^{-1}(\psi)) \left|\frac{\partial}{\partial \psi} g^{-1}(\psi)\right| = \pi_{\Theta}(\theta) \left|\frac{\partial \theta}{\partial \psi}\right| , \tag{11.3}$$

where $\theta = g^{-1}(\psi)$. It is intuitively reasonable to hope that $\pi_\psi$ represents the same degree of prior knowledge as $\pi_\Theta$. If this holds, we refer to $\pi_\Theta$ as **transformation-invariant**.

The prior specification suggested by **Jeffreys' rule** achieves transformation invariance by exploiting the mechanics of the change-of-variable formula. Consider $\mathcal{I}(\psi)$, the Fisher information with respect to the parameter $\psi$. We can write

$$\frac{\partial}{\partial\psi}\log f(\mathbf{y}|\theta) = \frac{\partial}{\partial\theta}\log f(\mathbf{y}|\theta)\frac{\partial\theta}{\partial\psi} \Rightarrow$$
$$-\left(\frac{\partial}{\partial\psi}\log f(\mathbf{y}|\theta)\right)^2 = -\left(\frac{\partial}{\partial\theta}\log f(\mathbf{y}|\theta)\right)^2\left(\frac{\partial\theta}{\partial\psi}\right)^2,$$

and take expectations on both sides to obtain

$$\mathcal{I}(\psi) = \mathcal{I}(\theta)\left(\frac{\partial\theta}{\partial\psi}\right)^2. \tag{11.4}$$

A **Jeffreys prior** for $\Theta$ is then specified as

$$\pi_\Theta(\theta) = \sqrt{|\mathcal{I}(\theta)|}. \tag{11.5}$$

Comparing equations (11.3) and (11.4), we can see that the implied prior for the transformed version of the parameters is also a Jeffreys prior, thus, $\pi_\Theta$ is transformation invariant. If $\boldsymbol{\Theta}$ is a vector of parameters, $|\mathcal{I}(\boldsymbol{\theta})|$ is the determinant of the Fisher information matrix, and expression (11.5) gives us the joint Jeffreys prior.

**Example 11.2.1** (Reference priors for normal parameters)
Let $\mathbf{Y} = (Y_1, \ldots, Y_n)^T$ be a random sample from an $\mathrm{N}(\mu, \sigma^2)$ distribution, and consider the problem of specifying reference priors for the parameters. It is reasonable to assume that $\mu$ and $\sigma$ are independent, that is, $\pi(\mu,\sigma) = \pi(\mu)\pi(\sigma)$. Consider the likelihood for $\mu$ first, treating $\sigma$ as known. Using the factorisation from Example 10.1.9, we can write

$$L(\mu) \propto \exp\left\{-\frac{n}{2\sigma^2}(\bar{y}-\mu)^2\right\},$$

which is in data-translated form, with $\psi = g(\mu) = \mu$ and $h(y) = \bar{y}$. Thus, a suitable prior for $\mu$ is

$$\pi(\mu) \propto \left|\frac{\partial\mu}{\partial\mu}\right| = 1,$$

a flat (uniform) density. This is an improper prior, as it cannot integrate to 1 over the interval $(-\infty,\infty)$. Now consider the likelihood for $\sigma$, treating $\mu$ as known:

$$L(\sigma) \propto \frac{1}{\sigma^n}\exp\left\{-\frac{1}{2\sigma^2}\sum_{i=1}^{n}(y_i-\mu)^2\right\}.$$

Define $s_\mu^2 = \frac{1}{n}\sum_{i=1}^n (y_i-\mu)^2$, which is not a function of $\sigma$, so we can multiply the expression by $s_\mu^n$ and write

$$L(\sigma) \propto \frac{s_\mu^n}{\sigma^n} \exp\left\{-\frac{n}{2}\frac{s_\mu^2}{\sigma^2}\right\}.$$

This expression is of the form $r\left(\frac{g(\sigma)}{h(y)}\right)$, where $g(\sigma) = \sigma$ and $h(y) = s_\mu$, thus our prior is

$$\pi(\sigma) \propto \left|\frac{g'(\sigma)}{g(\sigma)}\right| = \frac{1}{\sigma}.$$

This is also an improper prior, because a function of this form cannot integrate to 1 over the interval $(0, \infty)$. We can think of it as the limit of a Gamma$(\alpha, \lambda)$ density as $\alpha \to 0$ and $\lambda \to 0$. Finally, putting the two priors together gives

$$\pi(\mu, \sigma) \propto \frac{1}{\sigma}.$$

**Example 11.2.2** (Jeffreys' rule for exponential parameter)
Suppose that $\boldsymbol{Y} = (Y_1, \ldots, Y_n)^T$ is a random sample from an Exp$(\theta)$ distribution. We know from Example 9.2.7 that the Fisher information is $\mathcal{I}(\theta) = n/\theta^2$, thus, a Jeffreys prior density for the rate parameter is

$$\pi(\theta) \propto \sqrt{|\mathcal{I}(\theta)|} \propto \frac{1}{\theta},$$

an improper prior.

### *11.2.2 Conjugate priors*

For a given likelihood, $f(\boldsymbol{y}|\boldsymbol{\theta})$, we can sometimes specify a prior in such a way that both prior and posterior belong to the same parametric family. In this case, we say that $\pi(\boldsymbol{\theta})$ is a **conjugate** prior for $f(\boldsymbol{y}|\boldsymbol{\theta})$. The main motivation for this choice of prior is mathematical convenience, however, the use of conjugate priors often leads to intuitively appealing results. The prior $\pi(\boldsymbol{\theta})$ may depend on unknown parameters, which are called **hyperparameters**. We assume for now that these are known.

**Example 11.2.3** (Normal prior and normal likelihood)
Let $\boldsymbol{X} = (X_1, \cdots, X_n)^T$ be a random sample from N$(\mu, \sigma^2)$, with $\sigma^2$ known. Let $\mu \sim$ N$(\mu_0, \sigma_0^2)$ be the prior. Then, ignoring terms that do not depend on $\mu$, the posterior is

$$
\begin{aligned}
\pi(\mu|\boldsymbol{x}) &\propto f(\boldsymbol{x}|\mu)\pi(\mu)\\
&\propto \exp\left[-\frac{n}{2\sigma^2}(\bar{x}-\mu)^2\right]\exp\left[-\frac{1}{2\sigma_0^2}(\mu-\mu_0)^2\right]\\
&\propto \exp\left[-\frac{1}{2}\left\{(n/\sigma^2+1/\sigma_0^2)\mu^2-2(n\bar{x}/\sigma^2+\mu_0/\sigma_0^2)\mu\right\}\right]\\
&\propto \exp\left[-\frac{1}{2}(n/\sigma^2+1/\sigma_0^2)\left(\mu-\frac{n\bar{x}/\sigma^2+\mu_0/\sigma_0^2}{n/\sigma^2+1/\sigma_0^2}\right)^2\right],
\end{aligned}
$$

where the last expression is obtained by completing the square inside the curly braces. Thus, the posterior distribution is

$$
(\mu|\boldsymbol{x}) \sim \mathrm{N}\left(\frac{n\bar{x}/\sigma^2+\mu_0/\sigma_0^2}{n/\sigma^2+1/\sigma_0^2}, \frac{1}{n/\sigma^2+1/\sigma_0^2}\right),
$$

and $\mathrm{N}(\mu_0, \sigma_0^2)$ is a conjugate prior. Notice that, as $\sigma_0^2 \to \infty$, the prior becomes flat and the posterior converges to $\mathrm{N}(\bar{x}, \sigma^2/n)$ (Figure 11.1). This is identical to the posterior we would have obtained had we specified a reference prior for $\mu$, that is, $\pi(\mu) \propto 1$. Also, as $n \to \infty$, the data swamp the prior and the posterior becomes independent of the prior parameters $\mu_0, \sigma_0^2$.

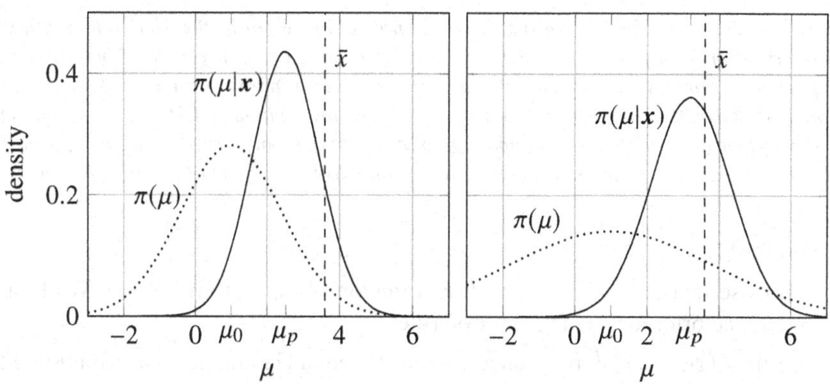


Figure 11.1 *An illustration of this example with $n = 7$, $\sigma^2 = 10$, and $\mu_0 = 1$. In the diagram on the left the prior is quite strong ($\sigma_0^2 = 1$), whereas on the right it is weak ($\sigma_0^2 = 8$). Suppose that we observe $\bar{x} = 3.6$. With a flatter prior, the mean of the posterior distribution, $\mu_p$, is closer to the MLE, $\bar{x}$.*

**Example 11.2.4** (Beta prior and binomial likelihood)
Let $(Y|\Theta = \theta) \sim \text{Bin}(n, \theta)$ and $\Theta \sim \text{Beta}(\alpha, \beta)$. The posterior density is

$$\begin{aligned}\pi(\theta|y) &\propto f(y|\theta)\pi(\theta) \\ &= \binom{n}{y}\theta^y(1-\theta)^{n-y}\frac{1}{B(\alpha,\beta)}\theta^{\alpha-1}(1-\theta)^{\beta-1} \\ &\propto \theta^{\alpha+y-1}(1-\theta)^{\beta+n-y-1}.\end{aligned}$$

Hence $(\Theta|Y = y) \sim \text{Beta}(\alpha + y, \beta + n - y)$, and thus $\text{Beta}(\alpha, \beta)$ is a conjugate prior for $\text{Bin}(n, \theta)$. As $n$ increases, our choice of hyperparameters $\alpha$ and $\beta$ becomes less important (Figure 11.2).

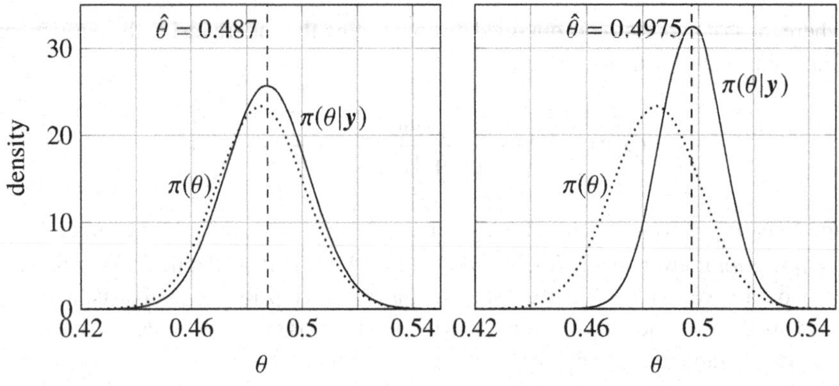


Figure 11.2 *An illustration of a simple Beta-binomial model for births. Here $\theta$ represents the proportion of female babies at birth. Suppose that on a given day 51% of babies born in a particular country were female. We use the weakly informative prior $\Theta \sim \text{Beta}(485, 515)$, which has mean 0.485 (the global average) and standard deviation 0.016. The diagrams show the posterior density of $\theta$ for a sample size of $n = 100$ (left) or $n = 1000$ (right). The dashed line is the posterior mean, $\hat{\theta}$. As the sample size increases, the effect of the prior is minimised.*

**Exercise 11.2**

1. Suppose that $\boldsymbol{X} = (X_1, \cdots, X_n)^T$ is a random sample from $\text{Pois}(\lambda)$. Work out a reference prior for $\lambda$ using Jeffreys' rule.
2. Let $\boldsymbol{Y} = (Y_1, \ldots, Y_n)^T$ be a random sample from a $\text{Gamma}(\alpha, \lambda)$ distribution. Find a reference prior for the parameters by expressing the likelihood in data-translated form.
3. Find a joint Jeffreys prior for the parameters of the $\text{N}(\mu, \sigma^2)$ distribution.
4. Show that the Gamma distribution is a conjugate prior for the Poisson.

## 11.3 Bayesian estimation

### *11.3.1 Point estimators*

In Bayesian analysis the key quantity used in inference is the **posterior distribution**, the conditional distribution of the parameters given the data. In practice, it is useful to be able to generate point estimators. For example, we may want to know where the posterior distribution is centred. Obvious choices of point estimators in Bayesian analysis are the posterior mean, median, and mode.

For example, the posterior mean is

$$\hat{\theta} = h(\boldsymbol{Y}) = \mathbb{E}(\theta|\boldsymbol{Y}),$$

where

$$h(\boldsymbol{y}) = \mathbb{E}(\theta|\boldsymbol{Y} = \boldsymbol{y}) = \int_{-\infty}^{\infty} \theta\pi(\theta|\boldsymbol{y})d\theta\,.$$

More generally, Bayesian parameter estimation is based on the concept of a **loss function**. The loss function is denoted by $L(\theta, a)$, where $\theta$ is the parameter to be estimated, and $a = a(\boldsymbol{Y})$ is an estimator. Intuitively, the loss represents the penalty incurred as a result of the difference between the estimated value $a$ and the true value $\theta$; it is a function of $\theta$, hence a random variable. The idea is to choose the estimator $\hat{\theta}$ which minimises the expected loss under the posterior distribution of $\theta$. This optimal estimator is known as the **Bayes estimator**, and can be expressed as $\hat{\theta} = g(\boldsymbol{Y})$, where

$$g(\boldsymbol{y}) = \arg\min_{a} \mathbb{E}[L(\theta, a)|\boldsymbol{y}] = \arg\min_{a} \int_{-\infty}^{\infty} L(\theta, a)\pi(\theta|\boldsymbol{y})d\theta\,.$$

We will now consider some of the most common loss functions and the resulting Bayes estimators.

*Quadratic loss*

If the chosen loss function is $L(\theta, a) = (\theta - a)^2$, then we need to minimise

$$\mathbb{E}[L(\theta, a)|\boldsymbol{y}] = \mathbb{E}[(\theta - a)^2|\boldsymbol{y}] = \mathbb{E}(\theta^2|\boldsymbol{y}) - 2a\,\mathbb{E}(\theta|\boldsymbol{y}) + a^2.$$

Differentiating with respect to $a$ yields

$$\frac{\partial}{\partial a}\mathbb{E}[L(\theta, a)|\boldsymbol{y}] = -2\,\mathbb{E}(\theta|\boldsymbol{y}) + 2a\,.$$

This is equal to 0 when $a = \mathbb{E}(\theta|\boldsymbol{y})$ and, furthermore, $\frac{\partial^2}{\partial a^2}\mathbb{E}[L(\theta, a)|\boldsymbol{y}] = 2 > 0$. The posterior mean $\hat{\theta} = \mathbb{E}(\theta|\boldsymbol{Y})$ is, thus, the Bayes estimator. We can readily show that, if our criterion for a good estimator is that it minimises posterior mean squared error, then the posterior mean is optimal. This is conceptually similar to least-squares regression, where we aim to minimise the sum of squares (loss function), and end up with $\hat{Y} = \mathbb{E}(Y|X)$ as our estimator (conditional mean).

**Proposition 11.3.1** (Posterior mean)
*Let $\hat{\theta} = \mathbb{E}(\theta|\boldsymbol{Y})$. Then, for any estimator $\tilde{\theta}$,*

$$\mathbb{E}[(\tilde{\theta} - \theta)^2] \geq \mathbb{E}[(\hat{\theta} - \theta)^2],$$

*where expectation is taken with respect to both $\boldsymbol{Y}$ and $\theta$.*

*Proof.*

We have

$$\begin{aligned}
\mathbb{E}[(\tilde{\theta} - \theta)^2] &= \mathbb{E}\left\{\mathbb{E}[(\tilde{\theta} - \theta)^2|\boldsymbol{Y}]\right\} \\
&= \int \left\{\int (\tilde{\theta} - \theta)^2 \pi(\theta|\boldsymbol{y})d\theta\right\} f_{\boldsymbol{Y}}(\boldsymbol{y})d\boldsymbol{y} \\
&\geq \int \left\{\int (\hat{\theta} - \theta)^2 \pi(\theta|\boldsymbol{y})d\theta\right\} f_{\boldsymbol{Y}}(\boldsymbol{y})d\boldsymbol{y} \qquad \text{as } \hat{\theta} \text{ minimises quad. loss} \\
&= \mathbb{E}\left\{\mathbb{E}[(\hat{\theta} - \theta)^2|\boldsymbol{Y}]\right\} \\
&= \mathbb{E}[(\hat{\theta} - \theta)^2],
\end{aligned}$$

hence the posterior mean is a minimum-mean-square estimator. This is in contrast to the classical frequentist approach, where the MMSE estimator does not usually exist, and we often resort to finding the minimum-variance unbiased estimator instead. □

### *11.3.2 Absolute loss*

Let the loss function be $L(\theta, a) = |\theta - a|$. We differentiate the expected loss with respect to $a$, recalling that $\frac{d}{dx}\int_{-\infty}^{x} g(u)du = g(x)$ and $\frac{d}{dx}\int_{x}^{\infty} g(u)du = -g(x)$, to obtain

$$\begin{aligned}
\frac{\partial}{\partial a}\mathbb{E}[L(\theta, a)|\boldsymbol{y}] &= \frac{\partial}{\partial a}\int_{-\infty}^{\infty} |\theta - a|\pi(\theta|\boldsymbol{y})d\theta \\
&= \frac{\partial}{\partial a}\left\{\int_{-\infty}^{a} (a - \theta)\pi(\theta|\boldsymbol{y})d\theta + \int_{a}^{\infty} (\theta - a)\pi(\theta|\boldsymbol{y})d\theta\right\} \\
&= \int_{-\infty}^{a} \pi(\theta|\boldsymbol{y})d\theta - \int_{a}^{\infty} \pi(\theta|\boldsymbol{y})d\theta,
\end{aligned}$$

with the remaining terms cancelling out. This derivative is equal to 0 when the two integrals are equal, that is, when $a$ is the median of $\theta|\boldsymbol{y}$. The second derivative is $2\pi(a|\boldsymbol{y}) > 0$, so this is a minimum. The Bayes estimator is, thus, $\hat{\theta} = \text{median}(\theta|\boldsymbol{Y})$, the posterior median.

### *11.3.3 0-1 loss*

Consider the loss function

$$L(\theta, a) = 1 - \mathbf{1}_{[a-c,a+c]}(\theta) = \begin{cases} 0 & \text{for } |\theta - a| \leq c\,, \\ 1 & \text{otherwise,} \end{cases}$$

where $c$ is a given constant. The expected loss is

$$\mathbb{E}[L(\theta,a)|\boldsymbol{y}] = \mathbb{E}\left[1 - \mathbf{1}_{[a-c,a+c]}(\theta)|\boldsymbol{y}\right] = 1 - \mathrm{P}(a-c \leq \theta \leq a+c|\boldsymbol{y}),$$

which is minimised when the probability in the final expression is maximised. We choose $a$ so that the interval $[a-c, a+c]$ has the highest posterior probability of any interval with length $2c$; this is known as the **modal interval** of length $2c$. We can take the midpoint of the modal interval as our Bayes estimator. Now let $c \to 0$ to obtain the loss function known as 0-1 loss,

$$L(\theta,a) = \begin{cases} 0 & \text{if } \theta = a, \\ 1 & \text{otherwise.} \end{cases}$$

If the posterior distribution is continuous, it is not possible to minimise the expected loss directly. However, if the distribution is unimodal, the endpoints of the modal interval always lie on opposite sides of the mode (see Example 8.3.9) so, as $c \to 0$, the modal interval converges to the mode. We conclude that the Bayes estimator is the posterior mode $\hat{\theta} = \text{mode}(\theta|\boldsymbol{Y})$, defined as

$$\text{mode}(\theta|\boldsymbol{Y}) = g(\boldsymbol{Y}),$$

where

$$g(\boldsymbol{y}) = \arg\max_{\theta} \pi(\theta|\boldsymbol{y}) = \arg\max_{\theta} f(\boldsymbol{y}|\theta)\pi(\theta).$$

If we choose a flat (non-informative) prior, the function $\pi(\theta)$ is constant and the expression simplifies to

$$g(\boldsymbol{y}) = \arg\max_{\theta} f(\boldsymbol{y}|\theta),$$

which, in the classical frequentist sense, is the maximum-likelihood estimator of an unknown (fixed) parameter $\theta$ based on a sample $\boldsymbol{y}$.

**Example 11.3.2** (Beta prior and binomial likelihood)
The posterior distribution is Beta$(\alpha + y, \beta + n - y)$. From the properties of the Beta distribution (see Exercise 7.4), the posterior mean is

$$\mathbb{E}(\theta|Y) = \frac{Y+\alpha}{n+\alpha+\beta} = \frac{n}{n+\alpha+\beta}\frac{Y}{n} + \left(1 - \frac{n}{n+\alpha+\beta}\right)\frac{\alpha}{\alpha+\beta}$$

which is a weighted sum of the prior mean, $\alpha/(\alpha+\beta)$, and the MLE, $Y/n$. This is a common occurrence in Bayesian estimation; we say that the MLE is **shrunk** towards the prior estimate (Figure 11.2). The posterior mode is

$$\begin{aligned} \text{mode}(\theta|Y) &= \frac{Y+\alpha-1}{n+\alpha+\beta-2} \\ &= \frac{n}{n+\alpha+\beta-2}\frac{Y}{n} + \left(1 - \frac{n}{n+\alpha+\beta-2}\right)\frac{\alpha-1}{\alpha+\beta-2}, \end{aligned}$$

a weighted sum of the prior mode, $(\alpha-1)/(\alpha+\beta-2)$, and the MLE. There is no closed form for the posterior median.

**Example 11.3.3** (Normal prior and normal likelihood)
The posterior distribution is

$$\mathrm{N}\left(\frac{n\bar{x}/\sigma^2 + \mu_0/\sigma_0^2}{n/\sigma^2 + 1/\sigma_0^2}, \frac{1}{n/\sigma^2 + 1/\sigma_0^2}\right),$$

so the posterior mean, median, and mode are all equal to

$$\begin{aligned}\hat{\mu} &= \frac{n\bar{X}/\sigma^2 + \mu_0/\sigma_0^2}{n/\sigma^2 + 1/\sigma_0^2} = \frac{n\sigma_0^2\bar{X} + \mu_0\sigma^2}{n\sigma_0^2 + \sigma^2} \\ &= \frac{n\sigma_0^2}{n\sigma_0^2 + \sigma^2}\bar{X} + \left(1 - \frac{n\sigma_0^2}{n\sigma_0^2 + \sigma^2}\right)\mu_0,\end{aligned}$$

a weighted sum of the prior mean, $\mu_0$, and the sample mean, $\bar{X}$, which is also the MLE. The weight given to the prior mean – the amount of shrinkage – is determined by $\sigma_0^2$ and $n$; it decreases as $\sigma_0^2$ increases (non-informative prior) and as $n$ increases (the data swamp the prior).

### *11.3.4 Interval estimates*

We can also use the posterior distribution to generate interval estimates for the parameter of interest. If $C$ is an interval such that

$$\mathrm{P}(\theta \in C | \boldsymbol{Y} = \boldsymbol{y}) = 1 - \alpha, \tag{11.6}$$

then we refer to $C$ as a $100(1-\alpha)\%$ **credible interval** for $\theta$ (also known as a **Bayesian confidence interval**). Notice that Bayesian interval estimates admit a far more natural probability interpretation than their frequentist counterparts: given the data, the probability that the value of the parameter falls in $C$ is $1-\alpha$. In addition, note that, for a given posterior, there is an infinite number of intervals satisfying equation (11.6). If the posterior is strictly unimodal, the interval $C = (a, b)$ with $\pi(a|\boldsymbol{y}) = \pi(b|\boldsymbol{y})$ is optimal, but (for convenience) we often choose an interval with equal tail probabilities, that is,

$$\int_{-\infty}^{a} \pi(\theta|\boldsymbol{y})d\theta = \int_{b}^{\infty} \pi(\theta|\boldsymbol{y})d\theta = \alpha/2\,.$$

**Example 11.3.4** (Bayesian CI for mean of normal – variance known)
Suppose that $\boldsymbol{Y}$ is a random sample from an $\mathrm{N}(\mu, \sigma^2)$ distribution, where $\sigma^2$ is known. If we choose a reference prior for $\mu$, the posterior distribution of $\mu$ is

$$(\mu | \boldsymbol{Y} = \boldsymbol{y}) \sim \mathrm{N}(\bar{y}, \sigma^2/n),$$

and the shortest $(1-\alpha)$ interval estimate for $\mu$ is then

$$\bar{y} \pm z_{1-\alpha/2}\frac{\sigma}{\sqrt{n}}\,.$$

This is identical to the frequentist interval in Example 8.3.2, although the interpretation is very different. As always, the result is dependent on our choice of prior; for example, using the conjugate (normal) prior from Example 11.2.3, we obtain

$$(\mu|\boldsymbol{Y} = \boldsymbol{y}) \sim \mathrm{N}\left(\frac{n\bar{y}/\sigma^2 + \mu_0/\sigma_0^2}{n/\sigma^2 + 1/\sigma_0^2}, \frac{1}{n/\sigma^2 + 1/\sigma_0^2}\right),$$

and the resulting confidence interval is

$$\frac{n\bar{y}/\sigma^2 + \mu_0/\sigma_0^2}{n/\sigma^2 + 1/\sigma_0^2} \pm z_{1-\alpha/2}\frac{1}{n/\sigma^2 + 1/\sigma_0^2}.$$

Comparing the two intervals, we see that using an informative prior reduces the length of the interval (the posterior variance is lower), and pulls the midpoint of the interval towards the prior mean (Figure 11.3).

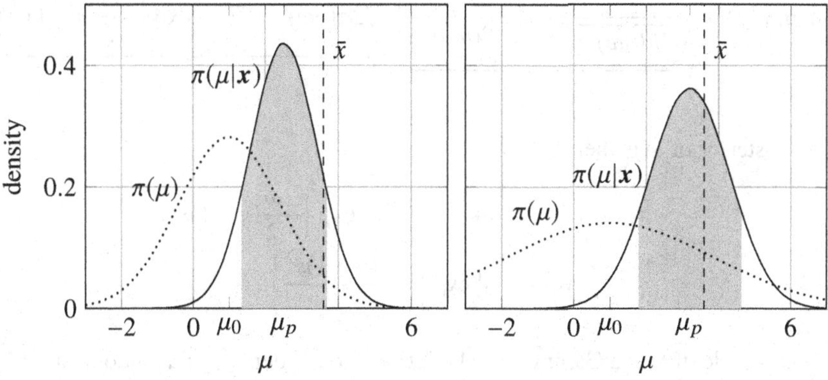


Figure 11.3 *Bayesian confidence intervals for the example from Figure 11.1. Notice that the 80% interval (shaded), which is centred at $\mu_p$, is wider when the prior is flatter, as there is more uncertainty about the parameter.*

**Example 11.3.5** (Bayesian confidence intervals for normal parameters)
Suppose that $\boldsymbol{Y}$ is a random sample from $\mathrm{N}(\mu, \sigma^2)$, but $\sigma^2$ is unknown. From Example 11.2.1, a suitable reference prior is $\pi(\mu, \sigma) \propto \sigma^{-1}$. Obtaining the posterior density of $\mu$ and $\sigma^2$ is somewhat more complicated, but we can simplify the calculations slightly by expressing them in terms of the **precision**, $\omega$, where $\omega = 1/\sigma^2$. Using the factorisation from Example 10.1.9, we can write

$$f(\boldsymbol{y}|\mu, \omega) = \left(\frac{\omega}{2\pi}\right)^{n/2} \exp\left\{-\frac{n\omega}{2}(\bar{y} - \mu)^2\right\} \exp\left\{-\frac{\omega}{2}(n-1)s^2\right\}.$$

The parameter $\omega$ is a monotonic function of $\sigma$, so a suitable reference prior is

$$\pi(\omega) = \pi(\sigma)\left(\frac{d\omega}{d\sigma}\right)^{-1} \propto \frac{1}{\sigma}\left(-\frac{2}{\sigma^3}\right)^{-1} \propto \sigma^2 = \frac{1}{\omega}.$$

The joint posterior density of $\mu$ and $\omega$ is then

$$\begin{aligned}\pi(\mu,\omega|\mathbf{y}) &\propto f(\mathbf{y}|\mu,\omega)\pi(\mu,\omega)\\ &= \left(\frac{\omega}{2\pi}\right)^{n/2}\exp\left\{-\frac{n\omega}{2}(\bar{y}-\mu)^2\right\}\exp\left\{-\frac{\omega}{2}(n-1)s^2\right\}\frac{1}{\omega}\\ &\propto \exp\left\{-\frac{(\mu-\bar{y})^2}{2(n\omega)^{-1}}\right\}\omega^{n/2-1}\exp\left\{-\frac{\omega}{2}(n-1)s^2\right\}. \qquad (11.7)\end{aligned}$$

This expression cannot be decomposed into separate (marginal) posterior densities for the parameters $\mu$ and $\omega$, as they are not independent. However, note that we can use the decomposition $\pi(\mu,\omega|\mathbf{y}) = \pi(\mu|\omega,\mathbf{y})\pi(\omega|\mathbf{y})$. The only term in the expression which involves $\mu$ is proportional to an $\mathrm{N}(\bar{y},(n\omega)^{-1})$ density function, which is unsurprising; conditioning on the precision takes us back to the known-variance case. Completing this density function yields

$$\pi(\mu,\omega|\mathbf{y}) \propto \underbrace{\frac{1}{\sqrt{2\pi(n\omega)^{-1}}}\exp\left\{-\frac{(\mu-\bar{y})^2}{2(n\omega)^{-1}}\right\}}_{\pi(\mu|\omega,\mathbf{y})}\underbrace{\sqrt{2\pi(n\omega)^{-1}}\,\omega^{n/2-1}\exp\left\{-\frac{\omega}{2}(n-1)s^2\right\}}_{\pi(\omega|\mathbf{y})}.$$

The posterior of $\omega$ is then

$$\begin{aligned}\pi(\omega|\mathbf{y}) &\propto \sqrt{2\pi(n\omega)^{-1}}\,\omega^{n/2-1}\exp\left\{-\frac{\omega}{2}(n-1)s^2\right\}\\ &\propto \omega^{\frac{n-1}{2}-1}\exp\left\{-\omega\frac{(n-1)s^2}{2}\right\},\end{aligned}$$

which we identify as a $\mathrm{Gamma}((n-1)/2,(n-1)s^2/2)$ density. If a random variable $U$ has distribution $\mathrm{Gamma}(k/2,\lambda/2)$, where $k$ is a positive integer, then $\lambda U \sim \chi^2_k$ (see Exercise 7.3). Thus, the posterior of $\omega$ can be rearranged to give $((n-1)s^2\omega|\mathbf{y}) \sim \chi^2_{n-1}$ or, equivalently,

$$\left(\left.\frac{(n-1)s^2}{\sigma^2}\right|\mathbf{y}\right) \sim \chi^2_{n-1}.$$

We can write

$$\mathrm{P}\left(\chi^2_{n-1,\alpha/2} < \frac{(n-1)s^2}{\sigma^2} < \chi^2_{n-1,1-\alpha/2}\,\middle|\,\mathbf{y}\right) = 1-\alpha,$$

which gives a $(1-\alpha)$ interval estimate for $\sigma^2$,

$$\left(\frac{(n-1)s^2}{\chi^2_{n-1,1-\alpha/2}},\frac{(n-1)s^2}{\chi^2_{n-1,\alpha/2}}\right).$$

The approach for constructing a confidence interval for $\mu$ is similar, except that we need to factorise the joint posterior density as $\pi(\mu, \omega|\mathbf{y}) = \pi(\omega|\mu, \mathbf{y})\pi(\mu|\mathbf{y})$. Rearranging (11.7), we have

$$\pi(\mu, \omega|\mathbf{y}) \propto \omega^{n/2-1} \exp\left\{-\omega \frac{n(\mu - \bar{y})^2 + (n-1)s^2}{2}\right\}.$$

The terms in $\omega$ belong to a Gamma$(n/2, (n(\mu - \bar{y})^2 + (n-1)s^2)/2)$ density function; the missing normalising constant is

$$\frac{1}{\Gamma(n/2)} \left(\frac{n(\mu - \bar{y})^2 + (n-1)s^2}{2}\right)^{n/2},$$

from which we deduce that

$$\pi(\mu|\mathbf{y}) \propto \left(\frac{n(\mu - \bar{y})^2 + (n-1)s^2}{2}\right)^{-n/2} \propto \left(1 + \frac{\left(\frac{\mu - \bar{y}}{s/\sqrt{n}}\right)^2}{n-1}\right)^{-n/2}.$$

Comparing this to the density of the $t$-distribution (Definition 8.1.3), we can see that

$$\left(\frac{\mu - \bar{y}}{s/\sqrt{n}}\middle|\mathbf{y}\right) \sim t_{n-1},$$

which results in the usual $(1-\alpha)$ interval estimate for $\mu$,

$$\bar{y} \pm t_{n-1,1-\alpha/2}\frac{s}{\sqrt{n}}.$$

**Exercise 11.3**

1. Consider the loss function

$$L(\theta, a) = \begin{cases} p|\theta - a| & \text{if } a < \theta, \\ (1-p)|\theta - a| & \text{if } a \geq \theta. \end{cases}$$

for some given $0 \leq p \leq 1$. What is the corresponding Bayes estimator? Verify your solution in the special case $p = 1/2$.

2. Let $\mathbf{X} = (X_1, \ldots, X_n)^T$ be a random sample from Unif$[\theta - 1, \theta + 1]$, where $\theta$ has prior Unif$[-1, 1]$. Show that the Bayesian estimator of $\theta$ under both squared error loss and absolute loss is

$$\frac{\min\{x_{(1)}, 0\} + \max\{x_{(n)}, 0\}}{2}.$$

3. Let $\mathbf{Y} = (Y_1, \ldots, Y_n)^T$ be a random sample from Pois$(\theta)$, where $\theta$ has prior Gamma$(\alpha, \lambda)$.
   (a) Find the Bayesian estimator of $\theta$ under
      i. squared error loss;
      ii. 0-1 loss.

(b) What would the estimator have been in each case if we had used the Jeffreys prior for $\theta$ instead?

4. Suppose that we have two normal random samples, $X_1, \cdots, X_n \sim \text{N}(\mu_X, \sigma^2)$ and $Y_1, \cdots, Y_n \sim \text{N}(\mu_Y, \sigma^2)$, where $\sigma^2$ is known. Using a sensible prior for $\mu_X$ and $\mu_Y$, construct a 90% confidence interval for the difference $\mu_X - \mu_Y$.

## 11.4 Hierarchical models and empirical Bayes

We will now focus on the hyperparameters in the prior $\pi(\boldsymbol{\theta})$, which we denote by $\boldsymbol{\eta}$. There are two main approaches to fitting a model where $\boldsymbol{\eta}$ is unknown. The first is to consider a fully Bayesian formulation where $\boldsymbol{\eta}$ is itself a random variable with prior density $h(\boldsymbol{\eta})$, which we call a **hyperprior**; this approach gives rise to **multistage** or **hierarchical models**. The second approach is to estimate $\boldsymbol{\eta}$ directly from the data and then treat this estimate as a fixed value. This is known as the **empirical Bayes** approach.

### *11.4.1 Hierarchical models*

In general, if the parameter $\boldsymbol{\theta}$ implicitly depends on a hyperparameter $\boldsymbol{\eta}$ with **hyperprior** $h(\boldsymbol{\eta})$, we can make the dependence explicit and express the joint posterior distribution of the parameters in the form

$$\pi(\boldsymbol{\theta}, \boldsymbol{\eta} | \boldsymbol{y}) \propto f(\boldsymbol{y} | \boldsymbol{\theta}, \boldsymbol{\eta}) \pi(\boldsymbol{\theta}, \boldsymbol{\eta}) = f(\boldsymbol{y} | \boldsymbol{\theta}) \pi(\boldsymbol{\theta} | \boldsymbol{\eta}) h(\boldsymbol{\eta}) . \tag{11.8}$$

We write $f(\boldsymbol{y}|\boldsymbol{\theta})$ instead of $f(\boldsymbol{y}|\boldsymbol{\theta},\boldsymbol{\eta})$ because $\boldsymbol{y}$ depends on $\boldsymbol{\eta}$ only through $\boldsymbol{\theta}$. This is a two-stage hierarchical Bayes model. It can easily be extended into a multistage model where the parameter $\boldsymbol{\eta}$ depends on another parameter $\boldsymbol{\phi}$, which may depend on another parameter $\boldsymbol{\omega}$, and so on. This approach allows us to construct more elaborate models but often leads to mathematically intractable integrals. These need to be evaluated numerically using Monte Carlo methods, described in the next chapter.

**Example 11.4.1** (Dependence on a common parameter)
Consider the case where we have a sample $Y_i \sim \text{Pois}(\theta_i)$ for $i = 1, \ldots, n$, and the parameters $\theta_1, \ldots, \theta_n$ are drawn from $\text{Exp}(\lambda)$. If $\lambda$ is a known constant, the posterior density is simply

$$\pi(\boldsymbol{\theta}|\boldsymbol{y}) \propto f(\boldsymbol{y}|\boldsymbol{\theta})\pi(\boldsymbol{\theta}) = \prod_{i=1}^{n} \left\{ \frac{e^{-\theta_i}\theta_i^{y_i}}{y_i!} \lambda e^{-\lambda\theta_i} \right\} \propto \prod_{i=1}^{n} \left\{ e^{-(\lambda+1)\theta_i} \theta_i^{y_i} \right\} .$$

From this we can deduce that $\theta_i|\boldsymbol{y} \sim \text{Gamma}(y_i + 1, \lambda + 1)$. Given the sample $\boldsymbol{y}$, each parameter $\theta_i$ is independent of the others, and will only depend on the corresponding observation $y_i$. For example, its posterior mean is $\hat{\theta}_i = (y_i + 1)/(\lambda + 1)$.

Now consider the case where the hyperparameter $\lambda$ is also a random variable; this is a hierarchical Bayes model. The hyperparameter in this case is the rate parameter of an exponential distribution, and an appropriate non-informative hyperprior is $h(\lambda) \propto 1/\lambda$, as previously established. This yields

$$\begin{aligned}\pi(\boldsymbol{\theta}, \lambda | \boldsymbol{y}) &\propto f(\boldsymbol{y}|\boldsymbol{\theta})\pi(\boldsymbol{\theta}|\lambda)h(\lambda) \\ &= \left\{\prod_{i=1}^{n}\left(\frac{e^{-\theta_i}\theta_i^{y_i}}{y_i!}\lambda e^{-\lambda\theta_i}\right)\right\}\frac{1}{\lambda} \\ &\propto \lambda^{n-1}e^{-\lambda\sum_i \theta_i}e^{-\sum_i \theta_i}\prod_{i=1}^{n}\theta_i^{y_i}.\end{aligned}$$

Noting that $\pi(\boldsymbol{\theta}, \lambda|\boldsymbol{y}) = h(\lambda|\boldsymbol{\theta}, \boldsymbol{y})\pi(\boldsymbol{\theta}|\boldsymbol{y})$, consider the posterior distribution of $\lambda$ first (conditional on $\boldsymbol{\theta}$). Ignoring terms that do not contain $\lambda$, we can identify this as a Gamma$(n, \sum_i \theta_i)$ distribution. Filling in the missing terms from this density yields

$$\pi(\boldsymbol{\theta}, \lambda|\boldsymbol{y}) \propto \frac{(\sum_i \theta_i)^n}{\Gamma(n)}\lambda^{n-1}e^{-\lambda\sum_i^n \theta_i}\left\{\frac{(\sum_i \theta_i)^n}{\Gamma(n)}\right\}^{-1}e^{-\sum_i \theta_i}\prod_{i=1}^{n}\theta_i^{y_i}.$$

Looking at the remaining terms, we can deduce that the posterior density of $\boldsymbol{\theta}$ is

$$\pi(\boldsymbol{\theta}|\boldsymbol{y}) \propto \left(\sum_{i=1}^{n}\theta_i\right)^{-n}e^{-\sum_i \theta_i}\prod_{i=1}^{n}\theta_i^{y_i},$$

which does not factorise into separate marginal densities for each of the $\theta_i$, because of the presence of the term $(\sum_i \theta_i)^{-n}$. Thus, the parameters $\theta_1, \ldots, \theta_n$ are no longer conditionally independent, and we need to draw inference on all of them simultaneously. In other words, we **pool** information from all the observations when estimating each $\theta_i$. This often leads to more sensible estimates.

This approach is particularly useful when there is natural structure in the data. Suppose that we observe $Y_{i,1}, \ldots, Y_{i,m_i} \sim \text{Pois}(\theta_i)$ for $i = 1, \ldots, n$, where each $i$ represents a different geographic location. In a frequentist setting, the MLE for $\theta_i$ is the group mean $\bar{Y}_i = \sum_j Y_{i,j}/m_i$, i.e. we only use information from the $i^{\text{th}}$ group (no pooling). A disadvantage with this approach is that the variance of the MLE is high when the corresponding sample size, $m_i$, is low.

At the other extreme, if we assume that $\theta_1 = \theta_2 = \ldots = \theta_n$, the MLE is the global mean $\bar{Y} = \sum_{i,j} Y_{i,j}/\sum_i m_i$, ignoring the groups (complete pooling). The issue here is that the groups may, in fact, have very different parameter values. If we introduce the prior $\theta_1, \ldots, \theta_n \sim \text{Exp}(\lambda)$ and draw inference on all the parameters simultaneously, we can benefit from *partial* pooling. The result is that the parameter estimates for the smaller groups are shrunk heavily towards the global mean.

### *11.4.2 Empirical Bayes*

In a Bayesian model, the data $\boldsymbol{Y}$ depend on a parameter $\boldsymbol{\theta}$, which may in turn depend on other parameters. The hierarchical Bayes approach allows us to construct such multistage models, but the final hyperparameter $\eta$ must follow a distribution that is fully known, such as the non-informative prior in Example 11.4.1. With the empirical Bayes method, we make no such assumption, but instead attempt to estimate $\eta$ from the data.

We consider a two-stage model, as before, but the same method can be extended to models with more stages. The technique we focus on is known as **parametric** empirical Bayes, because we assume that only the parameters of the density $\pi(\boldsymbol{\theta}|\boldsymbol{\eta})$, the vector of hyperparameters $\boldsymbol{\eta}$, are unknown. Treating the form of the function $\pi(\boldsymbol{\theta}|\boldsymbol{\eta})$ as unknown leads to **nonparametric** empirical Bayes methods, which are beyond the scope of this book.

The conditional density of $\boldsymbol{Y}$ given $\boldsymbol{\eta}$ can be written in the form

$$\begin{aligned} f(\boldsymbol{y}|\boldsymbol{\eta}) = \frac{f(\boldsymbol{y},\boldsymbol{\eta})}{h(\boldsymbol{\eta})} &= \int \frac{f(\boldsymbol{y},\boldsymbol{\theta},\boldsymbol{\eta})}{h(\boldsymbol{\eta})} d\boldsymbol{\theta} \\ &= \int \frac{f(\boldsymbol{y},\boldsymbol{\theta},\boldsymbol{\eta})}{f(\boldsymbol{\theta},\boldsymbol{\eta})} \frac{f(\boldsymbol{\theta},\boldsymbol{\eta})}{h(\boldsymbol{\eta})} d\boldsymbol{\theta} \\ &= \int f(\boldsymbol{y}|\boldsymbol{\theta},\boldsymbol{\eta}) h(\boldsymbol{\theta}|\boldsymbol{\eta}) d\boldsymbol{\theta} . \end{aligned} \tag{11.9}$$

If the integral is tractable, we can treat this density as a likelihood function for $\boldsymbol{\eta}$ and maximise it to obtain an estimator $\hat{\boldsymbol{\eta}}(\boldsymbol{y})$. This approach is known as **type-II maximum likelihood**. If the integral is intractable, or if direct maximisation of $f(\boldsymbol{y}|\boldsymbol{\eta})$ is not possible (such as in the Beta-Binomial model), a common approach is to calculate the moments of $\boldsymbol{Y}|\boldsymbol{\eta}$ by iterating expectations, and then use these to construct a method-of-moments estimator for $\boldsymbol{\eta}$. For example, the conditional mean is

$$\mathbb{E}(\boldsymbol{Y}|\boldsymbol{\eta}) = \mathbb{E}\left[\mathbb{E}(\boldsymbol{Y}|\boldsymbol{\eta},\boldsymbol{\theta})\right] = \mathbb{E}\left[\mathbb{E}(\boldsymbol{Y}|\boldsymbol{\theta})|\boldsymbol{\eta}\right] .$$

**Example 11.4.2** (Dependence on a common parameter – continued)
We now apply the empirical Bayes method to Example 11.4.1. The density of $\boldsymbol{Y}$ conditional on the hyperparameter $\lambda$ is

$$\begin{aligned} f(\boldsymbol{y}|\lambda) &= \int_{\mathbb{R}^n} f(\boldsymbol{y}|\boldsymbol{\theta})\pi(\boldsymbol{\theta}|\lambda) d\boldsymbol{\theta} \\ &= \int_0^\infty \cdots \int_0^\infty \prod_{i=1}^n \left( \frac{e^{-\theta_i}\theta_i^{y_i}}{y_i!} \lambda e^{-\lambda\theta_i} \right) d\theta_1 \dots d\theta_n \\ &= \prod_{i=1}^n \left( \int_0^\infty \frac{\lambda e^{-(\lambda+1)\theta_i}\theta_i^{y_i}}{y_i!} d\theta_i \right) \qquad \text{separating integrals} \\ &= \prod_{i=1}^n \left( \frac{\lambda}{(\lambda+1)^{y_i+1}} \int_0^\infty \frac{(\lambda+1)^{y_i+1}}{\Gamma(y_i+1)} e^{-(\lambda+1)\theta_i}\theta_i^{y_i} d\theta_i \right) \\ &= \frac{\lambda^n}{(\lambda+1)^{\sum_i y_i + n}} , \end{aligned}$$

because the final integrand is a Gamma$(y_i+1, \lambda+1)$ density. We now need to maximise the marginal density $f(\boldsymbol{y}|\lambda)$. We have

$$\log f(\boldsymbol{y}|\lambda) = n\log\lambda - \left(\sum_i y_i + n\right)\log(\lambda+1) \Rightarrow$$

$$\frac{\partial}{\partial\lambda}\log f(\boldsymbol{y}|\lambda) = \frac{n}{\lambda} - \frac{\sum_i y_i + n}{\lambda+1},$$

which we set equal to zero to obtain the marginal MLE,

$$\hat{\lambda} = \frac{n}{\sum_i Y_i} = \bar{Y}^{-1}.$$

Substituting this estimate back into the posterior density yields the estimated posterior,

$$\pi(\boldsymbol{\theta}|\boldsymbol{y},\hat{\lambda}) \propto \prod_{i=1}^{n}\left\{e^{-(\bar{y}^{-1}+1)\theta_i}\theta_i^{y_i}\right\},$$

which corresponds to a Gamma$(y_i+1, \bar{y}^{-1}+1)$ distribution for $\theta_i$.

### *11.4.3 Predictive inference*

Suppose that we have observed $X_1,\dots,X_n$ and would like to make a prediction for some $Y$ based on these observations. In Bayesian predictive inference, we generate the density for the random variable $Y|\boldsymbol{X}=\boldsymbol{x}$ using the posterior for the parameters. The distribution of $Y$ conditional on the observed data is known as the **posterior predictive distribution**. This is to distinguish it from the **prior predictive distribution**, which is just the marginal distribution of $\boldsymbol{X}$.

Using the same manipulations as in (11.9), we can write the posterior predictive density as

$$f(y|\boldsymbol{x}) = \int_{-\infty}^{\infty} f(y|\boldsymbol{x},\theta)\pi(\theta|\boldsymbol{x})d\theta.$$

The motivation behind this method is that the density $f(y|\boldsymbol{x},\theta)$ tends to be much simpler than $f(y|\boldsymbol{x})$. In fact, it is often the case that $f(y|\boldsymbol{x},\theta) = f(y|\theta)$, as in the following example.

**Example 11.4.3** (Normal sample)
Consider the setup of Example 11.2.3, where $X_1, X_2, \dots$ are IID $\mathrm{N}(\mu,\sigma^2)$, and $\mu \sim \mathrm{N}(\mu_0,\sigma_0^2)$. Assume that we are interested in the predictive distribution of the observation $Y = X_{n+1}$ conditional on the sample $\boldsymbol{X} = (X_1,\dots,X_n)^T$. We have already established that the posterior density of $\mu$ given $\boldsymbol{X}$ is $\mathrm{N}(\mu_p,\sigma_p^2)$, where

$$\mu_p = \frac{n\bar{x}/\sigma^2 + \mu_0/\sigma_0^2}{n/\sigma^2 + 1/\sigma_0^2} \quad \text{and} \quad \sigma_p^2 = \frac{1}{n/\sigma^2 + 1/\sigma_0^2}.$$

The predictive density is given by

$$f(y|\boldsymbol{x}) = \int_{-\infty}^{\infty} f(y|\boldsymbol{x},\mu)\pi(\mu|\boldsymbol{x})d\mu$$
$$= \int_{-\infty}^{\infty} \frac{1}{\sqrt{2\pi\sigma^2}} \exp\left\{\frac{(y-\mu)^2}{2\sigma^2}\right\} \frac{1}{\sqrt{2\pi\sigma_p^2}} \exp\left\{\frac{(\mu-\mu_p)^2}{2\sigma_p^2}\right\} d\mu.$$

This integral is straightforward (though tedious) to evaluate, by completing the square in $\mu$. It yields

$$f(y|\boldsymbol{x}) = \frac{1}{\sqrt{2\pi(\sigma^2+\sigma_p^2)}} \exp\left\{\frac{(y-\mu_p)^2}{2(\sigma^2+\sigma_p^2)}\right\} \int_{-\infty}^{\infty} \frac{1}{\sqrt{2\pi\sigma_q^2}} \exp\left\{\frac{(\mu-\mu_q)^2}{2\sigma_q^2}\right\} d\mu,$$

where

$$\sigma_q^2 = \left(\frac{1}{\sigma^2}+\frac{1}{\sigma_p^2}\right)^{-1} \quad \text{and} \quad \mu_q = \frac{y/\sigma^2+\mu_p/\sigma_p^2}{1/\sigma^2+1/\sigma_p^2}.$$

The integral in the above expression evaluates to 1, and we can identify the remaining part as a normal density. The predictive distribution of $Y$ given $\boldsymbol{X}$ is, thus,

$$\mathrm{N}(\mu_p, \sigma^2+\sigma_p^2).$$

Notice that, while the mean of the predictive distribution is just the posterior mean, $\mu_p$, the variance consists of two parts, as there are two sources of uncertainty: one is from the variance of the new observation ($\sigma^2$), and the other is from estimating the mean ($\sigma_p^2$).

**Example 11.4.4** (AR(1) process)
Consider the autoregressive time series model from section 6.6, where

$$y_t = \phi y_{t-1} + \varepsilon_t, \quad \{\varepsilon_t\} \sim \mathrm{N}(0,\sigma^2).$$

We assume that $-1 < \phi < 1$, and that there is a fixed starting value $y_0$. We want to predict $y_{n+1}$ given observations $y_1,\dots,y_n$. More formally, we wish to draw from the predictive distribution $f(y_{n+1}|\boldsymbol{y}_n)$, which depends on the parameters $\phi$ and $\sigma$. We take advantage of the fact that (if the parameters are known) the next observation $y_{n+1}$ depends on the past only through the present, $y_n$, so we can write

$$f(y_{n+1}|\boldsymbol{y}_n) = \int_{-\infty}^{\infty}\int_{-\infty}^{\infty} f(y_{n+1}|y_1,\dots,y_n,\phi,\sigma)\pi(\phi,\sigma|\boldsymbol{y}_n)d\phi d\sigma$$
$$= \int_{-\infty}^{\infty}\int_{-\infty}^{\infty} f(y_{n+1}|y_n,\phi,\sigma)\pi(\phi,\sigma|\boldsymbol{y}_n)d\phi d\sigma.$$

We make the substitution $\omega = 1/\sigma^2$, so that $(y_{n+1}|y_n,\phi,\omega) \sim \mathrm{N}(\phi y_n, 1/\omega)$. The density is

$$f(y_{n+1}|y_n,\phi,\omega) = \sqrt{\frac{\omega}{2\pi}} \exp\left\{-\frac{\omega}{2}(y_{n+1}-\phi y_n)^2\right\}.$$

The posterior density of the parameters is

$$\pi(\phi, \omega | \boldsymbol{y}_n) \propto f(\boldsymbol{y}_n | \phi, \omega) \pi(\phi, \omega).$$

We assume that the parameters $\phi$ and $\omega$ are independent and write $\pi(\phi, \omega) = \pi(\phi)\pi(\omega)$. The autoregressive parameter $\phi$ has finite support, so we can simply choose a uniform prior in the interval $(-1, 1)$.

As we found in Example 11.3.5, a suitable reference prior for $\omega$ is

$$\pi(\omega) \propto \frac{1}{\omega},$$

which gives

$$\pi(\phi, \omega) \propto \mathbf{1}_{(-1,1)}(\phi)\omega^{-1}.$$

The density of the sample conditional on the parameters can be worked out through successive conditioning (Proposition 5.7.1), so

$$\begin{aligned} f(y_n, \ldots, y_1 | \phi, \omega) &= \prod_{t=1}^{n} f(y_t | \phi, \omega, y_{t-1}) \\ &= \prod_{t=1}^{n} \left(\frac{\omega}{2\pi}\right)^{1/2} \exp\left\{-\frac{\omega}{2}(y_t - \phi y_{t-1})^2\right\} \\ &\propto \omega^{n/2} \exp\left\{-\frac{\omega}{2}\sum_{t=1}^{n}(y_t - \phi y_{t-1})^2\right\}. \end{aligned}$$

The posterior density of the parameters is, then,

$$\begin{aligned} \pi(\phi, \omega | \boldsymbol{Y}_n) &\propto \mathbf{1}_{(-1,1)}(\phi)\omega^{n/2-1} \exp\left\{-\omega \sum_{t=1}^{n} \frac{1}{2}(y_t - \phi y_{t-1})^2\right\} \\ &= \mathbf{1}_{(-1,1)}(\phi)\omega^{n/2-1} \exp\left(-\omega\, S_{YY}(\phi)/2\right), \end{aligned}$$

where

$$S_{YY}(\phi) = \sum_{t=1}^{n}(y_t - \phi y_{t-1})^2.$$

Treating everything else as fixed, the terms involving $\omega$ in the above expression form a Gamma$(n/2, S_{YY}(\phi)/2)$ density. Filling in the missing terms yields

$$\pi(\omega | \phi, \boldsymbol{Y}_n) = \frac{1}{\Gamma(n/2)}\left(\frac{S_{YY}(\phi)}{2}\right)^{n/2} \omega^{n/2-1} \exp\left\{-\omega \frac{S_{YY}(\phi)}{2}\right\},$$

which leaves us with

$$\pi(\phi | \boldsymbol{Y}_n) \propto \mathbf{1}_{(-1,1)}(\phi)\left(\frac{S_{YY}(\phi)}{2}\right)^{-n/2}$$

as the posterior density for $\phi$.

We will return to this example in the next chapter, in order to describe how to draw from the predictive distribution using a composition algorithm.

**Exercise 11.4**

1. Let $Y_1, Y_2, \ldots$ be IID Geometric($\theta$), where $\theta$ has prior Beta($\alpha, \beta$). We observe a random sample $\boldsymbol{Y} = (Y_1, \ldots, Y_n)^T$ and want to predict a new observation, $Y_{n+1}$.
   (a) Find the posterior distribution of $\theta$. What do you observe?
   (b) Find the posterior predictive density of $Y_{n+1}$ conditional on $\boldsymbol{Y}$.
   (c) Repeat the exercise, but with a Jeffreys prior for $\theta$.
2. Consider again the case where there are two normal populations and we take a random sample from each one, $X_1, \cdots, X_n \sim \mathrm{N}(\mu_X, \sigma^2)$ and $Y_1, \cdots, Y_n \sim \mathrm{N}(\mu_Y, \sigma^2)$, with known $\sigma^2$. The quantity of interest is the probability $\mathrm{P}(X_{n+1} > Y_{n+1})$, where $X_{n+1}$ and $Y_{n+1}$ denote a future pair of observations from the corresponding populations.
   (a) Give a frequentist estimate of this probability (i.e. using the MLEs of $\mu_X$ and $\mu_Y$).
   (b) Compute the same probability in a Bayesian setting, with the prior $\mu_X, \mu_Y \sim \mathrm{N}(\mu, \tau^2)$.
3. Suppose that $Y_1, Y_2, \ldots$ are IID Pois($\theta$), where $\theta$ has prior Gamma($\alpha, \lambda$). Assume that $\alpha$ and $\lambda$ are known.
   (a) Given a random sample $\boldsymbol{Y} = (Y_1, \ldots, Y_n)^T$, find the prior predictive density, $f(\boldsymbol{y})$.
   (b) By considering the case $n = 1$, show that the marginal distribution of each $Y_i$ is Polya($\alpha, \lambda/(\lambda + 1)$). Are the $Y_i$ independent?
   (c) Find the posterior predictive density of $Y_{n+1}$ conditional on $\boldsymbol{Y}$.

## 11.5 Further exercises

1. Let $Y_1, Y_2, \ldots, Y_n$ be a random sample from the distribution with density function

$$f_Y(y|\theta) = \begin{cases} e^{-(y-\theta)} & \text{if } y > \theta\,, \\ 0 & \text{otherwise,} \end{cases}$$

   where $\theta$ has prior Exp(1). What is the posterior distribution of $\theta$? What happens in the case where $n = 1$? Find the Bayesian estimator of $\theta$ under squared error loss, absolute loss, and 0-1 loss.
2. The random variable $Y$ follows a **scaled inverse chi-squared distribution** on $n$ degrees of freedom, denoted

$$Y \sim \chi^{-2}_{(n,\sigma_0^2)}, \tag{11.10}$$

   if $(\sigma_0^2 Y)^{-1} \sim \chi^2_n$. Work out the density function of this distribution. Now suppose that we have a random sample from $\mathrm{N}(\mu, \sigma^2)$, where $\mu$ is known and $\sigma^2$ is unknown. Show that the scaled inverse chi-squared is a conjugate prior for $\sigma^2$.

3. Suppose that $X_1, X_2, \dots$ are IID Exp($\theta$), where $\theta$ has prior Gamma($\alpha, \lambda$). We have observed a random sample $\boldsymbol{X} = (X_1, \dots, X_n)^T$ and are interested in predicting $Y = X_{n+1}$. Show that the posterior predictive density of $Y$ is of the form

$$f(y|\boldsymbol{x}) = \begin{cases} \frac{\beta}{\tau}\left(1 + \frac{y}{\tau}\right)^{-(\beta+1)} & \text{if } y > 0\,, \\ 0 & \text{otherwise,} \end{cases}$$

where $\beta$ and $\tau$ are to be determined. [This is the **Lomax distribution**, a close relative of the Pareto.]

4. Let $Y \sim \text{Bin}(n, p)$ represent the number of heads obtained in $n$ tosses of a possibly biased coin. For convenience, we use the variable $\theta$ to indicate whether the coin is biased. If $\theta = 0$ the coin is fair and we have $p = 0.5$, a degenerate distribution with all its mass on 0.5; if $\theta = 1$ the coin is biased and we have $p \sim \text{Unif}[0, 1]$. Our prior belief is $\text{P}(\theta = 0) = \text{P}(\theta = 1) = 1/2$. Now suppose that we perform the experiment and the coin comes up heads every time (so $Y = n$). Write down an expression for the posterior probability that the coin is fair. Verify that your result is correct for the cases $n = 0$ and $n \to \infty$. What is this probability equal to when $n = 1$? Can you explain why?
[Note: The prior distribution for $p$ is a mixture of an infinitely thin *spike* (a degenerate distribution) and a flat *slab* (a continuous uniform). This is known as a **spike-and-slab** prior and is useful in variable-selection problems.]

5. Let $\boldsymbol{Y} = (Y_1, \dots, Y_n)^T$ be a random sample from an exponential family with density

$$f(y|\boldsymbol{\theta}) = a(y)b(\boldsymbol{\theta})\exp\left[\sum_{i=1}^{k} c_i(\boldsymbol{\theta})h_i(y)\right].$$

Show that a density of the form

$$\pi(\boldsymbol{\theta}) \propto b(\boldsymbol{\theta})^m \exp\left[\sum_{i=1}^{k} c_i(\boldsymbol{\theta})d_i\right],$$

where $m, d_1, \dots, d_k$ are constants, is a conjugate prior for $\boldsymbol{\theta}$. Which of the conjugate priors encountered in this chapter (normal-normal, Gamma-Poisson, Beta-binomial, etc.) are special cases of this result?

6. Let $\boldsymbol{X} = (X_1, \dots, X_n)^T$ be a random sample from Lognormal($\theta, \sigma^2$), where $\sigma^2$ is known. The density function is

$$f(x|\boldsymbol{\theta}) = \begin{cases} \frac{1}{x\sqrt{2\pi\sigma^2}} \exp\left\{-\frac{(\log x - \theta)^2}{2\sigma^2}\right\} & \text{if } x > 0\,, \\ 0 & \text{otherwise.} \end{cases}$$

Find a conjugate prior for $\theta$. What are the parameters of the posterior distribution?

7. Let $\boldsymbol{Y} = (Y_1, \dots, Y_n)^T$ be a random sample from Unif$[0, \theta]$. The prior distribution for $\theta$ has density

$$\pi(\theta) = \frac{\gamma\alpha^\gamma}{\theta^{\gamma+1}}\mathbf{1}_{[\alpha,\infty)}(\theta)\,,$$

where $\gamma, \alpha > 0$ (this is known as the Type I Pareto).

(a) What is the posterior distribution of $\theta$?

(b) Find the Bayesian estimator of $\theta$ under squared error loss, absolute loss, and 0-1 loss.

(c) Construct a 95% Bayesian confidence interval for $\theta$.

8. For the Pareto-uniform example in question 7, estimate the hyperparameters $\gamma$ and $\alpha$ from the data with the empirical Bayes approach. Show that this is equivalent to using the improper prior $\pi(\theta) \propto 1/\theta$.

9. Let $\boldsymbol{Y} = (Y_1, \ldots, Y_n)^T$ be a random sample from Pois($\theta$), where $\theta$ has prior Gamma($\alpha, \lambda$).

   (a) Find the conditional density $f(\boldsymbol{y}|\boldsymbol{\eta})$, where $\boldsymbol{\eta} = (\alpha, \lambda)$.

   (b) Treating this density as a likelihood function, show that we cannot obtain a closed form solution for the type-II maximum-likelihood estimator of $\boldsymbol{\eta}$.

   (c) Construct a method-of-moments estimator for $\boldsymbol{\eta}$. [*Hint*: Use the law of iterated expectations and the corresponding result for the variance from Proposition 5.4.10]

## 11.6 Chapter summary

On completion of this chapter you should be able to:

- explain the fundamental differences between classical and Bayesian inference,
- define the sampling, prior, and posterior densities, and state the connection between them,
- choose an appropriate prior distribution,
- obtain, by identification or direct integration, the posterior distribution in simple cases,
- understand why different loss functions lead to different point estimators,
- construct Bayesian interval estimators for parameters,
- formulate a hierarchical (multistage) model,
- set up an empirical Bayes model and describe how to estimate the hyperparameters, and
- write down an expression for the predictive distribution of future observations from a model.

Chapter 12

# Simulation methods

In statistical inference, we are often interested in determining the properties of a statistic under a particular model. This may be in order to get some handle on the quality of a point estimator, or to construct a critical region for a test. Our approach of first choice is to use direct mathematical reasoning, however, in many instances, particularly for models of practical interest, the mathematics involved are intractable. We may be able to appeal to an asymptotic result but it is often unclear how good an approximation this will provide in finite samples. For example, the central limit theorem tells us that the limiting distribution of the sample mean is normal but gives no indication of how good an approximation the normal is for, say, a sample of size 50.

One solution to these problems is to use simulation. Probabilistic simulations are often referred to as Monte Carlo techniques, after the location made famous by its casinos. As we explain in section 12.2, Monte Carlo methods provide a simple and flexible mechanism for evaluating intractable integrals, based on techniques for generating random numbers. The application of Monte Carlo methods, exploiting the availability of fast, inexpensive computers, has been fundamental in boosting the practical applicability of the Bayesian approach.

This chapter is an overview of some of the main modern simulation methods used in statistics. The reader is strongly encouraged to attempt to implement some of these algorithms in R, Python, or another programming language.

## 12.1 Simulating independent values from a distribution

Recall that mathematical models fall into two categories: deterministic and stochastic. For a deterministic model, if we know the input, we can determine exactly what the output will be. This is not true for a stochastic model, as its output is a random variable. Remember that we are talking about the properties of models; discussions of the deterministic or stochastic nature of real-life phenomena often involve very muddled thinking and should be left to the philosophers.

Simple deterministic models can produce remarkably complex and beautiful behaviour. An example is the logistic map,

$$x_{n+1} = rx_n(1 - x_n), \tag{12.1}$$

for $n = 1, 2, \ldots$ and starting value $x_0 \in (0, 1)$. For some values of the parameter $r$, this map will produce a sequence of numbers $x_1, x_2, \ldots$ that look random. Put more precisely, we would not reject a null hypothesis of independence if we were to test these values. We call sequences of this sort **pseudo-random**.

Computer programs cannot produce random numbers; for a given input we can, in principle, predict exactly what the output of a program will be. However, we can use a computer to generate a sequence of pseudo-random numbers. The distinction is somewhat artificial since no finite sequence of numbers is random; randomness is a property that can only be attributed to mathematical models (random variables). From this point on we will drop the "pseudo", with the understanding that, whenever we use the term random numbers, what we mean is a sequence that is indistinguishable from instances of a random variable.

In this section, we describe algorithms for generating random samples from some of the distributions we have encountered thus far. All of these methods rely on us being able to obtain a random sample from the continuous uniform distribution on the interval $[0, 1]$, which is the usual output of a random-number generator.

### *12.1.1 Table lookup*

A simple way to generate instances of discrete random variables is with the following method. Let $X$ be a discrete random variable with distribution function $F_X(x)$. Start by generating a uniform random variable $U \sim \text{Unif}[0, 1]$ and let $u$ be the generated value.

Then, set $X = x_i$, where $x_i$ is such that $F_X(x_{i-1}) < u \leq F_X(x_i)$. This is known as the **table lookup** method, because we are effectively looking up the value of $U$ in the cumulative probability table of $X$.

The generated variable has the required distribution because

$$\text{P}(X = x_i) = \text{P}(F_X(x_{i-1}) < U \leq F_X(x_i)) = F_X(x_i) - F_X(x_{i-1}).$$

**Example 12.1.1** (Generating Poisson variables)
Assume we want to generate values from the Poisson distribution with $\lambda = 1.25$. This has the following probability table (to 4 d.p.):

We draw $U_1, U_2, \ldots \sim \text{Unif}[0, 1]$ and compare each $u_i$ to $F_X(x)$, starting at $F_X(0)$, until we find the smallest value for which $u_i < F_X(x_i)$. For example, if the first five values drawn are $(0.019, 0.968, 0.368, 0.160, 0.645)$, then the corresponding Poisson sample is $(0, 3, 1, 0, 2)$.

| $x$ | 0 | 1 | 2 | 3 | 4 | ... |
|---|---|---|---|---|---|---|
| $f_X(x)$ | 0.2865 | 0.3581 | 0.2238 | 0.0933 | 0.0291 | ... |
| $F_X(x)$ | 0.2865 | 0.6446 | 0.8685 | 0.9617 | 0.9909 | ... |

Table 12.1 *Probability table for the* Pois(1.25) *distribution*

In Example 12.1.1 the required search time will depend on the value of $\lambda$, since $\mathbb{E}(X) = \lambda$. The higher the mean, the larger the value of $X$ on average, so we will need to search further along the table for the correct value. This method also involves some setup time, because the table needs to be constructed in advance. In practice, the table lookup method is most commonly used for discrete distributions with finite support.

### *12.1.2 Probability integral*

Recall the probability integral transformation as described in Chapter 2. $X$ is a continuous random variable with CDF $F_X(x)$, and we will also assume (for the moment) that $F_X$ is a strictly increasing function over the support of $X$. We have shown that, if $U \sim \text{Unif}[0,1]$, then $F_X^{-1}(U) \sim F_X$. Thus, if $u$ is a draw from Unif[0, 1], we can treat $x = F_X^{-1}(u)$ as a draw from $F_X$ (Figure 12.1).

This result also holds if we relax the extra condition on $F_X$ and allow it to be a non-decreasing (as opposed to increasing) function. However, since there may be intervals where $F_X$ is constant, care is needed to ensure that $F_X^{-1}$ is a single-valued function. We define $F_X^{-1}(u) = \min\{x : F_X(x) = u\}$ for $0 \leq u \leq 1$. For every interval $[x_1, x_2]$ where $F_X$ is constant (say $F_X(x_1) = F_X(x_2) = u$), we set $F_X^{-1}(u) = x_1$, the left endpoint of the interval. Proof of the more general result is left as an exercise.

If the function $F_X(x)$ has discontinuities, we can still use the probability integral method by setting $F_X^{-1}(u) = min\{x : F_X(x) \geq u\}$. Note that this generalised version applied to a discrete random variable is identical to the table lookup method (Figure 12.2).

**Example 12.1.2** (Generating exponential variables (and related cases))
Let $X \sim \text{Exp}(\lambda)$, so that $F_X(x) = 1 - e^{-\lambda x}$ and $F_X^{-1}(u) = -\frac{1}{\lambda}\log(1-u)$. Equivalently, we can set $F_X^{-1}(u) = -\frac{1}{\lambda}\log(u)$, because $1 - U \sim \text{Unif}[0,1]$ as well. We will later discuss how this property can be exploited to reduce the variance of estimates based on this distribution.

If $X_1, X_2, \ldots$ are independent Exp($\lambda$) random variables then $\sum_{i=1}^{r} X_i \sim \text{Gamma}(r, \lambda)$, so we can use this method to draw samples from the Gamma distribution, as long as the shape parameter $r$ is an integer. Furthermore, if $\lambda = 0.5$ then $\sum_{i=1}^{r} X_i \sim \Gamma(r, 0.5)$, which corresponds to the chi-squared distribution with $2r$ degrees of freedom.

The probability integral approach fails when attempting to sample from distributions where the CDF does not have a closed form. Unfortunately, many important distributions, such as the normal or the chi-squared with odd degrees of freedom, fall in this category. However, for the particular case of the standard normal there is a remarkably elegant method which enables us to generate a sample directly.

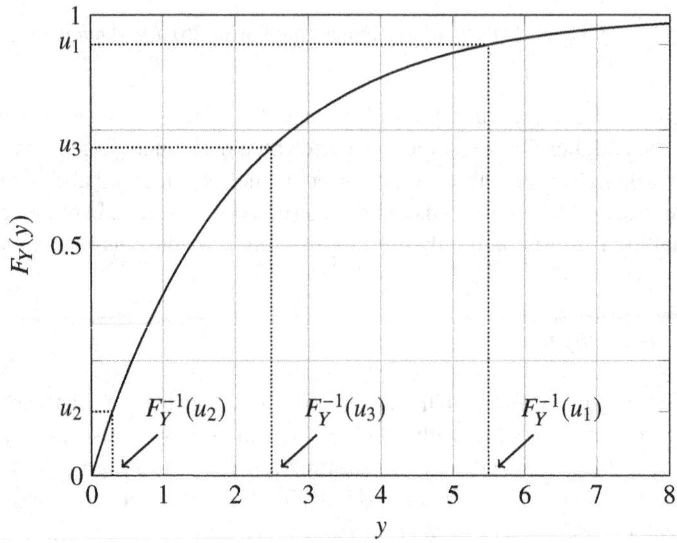


Figure 12.1 *Generating values from the* Exp(0.5) *distribution with the probability integral transformation. The values* $u_1, u_2, u_3$ *are drawn from* Unif[0.1]. *The corresponding* $F_Y^{-1}(u_1)$, $F_Y^{-1}(u_2)$, $F_Y^{-1}(u_3)$ *are draws from* Exp(0.5).

### 12.1.3 *Box-Muller method*

Let $U, V$ be independent Unif[0, 1] random variables and define $R = \sqrt{-2 \log U}$ and $\Theta = 2\pi V$. Now convert the polar coordinates $(R, \Theta)$ into cartesian coordinates $(X, Y)$ using the usual formulae, that is, $X = R\cos\Theta$ and $Y = R\sin\Theta$. The random variables $X, Y$ are independent standard normal. Proof of this proposition is left as an exercise.

The Box-Muller method can be also used to draw from the general normal distribution, since $\mu + \sigma X \sim \text{N}(\mu, \sigma^2)$.

### 12.1.4 *Accept/reject method*

Consider a situation where we have a distribution that is hard to simulate from directly. The **accept/reject** method works by simulating from another, simpler distribution (known as the **proposal** distribution), and keeping only a subset of the values initially simulated. The values that are kept are accepted and the others are rejected, hence accept/reject. By careful choice of the mechanism for acceptance, we can ensure that the resulting sequence represents simulated instances from the distribution of interest.

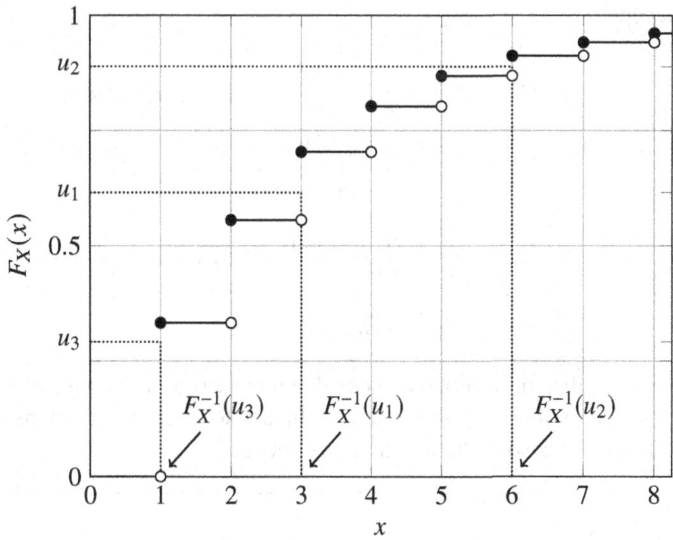


Figure 12.2 *Generating values from the* Geometric(1/3) *distribution with the probability integral transformation. The values* $u_1, u_2, u_3$ *are drawn from* Unif[0.1]. *The corresponding* $F_X^{-1}(u_1)$, $F_X^{-1}(u_2)$, $F_X^{-1}(u_3)$ *are draws from* Geometric(1/3).

Provided that the distribution of interest has bounded support, we can use sampling from the uniform distribution as the basis for a simple accept/reject algorithm. Consider a random variable $Y$ with density $f_Y$ and support $[a, b]$. Suppose in addition that $m = \max_y f_Y(y)$. A simple accept/reject scheme to generate the $i^{\text{th}}$ member of our simulated sample is:

*Repeat*: simulate independent $u_1 \leftarrow \text{Unif}[a, b]$ and $u_2 \leftarrow \text{Unif}[0, m]$
*Until*: $f_Y(u_1) \geq u_2$
*Set*: $y_i = u_1$

The algorithm has a neat intuitive interpretation. Imagine that we plot the density $f_Y$ and draw a rectangle of width $b - a$ and height $m$ that completely encompasses the density. We then throw points at random into the rectangle; these are the points $(u_1, u_2)$ from the algorithm above. Any that land under the density curve we keep and those that land over the density curve we reject. The name **hit-or-miss** that is commonly applied to this algorithm comes from this idea of throwing points into a rectangle.

We can readily calculate the probability that a value is accepted. Suppose that $U_1 \sim \text{Unif}[a, b]$ and $U_2 \sim \text{Unif}[0, m]$ and that $U_1$ and $U_2$ are independent, so the joint density of $U_1$ and $U_2$ is given by $f_{U_1,U_2}(u_1, u_2) = 1/m(b - a)$. The probability that a value is accepted is

$$\begin{aligned}
P(f_Y(U_1) \geq U_2) &= \int_a^b \int_0^{f_Y(u_1)} f_{U_1,U_2}(u_1,u_2)du_2du_1 \\
&= \int_a^b \int_0^{f_Y(u_1)} \frac{1}{m(b-a)}du_2du_1 \\
&= \int_a^b f_Y(u_1)\frac{1}{m(b-a)}du_1 \\
&= \frac{1}{m(b-a)}
\end{aligned}$$

Thus, the probability of acceptance goes down as $m$ (the maximum of the density) and $b-a$ (the width of the support) increase. In fact, for most distributions of practical interest, hit-or-miss is so inefficient that it is not useful.

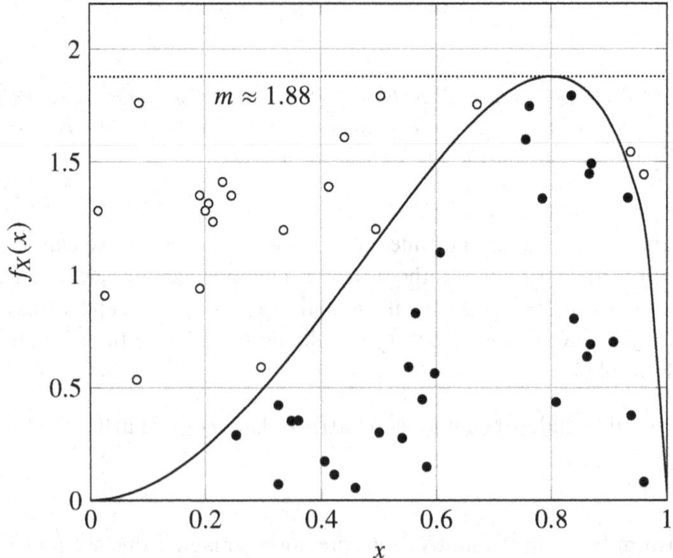


Figure 12.3 *Accept/reject sampling from the* Beta(3, 1.5) *distribution. The mode is* 0.8 *and the corresponding maximum, m, is approximately* 1.88. *The algorithm is equivalent to picking random points in a rectangle of width 1 and height m, and keeping them if they fall under the density curve. The area of the rectangle is m and the area under the curve is 1, so each point is accepted with probability* $1/m \approx 0.53$.

The problems with hit-or-miss are twofold. Firstly, it is restricted to cases where we have bounded support. Secondly, for most distributions, the probability of rejection is large; resources are wasted by simulating a lot of values from regions of the support that have low probability density. One obvious solution is to simulate from a density that more closely approximates our target. Suppose that we can find a function density,

$g$, that is easy to simulate from, and a constant $c$ such that

$$f_Y(y) \leq c\, g(y) \text{ for all } y.$$

If we were to draw the curve $f_Y(y)$ it would lie completely underneath the curve $c\, g(y)$. The $i^{\text{th}}$ member of our simulated sample is then generated by:

*Repeat*: simulate independent $x \leftarrow g$ and $u \leftarrow \text{Unif}[0,1]$
*Until*: $f_Y(x)/(c\, g(x)) \geq u$
*Set*: $y_i = x$

We can readily show that this method will generate a sequence of simulated instances of the random variable $Y$. For notational convenience, let $X$ be a random variable with density function $g$, and let $U \sim \text{Unif}[0,1]$. By independence of $X$ and $U$, their joint density is given by

$$f_{X,U}(x,u) = g(x)\mathbf{1}_{[0,1]}(u).$$

By construction, our accept/reject algorithm simulates instance of the random variable

$$X \,\Big|\, U \leq \frac{f_Y(x)}{c\, g(x)}.$$

This is the same as simulating instances of $Y$, since

$$\begin{aligned}
\mathrm{P}\left(X \leq y \,\Big|\, U \leq \frac{f_Y(x)}{c\, g(x)}\right) &= \frac{\mathrm{P}\left(X \leq y, U \leq \frac{f_Y(x)}{c\, g(x)}\right)}{\mathrm{P}\left(U \leq \frac{f_Y(x)}{c\, g(x)}\right)} \\
&= \frac{\int_{-\infty}^{y}\int_0^{\frac{f_Y(x)}{c\, g(x)}} g(x)dudx}{\int_{-\infty}^{\infty}\int_0^{\frac{f_Y(x)}{c\, g(x)}} g(x)dudx} \\
&= \frac{\frac{1}{c}\int_{-\infty}^{y} f_Y(x)dx}{\frac{1}{c}\int_{-\infty}^{\infty} f_Y(x)dx} \\
&= \int_{-\infty}^{y} f_Y(x)dx = F_Y(y).
\end{aligned}$$

For each proposed value $x$, the probability of acceptance is $1/c$. In order for the algorithm to be efficient, we should keep $c$ as small as possible.

### *12.1.5 Composition*

Consider the case where we wish to sample from a joint density $f(\boldsymbol{x}) = f(x_1, \ldots, x_p)$, but it is not possible to do so directly. We may still be able to construct a sample by successive conditioning, using the property

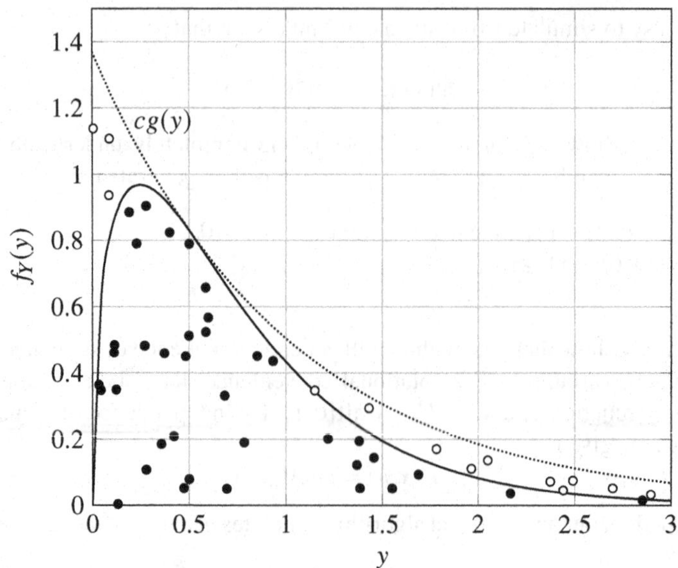


Figure 12.4 *Accept/reject sampling from the* Gamma(1.5,2) *distribution using the proposal density* $g(y) = e^{-y}$, $y > 0$, *which is an* Exp(1). *With a bit of calculus, we can find that the maximum of* $f_Y(y)/g(y)$ *occurs at* $y = 0.5$ *and is equal to* $c = 2e^{-1/2}/\Gamma(3/2) \approx 1.37$. *The algorithm is equivalent to picking random points under the curve* $cg(y)$ *and keeping them if they fall under the density curve* $f_Y(y)$. *The areas under* $cg(y)$ *and* $f_Y(y)$ *are, respectively,* $c$ *and* 1, *so each point is accepted with probability* $1/c \approx 0.73$.

$$f(\boldsymbol{x}) = f(x_p|x_{p-1},\ldots,x_1)f(x_{p-1}|x_{p-2},\ldots,x_1)\ldots f(x_1)\,. \tag{12.2}$$

The algorithm consists of the following steps:

Step 1. Draw $X_1$ from the marginal $f(x_1)$.

Step 2. Draw $X_2$ from $f(x_2|x_1)$.

$\vdots$

Step $p$. Draw $X_p$ from $f(x_p|x_{p-1},\ldots,x_1)$.

This approach, known as **composition sampling**, lends itself naturally to hierarchical Bayes models and predictive inference.

**Example 12.1.3** (Composition sampling for the AR(1) process)
Returning to Example 11.4.4, recall that we derived the following posterior distributions,

$$\pi(\phi|\mathbf{y}_n) \propto \mathbf{1}_{(-1,1)}(\phi)\left(\frac{S_{YY}(\phi)}{2}\right)^{-n/2},$$

$$\pi(\omega|\phi,\mathbf{y}_n) \propto \omega^{n/2-1}\exp\left\{-\omega\frac{S_{YY}(\phi)}{2}\right\} \Rightarrow \omega|\phi,\mathbf{y}_n \sim \text{Gamma}\left(\frac{n}{2},\frac{S_{YY}(\phi)}{2}\right)$$

$$f(y_{n+1}|\mathbf{y}_n,\phi,\omega) \propto \exp\left\{-\frac{\omega}{2}(y_{n+1}-\phi y_n)^2\right\} \Rightarrow y_{n+1}|\mathbf{y}_n,\phi,\omega \sim \text{N}(y_n,1/\omega),$$

where $S_{YY}(\phi) = \sum_{t=1}^{n}(y_t - \phi y_{t-1})^2$. We can thus draw from the predictive distribution of $y_{n+1}$ with a composition algorithm, for $k = 1,\ldots,m$,

Step 1. Draw $\phi^{(k)}$ from $\pi(\phi|\mathbf{y}_n)$.
Step 2. Draw $\omega^{(k)}$ from $\pi(\omega|\phi^{(k)},\mathbf{y}_n)$, a Gamma density.
Step 3. Draw $y_{n+1}^{(k)}$ from $f(y_{n+1}|\mathbf{y}_n,\phi^{(k)},\omega^{(k)})$, a normal density

Cycling through Steps 1–3, we generate a sample $y_{n+1}^{(1)},\ldots,y_{n+1}^{(m)}$, which can then be used to compute an approximation of the predictive density $f(y_{n+1}|\mathbf{y}_n)$.

The first step of the algorithm is quite difficult, because $\pi(\phi|\mathbf{y}_n)$ cannot be computed explicitly; we discuss a possible approach in the context of importance sampling (subsection 12.2.3).

### 12.1.6 Simulating model structure and the bootstrap

Many of the statistical models we have encountered so far are linear, in the sense that the output is a linear combination of the input variables and error terms. Examples include the linear regression models of section 6.3 and the time series models in section 6.6. Simulation allows us to assess the correctness of these models ('*Does the model behave as expected if its assumptions are met?*') and also their robustness ('*What is the behaviour of the model if its assumptions are not met?*').

Consider the simple linear regression model,

$$Y_i = \alpha + \beta X_i + \varepsilon_i \quad \text{where} \quad \{\varepsilon_i\} \sim \text{N}(0,\sigma^2),$$

for $i = 1,\ldots,n$. We can check that the confidence intervals for the model parameters have the correct coverage probability. If we specify values for the parameters ($\alpha$, $\beta$, and $\sigma^2$) and the explanatory variable, $\mathbf{X}$, we can simulate data for the response variable by setting

$$Y_i^* = \alpha + \beta X_i + \varepsilon_i^*,$$

where $\varepsilon_1^*,\ldots,\varepsilon_n^*$ are drawn independently from $\text{N}(0,\sigma^2)$. If we generate a large number of these $n$-member samples, we can check that, say, the 95% confidence interval for $\beta$ covers the true value of $\beta$ (which is known) in 95% of the generated samples.

Furthermore, we can check the coverage probability when some of the model assumptions are violated, for example, if the error terms

- are not independent, so $(\varepsilon_1^*, \ldots, \varepsilon_n^*)^T \sim \mathrm{N}(\mathbf{0}, \mathbf{\Sigma})$, where $\mathbf{\Sigma}$ is not diagonal.
- do not all have the same variance, say $\varepsilon_i^* \sim \mathrm{N}(0, u_i \sigma^2)$, where $U_i$ is drawn from Unif[0.5, 1.5].
- follow a different distribution, say $\varepsilon_i^* = \sigma(v_i - 1)$, where $V_i$ is drawn from Exp(1).

This approach can be powerful; in the linear regression case, we discover that the errors being independent is more important than them having equal variance or being normally distributed, in the sense that the model fit suffers more when the assumption of independence is violated (Figure 12.5).

A similar method we can use to assess the fit of a particular model is to generate new samples from the existing data set, a process is known as **bootstrapping**. In the context of the simple linear regression model, we could generate

$$Y_i^* = \hat{\alpha} + \hat{\beta} X_i + \varepsilon_i^*,$$

where $\varepsilon_1^*, \ldots, \varepsilon_n^*$ are drawn independently from $\mathrm{N}(0, \hat{\sigma}^2)$. We then refit the same model to the new sample of pairs $\{(X_1, Y_1^*), \ldots, (X_n, Y_n^*)\}$. As before, if we generate a large number of these $n$-member paired samples, we can approximate the distribution of the model parameter estimators. This particular application is an example of a **parametric bootstrap**, because we make distributional assumptions in resampling the new data; the errors are assumed to be normal.

A related approach is the **nonparametric** or **resampling bootstrap**. Suppose that we use the sample $\boldsymbol{Y} = (Y_1, \ldots, Y_n)^T$ to estimate a parameter of interest, $\theta$, but we cannot work out the distribution of the estimator $\hat{\theta} = h(\boldsymbol{Y})$ analytically. We generate new samples, $\boldsymbol{Y}_1^*, \boldsymbol{Y}_2^*, \ldots, \boldsymbol{Y}_m^*$, also of size $n$, by sampling with replacement from $\boldsymbol{Y}$ *itself*. For example, if $n = 5$, the first bootstrap sample could be

$$\boldsymbol{Y}_1^* = (Y_2, Y_5, Y_1, Y_3, Y_3)^T .$$

We can now compute the bootstrap estimates of $\theta$ as $\hat{\theta}_j^* = h(\boldsymbol{Y}_j^*)$, for $j = 1, \ldots, m$. These allow us to assess the properties of $\hat{\theta}$; for example, a bootstrap estimate of the variance of $\hat{\theta}$ is

$$\widehat{\mathrm{Var}(\hat{\theta})} = \frac{1}{m-1} \sum_{i=1}^{m} \left( \hat{\theta}_j^* - \widehat{\mathbb{E}(\hat{\theta})} \right)^2 , \tag{12.3}$$

where

$$\widehat{\mathbb{E}(\hat{\theta})} = \frac{1}{m} \sum_{j=1}^{m} \hat{\theta}_j^*$$

is the sample mean of the bootstrap estimates.

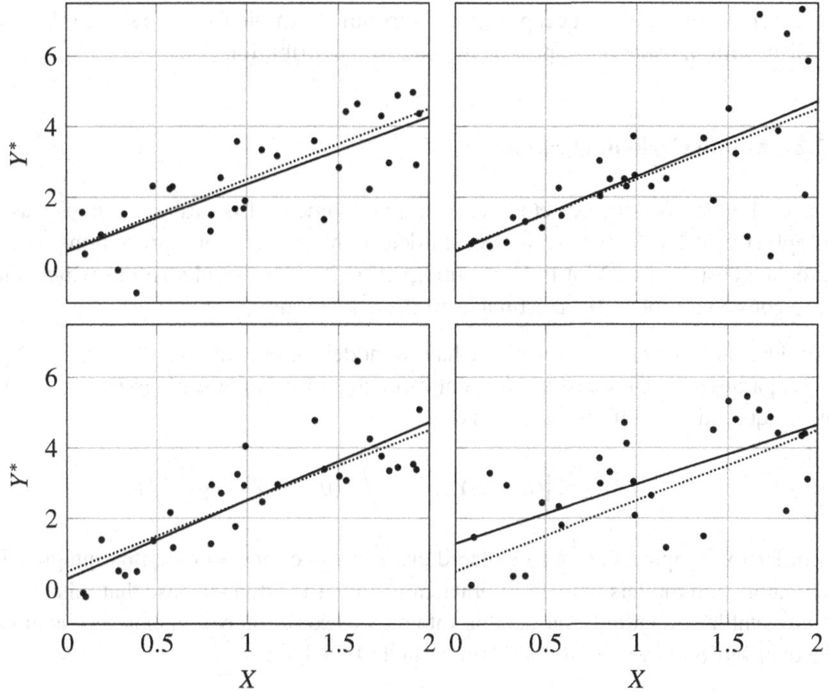


Figure 12.5 *Simulated data from the model* $Y_i^* = 0.5 + 2X_i + \varepsilon_i^*$, *where the error terms have variance 1. The dotted line represents the true model, whereas the solid line is the least-squares regression line. Top left: all assumptions are correct. Top right: the variance of* $\varepsilon_i^*$ *is proportional to* $x_i$, *i.e. not constant (the variance is equal to 1 on average). Bottom left:* $\varepsilon_i^* = v_i - 1$, *where* $V_i \sim \text{Exp}(1)$. *Bottom right: the X values are in ascending order and there is serial correlation in the errors,* $\text{Corr}(\varepsilon_i^*, \varepsilon_{i+1}^*) = 0.5$. *Notice that the difference between the true and the estimated line is greatest when the assumption of independence is violated (bottom right).*

**Exercise 12.1**

1. Recall from Example 6.5.6 that the Weibull distribution has cumulative distribution function

$$F_X(x) = \begin{cases} 1 - \exp(-\beta x^\alpha) & \text{for } 0 < x < \infty, \\ 0 & \text{otherwise.} \end{cases}$$

   Show how you can use the probability integral transform to draw samples from this distribution. Verify that in the special case where $\alpha = 1$ (when $X$ is exponentially distributed), the sampler reduces to the one derived in Example 12.1.2.

2. Describe an algorithm to simulate instances from a $\text{Bin}(n, p)$ distribution.
3. Prove that the random variables produced by the Box-Muller method are standard normal and independent.

4. Describe, in detail, an accept/reject algorithm to sample from the standard normal distribution, using an Exp(1) as the proposal distribution.

## 12.2 Monte Carlo integration

One of the factors that restricted the early development of Bayesian methods was the complexity of the integrals involved in determining posterior distributions and the associated estimators. **Monte Carlo integration** is a class of algorithms which allow us to construct probabilistic estimates of these integrals.

Consider the following situation: we have a model for our sample, $\boldsymbol{Y} = (Y_1, \dots, Y_n)^T$, and a parameter of interest, $\boldsymbol{\theta}$. A point estimator of $\boldsymbol{\theta}$ is the statistic $\hat{\boldsymbol{\theta}} = h(\boldsymbol{Y})$. The mean squared error of this statistic is

$$\underset{\boldsymbol{\theta}}{\mathrm{MSE}}(\hat{\boldsymbol{\theta}}) = \mathbb{E}\left[(\boldsymbol{\theta} - h(\boldsymbol{Y}))^2\right] = \int_{\mathbb{R}^n} (\boldsymbol{\theta} - h(\boldsymbol{Y}))^2 f_{\boldsymbol{Y}}(\boldsymbol{y}) d\boldsymbol{y} .$$

In order to compute the mean squared error, we need to evaluate this integral. For estimators and models that are of practical value, it is often the case that this integral is intractable. Potentially intractable integrals arise in many contexts. For example, the distribution function of a random variable $Y$ is defined as

$$F_Y(y) = \int_{-\infty}^{y} f_Y(u) du .$$

This integral may not have a closed form solution; the normal is an example of a distribution for which this is the case. It is worth noting at this point that we can write the distribution function as an expectation,

$$F_Y(y) = \mathbb{E}\left(\mathbf{1}_{(-\infty, y]}(Y)\right) .$$

In fact, expressing the integrals to be evaluated as expectations is the key to the Monte Carlo approach.

### 12.2.1 *Averaging over simulated instances*

The basic principles of Monte Carlo integration are very straightforward. Suppose that we need to evaluate the (univariate) integral,

$$J = \int_A g(y) dy. \tag{12.4}$$

Suppose in addition that $f_Y$ is a known density with support $A$. We can write

$$J = \int_A g(y) dy = \int_A \frac{g(y)}{f_Y(y)} f_Y(y) dy = \mathbb{E}\left[\frac{g(Y)}{f_Y(Y)}\right] .$$

The law of large numbers tells us that, for a reasonably well-behaved function $h$ and random sample $Y_1, \dots, Y_n$, we have

$$\frac{1}{n}\sum_{i=1}^{n} h(Y_i) \xrightarrow{p} \mathbb{E}(h(Y)) \text{ as } n \to \infty .$$

This is the basis for simple Monte Carlo integration. Consider a random sample $Y_1, \dots, Y_n$ of variables with density $f_Y$. If we take

$$\hat{J} = \frac{1}{n}\sum_{i=1}^{n} \frac{g(Y_i)}{f_Y(Y_i)},$$

then, by the law of large numbers, this estimator converges in probability to $J$ as $n \to \infty$. We generate an observed value of this estimator using values $y_1, \dots, y_n$ simulated from $f_Y$.

We can readily derive the properties of $\hat{J}$ as an estimator of $J$. The mean is

$$\mathbb{E}(\hat{J}) = \frac{1}{n}\sum_{i=1}^{n} \mathbb{E}\left(\frac{g(Y_i)}{f_Y(Y_i)}\right) = \frac{1}{n}\sum_{i=1}^{n} \int_A g(y)dy = \frac{1}{n}\sum_{i=1}^{n} J = J,$$

so $\hat{J}$ is unbiased. The variance is

$$\text{Var}(\hat{J}) = \frac{1}{n^2}\sum_{i=1}^{n} \text{Var}\left(\frac{g(Y_i)}{f_Y(Y_i)}\right) = \frac{1}{n}\text{Var}\left(\frac{g(Y)}{f_Y(Y)}\right),$$

which converges to 0 as $n \to \infty$, so $\hat{J}$ is consistent. In practice, we can estimate the variance by applying the Monte Carlo approach again. Let $V$ be the variance of $\hat{J}$, so

$$V = \frac{1}{n}\left\{\mathbb{E}\left[\left(\frac{g(Y)}{f_Y(Y)}\right)^2\right] - J^2\right\},$$

which we can estimate with

$$\hat{V} = \frac{1}{n}\left\{\frac{1}{n}\sum_{i=1}^{n}\left(\frac{g(Y_i)}{f_Y(Y_i)}\right)^2 - \hat{J}^2\right\}.$$

### 12.2.2 *Univariate vs. multivariate integrals*

If $A = (a, b)$ is an interval of finite length, an obvious choice of distribution is to take $Y_1, \dots, Y_n \sim \text{Unif}[a, b]$, so $f_Y(y) = 1/(b-a)$ for $y \in A$. This approach is known as **uniform sampling**, and results in the estimator

$$\hat{J}_U = \frac{b-a}{n}\sum_{i=1}^{n} g(Y_i).$$

An even simpler way to estimate $J$ would be to start by subdividing $(a, b)$ into $n$ equal intervals, $(a, a + \delta), (a + \delta, a + 2\delta), \ldots, (a + (n - 1)\delta, b)$, where $\delta = (b - a)/n$. We can then approximate the region corresponding to the $i^{\text{th}}$ interval as a rectangle with base $(b - a)/n$ and height $g(x_i)$, where $x_i$ is the midpoint of the interval. This yields

$$\tilde{J} = \frac{b-a}{n} \sum_{i=1}^{n} g(x_i),$$

which is very similar to $\hat{J}_U$. In fact, the only difference is that the $x_1, \ldots, x_n$ are evenly spaced, while the $Y_1, \ldots, Y_n$ are drawn randomly.

In the univariate case, there is no compelling reason to use the Monte Carlo approach. The simple method is no worse, and can be made a lot better with some simple modifications, such as approximating $g(x)$ in each region by a linear function (trapezoid rule) or a quadratic function (Simpson's rule). The advantage of Monte Carlo integration becomes apparent in the multivariate case.

Suppose that $g(\boldsymbol{y})$ is a scalar-valued function that takes a $p$-dimensional argument, and we wish to integrate the function over some region $\boldsymbol{A}$. To simplify this illustration, we assume that $\boldsymbol{A}$ is such that the $j^{\text{th}}$ component of $\boldsymbol{y}$ takes values in the interval $(a_j, b_j)$, so we can write $\boldsymbol{A} = (a_1, b_1) \times \cdots \times (a_p, b_p)$, which is a **hyperrectangle**. The integral can then be expressed as

$$J = \int_{\boldsymbol{A}} g(\boldsymbol{y}) d\boldsymbol{y} = \int_{a_p}^{b_p} \cdots \int_{a_1}^{b_1} g(\boldsymbol{y}) dy_1 \ldots dy_p .$$

This is analogous to integrating a bivariate density over a rectangular region of the plane (Example 4.2.10).

The Monte Carlo approach generalises very easily. If $\boldsymbol{Y}$ is a $p$-dimensional random vector with support $\boldsymbol{A}$ and density $f_{\boldsymbol{Y}}$, we draw a random sample $\boldsymbol{Y}_1, \ldots, \boldsymbol{Y}_n$ from $f_{\boldsymbol{Y}}$ and take

$$\hat{J} = \frac{1}{n} \sum_{i=1}^{n} \frac{g(\boldsymbol{Y}_i)}{f_{\boldsymbol{Y}}(\boldsymbol{Y}_i)} .$$

For example, if the $j^{\text{th}}$ component of $\boldsymbol{Y}$ is $\text{Unif}[a_j, b_j]$, and the different components are independent, we end up with

$$\hat{J}_U = \frac{\prod_{j=1}^{p} (b_j - a_j)}{n} \sum_{i=1}^{n} g(\boldsymbol{Y}_i),$$

With the simple approach, we need to subdivide each $(a_j, b_j)$ into $n_j$ equal intervals, which splits $\boldsymbol{A}$ into $\prod_{j=1}^{p} n_j$ equal hyperrectangles. The resulting estimate is

$$\tilde{J} = \left( \prod_{j=1}^{p} \frac{b_j - a_j}{n_j} \right) \sum_{i_1=1}^{n_1} \cdots \sum_{i_p=1}^{n_p} g(x_{i_1} \ldots x_{i_p}),$$

where $x_{i_j}$ is the midpoint of the $i^{\text{th}}$ interval in the $j^{\text{th}}$ dimension.

This highlights the difference between the two methods. The simple approach requires $p$ summations and $\prod_{j=1}^{p} n_j$ function evaluations, which quickly becomes very impractical as the dimension increases. By contrast, Monte Carlo integration requires only a single summation and $n$ evaluations of the function $g(\boldsymbol{y})$, regardless of the dimension. Even better, the variance of the Monte Carlo estimator remains

$$\text{Var}(\hat{J}) = \frac{1}{n}\,\text{Var}\left(\frac{g(\boldsymbol{Y})}{f_{\boldsymbol{Y}}(\boldsymbol{Y})}\right),$$

so we are not sacrificing accuracy in the process.

The Monte Carlo approach is also much better than the hit-or-miss algorithm, as the following example demonstrates.

**Example 12.2.1** (Area of the quarter circle)
Suppose that the integral to be evaluated is

$$J = \int_0^1 \sqrt{1 - y^2}dy\,.$$

This is the area of the quarter-circle with unit radius, so we know that it is equal to $\pi/4$.

Consider a simple hit-or-miss algorithm. The function of interest attains a maximum value of 1, so we generate a random sample $X_1, \ldots, X_n, Y_1, \ldots, Y_n$ from Unif[0, 1]. We define

$$H_i = \begin{cases} 1 & \text{if } Y_i \leq \sqrt{1 - (X_i)^2}\,, \\ 0 & \text{otherwise}\,, \end{cases}$$

so $H_i = 1$ indicates a hit. The estimator is then

$$\hat{J}_1 = \frac{1}{n}\sum_{i=1}^{n} H_i\,.$$

The probability of a hit is $\pi/4$, so $H_i \sim \text{Bernoulli}(\pi/4)$, and we conclude that

$$\text{Var}(\hat{J}_1) = \frac{1}{n}\frac{\pi}{4}\left(1 - \frac{\pi}{4}\right) \approx \frac{0.1685}{n}\,,$$

for a sample of size $2n$.

Now consider the Monte Carlo approach. Given the same uniform sample as before (which we relabel $Y_1, \ldots, Y_{2n}$ for convenience), we set

$$\hat{J}_2 = \frac{1}{2n}\sum_{i=1}^{2n}\sqrt{1 - Y_i^2}\,.$$

This estimator has variance

$$\mathrm{Var}(\hat{J}_2) = \frac{1}{2n}\mathrm{Var}\left(\sqrt{1-Y^2}\right) = \frac{1}{2n}\left[\int_0^1 \left(\sqrt{1-y^2}\right)^2 dy - \left(\frac{\pi}{4}\right)^2\right]$$
$$= \frac{1}{2n}\left[\frac{2}{3} - \frac{\pi^2}{4}\right] \approx \frac{0.0249}{n}.$$

Compared to the hit-or-miss algorithm, Monte Carlo integration reduced the variance by a factor of around 7.

### 12.2.3 *Importance sampling*

Consider the integral

$$J(y) = \mathbb{E}[g(y|X)] = \int_{D_X} g(y|x) f_X(x) dx,$$

where $D_X$ is the support of the density function $f_X$. If we can draw a random sample $X_1, \dots, X_n$ from $f_X$, we can construct a Monte Carlo estimator of this integral,

$$\hat{J}^{(M)}(y) = \frac{1}{n}\sum_{i=1}^{n} g(y|X_i). \tag{12.5}$$

In many cases of interest we cannot sample directly from $f_X$, for example, if $f_X$ is an improper prior. Now suppose that $f_Z$ is a density function that is easy to sample from, and which has support $D_Z$, where $D_X \subseteq D_Z$. We have

$$\mathbb{E}\left[g(y|Z)\frac{f_X(Z)}{f_Z(Z)}\right] = \int_{D_Z} g(y|x)\frac{f_X(x)}{f_Z(x)} f_Z(x) dx = \int_{D_Z} g(y|x) f_X(x) dx = J(y).$$

Using the approach of averaging over simulated instances, if $Z_1, \dots, Z_n$ is a random sample drawn from $f_Z$, then

$$\hat{J}(y) = \frac{1}{n}\sum_{i=1}^{n} g(y|Z_i)\frac{f_X(Z_i)}{f_Z(Z_i)}$$

is an unbiased estimator of $J(y)$, which converges to $J(y)$ in probability as $n \to \infty$. This method is known as **importance sampling**, and $f_Z$ is commonly referred to as the **biasing density**. If we express the estimator in the form

$$\hat{J}(y) = \frac{1}{n}\sum_{i=1}^{n} g(y|Z_i) W_i, \quad W_i = \frac{f_X(Z_i)}{f_Z(Z_i)},$$

where $W_i$ are the **weights**, we can see that importance sampling compensates for the use of the biasing density $f_Z$ instead of the "true" density $f_X$ by placing weights greater than 1 in the regions where $f_X(x) > f_Z(x)$, and smaller than 1 in the regions where $f_X(x) < f_Z(x)$. These correspond, respectively, to regions of high and low probability for $X$.

Importance sampling is a useful technique even when we can sample from $f_X$ directly. The estimators $\hat{J}^{(M)}(y)$ (from (12.5)) and $\hat{J}(y)$ are both unbiased, so we compare their variances. The variance of the Monte Carlo estimator is

$$\operatorname{Var}\left(\hat{J}^{(M)}(y)\right) = \frac{1}{n}\operatorname{Var}\left(g(y|X)\right) = \frac{1}{n}\left(\int_{D_X} g(y|x)^2 f_X(x)dx - J(y)^2\right),$$

while that of the importance sampling estimator is

$$\operatorname{Var}(\hat{J}(y)) = \frac{1}{n}\operatorname{Var}\left(g(y|Z)\frac{f_X(Z)}{f_Z(Z)}\right) = \frac{1}{n}\left[\int_{D_Z}\left(g(y|x)\frac{f_X(x)}{f_Z(x)}\right)^2 f_Z(x)dx - J(y)^2\right].$$

The importance sampling estimator has lower variance when

$$\begin{aligned} 0 &< \int_{D_X} g(y|x)^2 f_X(x)dx - \int_{D_Z}\left(g(y|x)\frac{f_X(x)}{f_Z(x)}\right)^2 f_Z(x)dx \\ &= \int_{D_Z} g(y|x)^2 f_X(x)\left(1 - \frac{f_X(x)}{f_Z(x)}\right)dx, \end{aligned}$$

thus, if we can choose a biasing density such that $f_Z(x) > f_X(x)$ when $g(y|x)^2 f_X(x)$ is large, and $f_Z(x) < f_X(x)$ when $g(y|x)^2 f_X(x)$ is small, the importance sampling estimator achieves **variance reduction**. For example, if $g(y|x)$ takes large values in the right tail of the distribution of $X$, we could use a transformation such as $Z = X + c$ with $c > 0$ (translation), or $Z = mX$ with $m > 0$ (scaling), to increase the density in the desired region.

**Example 12.2.2** (Variance reduction with importance sampling)
We can express the integral from Example 12.2.1 as

$$\mathbb{E}\left(\sqrt{1 - Y^2}\right) = \int_0^1 \sqrt{1 - y^2}dy = \int_0^1 g(y)f_Y(y)dy,$$

where $g(y) = \sqrt{1 - y^2}$ and $f_Y(y) = 1$ for $y \in [0, 1]$, that is, $Y \sim \text{Unif}[0, 1]$. To achieve variance reduction, we need a biasing density that is $> 1$ when $g(y)^2 f_Y(y) = 1 - y^2$ is large, and $< 1$ when it is small. Consider the simple density function

$$f_Z(y) = \begin{cases} \frac{3}{2} - y & 0 \leq y \leq 1, \\ 0 & \text{otherwise.} \end{cases}$$

The CDF of $Z$ is $F_Z(y) = (3y - y^2)/2$, which we can readily invert; given a Unif[0, 1] random sample, $Y_1, \ldots, Y_{2n}$, we can draw $Z_1, \ldots, Z_{2n}$ from $f_Z$ using the probability integral transformation,

$$Z_i = \frac{3 - \sqrt{9 - 8Y_i}}{2}, \text{ for } i = 1, \ldots, 2n.$$

The importance sampling estimator of $J$ is, thus,

$$\hat{J}_3 = \frac{1}{n}\sum_{i=1}^{n} g(Z_i)\frac{f_Y(Z_i)}{f_Z(Z_i)} = \frac{1}{2n}\sum_{i=1}^{2n}\sqrt{1-(Z_i)^2}\frac{1}{3/2-Z_i}.$$

Its variance is

$$\begin{aligned}\text{Var}(\hat{J}_3) &= \frac{1}{2n}\text{Var}\left(\frac{\sqrt{1-(Z)^2}}{3/2-Z}\right)\\ &= \frac{1}{2n}\left[\int_0^1\left(\frac{\sqrt{1-y^2}}{3/2-y}\right)^2(3/2-y)dy - \left(\frac{\pi}{4}\right)^2\right] \approx \frac{0.0049}{n}.\end{aligned}$$

Compared to Monte Carlo integration, importance sampling reduced the variance of the estimator by a factor of around 5.

Now suppose that the density $f_X$ is known only up to a constant, a very common situation in Bayesian inference. We write $f_X(x) = kh(x)$, where $h$ is a known function, and $k > 0$ is an unknown normalising constant. Although we cannot evaluate $\hat{J}(y)$ directly (the weights depend on $k$), we can compute an alternative estimator,

$$\hat{J}^*(y) = \frac{\frac{1}{n}\sum_{i=1}^n g(y|Z_i)W_i}{\frac{1}{n}\sum_{i=1}^n W_i} = \frac{\sum_{i=1}^n g(y|Z_i)W_i^*}{\sum_{i=1}^n W_i^*}, \quad \text{where } W_i^* = \frac{h(Z_i)}{f_Z(Z_i)},$$

where $k$ cancels out, as it appears in both the numerator and the denominator. The new estimator is biased, but we can show that it still converges to the correct value. We have

$$\mathbb{E}(W_i) = \mathbb{E}\left(\frac{f_X(Z_i)}{f_Z(Z_i)}\right) = \int_{D_Z}\frac{f_X(x)}{f_Z(x)}f_Z(x)dx = \int_{D_Z} f_X(x)dx = 1,$$

which implies that

$$\frac{1}{n}\sum_{i=1}^n W_i \xrightarrow{p} 1.$$

For the original estimator, we know that

$$\frac{1}{n}\sum_{i=1}^n g(y|Z_i)W_i \xrightarrow{p} J(y),$$

so we deduce that

$$\hat{J}^*(y) = \frac{\frac{1}{n}\sum_{i=1}^n g(y|Z_i)W_i}{\frac{1}{n}\sum_{i=1}^n W_i} \xrightarrow{p} J(y) \quad \text{as } n\to\infty.$$

**Example 12.2.3** (Importance sampling for the AR(1) process)
Returning to Example 12.1.3, recall that the only remaining problem is sampling from the posterior distribution of $\phi$,

$$\pi(\phi|\mathbf{Y}_n) \propto h(\phi|\mathbf{y}_n) = \mathbf{1}_{(-1,1)}(\phi)\left(\frac{S_{YY}(\phi)}{2}\right)^{-n/2}.$$

Let $f_Z$ be a density that is easy to sample from, and whose support contains the interval $(-1,1)$. An importance sampling algorithm is

Step 1. Draw $z_1,\dots,z_m$ from $f_Z$.

Step 2. Draw $\omega_i$ from Gamma$(n/2, S_{YY}(z_i)/2)$, for $i = 1,\dots,m$.

Step 3. Set $\hat{f}(y_{n+1}|\mathbf{y}_n) = \sum_{i=1}^m w_i^* f(y_{n+1}|\mathbf{y}_n, z_i, \omega_i)/\sum_{i=1}^m w_i^*$, where the weights are $w_i^* = h(z_i|\mathbf{y}_n)/f_Z(z_i)$.

### *12.2.4 Antithetic variates*

The simulation techniques we have encountered so far have all been based on drawing an IID sample from some probability distribution. We now demonstrate a different approach: selecting a non-independent sample in such a way that reduces the variance of the resulting estimators.

Suppose that the quantity to be estimated can be expressed as the expectation of a random variable $X$. We generate a sequence of IID random variables $X_1, X_2, \dots, X_n$, all with the same distribution as $X$, and use the sample mean $\bar{X}$ as our estimator. Now assume that for each $X_i$ we can generate another random variable $Y_i = g(X_i)$, also with the same distribution as $X_i$, but such that $\text{Corr}(X_i, Y_i) = \rho$. We thus obtain another sequence of IID random variables $Y_1, Y_2, \dots, Y_n$. Taking $\hat{\mu} = (\bar{X}+\bar{Y})/2$ as our estimator of $\mathbb{E}(X)$, we have

$$\mathbb{E}\left(\frac{\bar{X}+\bar{Y}}{2}\right) = \mathbb{E}(X),$$

and

$$\text{Var}\left(\frac{\bar{X}+\bar{Y}}{2}\right) = \frac{1}{4}\left[\text{Var}(\bar{X}) + \text{Var}(\bar{Y}) + 2\text{Cov}(\bar{X},\bar{Y})\right] = \frac{1}{4}\left[2\frac{\text{Var}(X)}{n} + 2\text{Cov}(\bar{X},\bar{Y})\right].$$

The covariance term is

$$\text{Cov}(\bar{X},\bar{Y}) = \text{Cov}\left(\frac{1}{n}\sum_{i=1}^n X_i, \frac{1}{n}\sum_{i=1}^n Y_i\right) = \frac{1}{n^2}\sum_{i=1}^n \text{Cov}(X_i, Y_i) = \frac{\text{Var}(X)\rho}{n},$$

so the variance of the new estimator is

$$\text{Var}(\hat{\mu}) = \frac{\text{Var}(X)}{2n}(1+\rho).$$

If $\rho = 0$, the variance of $\hat{\mu}$ is exactly what we would have obtained from an IID sample of size $2n$. However, if we construct each $Y_i$ so that it is negatively correlated with $X_i$, the variance of $\hat{\mu}$ will be lower, possibly considerably so. We refer to such samples as **antithetic variates**. Informally, the intuition behind this method is that, if $X_i$ takes a high value then $Y_i$ will take a low value, as $\rho < 0$, so the estimation errors cancel out.

For example, if we are sampling using the probability integral method, then for each sampled $U_i \sim \text{Unif}[0, 1]$ variable, we can define

$$X_i = F_X^{-1}(U_i) \quad \text{and} \quad Y_i = F_X^{-1}(1 - U_i).$$

As $1 - U_i \sim \text{Unif}[0, 1]$, each $Y_i$ will have the required distribution. The generated variables are negatively correlated because $X_i$ takes a high value when $Y_i$ takes a low value, and vice versa.

**Example 12.2.4** (Variance reduction with antithetic variables)
Consider again the setup of Example 12.2.1. Given the same random sample, $Y_1, \dots, Y_{2n}$, define $Y_i^* = 1 - Y_i$ for $i = 1, \dots, 2n$. Notice that $Y_i^* \sim \text{Unif}[0, 1]$, so the estimator

$$\hat{J}_2^* = \frac{1}{2n} \sum_{i=1}^{2n} \sqrt{1 - (Y_i^*)^2}$$

has the same distribution as $\hat{J}_2$. The $\{Y_i\}$ and the $\{Y_i^*\}$ have the same distribution and are negatively correlated, as required. Now define the estimator

$$\hat{J}_4 = \frac{\hat{J}_2 + \hat{J}_2^*}{2},$$

which is also unbiased and has variance

$$\text{Var}(\hat{J}_4) = \frac{1}{4} \left[ \text{Var}(\hat{J}_2) + \text{Var}(\hat{J}_2^*) + 2\text{Cov}(\hat{J}_2, \hat{J}_2^*) \right].$$

The covariance term in this expression is

$$\begin{aligned}
\text{Cov}(\hat{J}_2, \hat{J}_2^*) &= \text{Cov}\left( \frac{1}{2n} \sum_{i=1}^{2n} \sqrt{1 - Y_i^2}, \frac{1}{2n} \sum_{i=1}^{2n} \sqrt{1 - (Y_i^*)^2} \right) \\
&= \frac{1}{2n} \text{Cov}\left( \sqrt{1 - Y^2}, \sqrt{1 - (1 - Y)^2} \right) \\
&= \frac{1}{2n} \left[ \int_0^1 \sqrt{1 - y^2} \sqrt{1 - (1 - y)^2} dy - \frac{\pi}{4} \frac{\pi}{4} \right].
\end{aligned}$$

This integral is intractable, but we can integrate numerically to find

$$\text{Cov}(\hat{J}_2, \hat{J}_2^*) = \frac{1}{2n} \left[ 0.58075 - \left( \frac{\pi}{4} \right)^2 \right] \approx -\frac{0.0180}{n}.$$

The variance of the estimator is, thus,

$$\text{Var}(\hat{J}_4) = \frac{1}{2}\left(\frac{0.0249}{n} - \frac{0.0180}{n}\right) \approx \frac{0.00345}{n},$$

which is an improvement over $\hat{J}_2$ by a factor of around 7, and a slight improvement over $\hat{J}_3$.

**Exercise 12.2**

1. Consider the integral

$$\frac{1}{\sqrt{2\pi}}\int_0^1 e^{-x^2}\,dx,$$

which is difficult to compute analytically. Outline a Monte Carlo procedure to estimate the integral and find an expression for the standard error of your estimator.

2. Let $X$ be a continuous random variable and suppose that we want to estimate the probability $p_k = \text{P}(X > k)$, where $k$ is a value that is very far in the right tail of the distribution of $X$.

   (a) We generate a sample $x_1, \dots, x_n$ and compute the Monte Carlo estimator

   $$\hat{p}_k = \frac{1}{n}\sum_{i=1}^{n} \mathbf{1}_{(k,\infty)}(x_i).$$

   Explain why $\hat{p}_k$, despite being unbiased, is not a very useful estimator of $p_k$.

   (b) Now suppose that $X \sim \text{N}(0,1)$ and $k = 3.5$. Use the transformation $Z = X + k$ to construct an importance sampling estimator for $p_k$.

   (c) Show that the importance sampling estimator in part (b) has lower variance than the Monte Carlo estimator in part (a).

3. Suppose that we have a sequence of IID random variables, $X_1, X_2, \dots, X_n$, and want to estimate the mean of their distribution, $\mu_x$. Now assume that we can generate another sequence of IID random variables, $Y_1, Y_2, \dots, Y_n$, where $X_i$ and $Y_i$ are correlated, and $\mu_y = \mathbb{E}(Y_i)$ is known. Consider the estimator

$$\tilde{\mu}_x = \frac{1}{n}\sum_{i=1}^{n}\left(X_i + \beta(Y_i - \mu_y)\right),$$

where $\beta \in \mathbb{R}$ is a tuning parameter.

   (a) Show that $\tilde{\mu}_x$ is unbiased for any choice of $\beta$.

   (b) Establish a condition for the variance of $\tilde{\mu}_x$ to be lower than that of the obvious estimator, $\hat{\mu}_x = \bar{X}$.

   (c) What value of $\beta$ would you choose to minimise the variance of $\tilde{\mu}_x$? How big is the reduction in variance relative to $\hat{\mu}_x$? Explain how this should inform our choice of $Y_i$.

   (d) Can we find the optimal value of $\beta$ in practice?

(e) Suppose that we want to use this method to evaluate the integral

$$\frac{1}{\sqrt{2\pi}} \int_0^1 e^{-u^2} du .$$

Define $X_i$, suggest an appropriate $Y_i$, and explain how you would obtain a reasonable value for $\beta$.

[This is an example of the **control variates** method.]

## 12.3 Markov chain Monte Carlo

We now consider a class of algorithms known as **Markov chain Monte Carlo** (MCMC) methods. Their common characteristic is that, in order to draw samples from a desired probability distribution $f_X(x)$, we construct a Markov chain $\{X_t\}$ which has $f_X(x)$ as its limiting distribution. In practice, we choose some large value $N$ and treat $X_{N+1}, X_{N+2}, \ldots$ as an identically distributed (but not independent) draws from $f_X(x)$. The observations $X_1, X_2, \ldots, X_N$ are discarded; this is known as the **burn-in period**.

### *12.3.1 Discrete Metropolis*

Assume that we are trying to sample from a discrete distribution whose mass function $f_X(x)$ is known only up to a constant. The **discrete Metropolis** algorithm is based on the idea of defining a Markov chain $\{X_t\}$ with $f_X(x)$ as its limiting distribution.

Let $\{x_1, \ldots, x_M\}$ be the support of $f_X(x)$. Suppose that $M$ is finite and let $\boldsymbol{Q} = \{q_{i,j}\}$ be an arbitrary $M \times M$ symmetric transition matrix, so $q_{i,j} = q_{j,i}$ for all $i, j$. To ensure that the finite-state Markov chain defined by $\boldsymbol{Q}$ is regular, we may impose the additional condition $q_{i,j} > 0$ for all $i, j$. It is easy to verify that this chain converges to a discrete uniform distribution, that is, the limiting probability of each state is $1/M$, where $M$ is the number of states. The vector $\frac{1}{M}\mathbf{1}_M^T$ is an equilibrium distribution, as

$$\left[\frac{1}{M}\mathbf{1}_M \boldsymbol{Q}\right]_i = \sum_{j=1}^{M} \frac{1}{M} q_{j,i} = \frac{1}{M} \sum_{j=1}^{M} q_{i,j} = \frac{1}{M},$$

and the chain is regular, so this is the limiting distribution.

Using the notation $\pi_i = f_X(x_i)$, we now define a new Markov chain with transition matrix $\boldsymbol{P} = \{p_{i,j}\}$, where

$$\begin{aligned} p_{i,j} &= \min\{1, \pi_j/\pi_i\}\, q_{i,j}\,, \text{ for } j \neq i\,, \\ p_{i,i} &= 1 - \sum_{j \neq i} p_{i,j}\,. \end{aligned}$$

The new chain is also regular. The vector $\boldsymbol{\pi} = (\pi_1, \ldots, \pi_M)$ satisfies the so-called **reversibility condition**,

$$\pi_j p_{j,i} = \min\{\pi_j, \pi_i\} q_{j,i} = \min\{\pi_i, \pi_j\} q_{i,j} = \pi_i p_{i,j}, \tag{12.6}$$

which implies that

$$[\boldsymbol{\pi P}]_i = \sum_{j=1}^{M} \pi_j p_{j,i} = \sum_{j=1}^{M} \pi_i p_{i,j} = \pi_i \sum_{j=1}^{M} p_{i,j} = \pi_i ,$$

and we deduce that the chain with transition matrix $\boldsymbol{P}$ has the equilibrium distribution $\boldsymbol{\pi}$, so this is the unique limiting distribution.

Intuitively, if $f_X(x_i)$ is one of the larger probabilities ($f_X(x_i) > 1/M$) then $\pi_j/\pi_i$ will be $< 1$ for most values of $j$, and $p_{i,j}$ will be smaller than $q_{i,j}$. Thus, $p_{i,i}$ will be larger than $q_{i,i}$, and the chain will remain in state $i$ more often. This has the effect of increasing the equilibrium probability $\pi_i$ which, as shown, becomes $f_X(x_i)$.

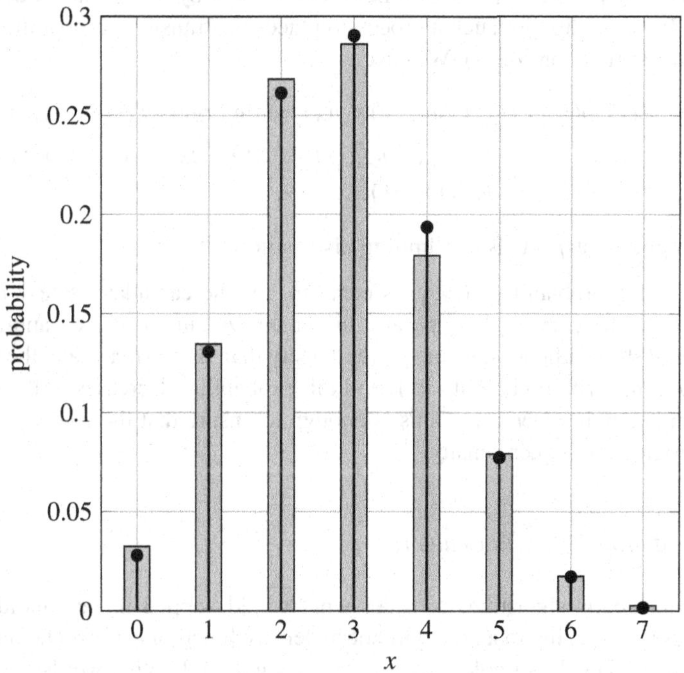


Figure 12.6 *A simple illustration of the discrete Metropolis algorithm. We want to sample from the* Bin(7, 0.4) *distribution using the chain with transition matrix* $\boldsymbol{Q} = \{q_{i,j}\}$, *where* $q_{i,j} = 1/7$ *for* $i, j \in \{0, 1, \ldots, 7\}$ *and* $i \neq j$ *(and 0 otherwise). Starting with* $X_1 = 0$, *we generate 5000 steps of the chain and discard the first 1000. The plot shows the histogram of the resulting values, along with the true distribution (black dots).*

### 12.3.2 Continuous Metropolis

The Metropolis algorithm can be adapted to the continuous case, where the distribution of interest has density function $f_X(x)$, known only up to a constant. We must first choose an (arbitrary) symmetric density function, that is, a PDF with the property $g(y|x) = g(x|y)$, which is known as the proposal density. In practice, $g(y|x)$ is usually a symmetric PDF with mean $x$. We now define a Markov chain as follows:

Step 1. Select an arbitrary initial state $X_1 = x_1$.

Step 2. Given the current state $X_t = x_t$, draw the **candidate state** $y$ from $g(y|x_t)$.

Step 3. Accept the candidate state with probability $q(x_t, y) = \min\{f_X(y)/f_X(x_t), 1\}$ and set the next state to $X_{t+1} = y$, or reject it with probability $1 - q(x_t, y)$ and set $X_{t+1} = x_t$.

The resulting process $\{X_t\}$ is a discrete-time continuous-valued Markov chain. Its transition function, $p(x, y) = q(x, y)g(y|x)$, corresponds to a distribution which is neither discrete nor continuous, because the chain will either remain in the same state with probability $> 0$, or jump to a new state chosen by a continuous density. The reversibility condition for such a process replaces the transition probability $p_{i,j}$ with the transition function $p(x, y)$. We have

$$\begin{aligned} f_X(x)p(x,y) &= f_X(x)q(x,y)g(y|x) = \min\{f_X(y), f_X(x)\}\, g(y|x) \\ &= \min\{f_X(x), f_X(y)\}\, g(x|y) = f_X(y)q(y,x)g(y|x) \\ &= f_X(y)p(y,x), \end{aligned} \tag{12.7}$$

which implies that $f_X(x)$ is the limiting distribution.

Intuitively, the probability $q(x_t, y)$ is equal to 1 if the candidate state $y$ has higher density than the current state $x_t$, and the chain will jump to the candidate state. However, if the candidate state has lower density than the current state, the chain will make the jump with probability $< 1$, and this probability decreases as the density of the candidate point decreases. This is enough to ensure that the chain spends more time in regions of higher density.

### 12.3.3 Metropolis-Hastings algorithm

In the Metropolis algorithm, consider what would happen if the candidate states were chosen according to a non-symmetric density function $g(y|x)$. The reversibility condition (12.7) will not hold because $g(y|x) \neq g(x|y)$. In other words, there will be values of $x, y$ for which

$$f_X(x)q(x,y)g(y|x) > f_X(y)q(y,x)g(x|y).$$

This implies that the chain will jump from state $x$ to state $y$ too often, and from state $y$ to state $x$ not often enough. One way we can restore balance is to modify the transition probability $q(x, y)$.

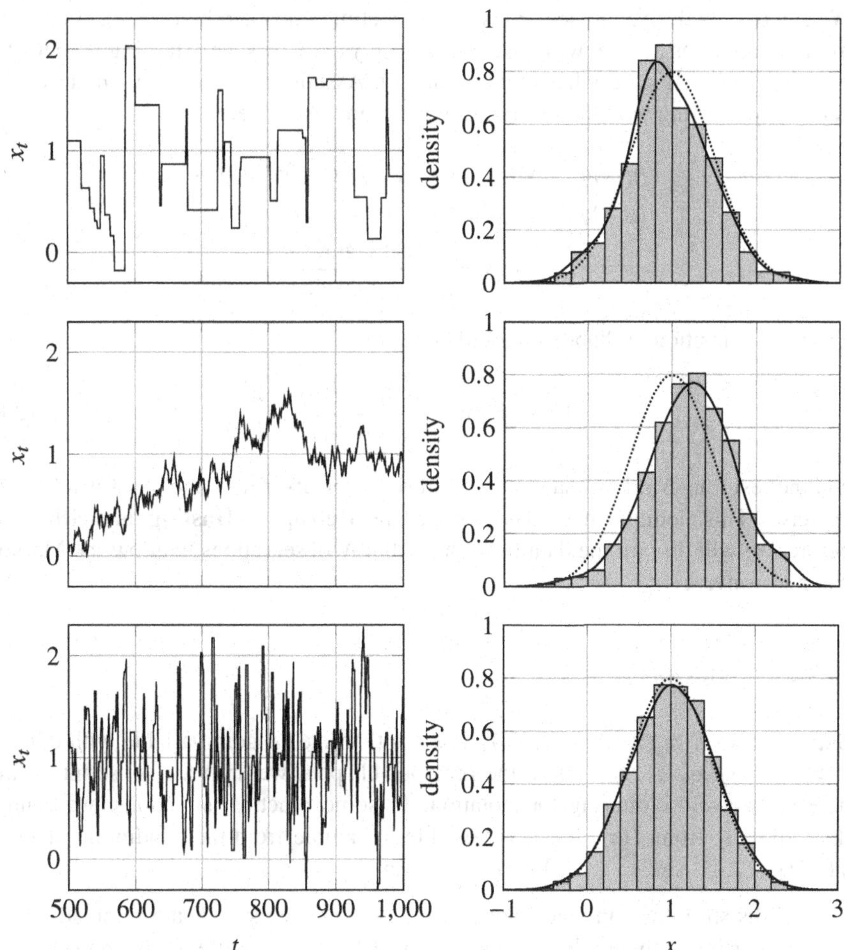


Figure 12.7 *Sampling from a* N(1, 0.25) *density with the continuous Metropolis algorithm, using various proposal densities. Top:* $g(y|x) = 0.05\exp(-10|y-x|)$*, a Laplace (double exponential) density shifted by x. Middle:* $(Y|X = x) \sim \text{Unif}[x - 0.1, x + 0.1]$*. Bottom:* $(Y|X = x) \sim \text{Logistic}(x, 1.5)$*. For each chain, we generate 10000 steps and discard the first 500. The plots on the left show the first 500 steps of each chain after the burn-in period. The plots on the right show the histogram of the resulting values and their smoothed density (dotted line), as well as the true density (solid line). Top left: the proposed steps are too big and the chain remains constant very often. Middle left: the steps are too small, so the chain does a poor job of exploring the space of the true density. Bottom left: the size of the steps is more reasonable, and the chain explores the space efficiently. It is a good idea to check the proportion of candidate states that are accepted; the proportions for these three chains are, respectively, 7%, 95%, and 56%.*

We can readily derive the new transition probability, denoted by $q'(x,y)$. As before, for any two states $x,y$ we will have either $q'(x,y) = 1$ (if state $y$ has higher density) or $q'(y,x) = 1$ (if state $x$ has higher density). Assume $q'(y,x) = 1$ without loss of generality. Starting with the reversibility condition, we have

$$\begin{aligned} f_X(x)q'(x,y)g(y|x) &= f_X(y)q'(y,x)g(x|y) \Rightarrow \\ f_X(x)q'(x,y)g(y|x) &= f_X(y)g(x|y) \Rightarrow \\ q'(x,y) &= \frac{f_X(y)g(x|y)}{f_X(x)g(y|x)}. \end{aligned}$$

Thus, the transition probability should be set to

$$q'(x,y) = \min\left\{\frac{f_X(y)g(x|y)}{f_X(x)g(y|x)}, 1\right\}, \tag{12.8}$$

and the resulting Markov chain has equilibrium density $f_X(x)$ (the proof is left as an exercise). This modified method is known as the **Metropolis-Hastings algorithm**. As before, we will discard the data from the initial $N$ observations to allow the Markov chain to converge.

### *12.3.4 Gibbs sampler*

Suppose that $\boldsymbol{X} = (X_1, X_2, \ldots, X_p)^T$ is a multivariate random variable with density $f(\boldsymbol{x}) = f(x_1, x_2, \ldots, x_p)$. As in the previous chapter, we drop the subscript on the density to avoid cluttering the notation. In some practical situations, we cannot draw directly from $f(\boldsymbol{x})$, but it is possible to sample from the conditional density $f(x_i|x_1, \ldots, x_{i-1}, x_{i+1}, \ldots, x_p)$ for $i = 1, \ldots, p$.

The **Gibbs sample** is an algorithm in which the elements of $\boldsymbol{X}$ are sampled one at a time. Starting with an arbitrary initial state $\boldsymbol{x}^{(0)}$, we repeat the following steps, for $t = 1, 2, \ldots,$

Step 1. Draw $x_1^{(t)}$ from $f(x_1|x_2^{(t-1)}, \ldots, x_p^{(t-1)})$

Step 2. Draw $x_2^{(t)}$ from $f(x_2|x_1^{(t)}, x_3^{(t-1)}, \ldots, x_p^{(t-1)})$

⋮

Step $i$. Draw $x_i^{(t)}$ from $f(x_i^{(t)}|x_1^{(t)}, \ldots, x_{i-1}^{(t)}, x_{i+1}^{(t-1)}, \ldots, x_p^{(t-1)})$

⋮

Step $p$. Draw $x_p^{(t)}$ from $f(x_p|x_1^{(t)}, \ldots, x_{p-1}^{(t)})$.

The Gibbs sampler can be viewed as a special case of the Metropolis-Hastings algorithm, as we will illustrate. For a vector $\boldsymbol{v}$, define $\boldsymbol{v}_{-i}$ to be the vector with its $i^{\text{th}}$ component removed. Using the notation of the Metropolis-Hastings algorithm, let the current state be $\boldsymbol{x} = (x_1^{(t)}, \ldots, x_{i-1}^{(t)}, x_i^{(t-1)}, x_{i+1}^{(t-1)}, \ldots, x_p^{(t-1)})$ and the next candidate state be $\boldsymbol{y} = (x_1^{(t)}, \ldots, x_{i-1}^{(t)}, x_i^{(t)}, x_{i+1}^{(t-1)}, \ldots, x_p^{(t-1)})$. The vectors $\boldsymbol{x}, \boldsymbol{y}$ differ only in their $i^{\text{th}}$ components, and we can write $\boldsymbol{x}_{-i} = \boldsymbol{y}_{-i}$. Consider the proposal density $g(\boldsymbol{y}|\boldsymbol{x})$ by which the candidate state is generated. By construction of the Gibbs sampler, only the $i^{\text{th}}$ component of $\boldsymbol{y}$ needs to be updated; we can write

$$g(\boldsymbol{y}|\boldsymbol{x}) = f(y_i|\boldsymbol{x}_{-i}) = f(x_i^{(t)}|\boldsymbol{x}_{-i}).$$

The only difference is that the Gibbs sampler always moves to the candidate state, whereas the Metropolis-Hastings algorithm does so with a probability that is possibly $< 1$. However, the transition probability for the Gibbs sample is

$$\begin{aligned}
p'_{GIBBS}(x,y) &= \min\left\{\frac{f(\boldsymbol{x})g(\boldsymbol{y}|\boldsymbol{x})}{f(\boldsymbol{y})g(\boldsymbol{x}|\boldsymbol{y})}, 1\right\} = \min\left\{\frac{f(x_i, \boldsymbol{x}_{-i})f(y_i|\boldsymbol{x}_{-i})}{f(y_i, \boldsymbol{y}_{-i})f(x_i|\boldsymbol{y}_{-i})}, 1\right\} \\
&= \min\left\{\frac{f(x_i|\boldsymbol{x}_{-i})f(\boldsymbol{x}_{-i})f(y_i|\boldsymbol{x}_{-i})}{f(y_i|\boldsymbol{y}_{-i})f(\boldsymbol{y}_{-i})f(x_i|\boldsymbol{y}_{-i})}, 1\right\} \\
&= \min\left\{\frac{f(x_i|\boldsymbol{x}_{-i})f(\boldsymbol{x}_{-i})f(y_i|\boldsymbol{x}_{-i})}{f(y_i|\boldsymbol{x}_{-i})f(\boldsymbol{x}_{-i})f(x_i|\boldsymbol{x}_{-i})}, 1\right\} \\
&= \min\{1, 1\} = 1,
\end{aligned}$$

which implies that the chain should always jump to state $\boldsymbol{y}$, regardless of the value of $x_i^{(t)}$. Thus, the Gibbs sampler is equivalent to the Metropolis-Hastings algorithm for a particular choice of proposal density $g(\boldsymbol{y}|\boldsymbol{x})$.

**Example 12.3.1** (Poisson change-point model)
In Example 5.3.1, we used the Poisson distribution to model the number of hurricanes that form in the Atlantic Basic in a given year. We suspect there was an increase in the Poisson parameter at some point in the past. If $\boldsymbol{Y} = (Y_1, \ldots, Y_n)^T$ represent consecutive yearly observations, and $m$ is the year after which the parameter changes (the change-point), we can write down the model

$$\begin{aligned}
Y_1, \ldots Y_m|\lambda_1, \lambda_2, m &\sim \text{Pois}(\lambda_1), \\
Y_{m+1}, \ldots Y_n|\lambda_1, \lambda_2, m &\sim \text{Pois}(\lambda_2),
\end{aligned}$$

where $m \in \{1, 2, \ldots, n-1\}$. A Jeffreys prior for $\lambda_1$ and $\lambda_2$ is

$$\pi(\lambda_i) \propto \frac{1}{\lambda_i} \quad \text{for } i = 1, 2.$$

We do not know the location of the change-point, so we give $m$ a discrete uniform prior. We assume that the parameters $\lambda_1$, $\lambda_2$, and $m$ are independent; the joint likelihood is

$$\begin{aligned}
f(\boldsymbol{y}, \lambda_1, \lambda_2, m) &= f(\boldsymbol{y}|\lambda_1, \lambda_2, m)\pi(\lambda_1)\pi(\lambda_2)\pi(m) \\
&= \frac{1}{\prod_{i=1}^m y_i!}\lambda_1^{\sum_{i=1}^m y_i} e^{-m\lambda_1} \frac{1}{\prod_{i=m+1}^n y_i!}\lambda_2^{\sum_{i=m+1}^n y_i} e^{-(n-m)\lambda_2} \frac{1}{\lambda_1}\frac{1}{\lambda_2}\frac{1}{n-1} \\
&= \frac{1}{\prod_{i=1}^n y_i!}\lambda_1^{\sum_{i=1}^m y_i - 1} e^{-m\lambda_1} \lambda_2^{\sum_{i=m+1}^n y_i - 1} e^{-(n-m)\lambda_2} \frac{1}{n-1}.
\end{aligned}$$

To construct a Gibbs sampler from the posterior joint density of the parameters, we must first work out the conditional distribution of each one. We have

$$\pi(\lambda_1|\boldsymbol{y},\lambda_2,m) \propto \lambda_1^{\sum_{i=1}^m y_i - 1} e^{-m\lambda_1} \Rightarrow (\lambda_1|\boldsymbol{y},\lambda_2,m) \sim \text{Gamma}\left(\sum_{i=1}^m y_i, m\right),$$

and, similarly,

$$\pi(\lambda_2|\boldsymbol{y},\lambda_1,m) \propto \lambda_2^{\sum_{i=m+1}^n y_i - 1} e^{-(n-m)\lambda_2} \Rightarrow (\lambda_2|\boldsymbol{y},\lambda_1,m) \sim \text{Gamma}\left(\sum_{i=m+1}^n y_i, n-m\right).$$

Finally, the conditional mass function of $m$ is

$$\pi(m|\boldsymbol{y},\lambda_1,\lambda_2) \propto e^{-m\lambda_1-(n-m)\lambda_2} \lambda_1^{\sum_{i=1}^m y_i - 1} \lambda_2^{\sum_{i=m+1}^n y_i - 1}.$$

This does not belong to any known mass function, but we can readily sample from it. For given values of the other parameters, we find the normalizing constant by simply computing $\pi(m|\boldsymbol{y},\lambda_1,\lambda_2)$ for $m = 1,2,\dots,n$.

We initialise the Gibbs sampler by drawing $m^{(0)}$ from the discrete uniform on $\{1,2,\dots,n-1\}$. The algorithm then cycles through the following steps, for $t = 1,2,\dots$,

Step 1. Draw $\lambda_1^{(t)}$ from Gamma $\left(\sum_{i=1}^{m^{(t)}} y_i, m^{(t)}\right)$.

Step 2. Draw $\lambda_2^{(t)}$ from Gamma $\left(\sum_{i=m^{(t)}+1}^{n} y_i, n - m^{(t)}\right)$.

Step 3. Draw $m^{(t)}$ from $\pi(m|\boldsymbol{y},\lambda_1^{(t)},\lambda_2^{(t)})$.

**Exercise 12.3**

1. Show that the Markov chain defined in the Metropolis-Hastings algorithm converges to the desired distribution. You may use the continuous-valued version of the reversibility condition.
2. Suppose that we want to use the discrete Metropolis algorithm to sample from a Polya$(r,p)$ distribution. The mass function is

$$f_X(x) \propto \frac{\Gamma(x+r)}{x!}(1-p)^x \quad \text{for } x = 0,1,2,\dots$$

We define a Markov chain on $\{0,1,2,\dots\}$ with transition matrix $\boldsymbol{Q}$, where

$$[\boldsymbol{Q}]_{i,j} = \begin{cases} 1/2 & \text{for } j = 0 \text{ or } j = i+1 \\ 1/2 & \text{otherwise.} \end{cases}$$

(a) Describe, in words, the Markov chain defined by $\boldsymbol{Q}$.

(b) Prove that this chain is ergodic. What is the mean recurrence time $\mu_0$?

(c) Work out the transition probabilities for a Markov chain whose equilibrium distribution is a Polya$(r,p)$.

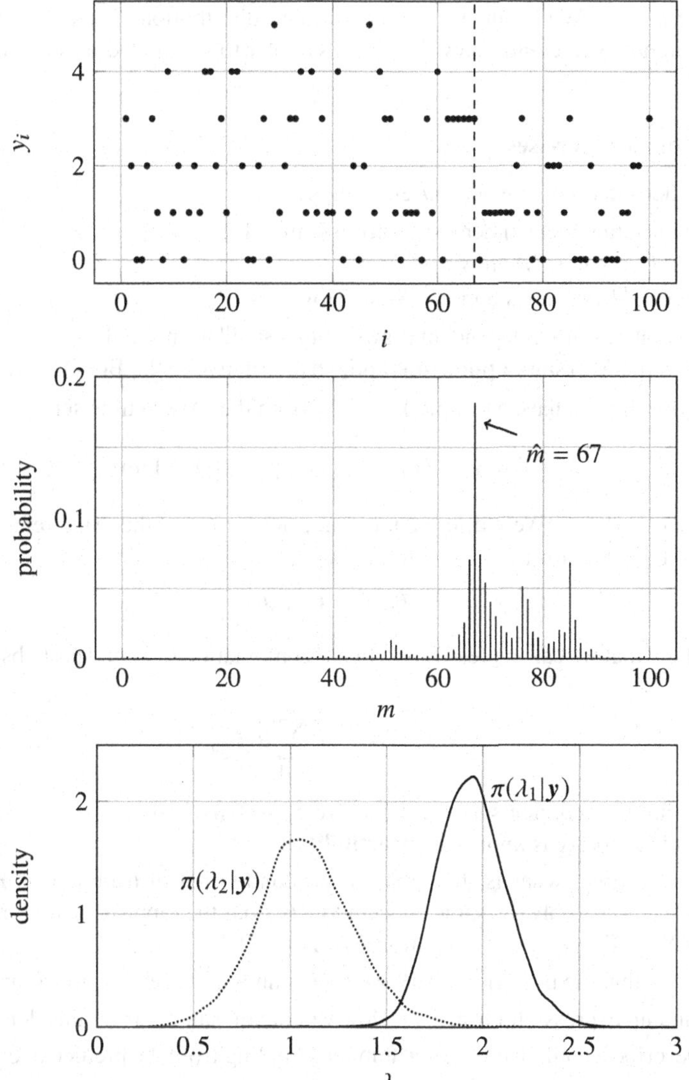


Figure 12.8 *Gibbs sampler for the Poisson change-point model. We generated* $n = 100$ *observations (top) using* $\lambda_1 = 2$, $\lambda_2 = 1$, *and* $m = 67$ *(dashed line). The middle plot shows the posterior mass function of* $m$. *The posterior density of* $\lambda_1$ *and* $\lambda_2$ *are in the bottom plot.*

3. Let $X, Y$ be random variables with joint mass/density function

$$f_{X,Y}(x, y) \propto \frac{n!}{x!(n-x)!} y^{x+a-1}(1-y)^{n-x+b-1}, \quad x = 0, 1, \ldots, n, \quad 0 \leq y \leq 1,$$

and $a, b > 0$. Work out the two conditional distributions ($f_{X|Y}$ and $f_{Y|X}$), and explain how you could apply the Gibbs sampler to simulate draws from $f_{X,Y}$.

## 12.4 Further exercises

1. Consider the following accept/reject algorithm:
   *Repeat*: draw independent $u_1, u_2$ from Unif$[-1, 1]$
   *Until*: $0 < s < 1$, where $s = u_1^2 + u_2^2$
   *Set*: $x = \frac{u_1}{s}\sqrt{-2\log s}$ and $y = \frac{u_2}{s}\sqrt{-2\log s}$
   Show that $x, y$ are independent draws from a standard normal.
   [This is the **Marsaglia polar method**, an alternative to the Box-Muller method.]
2. Suppose that we have a sample $Y_1, \dots, Y_n$ from the MA(1) time series model
$$Y_t = \varepsilon_t + \theta\varepsilon_{t-1}, \qquad \{\varepsilon_t\} \sim \text{NID}(0, 1/\omega),$$
   where $0 < \theta < 1$. We will use a uniform prior for $\theta$ and the usual reference prior for $\omega$ (the precision),
$$\pi(\theta, \omega) \propto \mathbf{1}_{(0,1)}(\theta)\,\frac{1}{\omega}\,.$$
   (a) Show that we can express $Y_t$ as a linear combination of $\varepsilon_t$ and past observations, that is,
$$Y_t = \varepsilon_t + \sum_{j=1}^{\infty} a_j Y_{t-j}\,,$$
   for some coefficients $a_1, a_2, \dots$, where $a_j \to 0$ as $j \to \infty$.
   [This property is known as **invertibility**.]
   (b) If $t$ is large, what is the approximate conditional distribution of $Y_{t+1}$ given $Y_1, \dots, Y_t$? Apply the same concept to work out the (approximate) joint density of $\boldsymbol{Y} = (Y_{t+1}, \dots, Y_n)^T$ conditional on $Y_1, \dots, Y_t$.
   (c) Show that the posterior $\pi(\omega|\theta, \boldsymbol{y})$ is a Gamma$(\alpha, \lambda)$ density, for some $\alpha, \lambda > 0$.
   (d) Find an expression for $\pi(\theta|\boldsymbol{y})$. How would you sample from this density?
   (e) Describe a composition algorithm for sampling from the predictive distribution of $Y_{n+1}$.
3. Let $\boldsymbol{X} = (X_1, \dots, X_n)^T$ be a random sample from a distribution with density function $f_X(x; \theta)$. Suppose that $\hat{\theta}$, our estimator of $\theta$, has bias
$$\eta = \mathbb{E}(\hat{\theta}) - \theta\,.$$
   In many practical cases, $\eta$ is a function of $\theta$ that is difficult to compute analytically.
   (a) We use the parametric bootstrap approach to generate new samples $\boldsymbol{X}_1^*, \dots, \boldsymbol{X}_m^*$, also of size $n$, with density $f_X(x; \hat{\theta})$. How could you estimate $\eta$ from the bootstrap samples?
   (b) Use the estimate of $\eta$ to construct a bias-corrected estimator of $\theta$.

(c) Explain why we might expect this procedure to work well if $\mathbb{E}(\hat{\theta}) - \theta$ is approximately pivotal for $\theta$.

4. For the Markov chains defined in section 12.3.1, why is the chain with transition matrix $\boldsymbol{P}$ regular? Prove that this chain is also regular in the case where the chain with transition matrix $\boldsymbol{Q}$ is regular, but $q_{i,j}$ is not necessarily positive for all $i, j$.

5. Suppose that we want to simulate from the N(0, 1) distribution using the continuous Metropolis algorithm. Our proposal density is uniform, $g(y|x) = \mathbf{1}_{(x-c,x+c)}(y)$, where $c$ is a tuning parameter.
   (a) Describe, in detail, the steps of the algorithm.
   (b) You implement this algorithm and find that the average acceptance probability, $q(x_t, y)$, decreases as $c$ increases, and increases as $c$ decreases. Explain why this is the case. Does this mean that we should choose a very small value of $c$?

6. The random variables $\boldsymbol{Y} = (Y_1, \dots, Y_n)^T$ are IID with density function

$$f_{Y_i}(y_i|\lambda, p) = (1-p)\mathbf{1}_{\{0\}}(y_i) + p\frac{e^{-\lambda}\lambda^{y_i}}{y_i!}, \quad i = 1, \dots, n,$$

where $\mathbf{1}_{\{0\}}(y_i)$ denotes an indicator function that takes the value 1 when $y_i = 0$, and 0 otherwise. This is a mixture model which can viewed in terms of the following generative process:

- With probability $(1-p)$ draw $Y_i$ from a distribution with all its mass at 0.
- With probability $p$ draw $Y_i$ from a Poisson distribution with parameter $\lambda$.

Suppose that the prior distribution for $p$ is Beta$(\alpha, \beta)$, while the prior for $\lambda$ is Gamma$(a, b)$. By introducing a latent random variable $Z_i \sim$ Bernoulli(p), construct a Gibbs sampler to obtain an approximation to the posterior distribution, $\pi(p, \lambda|\boldsymbol{y})$.

7. A credit card company classifies its customers into 4 categories with proportions

$$\left(\frac{\theta}{8}, \frac{1}{2} - \frac{\theta}{4}, \frac{1+\theta}{4}, \frac{1}{4} - \frac{\theta}{8}\right),$$

where $\theta \in (0, 1)$. We obtain a random sample of 100 customers from this population; the counts for each category are

$$(6, 43, 30, 21).$$

Describe how you would use the Metropolis algorithm to compute the posterior mean of $\theta$ given this sample. You may use a flat (uniform) prior for $\theta$.

8. Let $X_1, \dots, X_n$ be independent geometrically distributed random variables, with $X_i \sim$ Geometric$(p_1)$ for $1 \le i \le k$, and $X_j \sim$ Geometric$(p_2)$ for $k < j \le n$, where $1 \le k < n$. The parameters all have flat priors, so the full prior density/mass function is

$$\pi(p_1, p_2, k) = \pi(p_1)\pi(p_2)\pi(k) = \mathbf{1}_{(0,1)}(p_1)\ \mathbf{1}_{(0,1)}(p_2)\ \mathbf{1}_{\{1,2,\dots,n-1\}}(k)\ \frac{1}{n-1}.$$

Construct a Gibbs sampler for the posterior distribution of the change-point $k$.

## 12.5 Chapter summary

On completion of this chapter you should be able to:

- use a (uniform) random-number generator to draw random samples from common distributions,
- set up a simple accept/reject sampling scheme,
- construct a random sample from a model with multiple parameters using the composition algorithm,
- apply simulation techniques to determine the properties of a model,
- understand the difference between the parametric and nonparametric bootstrap,
- estimate a univariate or multivariate integral using the Monte Carlo approach, and compute the variance of your estimator,
- use an importance sampling scheme to draw samples from a given density,
- explain how antithetic variates can be used to achieve variance reduction, and
- describe the different versions of MCMC and prove that the chains converge to the desired distribution.

Appendix A

# Proof of Proposition 5.7.2

Before we attempt this proof, consider the following matrix algebra result.

**Lemma A.1** (Matrix inversion lemma)

Suppose that the matrix $\boldsymbol{A}$ can be written in block matrix form as

$$\boldsymbol{A} = \begin{bmatrix} \boldsymbol{A}_{11} & \boldsymbol{A}_{12} \\ \boldsymbol{A}_{21} & \boldsymbol{A}_{22} \end{bmatrix}.$$

If the submatrices $\boldsymbol{A}_{11}$ and $\boldsymbol{A}_{22}$ are non-singular, the inverse of $\boldsymbol{A}$ is given by

$$\boldsymbol{A}^{-1} = \boldsymbol{B} = \begin{bmatrix} \left(\boldsymbol{A}_{11} - \boldsymbol{A}_{12}\boldsymbol{A}_{22}^{-1}\boldsymbol{A}_{21}\right)^{-1} & -\left(\boldsymbol{A}_{11} - \boldsymbol{A}_{12}\boldsymbol{A}_{22}^{-1}\boldsymbol{A}_{21}\right)^{-1}\boldsymbol{A}_{12}\boldsymbol{A}_{22}^{-1} \\ -\left(\boldsymbol{A}_{22} - \boldsymbol{A}_{21}\boldsymbol{A}_{11}^{-1}\boldsymbol{A}_{12}\right)^{-1}\boldsymbol{A}_{21}\boldsymbol{A}_{11}^{-1} & \left(\boldsymbol{A}_{22} - \boldsymbol{A}_{21}\boldsymbol{A}_{11}^{-1}\boldsymbol{A}_{12}\right)^{-1} \end{bmatrix}$$

*Proof.*

We can easily prove this result by writing out the product of $\boldsymbol{A}$ and $\boldsymbol{B}$. For example, the top-left block of $\boldsymbol{BA}$ is

$$\begin{aligned}
&\left[\left(\boldsymbol{A}_{11} - \boldsymbol{A}_{12}\boldsymbol{A}_{22}^{-1}\boldsymbol{A}_{21}\right)^{-1} \quad -\left(\boldsymbol{A}_{11} - \boldsymbol{A}_{12}\boldsymbol{A}_{22}^{-1}\boldsymbol{A}_{21}\right)^{-1}\boldsymbol{A}_{12}\boldsymbol{A}_{22}^{-1}\right]\begin{bmatrix} \boldsymbol{A}_{11} \\ \boldsymbol{A}_{21} \end{bmatrix} \\
&= \left(\boldsymbol{A}_{11} - \boldsymbol{A}_{12}\boldsymbol{A}_{22}^{-1}\boldsymbol{A}_{21}\right)^{-1}\boldsymbol{A}_{11} - \left(\boldsymbol{A}_{11} - \boldsymbol{A}_{12}\boldsymbol{A}_{22}^{-1}\boldsymbol{A}_{21}\right)^{-1}\boldsymbol{A}_{12}\boldsymbol{A}_{22}^{-1}\boldsymbol{A}_{21} \\
&= \left(\boldsymbol{A}_{11} - \boldsymbol{A}_{12}\boldsymbol{A}_{22}^{-1}\boldsymbol{A}_{21}\right)^{-1}\left(\boldsymbol{A}_{11} - \boldsymbol{A}_{12}\boldsymbol{A}_{22}^{-1}\boldsymbol{A}_{21}\right) = \boldsymbol{I}.
\end{aligned}$$

□

**Proposition 5.7.2** (Conditional distribution for multivariate normal)

Suppose that the random vectors $\boldsymbol{X} = (X_1, \dots, X_n)^T$ and $\boldsymbol{Y} = (Y_1, \dots, Y_m)^T$ (for some integers $n$ and $m$) are jointly normal, with $\boldsymbol{X} \sim \text{N}(\boldsymbol{\mu}_X, \boldsymbol{\Sigma}_X)$ and $\boldsymbol{Y} \sim \text{N}(\boldsymbol{\mu}_Y, \boldsymbol{\Sigma}_Y)$. If, in addition, $\text{Cov}(\boldsymbol{X}, \boldsymbol{Y}) = \boldsymbol{\Sigma}_{XY} = \boldsymbol{\Sigma}_{YX}^T$, then

$$\begin{aligned}
\mathbb{E}(\boldsymbol{Y}|\boldsymbol{X}) &= \boldsymbol{\mu}_Y + \boldsymbol{\Sigma}_{YX}\boldsymbol{\Sigma}_X^{-1}(\boldsymbol{X} - \boldsymbol{\mu}_X), \\
\text{Var}(\boldsymbol{Y}|\boldsymbol{X}) &= \boldsymbol{\Sigma}_Y - \boldsymbol{\Sigma}_{YX}\boldsymbol{\Sigma}_X^{-1}\boldsymbol{\Sigma}_{XY},
\end{aligned}$$

and the conditional distribution of $\boldsymbol{Y}$ given $\boldsymbol{X} = \boldsymbol{x}$ is multivariate normal.

*Proof.*

Assume for now that $\boldsymbol{\mu}_X = \boldsymbol{\mu}_Y = \boldsymbol{0}$, so the joint distribution of the random vectors is

$$\begin{bmatrix} \boldsymbol{X} \\ \boldsymbol{Y} \end{bmatrix} \sim \mathrm{N}(\boldsymbol{0}, \boldsymbol{\Sigma}), \text{ where } \boldsymbol{\Sigma} = \begin{bmatrix} \boldsymbol{\Sigma}_X & \boldsymbol{\Sigma}_{XY} \\ \boldsymbol{\Sigma}_{YX} & \boldsymbol{\Sigma}_Y \end{bmatrix}.$$

The conditional density of $\boldsymbol{Y}$ given $\boldsymbol{X}$ is

$$f_{\boldsymbol{Y}|\boldsymbol{X}}(\boldsymbol{y}|\boldsymbol{x}) = \frac{f_{\boldsymbol{X},\boldsymbol{Y}}(\boldsymbol{x},\boldsymbol{y})}{f_{\boldsymbol{X}}(\boldsymbol{x})} = \frac{(2\pi)^{-(n+m)/2}|\boldsymbol{\Sigma}|^{-1/2} \exp\left\{-\frac{1}{2}[\boldsymbol{x}^T \ \boldsymbol{y}^T]\boldsymbol{\Sigma}^{-1}\begin{bmatrix}\boldsymbol{x}\\ \boldsymbol{y}\end{bmatrix}\right\}}{(2\pi)^{-n/2}|\boldsymbol{\Sigma}_X|^{-1/2}\exp\left\{-\frac{1}{2}\boldsymbol{x}^T\boldsymbol{\Sigma}_X^{-1}\boldsymbol{x}\right\}}$$
$$\propto \exp\left\{-\frac{1}{2}\left([\boldsymbol{x}^T \ \boldsymbol{y}^T]\boldsymbol{\Sigma}^{-1}\begin{bmatrix}\boldsymbol{x}\\ \boldsymbol{y}\end{bmatrix} - \boldsymbol{x}^T\boldsymbol{\Sigma}_X^{-1}\boldsymbol{x}\right)\right\} = \exp\left\{-\frac{1}{2}\boldsymbol{U}(\boldsymbol{x},\boldsymbol{y})\right\}$$

where $\boldsymbol{U}(\boldsymbol{x},\boldsymbol{y})$ is the matrix we need to evaluate. Applying the matrix inversion lemma to $\boldsymbol{\Sigma}$, we have

$$\boldsymbol{\Sigma}^{-1} = \begin{bmatrix} \left(\boldsymbol{\Sigma}_X - \boldsymbol{\Sigma}_{XY}\boldsymbol{\Sigma}_Y^{-1}\boldsymbol{\Sigma}_{YX}\right)^{-1} & -\left(\boldsymbol{\Sigma}_X - \boldsymbol{\Sigma}_{XY}\boldsymbol{\Sigma}_Y^{-1}\boldsymbol{\Sigma}_{YX}\right)^{-1}\boldsymbol{\Sigma}_{XY}\boldsymbol{\Sigma}_Y^{-1} \\ -\left(\boldsymbol{\Sigma}_Y - \boldsymbol{\Sigma}_{YX}\boldsymbol{\Sigma}_X^{-1}\boldsymbol{\Sigma}_{XY}\right)^{-1}\boldsymbol{\Sigma}_{YX}\boldsymbol{\Sigma}_X^{-1} & \left(\boldsymbol{\Sigma}_Y - \boldsymbol{\Sigma}_{YX}\boldsymbol{\Sigma}_X^{-1}\boldsymbol{\Sigma}_{XY}\right)^{-1} \end{bmatrix}$$

As the inverse of a symmetric matrix, $\boldsymbol{\Sigma}^{-1}$ must also be symmetric, which implies that the top-right and bottom-left blocks are transposes of each other. We define the matrices $\boldsymbol{\Sigma}_{Y|X} = \boldsymbol{\Sigma}_Y - \boldsymbol{\Sigma}_{YX}\boldsymbol{\Sigma}_X^{-1}\boldsymbol{\Sigma}_{XY}$ and $\boldsymbol{\Sigma}_{X|Y} = \boldsymbol{\Sigma}_X - \boldsymbol{\Sigma}_{XY}\boldsymbol{\Sigma}_Y^{-1}\boldsymbol{\Sigma}_{YX}$ (both of them symmetric), so we can write

$$\boldsymbol{\Sigma}^{-1} = \begin{bmatrix} \boldsymbol{\Sigma}_{X|Y}^{-1} & \left(-\boldsymbol{\Sigma}_{Y|X}^{-1}\boldsymbol{\Sigma}_{YX}\boldsymbol{\Sigma}_X^{-1}\right)^T \\ -\boldsymbol{\Sigma}_{Y|X}^{-1}\boldsymbol{\Sigma}_{YX}\boldsymbol{\Sigma}_X^{-1} & \boldsymbol{\Sigma}_{Y|X}^{-1} \end{bmatrix} = \begin{bmatrix} \boldsymbol{\Sigma}_{X|Y}^{-1} & -\boldsymbol{\Sigma}_X^{-1}\boldsymbol{\Sigma}_{XY}\boldsymbol{\Sigma}_{Y|X}^{-1} \\ -\boldsymbol{\Sigma}_{Y|X}^{-1}\boldsymbol{\Sigma}_{YX}\boldsymbol{\Sigma}_X^{-1} & \boldsymbol{\Sigma}_{Y|X}^{-1} \end{bmatrix}$$

We now evaluate the matrix expression in the joint density,

$$\begin{aligned}
\boldsymbol{U}(\boldsymbol{x},\boldsymbol{y}) &= [\boldsymbol{x}^T \ \boldsymbol{y}^T]\begin{bmatrix} \boldsymbol{\Sigma}_{X|Y}^{-1} & -\boldsymbol{\Sigma}_X^{-1}\boldsymbol{\Sigma}_{XY}\boldsymbol{\Sigma}_{Y|X}^{-1} \\ -\boldsymbol{\Sigma}_{Y|X}^{-1}\boldsymbol{\Sigma}_{YX}\boldsymbol{\Sigma}_X^{-1} & \boldsymbol{\Sigma}_{Y|X}^{-1} \end{bmatrix}\begin{bmatrix}\boldsymbol{x}\\ \boldsymbol{y}\end{bmatrix} - \boldsymbol{x}^T\boldsymbol{\Sigma}_X^{-1}\boldsymbol{x} \\
&= [\boldsymbol{x}^T \ \boldsymbol{y}^T]\begin{bmatrix} \boldsymbol{\Sigma}_{X|Y}^{-1}\boldsymbol{x} - \boldsymbol{\Sigma}_X^{-1}\boldsymbol{\Sigma}_{XY}\boldsymbol{\Sigma}_{Y|X}^{-1}\boldsymbol{y} \\ -\boldsymbol{\Sigma}_{Y|X}^{-1}\boldsymbol{\Sigma}_{YX}\boldsymbol{\Sigma}_X^{-1}\boldsymbol{x} + \boldsymbol{\Sigma}_{Y|X}^{-1}\boldsymbol{y} \end{bmatrix} - \boldsymbol{x}^T\boldsymbol{\Sigma}_X^{-1}\boldsymbol{x} \\
&= \boldsymbol{x}^T\boldsymbol{\Sigma}_{X|Y}^{-1}\boldsymbol{x} - \boldsymbol{x}^T\boldsymbol{\Sigma}_X^{-1}\boldsymbol{\Sigma}_{XY}\boldsymbol{\Sigma}_{Y|X}^{-1}\boldsymbol{y} - \boldsymbol{y}^T\boldsymbol{\Sigma}_{Y|X}^{-1}\boldsymbol{\Sigma}_{YX}\boldsymbol{\Sigma}_X^{-1}\boldsymbol{x} + \boldsymbol{y}^T\boldsymbol{\Sigma}_{Y|X}^{-1}\boldsymbol{y} - \boldsymbol{x}^T\boldsymbol{\Sigma}_X^{-1}\boldsymbol{x} \\
&= (\boldsymbol{y} - \boldsymbol{\Sigma}_{YX}\boldsymbol{\Sigma}_X^{-1}\boldsymbol{x})^T\boldsymbol{\Sigma}_{Y|X}^{-1}(\boldsymbol{y} - \boldsymbol{\Sigma}_{YX}\boldsymbol{\Sigma}_X^{-1}\boldsymbol{x}) + \boldsymbol{C}(\boldsymbol{x}),
\end{aligned}$$

where $\boldsymbol{C}(\boldsymbol{x})$ includes only terms in $\boldsymbol{x}$. Substituting this expression into the conditional density (ignoring the term $\boldsymbol{C}(\boldsymbol{x})$, which is another multiplicative constant) gives

$$f_{\boldsymbol{Y}|\boldsymbol{X}}(\boldsymbol{y}|\boldsymbol{x}) \propto \exp\left\{-\frac{1}{2}\left[(\boldsymbol{y} - \boldsymbol{\Sigma}_{YX}\boldsymbol{\Sigma}_X^{-1}\boldsymbol{x})^T\boldsymbol{\Sigma}_{Y|X}^{-1}(\boldsymbol{y} - \boldsymbol{\Sigma}_{YX}\boldsymbol{\Sigma}_X^{-1}\boldsymbol{x})\right]\right\}.$$

We identify this as a multivariate normal density with mean $\boldsymbol{\mu}_{Y|X} = \boldsymbol{\Sigma}_{YX}\boldsymbol{\Sigma}_X^{-1}\boldsymbol{x}$ and variance $\boldsymbol{\Sigma}_{Y|X}$.

Finally, we make the transformation $\boldsymbol{X}^* = \boldsymbol{X} + \boldsymbol{\mu}_X$ and $\boldsymbol{Y}^* = \boldsymbol{Y} + \boldsymbol{\mu}_Y$. The random vectors $\boldsymbol{X}^*$ and $\boldsymbol{Y}^*$ are jointly normal with the desired mean and variance, and they too have covariance $\boldsymbol{\Sigma}_{XY}$. The Jacobian of the transformation is the identity matrix, so the resulting conditional density is

$$f_{\boldsymbol{Y}^*|\boldsymbol{X}^*}(\boldsymbol{y}^*|\boldsymbol{x}^*) \propto \exp\left\{-\frac{1}{2}\left[(\boldsymbol{y}^* - \boldsymbol{\mu}^*_{Y|X})^T\boldsymbol{\Sigma}^{-1}_{Y|X}(\boldsymbol{y}^* - \boldsymbol{\mu}^*_{Y|X})\right]\right\}$$

where $\boldsymbol{\mu}^*_{Y|X} = \boldsymbol{\mu}_Y + \boldsymbol{\Sigma}_{YX}\boldsymbol{\Sigma}_X^{-1}(\boldsymbol{x}^* - \boldsymbol{\mu}_X)$. This completes the proof. □

# Index


**A**

Absolute loss, 364
Absorption, Markov chains, 211–213
Accelerated-life models, 187
Accept/reject method, 384–387
Alternative hypothesis, 280
Analysis of variance (ANOVA), 177, 217
Antithetic variates, 399–401
Aperiodic Markov chains, 213, 214
Arithmetic mean, 61
Arrival and interarrival times, 203–204
Autocorrelation, 192–197
Autocorrelation function, 194
Autocovariance, 193
Autocovariance function, 194
Autoregressive model, 191
Averaging over simulated instances, 392–393

**B**

Bayes estimator, 363
Bayesian confidence interval, 366
Bayesian estimation, 363–370
- interval, 366–367
- point, 363–366

Bayesian inference, 355–378
- choosing a prior, 357–362
- empirical Bayes, 372–373
- hierarchical models, 370–371
- predictive inference, 373–374
- prior and posterior distributions, 355–357

Bayes' theorem, 25–27
Bernoulli distribution, 45
Bernoulli response, 179–180
Bernoulli trials, 82
Best linear unbiased estimator (BLUE), 266
Beta distribution, 247
Beta function, 247
Bias, 257–258
Biasing density, 396
Bias-variance tradeoff, 258
Binomial coefficient, 21–22
Binomial distribution, 44–45
Bivariate transformations, 130–133
Boole's inequality, 14–15
Bootstrap/bootstrapping, 389–392
- nonparametric, 390
- parametric, 390

Borel-Cantelli lemma
- first, 91–92
- second, 92

Borel $\sigma$-algebra, 93–96
- characterisation of a valid density, 95–96
- identity function, 94–95
- measurable function, 93
- random variable, 94–96
- Borel set, 93

Box-Muller method, 384

**C**

Canonical link function, 179
Categorical distribution, 218
Cauchy distribution, 97
Censoring of time-to-event data, 186
Central limit theorem, 225–227
Central sample moments, 228–229
Central tendency, 61
Certain event, 13
Change-of-variables formula, 131
Chebyshev inequality, 67–68

Chi-squared distribution, 236–240
Cholesky decomposition, 142
Combinations, *see* Permutations and combinations
Complement, 8, 13
Composite hypothesis, 279–280
Composition sampling, 387–389
Compound Poisson process, 205
Conditional central moment, 158–159
Conditional cumulative distribution function, 149
Conditional density, 150–151
Conditional distributions, 147–169
  conditional expectation and conditional moments, 154–162
  conditioning for random vectors, 166–167
  continuous, 150–152
  discrete, 147–149
  hierarchies and mixtures, 162–163
  random sums, 164–166
  relationship between joint, marginal, and conditional, 153–154
Conditional expectation, 154–162; *see also* Conditional moments
Conditional mass function, 148
Conditional moment-generating functions, 161–162
Conditional moments, 158–162
Conditional probability, 23–25
  conditioning notation, 25
  definition, 23–24
  multiplication law, 25
  properties, 24–25
Conditional variance, 159–160
Conditioning for random vectors, 166–167
Confidence coefficient, 269–271
Confidence interval, 269, 271
Confidence level, 269
Confidence set, 267
Conjugate priors, 360–362
Constrained model, 318
Continuous conditional distributions, 150–152
Continuous Metropolis, 404
Continuous random variables, transformations of, 130–134
Continuous-time Markov chains, 216–217
Continuous uniform distribution, 53
Control variates method, 402
Convergence, 88–93
  almost surely, 89
  in distribution, 88–89, 90
  in mean square, 89–90
  in probability, 89
  of real sequence, 88
Convex functions, 68
Convolution, 136
Correlation, 117, 118–120; *see also* Covariance and correlation
  in terms of cumulants, 122–123
Counting processes, 197–199
Covariance and correlation, 117–120
Covariance matrix, 128, 129
Covariates in time-to-event models, 187–188
  accelerated-life models, 187
  proportional-hazard models, 187
Coverage probability, 269–271
Cramér-Rao lower bound, 341–344
for a biased estimator, 352
for a vector of parameters, 344
Credible interval, *see* Bayesian confidence interval
Cumulant-generating functions and cumulants, 79–81
Cumulative distribution functions, 37–40
  probabilities from, 39–40
  properties, 38–39
  right continuity, 38
Cumulative hazard function, 183–184

**D**

Data sets, 4
Data-translated form, 358

Decision rules, hypothesis testing, 280–281
Density functions, continuous random variables and, 51–60
arguments of functions, 60
cumulative distribution function, 51
exponential family, 58–60
probability, 52–53
properties, 51
simple polynomial density, 55–56
Deterministic model, 381, 382
Discrete and continuous random variables, 41–61
appropriateness of, 41–42
density functions, 51–60
mass functions, 43–50
parameters and families of distributions, 60
support, 42
Discrete conditional distributions, 147–149
Discrete Metropolis, 402–403
Discrete uniform distribution, 50
Distributional assumptions, 172
Distribution family, 60
Distributions, *see* Probability distributions
Double exponential distribution, 98

**E**

EM algorithm, 310–315
Empirical Bayes, 370, 372–373
Empty set, 12–13
Ergodic chain, 216
Error sum of squares, 263
Event, 8
Expectation, 116–120
conditional, 154–162
Expectation operator, 63–64
Expected length, 269–271
Expected value, 62
Experiment, 7
Exponential dispersion family, 178–179
Exponential distribution, 53–55
Exponential family, 58–60
Extreme-value distribution, 248

**F**

*F*-distribution, 290
*F*-statistic, 290-291
First Borel-Cantelli lemma, 91–92
First-passage time, 215, 216
Fisher information, 297–299
Flat priors, 358

**G**

Gambler's ruin, 220
Gamma distribution, 57–58
Gauss-Markov theorem, 266
Generalised linear models, 177–181
link function, 178–181
motivation, 177–178
Geometric distribution, 45–46
Gibbs inequality, 311–312
Gibbs sampler, 406–410
Gompertz distribution, 219
Gumbel distribution, 248

**H**

Hat matrix, 265
Hazard function, 182–185
continuous case, 183–184
cumulative, 183–184
discrete case, 182–183
Hierarchical model, 162–163
Bayesian, 370-371
Higher-order sample moments, 228–236
joint sample moments, 233–236
sample moments and central sample moments, 228–229
sample variance, 229–233
Homogeneity, Markov chains, 207–208
Hyperparameters, 360
Hyperprior, 370
Hyperrectangle, 394
Hypothesis testing, 3, 278–285
composite hypothesis, 279–280
constructing, basic ideas in, 282–283
decision rules, 280–281

error types, 281–282
power function, 282
$p$-value, 283–285
significance level and size, 281–282
simple hypothesis, 279–280
statistical hypotheses, 279–280

**I**

Idempotent, 265
Identical distributions, 126–127
Importance sampling, 396–399
Impossible event, 13
Inclusion-exclusion principle, 14
Increments, 198
independent, 198
stationary, 199
Independence, 27–30
definition, 27–28
examples of, 28–29
mutual, 28
pairwise, 30
properties, 28
Independent and identically distributed (IID), 172, 173–174
error sequences, 173–174
random sample, 173
Independent increments, 198
Independent random variables, 124–127
Indicator function, 10–11
Induced likelihood, 307
Inequalities involving expectation, 67–70
Inferential theory, 327–352
most powerful tests, 345–350
sufficiency, 327–341
variance of unbiased estimators, 341–345
Information, 296–302; *see also* Score
Fisher, 297–299
observed, 299
Information matrix, 300
Interarrival time, *see* Arrival and interarrival times
Interquartile range, 240
Interval estimates, 366–367
Interval estimation, 3, 267–278
Bayesian, *see* Bayesian estimation, interval
confidence sets, 278
coverage probability and length, 269–271
order statistics, 275–277
pivotal functions, 272–275
Interval estimators, 267–268
Intuitive probability, 7–9
Inverse image, 82
Invertibility, 410

**J**

Jacobian, 131
Jeffreys prior, 359
Jeffreys' rule, 359
Jensen's inequality, 68–70
Joint central moment, 120
Joint central sample moment, 233
Joint cumulant-generating function, 122
Joint cumulative distribution function, 101–104
bivariate, 102
properties, 102
Joint mass and joint density, 104–15
Joint moment-generating function, 121–122
Joint moments, 120–123
Joint sample moments, 233–236
sample correlation, 234–235
sample covariance, 233–234

**K**

Kullback-Leibler divergence, 311–312
for discrete distributions, 324
Kurtosis
coefficient of, 71, 72
excess, 72

**L**

Laplace distribution, 98
Law (of a random variable), 94
Law of iterated expectations, 155–156
Law of large numbers, *see* Strong law of

large numbers / Weak law of large numbers
Law of total probability, 25–27
Likelihood-based inference, 293–325
likelihood and log-likelihood function, 293–296
likelihood-ratio test, 316–323
maximum-likelihood estimation (MLE), 302–315
score and information, 296–302
Likelihood-ratio test, 316–323
nuisance parameters, 317–318
properties, 318–320
score and Wald tests, 320–323
Limit inferior, 92
Limiting distribution, Markov chains, 214–215
Limit superior, 91
Linear models, 174–181
applications, 176–177
generalised, 177–181
multiple linear regression, 175–176
simple linear regression, 174–175
Linear predictor, 178
Linear regression
multiple, 175–176
simple, 174–175
Link function, 178
Location-scale families, 87
Logistic distribution, 218, 315
Logistic regression, 179–180
Logit, 179
Log-likelihood function, 293–296
Log-likelihood ratio, 310–311
Log-logistic distribution, 219
Lognormal distribution, 88, 378
Lomax distribution, 377
Loss function, 363

**M**

Maclaurin series, 75
Marginal distribution from joint distribution, 103–104
Markov chain Monte Carlo, 402–410
continuous Metropolis, 404
discrete Metropolis, 402–403
Gibbs sampler, 406–410
Metropolis-Hastings algorithm, 404–406
Markov chains, 207–217
absorption, 211–213
classification of states and chains, 210–211
continuous-time, 216–217
fundamental theorem of, 215–216
homogeneity, 207–208
limiting distribution, 214–215
periodicity, 213–214
recurrence and transience, 215–216
Markov inequality, 67
Markov property, 207
Marsaglia polar method, 410
Mass functions, discrete random variables and, 43–50
distribution function and, 43–44
properties, 43
Mathematical probability, 9–16; *see also* Measure
Maximum-likelihood estimate (MLE), 302–303
Maximum-likelihood estimation, 302–315
EM algorithm, 310–315
numerical maximisation of likelihood, 308–310
properties of, 304–308
Maximum-likelihood estimators (MLE), 302–304
asymptotic normality of, 305–306
consistency of, 305
large-sample distribution of, 306–307
Mean of random variables, 61–63
Mean recurrence time, 215, 216
Mean squared error, 258
Mean-value function, 206
Measurable space, 10
Measure, 9–11
indicator function as, 10–11
probability, 11–15

Measure space, 10
Median, 61
Memoryless property, 184–185
Method-of-moments estimators, 261–263
Method of scoring, 309
Methods for counting outcomes, 16–23; *see also* Permutations and combinations
Metropolis-Hastings algorithm, 404–406
Minimum-mean-square estimator (MMSE), 287–289, 364
Minimum-mean-square linear estimator (MMSLE), 288–289
Minimum-variance unbiased estimator (MVUE), 341
Mixture distribution, 162–163
estimation for, 313-314
Modal interval, 365
Mode, 61
Moment-generating function, 74–78
  derivatives at zero, 76
  as a polynomial, 75
  of a sum, 136-137
  uniqueness of, 77
Moments, 70–74
Monotone functions of random variables, 84–88
Monte Carlo integration, 381, 392–402
  antithetic variates, 399–402
  averaging over simulated instances, 392–393
  importance sampling, 396–399
  univariate *vs.* multivariate integrals, 393–396
Most powerful tests, 345–350
Moving-average models, 191–192
Multinomial coefficients, 20–23
Multinomial distribution, 145
Multiple linear regression, 175–176
Multistage models, *see* Hierarchical models
Multivariate distributions, 101–145
  expectation and joint moments, 116–123
  independent random variables, 124–127
  joint and marginal distributions, 101–104
  joint mass and joint density, 104–115
  random vectors and random matrices, 127–134
  sums of random variables, 134–138
Multivariate normal distribution, 138–142
Multivariate transformations, 133–134
Multivariate *vs.* univariate integrals, 393–396
Mutual independence, 28, 125–126
Mutually exclusive events, 26

**N**

Natural exponential family, 59
Negative binomial distribution, 46–48
Newton-Raphson scheme, 309–310
Neyman-Pearson lemma, 346–348
Non-homogeneous Poisson processes, 205–206
Nonparametric bootstrap, 390
Nonparametric empirical Bayes, 372
Nonparametric test, 279
Normal distribution, 56–57
  bivariate, 139–140
  multivariate, 138–143
Nuisance parameters, 317–318
Null hypothesis, 278, 280
Null recurrent, 216

**O**

Observed information, 299
Observed sample, 171–172
Order statistics, 240–247
  distribution of, 243–246
  interval estimation, 275–277
  sample quantiles and, 241
Ordinary least squares (OLS), 263–267

**P**

Pairwise independent, 125

Parameters, 172
  of distribution, 44, 60
Parametric bootstrap, 390
Parametric empirical Bayes, 372
Parametric test, 279
Pareto distribution, 97, 248, 296, 378
Parsimony, 318
Partition, 25–26
Pascal's triangle, 21
Periodicity, Markov chains, 213–214
Permutations and combinations, 17–23
  difference between, 18–19
  multinomial coefficients, 20–23
  number of, 18
Pivotal functions, 252–255
  interval estimation, 272–275
Plug-in estimator, 306
Point estimation, 3, 255–267
  bias, 257–258
  bias-variance tradeoff, 258
  consistency, 258–261
  mean squared error, 258
  method of moments, 261–263
  ordinary least squares (OLS), 263–267
Point estimators, 363–364
Poisson distribution, 48–50
  cumulant-generating function, 80
Poisson processes, 197–206
  arrival and interarrival times, 203–204
  compound, 205
  definitions of, 199–202
  elementary properties of, 200
  from first principles, 200–202
  non-homogeneous, 205–206
  stochastic processes and counting processes, 197–199
  thinning and superposition, 203
Poisson regression, 180
Pooling, 371
Population, 3–4
Positive recurrent, 216
Posterior distribution, 355–357, 363
Posterior predictive distribution, 373
Power function, 282
Prediction, 285–289
Prior distribution, 355–357
Prior predictive distribution, 373
Priors, 357–362
  conjugate, 360–362
  improper, 357
  non-informative, 357
  reference, 357–360
Probability, 7–31
  Bayes' theorem, 25–27
  conditional, 23–25
  independence, 27–30
  intuitive, 7–9
  law of total probability, 25–27
  mathematical, 9–16
  methods for counting outcomes, 16–23
Probability density function, *see* Density functions
Probability distributions
Bernoulli, 45
Beta, 247
binomial, 44–45
categorical, 218
Cauchy, 97
chi-squared, 236–237
continuous uniform, 53
discrete uniform, 50
double exponential, 98
exponential, 53–55
extreme value, 248
*F*, 290
Gamma, 57–58
geometric, 45–46
Gompertz, 219
Gumbel, 248
Laplace, 98
logistic, 218, 315
log-logistic, 219
lognormal, 88, 378
Lomax, 377
multinomial, 145
multivariate normal, 138–142
negative binomial, 46–48

normal, 56–57
Pareto, 97, 248, 296, 378
Poisson, 48–50
scaled inverse chi-squared, 377
*t*, 252-253
Weibull, 97, 185
Probability integral, 383–384
Probability mass function, *see* Mass functions
Probability measure, 11–15
Probability space, 11, 12–13
Proportional-hazard models, 187
Pseudo-random numbers, 382
*p*-value, 283–285

**Q**

Quantile, 240
Quantile-based statistics, 240
Quasi-likelihood, 324

**R**

Random matrices, random vectors and, 127–130
Random sums, 164–166
Random variables, 33–98
  convergence, 88–93
  cumulant-generating functions and cumulants, 79–81
  cumulative distribution functions, 37–40
  definition of, 33, 35
  discrete and continuous, 41–61
  expectation operator, 63–64
  formal definition of, 93–96
  functions of, 35, 81–88
  inequalities involving expectation, 67–70
  mean of, 61–63
  moment-generating functions, 74–78
  moments, 70–74
  positive, 36
  real numbers with outcomes, 33–37
  sums of, 134–138
  variance of, 65–66
Random vectors
  conditioning for, 166–167
  random matrices and, 127–130
Rao-Blackwell theorem, 338–340
Recurrence and transience, 215–216
Reference priors, 357–360
Reliability function, *see* Survival function
Resampling bootstrap, *see* Nonparametric bootstrap
Reversibility condition, 403

**S**

$\sigma$-algebra, 9–10
Sample, 172
  functions of, 251–255
    pivotal function, 252–255
    statistics, 251–252
Sample correlation, 234–235
Sample covariance, 233–234
Sample mean, 223–227
  central limit theorem, 225–227
  mean and variance of, 224
  properties of, 224
  sample variance and, 232–233
  and variance for a normal population, 236–240
Sample minimum and sample maximum, 242–243
Sample moments, 228–229
  joint, 233–236
Sample quantiles, 240–248
  order statistics and, 240–248
Sample space, 8
  empty set and, 12–13
Sample variance, 229–233
  properties of, 230–232
  representations of, 229–230
  sample mean and, 232–233
Sampling distribution, 252
Scaled inverse chi-squared distribution, 377
Score, 296–302; *see also* Information
Score test, 320–323
Score vector, 300

Second Borel-Cantelli lemma, 92
Significance level, 281–282
Simple hypothesis, 279–280
Simple linear regression, 174–175
Simulation methods, 381–412
Size of test, 281–282
Skewness, 71–72
Spike-and-slab prior, 377
Square of a continuous random variable, 83–84
Stationary distribution, of Markov chains, 214
Stationary increments, 199
Statistical hypotheses, 279–280
Statistical inference, 1–5
Step function, 44
Stochastic model, 381
Stochastic processes, 197–199
Strict stationarity, 196
Strong law of large numbers, 261
Structural assumptions, 172
Sufficiency, 327–341
 application in point estimation, 338–341
 factorisation theorem, 329–334
 minimal, 334–338
Sufficiency principle, 328
Sums of random variables, 134–138
 $n$ independent random variables, 137
 two random variables, 135–137
Superposition, 203
Survival analysis, 181
Survival function, 182–185
Survival time, 181

**T**
Table lookup method, 382–383
$t$-distribution, 252–253
Thinning, 203
Time series models, 188–197
 autocorrelation, 192–197
 autocovariance, 193
 autoregressive models, 191
 covariance stationary, 193–195
 invertibility, 410
 moving-average models, 191–192
 strictly stationary, 196
Time-to-event models, 181–188
 censoring of, 186
 covariates in, 187–188
 survival function and hazard function, 182–185
Total Fisher information, 297
Transformation invariance, 358–359
Transformations of continuous random variables, 130–134
Transition matrix, 208–210
Type-II maximum likelihood, 372

**U**
Unbiased estimator, 257
Unconstrained model, 318
Uniformly most powerful test (UMPT), 348–350
Uniform sampling, 393
Unimodality, 272
Union, 13
Univariate, 4

**V**
Variance of a random variable, 65–66
Variance of unbiased estimators, 341–345
Variance reduction, 397–398

**W**
Waiting-time paradox, 204
Wald test, 320, 321–323
Weak law of large numbers, 261
Weibull distribution, 97, 185
Weights, 396–397

**Y**
Yang Hui's triangle, 21

**Z**
0-1 loss, 364–366
Zero mean, 228–229

Printed in the United States
by Baker & Taylor Publisher Services

Printed in the United States
by Baker & Taylor Publisher Services